OXFORD MEDICAL PUBLICATIONS

Treatment of Early Breast Cancer
Volume 1

Worldwide Evidence 1985–1990

Early Breast Cancer Trialists' Collaborative Group
ICRF/MRC Clinical Trial Service Unit
Nuffield Department of Clinical Medicine
Radcliffe Infirmary
Oxford OX2 6HE
UK

Treatment of Early Breast Cancer

VOLUME 1

Worldwide Evidence 1985–1990

A systematic overview of all available randomized trials of adjuvant endocrine and cytotoxic therapy

Early Breast Cancer Trialists'
Collaborative Group (EBCTCG)

Oxford   New York   Tokyo
OXFORD UNIVERSITY PRESS
1990

Oxford University Press, Walton Street, Oxford OX2 6DP
Oxford New York Toronto
Delhi Bombay Calcutta Madras Karachi
Petaling Jaya Singapore Hong Kong Tokyo
Nairobi Dar es Salaam Cape Town
Melbourne Auckland
and associated companies in
Berlin Ibadan

Oxford is a trade mark of Oxford University Press

Published in the United States
by Oxford University Press, New York

© Oxford University Press 1990

All rights reserved. No part of this publication may be reproduced,
stored in a retrieval system, or transmitted, in any form or by any means,
electronic, mechanical, photocopying, recording, or otherwise, without
the prior permission of Oxford University Press.

This book is sold subject to the condition that it shall not, by way
of trade or otherwise, be lent, re-sold, hired out, or otherwise circulated
without the publisher's prior consent in any form of binding or cover
other than that in which it is published and without a similar condition
including this condition being imposed on the subsequent purchaser.

British Library Cataloguing in Publication Data
Treatment of early breast cancer.
Vol. 1, Worldwide evidence, 1985-1990
1. Women. Breasts. Cancer. Therapy. Clinical trials
I. Early Breast Cancer Trialists' Collaborative Group
616.9944906

ISBN 0-19-262015-0
0 19 2620142 (pb)

Library of Congress Cataloging in Publication Data
(Data available)

ISBN 0 19 262015 0
0 19 2620142 (pb)

Typeset by the authors
Printed in Great Britain by
Butler & Tanner Ltd, Frome and London

SUMMARY

This monograph reviews in detail all available randomized trials that began before 1985 of the treatment of early breast cancer with adjuvant tamoxifen, cytotoxic chemotherapy, radiotherapy or ovarian ablation. Coverage was reasonably complete for most countries, and involved a total of about 40,000 women in about 100 randomized trials. Of 16,513 women in 28 trials of tamoxifen, over 5000 had relapsed and almost 4000 had died. Of 13,442 women in 40 trials of chemotherapy, similar numbers had relapsed and had died. Of 9933 women in 19 trials of radiotherapy, almost 4000 had died, and of 2180 in 10 trials of ovarian ablation, over 1000 had died. Systematic overviews of the results of these trials can help limit the selective biases that might be produced by undue emphasis on particular studies, and can also help limit purely random errors. Such overviews demonstrated that medical treatment could produce moderate, but highly significant, improvements in 5-year survival. The overall reductions in the odds of death during the first 5 years were 16% SD 3 for tamoxifen versus no tamoxifen ($P<0.00001$), and 11% SD 4 for chemotherapy versus no chemotherapy ($P=0.003$). Radiotherapy did not appear to produce any clear improvement in survival, and ovarian ablation produced results that appeared promising but were based on more limited numbers.

In tamoxifen trials, there was a clear reduction in mortality only among women aged 50 or older, for whom allocation to tamoxifen reduced the annual odds of death during the first five years by about one-fifth. In chemotherapy trials, there was a clear reduction only among women aged under 50, for whom allocation to polychemotherapy reduced the annual odds of death during the first five years by about one-quarter. The improvements in recurrence-free survival produced by tamoxifen and by cytotoxic therapy were, however, statistically definite both for older and for younger women. Direct comparisons showed that combination chemotherapy was significantly more effective than single-agent therapy ($P=0.001$ both for recurrence and for mortality), but did not provide any evidence that chemotherapy durations of a year or more were any more effective than durations of about 6 months. No significant "interaction" between nodal status and the response to treatment was apparent, although only for tamoxifen was the improvement in recurrence-free survival clearly significant ($P<0.0001$) even for "node-negative" women. Likewise, in these trials the estrogen receptor measurements then available (which were less reliable than the best such measurements now available) could not identify a group of women wholly unresponsive to tamoxifen.

Because it involved several thousand women, this overview was able to demonstrate particularly clearly that both tamoxifen and cytotoxic therapy can reduce five-year mortality. Further collaborative follow-up will be needed to discover whether the definite differences at 5 years are likely to persist long-term.

1. CONTENTS

1 CONTENTS — page 2
1.1 General structure of the present report
1.2 Tables and Figures in the text, appendix & "short reports"

2 FOREWORD — page 5

3 INTRODUCTION — page 6
3.1 Systematic public availability of randomized trial results
3.2 Informative overviews even of somewhat different trials, despite heterogeneity of design and of size of treatment effect
3.3 Importance of reliably assessing MODERATE treatment effects
3.4 Two main reasons for using overviews of properly randomized trials to assess MODERATE treatment effects: avoiding selection biases and reducing random errors
3.5 Review of many trials, with no data-dependent omissions, to limit selection bias in assessment of treatment effects
3.6 Other selection biases either in design or in analysis of trials
3.7 Proper randomization (with no subsequent exclusions) to limit selection bias in design of trials of moderate treatment effects
3.8 Selection biases from subgroup analyses: statistical difficulty in the assessment of qualitative "interactions" and of quantitative "interactions" **(Table 1)**
3.9 Specific categories of patient or of treatment: data-dependent emphasis on subgroup analyses *versus* indirect extrapolation of overall analyses
3.10 Use of recurrence data, as well as mortality data, to study interactions unbiasedly
3.11 Use of overviews to assess effects of treatment on other specific causes of early death, or other rare endpoints
3.12 Use of overviews to improve the reliability of data from particular trials

4 DATA COLLECTION METHODS — page 10
4.1 Summary of the inclusion criteria for trials in the present overviews
4.2 Principles of identification of trials: selection bias may be limited even without absolute completeness
4.3 Practice of identification of trials: multiple sources of information
4.4 Methods of seeking data from individual trials
4.5 Methods of checking data from individual trials
4.6 Methods of publishing data from individual trials

5 STATISTICAL METHODS — page 12
5.1 Medical assumptions: (1) modest sizes of treatment effects, and (2) differences of size but not of direction between effects in different circumstances
5.2 Comparisons only of like with like, based on "Observed minus Expected "(O-E) differences within each separate trial **(Table 2)**
5.3 Routine stratification of all main analyses by age (<50, 50+) and by year (1, 2, 3, 4, 5+) of follow-up
5.4 Additional stratification of selected analyses for the available information on nodal status or Estrogen Receptor status
5.5 Arithmetic procedures for calculation of Observed minus Expected (O-E) numbers of events among treatment-allocated patients, and for obtaining "two-sided" significance levels (2P)
5.6 Interpretation of P-values: statistical significance and medical judgement
5.7 **RESULTS OF RADIOTHERAPY TRIALS (Table 3, Figure 1)** as an example of the summation of (O-E) values from different trials to provide an overall test of the "null hypothesis" of no treatment effect
5.8 Use of (O-E) values to provide a description of the typical reduction in the odds of treatment failure (i.e. to describe the "alternative" hypothesis)
5.9 Graphical display methods for separate trial results, and for an overview
5.10 Similarities between risk ratios, death rate ratios and odds ratios when event rates are low
5.11 Logrank "year of death" analyses for statistical significance tests, and for estimation of reductions in annual odds of death
5.12 Life-table estimation for descriptive purposes
5.13 Test of heterogeneity between several different trial results
5.14 Arithmetic details of tests for trend, for heterogeneity and for interaction
5.15 Practical meaning of a clear effect of treatment in a single large trial
5.16 Practical meaning of a clear effect of treatment in an overview of many trials
5.17 Fixed effect "assumption-free" methods, and random-effect "assumed representativeness" methods

6 MATERIALS AVAILABLE FOR REVIEW — page 19

7 STRUCTURE OF RESULTS AND DISCUSSION — page 20
7.1 Overall analyses followed by subgroup analyses
7.2 Mortality analyses followed by recurrence analyses

8 RESULTS OF TAMOXIFEN TRIALS — page 21
(Tables 4-10, Figs 2-11)
8.1 Overall analysis of MORTALITY in tamoxifen trials
8.2 Overall analysis of RECURRENCE in tamoxifen trials
8.3 Heterogeneity between tamoxifen TRIALS, and heterogeneity between WOMEN
8.4 Heterogeneity of MORTALITY results in tamoxifen trials
8.5 Heterogeneity of RECURRENCE results in tamoxifen trials
8.6 Influence of age on the effect of tamoxifen: MORTALITY data
8.7 Influence of age on the effect of tamoxifen: RECURRENCE data
8.8 Description of the SIZE of the effect of tamoxifen: MORTALITY data
8.9 Description of the SIZE of the effect of tamoxifen: RECURRENCE data
8.10 Further indirect comparisons: factors other than age
8.11 Effects of different durations of tamoxifen treatment: MORTALITY data
8.12 Effects of different durations of tamoxifen treatment: RECURRENCE data
8.13 Interactions of tamoxifen with chemotherapy: MORTALITY data
8.14 Interactions of tamoxifen with chemotherapy: RECURRENCE data

8.15 Effects of different tamoxifen doses: MORTALITY data
8.16 Effects of different tamoxifen doses: RECURRENCE data
8.17 Effects of tamoxifen in different nodal categories: MORTALITY data
8.18 Effects of tamoxifen in different nodal categories: RECURRENCE data
8.19 Effects of tamoxifen in different Estrogen Receptor categories: MORTALITY data
8.20 Effects of tamoxifen in different Estrogen Receptor categories: RECURRENCE data
8.21 Age-specific effects of ER assay results

9 RESULTS OF CYTOTOXIC TRIALS page 50
(Tables 11-16, Figures 12-23)

9.1 Overall analysis of MORTALITY in chemotherapy trials
9.2 Overall analysis of RECURRENCE in chemotherapy trials
9.3 Heterogeneity between chemotherapy TRIALS, and heterogeneity between WOMEN
9.4 Heterogeneity of MORTALITY results in chemotherapy trials
9.5 Heterogeneity of RECURRENCE results in chemotherapy trials
9.6 Influence of age on the effect of chemotherapy: MORTALITY data
9.7 Influence of age on the effect of chemotherapy: RECURRENCE data
9.8 Further comparisons: factors other than age
9.9 DIRECT and INDIRECT comparisons of the effects of multiple-agent and single-agent chemotherapeutic regimens: MORTALITY data
9.10 DIRECT and INDIRECT comparisons of the effects of multiple-agent and single-agent chemotherapeutic regimens: RECURRENCE data
9.11 Interactions of chemotherapy with tamoxifen: MORTALITY data
9.12 Interactions of chemotherapy with tamoxifen: RECURRENCE data
9.13 DIRECT comparisons of the effects of different duration of therapy: MORTALITY data
9.14 DIRECT comparisons of the effects of different duration of therapy: RECURRENCE data
9.15 Effects of chemotherapy in different nodal categories: MORTALITY data
9.16 Effects of chemotherapy in different nodal categories: RECURRENCE data
9.17 Description of the SIZE of the effect of chemotherapy: MORTALITY data
9.18 Description of the SIZE of the effect of chemotherapy: RECURRENCE data

10 RESULTS OF THE OVARIAN ABLATION TRIALS
(Table 17) page 82

10.1 MORTALITY in ovarian ablation trials
10.2 RECURRENCE in ovarian ablation trials

11 SUMMARY OF RESULTS (Tables 18-19) page 84

11.1 General principles
11.2 Subgroup analyses of mortality and of recurrence
11.3 Indirect comparisons: (a) effects of different treatments, (b) effects in different patient categories
11.4 Effects of tamoxifen
11.5 Effects of chemotherapy
11.6 Duration of benefits
11.7 Effects of full compliance with allocated treatment
11.8 Generalizability of trial results to clinical practice

12 IMPLICATIONS page 88

12.1 Implications for patient management
12.2 Implications for the conduct of future trials
12.3 The "Uncertainty Principle" as the fundamental eligibility criterion for clinical trials
12.4 Numbers needed in future randomized trials
12.5 Implications for future overviews in breast cancer

13 ACKNOWLEDGEMENTS page 90

14 REFERENCES page 91

APPENDIX TABLES AND FIGURE

SHORT REPORTS, ONE ON EACH RELEVANT TRIAL

1.1 General structure of the present report

This report is long because it describes the first worldwide collaborative overview of many cancer trials, and the medical principles, statistical principles, statistical details and organizational details are all discussed at length before the main results (which themselves include both analyses and commentary) are presented. But, many readers may need convenient access to the main results without scrutiny of a mass of ancillary detail, or may need to skip selected parts when looking up particular methods or results. For those who do wish to locate particular parts quickly, or to skip-read much of the report, the text is broken into many short sections with long subheadings. These subheadings are brought together here to form a contents list that indicates the structure of the whole report.

1.2 Figures and Tables in the text, appendix and "short reports"

Each trial has been given a unique code consisting of the year it began plus a letter (e.g. 78G: 78G3 would refer to just a part of trial 78G), and such codes are used in some figures or tables to refer to particular trials or parts of trials. Where figures or tables involve only Mortality (M) or only Recurrence (R), this is usually indicated by M or by R in their numbering. Where figures or tables involve only women who were under 50 [<50] years of age at entry, or 50 or more [≥50] years of age at entry, this too is usually indicated in their numbering (e.g. Fig. 2M[≥50]). **MORTALITY and RECURRENCE analyses will, respectively, generally be on the left and on the right of two facing pages**, e.g.

Fig. 2M[all ages] | Fig. 2R[all ages]

Fig.2M[<50]   Fig. 2M[≥50] | Fig. 2R[<50]   Fig. 2R[≥50]

FIGURES (in the text and appendix)

Fig. 1M. MORTALITY (Example of graphic display of an overview of 19 trials): Effects of post-mastectomy radiotherapy

Figs. 2M (& 2R) MORTALITY (& RECURRENCE). Effects of tamoxifen in all available unconfounded randomized trials

Figs. 3M (& 3R). Effects of tamoxifen among women of DIFFERENT AGES

Figs. 4M (& 4R). Effects of ANY DURATION of tamoxifen, by year from entry

Figs. 5M (& 5R). Effects of 2 OR MORE YEARS of tamoxifen, by year from entry

Figs. 6M (& 6R): Outcome by year from entry, subdivided by NODAL STATUS but not by treatment

Figs. 7M (& 7R). Details of effects of tamoxifen in various NODAL categories: [N0/N-], [N1-3], [N4+] or [N+/N?]

Figs. 8M (& 8R): Summary of effects of tamoxifen in various NODAL categories

Figs. 9M (& 9R). Outcome by year from entry, subdivided by ER STATUS but not by treatment

Figs. 10M (& 10R). Details of effects of tamoxifen in various ER categories: [ER-poor], [ER+], [ER++] or [ER?]

Figs. 11M (& 11R). Summary of effects of tamoxifen in various ER categories

Figs. 12M (& 12R). Effects of cytotoxic in all available unconfounded randomized trials

Figs. 13M (& 13R). Effects of cytotoxic among women of DIFFERENT AGES at entry

Figs. 14M (& 14R). Direct comparisons of MULTIPLE-AGENT cytotoxic versus SINGLE-AGENT cytotoxic

Figs. 15M (& 15R). Direct comparisons of MULTIPLE-AGENT cytotoxic versus SINGLE AGENT, OR NO, cytotoxic

Figs. 16M (& 16R). Direct comparisons of DIFFERENT DURATIONS of the same cytotoxic

Figs. 17M (& 17R). Outcome by year from entry, subdivided by NODAL STATUS

Figs. 18M (& 18R). Details of effects of cytotoxic in various NODAL categories: [N0/N-], [N1-3], [N4+] or [N+/N?]

Figs. 19M (& 19R). Summary of effects of cytotoxic in various NODAL categories

Figs. 20M (& 20R). Effects of ANY TYPE of prolonged cytotoxic, by year from entry

Figs. 21M (& 21R). Effects of MULTI-AGENT prolonged cytotoxic, by year from entry

Figs. 22M (& 22R). Effects of CMF-BASED prolonged cytotoxic, by year from entry

Figs. 23M (& 23R). Direct comparison of MULTIPLE versus NO/SINGLE-AGENT cytotoxic

Appendix Figs. 1F (& 1B). FRONT (& BACK) of 1985 data form

TABLES (in the text, appendix and short reports)

Table 1. Three examples, one hypothetical and two real, of apparent interactions produced merely by the play of chance
  (a) Hypothetical trial:
      false negative, and exaggeratedly positive, results
  (b) Real trial (ISIS-2):
      false negative result by astrological subgroups
  (c) Real trial (ISIS-1):
      exaggeratedly positive result by astrological subgroups

Table 2. Statistical methods
  (a) Calculation of O-E and its variance
  (b) Principle of unbiased comparison of trial results

Table 3M. MORTALITY (Example of tabular display of an overview of 19 trials): Effects of radiotherapy in all available unconfounded randomized trials

Tables 4M (& R). Trials available for evaluation of adjuvant tamoxifen in early breast cancer

Tables 5M (& 5R). Effects of tamoxifen in four large trials, and in four categories of smaller trial

Tables 6M (& 6R). Effects of tamoxfen, by year from entry

Tables 7M (& 7R). Indirect comparisons between the effects of different tamoxifen regimens

Tables 8M (& 8R). Effects of tamoxifen, by categories of nodal status

Tables 9M (& 9R). Effects of tamoxifen, by categories of estrogen receptor status

Tables 10M (& 10R). Trials available for evaluation of adjuvant cytotoxic in early breast cancer

Table 11M (& 11R). Indirect and direct comparisons between the effects of different cytotoxic regimens

Tables 12M (& 12R). Effects of cytotoxics, by categories of nodal status

Tables 13M (& 13R). Effects of prolonged cytotoxic, by year from entry

Tables 14M (& 14R). Effects of multi-drug cytotoxic versus no cytotoxic, by year from entry

Tables 15M (& 15R). Effects of CMF-containing cytotoxic versus no cytotoxic, by year from entry

Tables 16M (& 16R). Effects of multi-drug versus single, or no, cytotoxic, by year from entry

Tables 17M (& 17R). Effects of ovarian ablation in all available unconfounded randomized trials

Tables 18M (& 18R). 5-year survival (& 5-year recurrence-free survival): absolute improvement expected per 100 tamoxifen-allocated women

Tables 19M (& 19R). 5-year survival (& 5-year recurrence-free survival): absolute improvement expected per 100 CMF-allocated women

Appendix Table 1. List of all relevant randomized trials in the overview

Appendix Tables 2M (& 2R). Effects of tamoxifen in all available unconfounded randomized trials of adjuvant tamoxifen versus control with no tamoxifen

Appendix Tables 3M (& 3R). Effects of cytotoxic in all available unconfounded randomized trials of one or more types of chemotherapy versus no cytotoxics

"SHORT-REPORT" Tables. Basic data for each separate trial (or part of a trial)
  (a) Events in each separate year from randomization
  (b) Outcome in selected subgroups of patients

2. FOREWORD

By Sir Walter Bodmer, FRS
ICRF Laboratories, London

Early breast cancer is a common diagnosis, made in several hundreds of thousands of women worldwide each year. When the disease is at an "early" stage, the obvious deposits of cancerous tissue at the primary site of the cancer in the breast or in the local lymph nodes can be removed by surgery. Even at this early stage, however, undetectably small deposits may already have been established elsewhere in the body that can, after a delay of some years, eventually cause a recurrence that may not be curable. It is for this reason that, in addition to surgery for early stage breast cancer, various "adjuvant" treatments whose aim is to eradicate these undetectably small deposits by some form of systemic therapy have been developed over the last few decades.

These adjuvant treatments have now been tested in a succession of clinical trials throughout the world. The trials have involved, in particular, use of the "anti-estrogen" tamoxifen and of cytotoxic chemotherapy using combinations such as "CMF". The effects of these treatments are unfortunately still comparatively small with, for example, proportional reductions in mortality over a 5-year period of only about 15%. As a result, individual trials rarely give clearly significant results. The individual trials have, moreover, become so numerous that their real implications for patient care are difficult to assess reliably without the systematic overview of their results that this monograph provides. Until more specific and effective treatments of breast cancer become available that can achieve greater reductions in 5-year mortality, such overviews are the only approach by which it is possible to assess reliably any small but significant effects of the currently available treatments.

But, any widely practicable treatment for early breast cancer that can produce just a **moderate** reduction in the risk of death (for example, from 35% down to 30%) could well save ten thousand lives a year. This monograph, which summarizes evidence from almost 100 randomized trials among almost 40,000 women studied in many different countries, shows that such reductions can indeed be achieved if its conclusions are widely disseminated. By their unique worldwide collaboration, the EBCTCG have provided reliable information that is relevant not just to doctors engaged in clinical trials, but to all those concerned with treating or advising women with breast cancer.

3. INTRODUCTION

3.1 Systematic public availability of randomized trial results

In early breast cancer all clinically apparent disease can, by definition, be removed surgically. Following such surgery, adjuvant systemic treatments involving various cytotoxic, hormonal or other therapies may be considered. Despite numerous clinical trials of adjuvant therapy, however, there has been uncertainty as to the net effects of such treatments, particularly on survival. The present report arises from a collaboration that started in 1984[1,2] between the principal investigators of these randomized trials of adjuvant therapy for early breast cancer that included some evaluation of tamoxifen, of cytotoxic therapy, of radiotherapy[3] or of ovarian ablation that was "unconfounded" (i.e. not mixed up with other evaluations: see below).

There have already been over 300 randomized trials assessing various different primary treatments and different adjuvant treatments for early breast cancer, and a few of the main therapeutic questions have already been addressed by several dozen different trials, some published and some not. The purpose of the present report is to bring together the principal results from as many as possible of the trials that evaluated certain adjuvant treatments (particularly tamoxifen or cytotoxic therapy), so that reviews of the trial evidence on those questions can be more reliable and informative than a review of just the published literature.[4] The methods and data for each of the trials involved are presented separately by each trial group in the brief trial summaries that form an Appendix to this report.

As examples illustrating some of the uses to which such data can be put, systematic overviews of the principal results from certain categories of these trials are also presented using statistical methods that are particularly suited to trial overviews.[5,6] The purpose of these overview analyses is not, however, to impose any particular interpretation or method of analysis of the trial results, but merely to demonstrate the ability of an appropriate overview of many trial results to yield informative conclusions. Different readers may prefer to review only a limited number of the trials: if so, the data for such sub-analyses are available in this report. Similarly, different readers may prefer to review only certain subcategories of all women, or to use different statistical methods of review: if so, the data from each separate trial are available in sufficient detail to permit various alternative approaches. Thus, one of the chief functions of the present report is just to make available unbiased data from all (or at least from as many as initially possible) of the relevant randomized trials in early breast cancer, so as to facilitate the construction, discussion and publication of various different interpretations of the trial evidence. The main biases to be avoided are discussed below.

3.2 Informative overviews even of somewhat different trials, despite heterogeneity of design and of size of treatment effect

To illustrate the circumstances to which overviews of different trial results may be relevant, the question of the net effects on mortality of adjuvant treatment with tamoxifen can be considered. There are already a few dozen randomized trials of adjuvant tamoxifen, but they are heterogeneous in their entry criteria, their treatment schedules, their follow-up procedures and their methods for treating relapses. At first sight such heterogeneity may appear to be an important reason not to perform overviews. A converse view, however, is that if a certain type of treatment is of some benefit in one trial then a somewhat similar type of treatment will probably also be of at least some benefit in another trial, and that even though these benefits are probably not the same **size** in each trial they will, but for the play of chance, probably tend to point in the same **direction** in each study. If so, then an overview of the separate tamoxifen trials might provide a useful test of whether or not such treatment has any therapeutic effect, together with at least some indication of the sizes of therapeutic effect that are "typical" in the circumstances of these particular trials.

At one extreme, therefore, each tamoxifen trial might be considered in virtual isolation from all others, while at the opposite extreme all tamoxifen trials might be considered together. Both of these extreme views have some merit, and the pursuit of each by different people will perhaps prove more illuminating than too definite an insistence on any one particular approach. This will be still more the case for the chemotherapy trial results, since several very different regimens have been tested, ranging from a few days of cyclophosphamide to some months or years of fairly intensive multiple-agent chemotherapy. Moreover, even chemotherapy schedules that nominally involve the same agents — cyclophosphamide, methotrexate, and 5-fluorouracil (CMF), for example — may differ so substantially in intensity or duration that their effects are substantially different.

3.3 Importance of reliably assessing MODERATE treatment effects

Opinions differ as to the likelihood of large (e.g. 50% or more) differences in long-term survival being produced by the types of treatment for early breast cancer that are usually compared in randomized trials. Some, remembering past experience with childhood leukemia[7] and Hodgkin's disease,[8] expect large improvements to emerge, while others regard the expectation of large differences in mortality to be somewhat unrealistic.[9] There appears, however, to be much wider agreement that **moderate** (e.g. 15% or 25%) differences in mortality might well exist and that reliable knowledge of such effects could be humanly important when choosing what treatments to recommend for an individual woman. For example, "just" a 20% reduction in a 50% risk of death would represent avoidance of death in 1 out of every 10 women treated, which is substantial.

If large mortality differences exist, then discovering them would in general be more important than discovering moderate differences. But, even in the absence of large differences the reliable demonstration, or reliable refutation, of any moderate differences may still be important. For example, each year in North America alone early breast cancer is diagnosed in well over 100,000 women.[10] Although the primary treatment can eliminate all clinically apparent disease, more than 30,000 of these women who originally presented with only early breast cancer will eventually suffer disease recurrence and die. Early breast cancer is also common in other parts of the world, so a widely practicable treatment that produced a reduction in mortality as small as 15% could avoid or delay several thousand deaths a year, provided it was adopted widely by the medical profession.

3.4 Two main reasons for using overviews of properly randomized trials to assess MODERATE treatment effects: avoiding selection biases and reducing random errors

Reliable detection (or refutation) of treatment effects that are only moderate in size requires the reliable exclusion of (i) moderate **biases** and (ii) moderate **random errors**, either of which might obscure (or mimic) moderate treatment effects. Each of these requirements may be difficult to meet adequately without a proper overview of the unconfounded randomized trials.[4]

First, without a systematic search for all relevant randomized trials an unrepresentative selection of the trial results may be reviewed. The selection biases caused by this might not matter much if treatment had a large effect on long-term mortality, but these and other selection biases do matter if moderate treatment effects are to be assessed reliably, or if an ineffective treatment is to be reliably recognized as such.

Second, unless comparisons of the effects of treatment are based on trials that together include several hundred (or, preferably, several thousand) **deaths**, the random play of chance can produce favorable or unfavorable random errors that are comparable in size with any moderate mortality reductions that might exist. Such large numbers of deaths are difficult to achieve in individual trials but may be achievable in an overview of the results of many different trials.

3.5 Review of many trials, with no data-dependent omissions, to limit selection bias in assessment of treatment effects

Even after a complete and exhaustive review just of the **published** literature there may be some biases in the selection of the trial results that could then be reviewed, since detailed results of particular trials may remain unpublished unless those results are exceptional.[11] Hence, overviews based only on the published literature may well produce moderately biased results.[12] Moreover, even a complete review of nothing but the published literature might be such a time-consuming task (especially if additional information needs to be sought by correspondence with some authors) that only fairly determined investigators would be likely to achieve it. Others may be satisfied with an incomplete review that excludes some of the lesser known publications, but this might introduce some further selection bias since even among published studies the favorable or unfavorable play of chance in the results of a trial may substantially influence how well known that trial becomes. Thus, trials that appear to have particularly promising results (or, for some treatments, particularly unpromising results) are likely to be among the best known. Even if any selective biases that this produces are only moderate in size, moderate biases may still make realistically moderate treatment effects impossible to assess reliably. The biases that can be introduced just by selective exclusion of certain randomized trial results can, however, be avoided by systematic review of all (or of an unbiased subset) of the randomized trials ever undertaken.

3.6 Other selection biases either in design or in analysis of trials

Overviews may also be of some limited help in controlling other selection biases in the design or in the analysis of trials. Selection biases in design can be produced not only by failure to allocate treatment properly at random, but also by post-randomization withdrawal of selected patients (see below: "Proper randomization"). Selection bias in analysis can be produced not only by undue emphasis on just a limited number of trials, but also by unduly data-dependent emphasis on the apparent effects of treatment in particular subcategories of patient (for real examples, see Table 1 and the associated discussion on "Selection bias from subgroup analyses").

3.7 Proper randomization (with no subsequent exclusions) to limit selection bias in design of trials of moderate treatment effects

Comparisons using historical controls,[13] data-base "efficacy" analyses[14] or other non-randomized methods may, no matter what precautions are taken, be subject to moderate biases, the exact size of which cannot be predicted reliably. For example, in a recent United Kingdom Medical Research Council leukemia trial, a significant improvement (P=0.003) in mortality was seen between the first and second half of the same supposedly homogeneous trial.[13] Hence, if some completely ineffective additional treatment had been specifically introduced halfway through that study, then a historically controlled "evaluation" of that treatment might have misleadingly concluded that that treatment worked.

Knowledge of the next treatment allocation **before** patient entry is confirmed (for example, where randomization lists are publicly available, allowing foreknowledge, or where allocation is alternate or based on odd/even dates or record numbers) can produce biases that are uncorrectable even if the original sequence of treatments was completely random. Trials that permit foreknowledge to bias patient entry are not, in fact, properly randomized, although they are often mistakenly described as such.

Non-randomized methods may sometimes suffice as a crude means of deciding whether or not large therapeutic effects exist, but they are generally of little value for reliable detection, or refutation, of **moderate** therapeutic effects. Hence, the present overviews are restricted to properly randomized trials.

3.8 Selection biases from subgroup analyses: statistical difficulty in the assessment of qualitative "interactions" and of quantitative "interactions"

Patients with early breast cancer may be very different from each other, and the treatment appropriate for one may not be appropriate for another. Ideally, therefore, what is wanted is not only an answer to the question "Is this treatment good on average for a wide range of patients?", but also an answer to the question "For which recognizable categories of patient is this treatment good?". In other words, the ideal would be a reliable description of the categories most likely to benefit from treatment. This ideal is, however, difficult to attain.

"Interactions" — that is, differences between the effects of treatment in different categories of women — may be of two types that can have quite different practical medical consequences. If treatment improves the prognosis appreciably in one category of women but does so to a negligible extent, or not at all, in another category then this is a **qualitative** interaction. If, however, treatment improves the prognosis appreciably in both categories but the improvement is just somewhat bigger in one category than in the other then this is **a quantitative** interaction. Unfortunately, the direct use of clinical trial results in particular subgroups of women to refute or to demonstrate any type of interaction is often extremely difficult — and, even if statistically significant evidence of an interaction is found, this may still fall far short of providing reliable evidence of a qualitative interaction.[15-18]

One possible determinant of the size of the absolute benefit of any therapy in some particular category of patient is the absolute risk of death (or recurrence) without treatment. For example, the number of regional lymph nodes containing breast cancer deposits, divided into three standard categories (N0, N1-3, N4+), is an important prognostic feature. If, therefore, some treatment reduces the risk of death by a similar **proportion** (e.g. one-quarter) in all patients, the **absolute** benefit during the first few years after treatment may be greater in poor-prognosis patients (e.g. 40% dead reduced to 30%) than in good-prognosis patients (e.g. 8% dead reduced to 6%). It might, therefore, be useful to know whether the proportional risk reduction (i.e. the "relative risk") produced by a particular treatment really is approximately similar for good-prognosis and for poor-prognosis women.

These and other questions about "interactions" between patient characteristics and treatment effects are easy to ask but surprisingly difficult to answer reliably. This is because quite striking-looking interactions can often be produced just by the play of chance, and these can mimic or obscure some of the moderate treatment effects that one might realistically expect. For example, if the patients in a trial with a highly significant (2P<0.002) overall benefit of treatment were categorized by some completely absurd criterion (e.g. their astrological birth signs[17,19]) then the effects of treatment in different patient categories may, just by the play of chance, appear to be quite large and highly significant in some categories, but small and non-significant in others (Table 1). For treatments that have a smaller proportional effect on the odds of survival than on the odds of recurrence-free survival, such chance fluctuations between the apparent effects of treatment in different categories of women particularly affect the mortality analyses. Thus, just from direct analyses of data on overall mortality it may be difficult to determine reliably which categories of patient can expect the greatest proportional risk reductions. This is so even when only a small number of categories are considered (e.g. N0, N1-3, N4+), and is still more so when the patients are divided into many categories, based perhaps on more than one criterion (e.g. on both age and nodal status).

In principle the statistical difficulties might be limited by restricting subgroup analyses to those based on biological or clinical plausibility, but in practice this may offer little protection, for some sort of plausible-sounding rationale can usually be invented for almost any result.

There are two main remedies for these statistical difficulties, but the extent to which each is helpful is a matter on which informed judgements differ. The first is to use the **overall** mortality results as a guide (or at least a context for speculation) as to the qualitative effects of treatment in each particular category,[15] and to give proportionately less weight to the actual mortality results observed in that category than to extrapolation of the overall results (see below: "Specific categories of patient"). The second is to be influenced not only by mortality data but also by data on recurrence-free survival (see section 3.10 below: "Use of recurrence data").

3.9 Specific categories of patient or of treatment: data-dependent emphasis on subgroup analyses versus indirect extrapolation of overall analyses

In addition to the overall analysis of all the unconfounded randomized trials of some particular type of treatment, such as tamoxifen or chemotherapy, many different subgroup analyses will generally be presented. Among these, however, the play of chance alone is likely to yield several false negative or false positive results. So, for example, if the overall analysis yields results that are not clearly significant, data-dependent emphasis on just a few subgroup analyses that happened to be conventionally significant might well be misleading since these might be "false positive" results generated by chance alone (or at least considerably magnified by the play of chance). Conversely, if the overall analysis yields highly significant evidence that some treatment does, overall, produce a moderate reduction in mortality, subgroup analyses could well generate "false negative" results merely by chance. In the latter instance, data-dependent emphasis on subgroup analyses that are not conventionally significant (or that point in a direction opposite to that of the overall analysis) might again be misleading.

Because direct analyses just among specific categories of women or of trials can give misleading results, data-dependent emphasis on particular such analyses may lead to importantly biased conclusions. Paradoxically, therefore, even effects among specific categories of women or of trial may best be assessed indirectly by approximate extrapolation from the apparent effects of treatment among all women in a wide class of trials. This would not be true if the treatment effects in the specific category of interest were qualitatively different from the overall

Table 1. Three examples, one hypothetical and two real, of apparent "interactions" produced merely by the play of chance

Table 1(a). Hypothetical trial: FALSE NEGATIVE and EXAGGERATEDLY POSITIVE mortality effects in subgroups defined only by whether day of birth was even or odd

Birthdate category	Treatment group	Control group	Statistical significance
EVEN* birthdate only (almost no mortality reduction apparent)	18/500	22/500	NS
ODD* birthdate only (mortality appears to be almost halved)	22/500	38/500	2P<0.05
ANY birthdate (appropriate **overall** analysis)	40/1000	60/1000	2P<0.05

* The apparent discrepancy between these two results is not particularly unusual (15), as there would be about a one-in-eight probability of chance alone producing a contrast at least as extreme as (or, probably, more extreme than) this between the **apparent** effects of treatment in two categories of women. Much greater discrepancies could easily arise in **data-dependent** subgroup analyses, i.e. where the subgroups are selected for special emphasis in presentation of trial results partly because the apparent discrepancy between them is striking: see the two real examples that follow.

Table 1(b). Real trial (ISIS-2): FALSE NEGATIVE mortality effect in a subgroup defined only by astrological "birth sign"

Astrological "birth sign"	Aspirin effect on day 0-35 mortality in acute myocardial infarction	
	Nos. of deaths by treatment group* ASPIRIN vs PLACEBO	Statistical significance (2P)
Libra or Gemini (taken together)	150 vs 147	0.5 (NS adverse)
All other signs (taken together)	564 vs 869	<0.000 0001
Any birth sign (appropriate **overall** analysis)	804 vs 1016	<0.000 001

* In ISIS-2, 8587 were allocated active aspirin and 8600 were allocated placebo (17).

Table 1(c). Real trial (ISIS-1): EXAGGERATEDLY POSITIVE mortality effect in a subgroup defined only by astrological "birth sign"

Astrological "birth sign"	Atenolol effect on day 0-1* mortality in acute myocardial infarction	
	Mortality reduction comparing Atenolol with Control group	Statistical significance (2P)
Leo (i.e. born between July 24 & August 23)	71% ± 23	<0.01
11 other birth signs (taken separately)	Mean 24%	Each >0.1 (NS)
Any birth sign (appropriate **overall** analysis)	30% ± 10	<0.004

* For day 0-7 mortality, the overall mortality difference was 313/8037 vs 365/7990 (2P<0.04), and analysis of each separate birth sign "revealed" significant benefit only for those born under the sign of Scorpio (18).

effects. But, it might well be true (even though the proportional risk reductions are not exactly the same size in different circumstances) as long as the two treatment effects do at least point in the same direction as each other (see section 5.16 below: "Practical meaning of a clear effect of treatment in an overview of many trials").

3.10 Use of recurrence data, as well as mortality data, to study interactions unbiasedly

Trials usually observe more recurrences than deaths and, more importantly, the proportional effects of treatment on the odds of an unfavorable outcome may be larger for disease recurrence than for death, especially if breast cancer deaths are diluted by other causes of death that are largely unaffected by treatment. These two factors (more recurrences than breast cancer deaths and, particularly, larger treatment effects on recurrence) have meant that several trials in early breast cancer have been large enough just on their own to demonstrate statistically significant evidence for an effect of treatment on recurrence. They also mean that, from a purely statistical viewpoint, the relative effects of treatment in different subcategories of women or of treatments may be measured more accurately for recurrence than for death.

For example, subgroup analyses based on subcategorization of a 10 standard deviation difference in recurrence among all women will be less subject to the play of chance than are subgroup analyses based on subcategorization of a statistically significant (2P<0.0001) difference in mortality of "only" 4 standard deviations among all women. The statistical difficulties that dominate the assessment of interactions may therefore be considerably less severe for recurrence than for mortality analyses. Hence, it may be useful to analyze not only death but also recurrence when trying to determine whether the proportional effects of treatment are substantially different in different categories of patient (e.g. N0, N1-3, N4+). This remains equally true, of course, when considering any other subsets of women or of trials — and, in particular, when considering indirect comparisons between the apparent effects in different trials of two subcategories of a particular class of treatments.

A delay of recurrence is not, of course, necessarily equivalent to a delay of death. Indeed, there are instances of treatments (radiotherapy, perhaps) that substantially delay local recurrence without any substantial effect on early mortality (or on distant recurrence) in most women.[3] However, for treatments that definitely can reduce breast cancer mortality, assessment of whether their effect on mortality is likely to be markedly different, either between different categories of patient or between different categories of trial, may be assisted by separate analyses in each of those different categories of the apparent effects of treatment on recurrence (or, for localized therapies, on distant recurrence). The extent to which information from recurrence analyses should influence the interpretation of mortality analyses is, however, a matter of judgement on which opinions differ. The aim of the present report is, therefore, merely to make both survival and recurrence-free survival analyses separately available for consideration.

3.11 Use of overviews to assess effects of treatment on other specific causes of early death, or other rare endpoints

In large populations there is observational epidemiologic evidence that ovarian ablation decreases the incidence of breast cancer but increases the incidence of myocardial infarction (MI),[20,21] that postmenopausal hormones decrease the incidence of MI but increase the incidence of endometrial cancer,[22,23] that various contraceptive pills increase the incidence of MI, pulmonary embolus and stroke but decrease the incidence of ovarian cancer,[24] and that estrogenic treatment of prostate cancer is usefully palliative but increases the incidence of MI.[25] Thus, the effects of hormonal factors on the primary incidence rates of various vascular and neoplastic diseases can be substantial but difficult to predict. If, therefore, ovarian ablation, tamoxifen or other hormone-related treatments are to be used in breast cancer — and particularly if prolonged treatment is envisaged — it is important to evaluate their effects not only on breast cancer but also on other causes of death. Likewise, chemotherapy[26] and radiation[27,28] may induce leukemia or various solid tumors, though they may decrease the immediate likelihood of developing a second primary breast cancer that is large enough to be detectable.

A principal goal of therapy is to reduce overall mortality, irrespective of the actual cause of death, and it might appear that an analysis of the effects of treatment on uncommon other causes of early death would not be of much relevance to this particular goal. But, that may not necessarily be the case. For example, if a particular treatment reduces the risk of death from breast cancer but increases the risk of death from myocardial infarction, then (as in prostate cancer[25]) the balance of cancer risk factors and coronary risk factors could importantly influence the choice of treatment for many patients, particularly those at low risk of death from cancer. Moreover, even if total mortality is reduced in the first few years by a particular treatment for breast cancer, specific adverse effects may become more important later (especially if, in the above example, the annual incidence of myocardial infarction eventually begins to exceed the breast cancer recurrence rate). In any such circumstances, analyses of overall early mortality could be seriously misleading, whereas cause-specific time-to-death analyses may help interpret the implications of the randomized trials appropriately.

In absolute terms, any effects of treatment on rare causes of death may be too small to be reliably detected even by cause-specific analyses of individual trials, although they might be assessed reliably by an overview of many trials. But if, as is the case with the present data, information about cause-specific mortality is available from only a limited number of all trials, it is particularly important to avoid selective emphasis on trials where cause-specific mortality is available just because it indicates something peculiar. For this reason, the present mortality analyses do not distinguish between different causes of death, which means that any fatal side-effects cannot be studied properly. (It also reduces the statistical power of the analyses, especially among older women, a number of whom may have died of unrelated causes.) This will, however, be rectified in future overviews.

3.12 Use of overviews to improve the reliability of data from particular trials

Another contribution of overviews to the correct interpretation of trial evidence is that they involve the scrutiny of each study, which may lead to recognition and, in some cases, correction of methodological errors in certain trials (or to the exclusion of seriously flawed studies). For example, biases may be produced by the loss, or withdrawal some time after randomization, of some patients who deviate from their allocated treatment or scheduled follow-up (or by replacement of non-compliers with new patients), but these biases can be corrected by restoring any excluded randomized patients (and excluding inappropriate replacements whose treatment allocation was not properly random). Similarly, biases produced by differences in the completeness of follow-up between treatment groups can sometimes be reduced or eliminated by imposing a common cut-off date on all follow-up, or by using national mortality records to get unbiased data. In the present study, exhaustive efforts have been made to seek additional data from the trials reviewed so that any such biases may be minimized.

4. DATA COLLECTION METHODS

4.1 Summary of the inclusion criteria for trials in the present overviews

Trials were to be included if, and only if, the following four criteria were all satisfied:

(i) **The trial started to enter patients before 1.1.1985** (see section 4.3 below)

(ii) **The trial contained some properly randomized comparison:** Trials were excluded if the stated methods of treatment allocation used might well be subject to potentially serious biases due to prior knowledge of the likely treatment allocation (e.g. widespread foreknowledge of randomization lists, or use of alternate allocation, allocation by odd/even dates of birth or record numbers, historical controls, etc). Trials were also excluded if imbalances between the treatment groups with respect to numbers of patients (either overall or in particular categories) or with respect to follow-up were demonstrated by further investigation to be due to lack of proper randomization.

(iii) **The trial included at least two treatment groups that provided an unconfounded* concurrently randomized comparison of any of the following (a-f):**
(a) radiotherapy versus no radiotherapy.
(b) tamoxifen versus no tamoxifen (including trials of tamoxifen plus chemotherapy versus the same regimen without tamoxifen);
(c) chemotherapy versus no chemotherapy (including trials of chemotherapy plus tamoxifen versus the same regimen without chemotherapy, but excluding trials of perioperative chemotherapy);
(d) polychemotherapy versus single-agent chemotherapy;
(e) short duration versus long duration of the same chemotherapy;
(f) ovarian ablation versus no ablation;

(iv) **The trial was not conducted in the USSR or Japan** (see section 4.3 below)

4.2 Principles of identification of trials: selection bias may be limited even without absolute completeness

In practice, it may never be possible to identify absolutely all relevant randomized trials and to obtain absolutely all the data from all such trials. Some randomized trials may, despite extensive efforts, be overlooked or otherwise unavailable, and in the many trials where national mortality records cannot or have not been used to complement other sources of information several patients may be lost to mortality follow-up. So, for almost any particular result from an overview of the material that is available, absolute completeness cannot be guaranteed. The question to ask of an overview of many trials, therefore, is not whether it can be guaranteed to be complete, nor even whether absolutely all selective bias can be excluded, but instead whether or not any remaining biases due to missing trials or missing patients could reasonably be thought to cause any serious problems of interpretation. Some readers may initially suppose that no trustworthy inferences can be drawn from a review that is at all incomplete. This would be too rigid, however, for if taken literally it would mean that once some information from some randomized trial had been permanently lost then no amount of evidence from subsequent trials could ever suffice to answer the question of interest. Moreover, even though the editorial columns of many journals and the discussion sections of many papers often do not involve a complete review of the trial evidence, they have already led to many well-founded conclusions (although they have also led to the perpetuation of some differences of opinion that might have been resolved by a more systematic approach to the randomized evidence).

Clear evidence of an effect of therapy is provided when an overview yields a result with random errors small enough to make that result significantly different not just from zero but also from the size of selective bias that could plausibly be attributed to any incompleteness of the overview. Greater completeness, therefore, serves two complementary purposes: first it generally reduces the size of selective bias that can plausibly be ascribed to incompleteness, and second it reduces the size of the random error.

The present overview has sought to make available information that is as complete as possible, and then to address particular questions using as much of this information as is possible without serious bias. This overview is based not on all data from all the relevant randomized trials, but instead on reasonably unbiased data from a reasonably unbiased selection of these trials. (See, for example, the practical reasons for exclusion of data from the USSR and Japan that are discussed below.)

4.3 Practice of identification of trials: multiple sources of information

Several avenues of enquiry were pursued to locate as many of the relevant trials as possible. A first attempt involved discussion with trialists and scrutiny of review articles. This was supplemented by scrutiny of the systematic lists of trials that have been prepared by the UICC (Geneva), NCI (Bethesda) and UKCCCR (London), and by scrutiny of all current and previous proceedings of ASCO, AACR and UICC meetings. It was further augmented by a formal computer-aided literature search, by enquiry of the manufacturers of tamoxifen and by discussion with at least one person from each major trial organization to seek more exact details of all the randomized trials in early breast cancer ever performed by those (or any other) trial organizations.

The list of trials identified by September 1984 was circulated among trialists invited to attend a meeting at Heathrow Airport, London, at that time.[1] The smallness of the numbers of additions made then and subsequently suggests that the large majority of all patients in randomized trials of tamoxifen or chemotherapy that started before 1984 have been identified, except those from trials in the USSR and Japan, where much relevant research was

---

* Only trials in which such comparisons are "unconfounded" are included. By definition, in unconfounded trials one group differs from another only in the treatment of interest: thus, the tamoxifen overview includes trials of tamoxifen plus chemotherapy versus the same chemotherapy alone, but not trials of tamoxifen plus prednisone versus no treatment. (In trials of chemotherapy, however, prednisone was considered an integral part of the chemotherapy regimen being evaluated, so trials of chemotherapy plus prednisone versus no treatment were included in the chemotherapy overview.)

found to be in progress. Visits to the USSR and Japan have helped clarify the situations in those countries, but although investigators in these countries are willing to collaborate, sufficient details are not yet available. To limit the possibility of bias, therefore, all trial results from the USSR and Japan have been excluded from the present (though, it is to be hoped, not from future) analyses.

Similarly, to limit the possibility of bias due to selectively incomplete inclusion of recent trials, all results from trials that started to randomize patients on or after 1.1.1985 have been excluded from the present report.

4.4 Methods of seeking data from individual trials

At the 1984 Heathrow meeting many investigators agreed that data on each individual patient should be sought, so that analyses of the duration of overall survival and of recurrence-free survival could be undertaken. Some trial groups provided only grouped tabulations of results, but most provided individual patient data either on special forms completed according to standard instructions (Appendix Figure 1) or, more commonly, on magnetic tape or computer listings. Instructions defined the exact information sought and divided it into two categories: "essential minimum data", which included identifiers, allocated treatments and duration of survival or follow-up, and "optional extra data", which included information on recurrence and the date of recurrence. The "essential minimum data" columns were completed on virtually all patients and the recurrence data columns were completed on about 94% of the patients for whom individual data profiles were provided. Nine data categories were marked with an asterisk indicating that these items need not be provided if it was not convenient to do so. Because the information in some of these categories is seriously incomplete (e.g. information on contralateral breast cancers) little use will be made of it in the present report, although more complete information is being sought for future reports.

Trials of tamoxifen and trials of cytotoxic chemotherapy were sought assiduously and checked extensively, and their results are the principal object of the present report. Some types of trials (e.g. those evaluating adjuvant radiotherapy and ovarian ablation) were sought but were not as extensively checked, so the analyses of them in the present report will be brief. Other types of trials (e.g. those evaluating immunotherapy, or different surgical procedures) were not thoroughly sought, so no analysis of them is presented; again, however, more complete information on them is being sought for future reports.

Trial results can affect both trial closure and trial publication, so the exclusion of ongoing or unpublished trials could introduce the sort of data-dependent selection bias that the overview is designed to avoid. But, public availability of interim results from an ongoing, or unpublished, trial might interfere with recruitment into the study or prejudice its subsequent publication. In order to limit this problem, the results of some ongoing trials were provided for the overview on the condition that their individual results were not to be presented separately while randomization continued. In this way, confidential results could contribute to the overview without being explicitly reported in the tabular analyses.

4.5 Methods of checking data from individual trials

The availability of data on individual patients permitted a wide range of fairly obvious consistency checks, many of which revealed some errors or omissions in at least some trials that could be rectified by correspondence with the principal investigators. For example, where investigators identified patients by sequential numbers, checks were made for any breaks in the sequence that might indicate improper exclusion of some randomized patients. Similarly, where the original treatment group sizes differed by more than one standard deviation (e.g. 110 allocated one treatment compared with only 90 allocated another treatment), checks were again made to see whether there had been any improper exclusions or undocumented changes in the treatment allocation proportions.

The chief purpose of the present report is to compare one treatment with another, rather than to estimate the absolute risks of death or recurrence in women with early breast cancer. For such comparisons to be unbiased the fundamental requirement is not that follow-up should be absolutely complete but merely that there should be no systematic differences in the completeness of follow-up between treatment-allocated and control-allocated patients who are not known to be dead. Consequently, for each trial from which individual patient data were available the duration of follow-up was checked, with particular attention to the proportions of those not known to be dead whose last follow-up was before 1.1.1984. (By the time of the main data collection in 1985 this would have represented a delay of well over a year.) In the few instances where significant differences in follow-up duration were observed, attempts were made to rectify this by seeking additional information, either from the trialists or from national mortality records. The balance between groups with respect to entry date, age, menopausal status, nodal status and, where available, estrogen receptor (ER) status was also checked, and any imbalances were pursued. Other checks were instituted on the internal consistency of individual patient records.[19]

Finally, trialists were supplied with the checked and corrected records of each individual patient (when individual patient data had been supplied) and with summary tables in standard format computed from these checked records (as in the brief trial summaries that are appended to this report). Minor corrections in many of the trial results were engendered by these checks, but the arithmetic conclusions of the overviews were not materially altered by them. This does not mean, however, that these extensive checks had little effect on the interpretation of the results. An overview can yield clear evidence of an effect of treatment only if the size of the effect is clearly different from the size of bias that could plausibly be postulated. So, even if such checks do not materially alter the estimate of the treatment effect, they can still help limit the plausible size of the bias and thereby considerably strengthen the confidence that can be attached to the principal conclusions. **The principal analyses are of all events reported as occurring before September 1, 1985, which is when the main data checks were first completed.** (The exhaustive mass of data checks that continued for long after that date produced numerous minor corrections, but no substantial changes in the results of any trial.)

4.6 Methods of publishing data from individual trials

The data obtained on outcome by allocated treatment in each trial belong not to the central organizers of the collaboration but to those responsible for conducting that trial. It is only the trialists who really know the strengths, limitations and unusual features of their trial, and who can vouch for the reliability of the data from it. For both these reasons, this overview is accompanied by separate short reports, prepared in collaboration with the individual trialists, describing their trial methods and the checked results from those trials that are used in the overview. The standard summary tables in those reports generally include enough data to permit analyses of time to death and to recurrence, stratified by age and by various other prognostic features. The present overview is based only on the data in the accompanying short reports — with the exception of the few trials that have, as yet, provided only confidential results — and thus derives almost entirely from results that are now publicly available. The names of the trials that did make this available, together with their sizes and a few other brief details, are given in Appendix Table 1, which includes information on a total of about 40,000 women randomized over the years into about 100 trials.

5. STATISTICAL METHODS

5.1 Medical assumptions: (1) modest sizes of treatment effects, and (2) differences of size but not of direction between effects in different circumstances

The two fundamental assumptions that underlie an overview of many trials are not statistical, but medical. One — the human importance of mortality differences that are only moderate in size — has already been discussed at length. The other fundamental assumption is more abstract: although the same type of treatment would probably not produce exactly the same size of therapeutic effect in different circumstances, it would probably at least produce effects that tended to point in the same direction. For example, if two trials have tested approximately similar treatments then, but for the play of chance, they should yield results on specific endpoints that, although perhaps not the same size, at least point in the same direction. Likewise, if a particular treatment is of some value in one category of women then it is probably also of some value in other categories. (More formally, this second medical assumption is that although "quantitative" interactions may be common, there are not likely to be any unanticipated "qualitative" interactions between the effects of treatment on particular modes of death and the sort of information that is commonly available at the start of a trial.)

Consider, for example, two properly randomized trials of tamoxifen in early breast cancer that differ somewhat in the type of breast cancer patient, the type of primary treatment, the proportions also given adjuvant chemotherapy, the tamoxifen dose and duration, the type of treatment on relapse, the definition of relapse, the completeness and duration of follow-up, the proportion of deaths attributable to breast cancer, etc. Because of this heterogeneity, it is unlikely that the real difference in risk between treated and control patients in both of these trials will be exactly the same **size**. Despite this, it is still quite likely that any such differences will tend to point in the same **direction**. In other words, if tamoxifen is of some real value in the circumstances of one trial then it is probably also of some real value in the circumstances of the other, although of course in the **actual** results of one particular trial the play of chance may well be at least as big as the treatment effect that is to be measured, and could therefore obscure this similarity of direction. The same is true when the real treatment effects in each of several dozen tamoxifen trials are considered: common sense suggests that although some differences must exist between the real sizes of the effects in different trials, the real directions of those effects would probably be the same. Although this underlying direction may well be obscured by the play of chance in some of the individual trial results, it is more likely to stand out clearly when many different results are reviewed together.

5.2 Comparisons only of like with like, based on "Observed minus Expected" (O-E) differences in each separate trial

The statistical methods to be used in the overview analyses reflect these medical assumptions. They are of maximal statistical sensitivity for the detection of certain modest treatment effects; and, they do not implicitly assume that the real risk reductions in different trials are the same size, but merely that they will tend to point in the same direction. Nor do they require the obviously unjustified assumption that patients in one trial can be compared directly with patients in other trials. The basic principle involved is to make comparisons of treatment with control only within one trial and to avoid completely any direct comparisons of patients in one trial with patients in another. This is achieved by calculating, within each separate trial, the standard quantity "Observed minus Expected" (O-E) for the number of deaths among treatment-allocated patients.[29-31] O-E is **negative** if the treated group fared **better** than the controls (and positive if it fared worse). In an evenly balanced trial O-E is approximately equal to half the number of deaths avoided: so, for example, an O-E of -7.5 would suggest avoidance of about 15 deaths (see Table 2(a) and its footnotes).

These O-E values, one from each trial, can then simply be added up[31] (with the variance of the grand total of the individual O-E values given by the sum of the individual variances of each O-E value; this effectively leads to the results of each trial being given a "weight" in the overall assessment that depends appropriately on the amount of statistical information provided by it). If treatment did nothing, then each of the individual O-E values could equally well be positive or negative, and their grand

Table 2. Statistical methods

Table 2(a). Example of calculation of O-E and its variance: hypothetical data

	Allocated treatment	Allocated control	Both together
Dead	Obs = 25 Exp = 32.5	Obs = 40 Exp = 32.5	65
Alive			135
Total	100	100	200

O = "Observed" number of deaths in the treatment group = 25

If a total of 65 die and treatment has no effect, then
E = "Expected" no. of deaths in treatment group = half of 65 = 32.5

Statistical calculation (treatment group only): O-E = 25 - 32.5 = -7.5
N.B. **Minus** denotes **benefit**, and **-7.5** suggests about **15** deaths avoided.

Finally, the "variance" of O-E = 32.5 x (100/200) x 135/(200-1) = 11.0

Table 2(b). Principle of unbiased combination of randomized trial results

Trial 1	Result 1
Trial 2	Result 2
Trial 3	Result 3
Sum of **separate*** results	Overview result = **grand total**, i.e. Result 1 + Result 2 + Result 3

* If treatment had no effect on outcome in any trial then each of the results, considered separately, would differ only randomly from zero, and so too would their grand total *(5,31)*. (An overall test of whether the grand total differs from zero does not depend on the unjustified assumption that any real effects in different trials must be of similar size.)

total would likewise differ only randomly from zero: Table 2(b). If, on the other hand, treatment reduced the risk of death to some extent in most or all of the trials, then any individual O-E value would be likely to be somewhat negative (i.e. favoring treatment), so when they are all added up their grand total may be clearly negative. Such arguments do not, of course, assume that the size of treatment effect is the same in all patients, or in all trials; indeed, it is probably not.

5.3 Routine stratification of all main analyses by age (<50, 50+) and by year (1, 2, 3, 4, 5+) of follow-up

As a second step towards the ideal of comparing only like with like, the "expected" numbers of events can be calculated separately in different patient categories within each trial, and the separate O-E values for each category can then be added up to get the overall O-E value for that trial. So, for example, in the present report, expected numbers of events have been calculated separately for women aged under 50 and for women aged 50 or older (see, for example, Table 3M). Moreover, within each age group they have first been calculated separately for each separate year of follow-up and then added up to get the overall O-E for that particular trial. (This yields a "logrank" year-of-death analysis.[31]) In principle, such "stratified" comparisons may be slightly preferable. In practice, however, the overall result that they yield for each major trial may differ only slightly from that yielded by a crude calculation based simply on comparison within each trial of total mortality in all treatment-allocated patients with total mortality in all control patients.* Stratification by these age groups corresponds approximately with menopausal status. (Two trials that included only postmenopausal women did not provide a subdivision of their results by age: their patients are classified as "50+".)

5.4 Additional stratification of selected analyses for the available information on nodal status or Estrogen Receptor status

For a substantial majority of all patients some information on what was recorded about their axillary lymph node status was available that could be used to classify patients into four main categories (the last of which included "nodal status not reported"). In addition, for a substantial minority of those in tamoxifen trials some information about Estrogen Receptor (ER) measurements in the excised primary tumor was available that could likewise be used to classify patients into four main categories (the last being "ER measurement not reported"). These nodal or ER categories could be used in a cautious search for interactions. Alternatively, they could be used for the "retrospective stratification" of analyses of the overall effects of treatment. But, in such a large data set retrospective stratification of the overall analyses is likely to make no material difference to the overall results (unless, by mistake, a stratum of "unknown" values for the stratifying factor had failed to be defined and used).

5.5 Arithmetic procedures for calculation of Observed minus Expected (O-E) numbers of events among treatment-allocated patients, and for obtaining "two-sided" significance levels (2P)

Suppose that a total number N of patients are randomized into a study with a fraction f allocated active treatment (and hence 1-f allocated control), and suppose that in both groups combined a total number D of these N patients die. If there is no difference in mortality between the treatment and control groups, the "Expected" number of deaths among the treatment-allocated patients is given by fD. Let O denote the Observed number of deaths among these treatment-allocated patients. The quantity Observed minus Expected (O-E) then differs only randomly from zero, its variance can be shown[5] to be given by the formula $fD(1-f)(N-D)/(N-1)$, and its standard deviation (sd) is the square root of this variance. An example of the calculation of O-E and its variance is given in Table 2.

As is usually the case in statistics, (O-E) values that differ from zero by 1 sd or less could easily arise just by the play of chance, and although the play of chance provides a less plausible (two-sided P-value* = 0.05) explanation for differences of about 2 sd, such differences can also arise just by chance, especially when many different comparisons are scrutinized.

5.6 Interpretation of P-values: statistical significance and medical judgement

In an unbiased overview of many properly randomized trial results the number of patients involved may be large, so even a realistically moderate treatment difference may produce a really extreme P-value (e.g. 2P<0.0001). When this happens, the P-value alone is generally sufficient to provide proof beyond reasonable doubt that a real treatment effect does exist. In contrast, P-values that are not conventionally significant, or that are only moderately significant (e.g. 2P=0.01), may be much more difficult to interpret appropriately. There may be circumstances (particularly in sub-analyses of an overall analysis that is not clearly significant) when the existence of a real treatment benefit should not be accepted despite an apparent difference of 2 or 2.5 (or even, perhaps, almost 3) standard deviations between treatment and control. Conversely, there may be circumstances (particularly in sub-analyses of a highly

---

* For example, in the NATO trial of tamoxifen, young women had a worse prognosis than older women and the play of chance happened to put somewhat more young women into the tamoxifen group than into the control group (69 vs 57). Correction for this imbalance, by calculating the O-E values separately in each age group and then adding them up, will increase the statistical significance of the overall result, but the change is small. (An uncorrected difference of 2.79 standard deviations becomes an "age-corrected" difference of 2.91 standard deviations.) Likewise, because treated patients tended to survive longer than control patients, they constituted somewhat more than half the survivors later on in the trial. So, if treatment had no effect on death in year 5, slightly more than half the deaths in year 5 would be expected among treatment-allocated patients. Calculations of the O-E values separately for each year of the trial corrects for this, but again this Logrank "year-corrected" analysis (see below) changes the overall result only slightly (from 2.91 sd to 2.94 sd).

---

* Two-sided P-values (2P) are used throughout, estimated from the "normal" approximation that if treatment had no effect whatever on outcome then (O-E)/sd would be distributed approximately like the standard normal (bell-shaped) distribution. In the standard normal distribution the probability of a result being, just by chance, less than -2 is about 0.025. The probability of it being bigger than +2 is also about 0.025, so the total probability (which is written "2P") of it differing from zero by more than 2 is about 0.05 (i.e. 0.025 + 0.025). If, therefore, (O-E) is negative, indicating a favorable effect of treatment, and is about equal to -2 sd then the two-sided P-value is about 0.05. (Values of -2.6 sd, -3.3 sd and -3.9 sd would correspond to 2P=0.01, 2P=0.001 and 2P=0.0001). Hence, "2P=0.05" means that if treatment does nothing at all then 0.05 is the approximate probability of getting, just by chance, a result at least as **extreme** as that actually observed (i.e. at least as good as -2 sd in favor of treatment **or** at least as bad as +2 sd against treatment). In the Figures, the estimated significance levels are printed to 2, 3, 4 or 5 decimal places, according to whether 2P<0.1, 0.01, 0.001 or 0.0001. The abbreviation NS (i.e. Not Significant) is used to denote 2P>0.1.

Table 3M. **MORTALITY in all available unconfounded randomized trials of post-mastectomy radiotherapy (logrank analyses of yearly rates, excluding deaths after 1 Sep 1985)**

Study No.	Study name	Systemic adjuvant, both arms	Age <50 at entry Deaths entered XRT	Control	Statistics O-E	Var	Age 50+ at entry Deaths/entered XRT	Control	Statistics O-E	Var	Sum of statistics for young & old O-E	Var	Chi-sq.
64B1	Oslo X-ray	—	37/111	38/108	-0.7	15.8	105/174	95/159	0.5	31.3	*	*	
64B2	Oslo Co-60	—	33/ 90	32/101	1.6	13.6	78/188	71/184	1.9	28.3	3.3	89.0	0.1
69A	Heidelberg XRT	—	12/ 20	12/ 17	-0.8	4.8	51/ 64	28/ 41	5.5	13.2	4.7	17.9	1.2
70A1	Manchester RBS1	—	44/132	55/137	-5.4	21.4	125/223	128/222	-1.8	46.1	-7.2	67.5	0.8
70B	King's/Cambridge	—	209/496	201/482	1.4	86.8	467/880	476/942	13.1	182.4	14.6	269.2	0.8
71B	Stockholm A	—	24/ 91	31/ 83	-5.3	12.3	95/232	101/238	-1.4	41.1	-6.7	53.4	0.8
71C	NSABP B-04**	—	43/117	50/118	-6.7	19.1	102/235	120/247	1.3	47.4	-5.4	66.5	0.4
72A	WSSA Glasgow	—	9/ 31	24/ 48	-5.3	7.0	39/ 63	43/ 75	1.4	14.3	-3.9	21.3	0.7
73A	Wessex	—	9/ 26	15/ 24	-4.1	4.9	18/ 45	25/ 51	-2.8	9.0	-6.9	13.9	3.5
73C1	Mayo Clinic	—	0/ 0	1/ 3	0.0	0.0	1/ 5	3/ 5	-1.2	0.8	*	*	
73C1	Mayo Clinic	CFPr	1/ 3	2/ 2	-0.9	0.6	4/ 8	6/ 8	-1.6	1.7	*	*	
73C2	Mayo Clinic	CFPr	2/ 5	2/ 2	-1.0	0.7	30/ 62	23/ 56	1.8	11.3	*	*	
73C3	Mayo Clinic	CFPr	5/ 12	4/ 15	1.1	2.0	3/ 4	0/ 2	1.1	0.6	*	*	
73C4	Mayo Clinic	CFPr	7/ 17	10/ 23	-0.9	3.8	3/ 5	2/ 4	0.6	0.9	-0.9	22.3	0.0
74B	Edinburgh I	—	18/ 73	15/ 65	0.8	7.6	32/100	35/110	1.0	14.6	1.8	22.2	0.2
75K	Piedmont OA	Mel	11/ 25	8/ 29	1.9	4.3	23/ 51	22/ 41	-1.9	9.6	*	*	
75K	Piedmont OA	CMF	15/ 26	6/ 24	4.4	4.5	22/ 43	20/ 42	0.5	8.9	4.9	27.3	0.9
76A2	SECSG 1	CMF	6/ 21	13/ 27	-2.6	4.2	12/ 28	17/ 34	-0.6	5.9	*	*	
76A3	SECSG 1	CMF	13/ 33	8/ 28	3.0	4.4	8/ 45	14/ 40	-3.9	4.9	-4.0	19.4	0.8
76C	Glasgow	CMF	16/ 47	9/ 33	0.6	5.3	28/ 65	33/ 69	-0.6	13.0	0.0	18.3	0.0
78A1	S Swedish BCG	C	17/ 98	16/ 91	0.4	7.9	9/ 49	11/ 48	-1.0	4.7	*	*	
78A2	S Swedish BCG	Tam	0/ 1	0/ 0	0.0	0.0	43/238	51/243	-5.6	22.0	-6.3	34.6	1.1
78D4	Scottish D	Tam	2/ 8	1/ 7	0.4	0.7	5/ 16	2/ 16	1.4	1.7	*	*	
78D4	Scottish D	—	0/ 6	0/ 7	0.0	0.0	6/ 17	2/ 16	2.0	1.9	3.8	4.3	3.3
78G1	CCABC Canada	CMF	6/ 23	4/ 29	0.9	2.3	0/ 8	2/ 6	-1.0	0.5	*	*	
78G1	CCABC Canada	CMF+Ooph	3/ 30	3/ 27	0.0	1.4	0/ 5	0/ 6	0.0	0.0	*	*	
78G2	CCABC Canada	CMF	0/ 12	1/ 9	-0.7	0.2	0/ 4	0/ 0	0.0	0.0	*	*	
78G3	CCABC Canada	CMF	14/ 61	10/ 54	1.2	5.5	2/ 11	4/ 6	-2.4	1.0	-2.1	11.0	0.5
82B	Danish BCG 82b	CMF	10/165	7/144	1.3	4.1	6/ 59	8/ 68	-0.5	3.3	0.7	7.3	0.1
82C1	Danish BCG 82c	Tam	0/ 3	0/ 1	0.0	0.0	19/229	18/232	0.5	8.8	0.5	8.8	0.0
84A2	BMFT 03 Germany	CMF	0/ 7	0/ 4	0.0	0.0	0/ 15	0/ 19	0.0	0.0	0.0	0.0	0.0
TOTAL: 19 TRIALS*** (Odds reduction ± SD)		+/-	566/1790	578/1742	-15.4 (6% ± 6)	245.0	1336/3171	1360/3230	6.4 (-1% ± 4)	529.2	-9.0 (1% ± 4)	774.2	0.1

Approximate chi-squared test for heterogeneity on 17 degrees of freedom = 15.2; 2P>0.1; NS.
* Items marked with an asterisk contribute to totals directly below them.   **Published results *(3)*, since individual patient data not available.
*** Data from about 10 randomized radiotherapy trials that began before 1.1.1985 were not available in 1985, and are not included here. Data from one large trial (Manchester Christie) have been excluded because of non-standard randomization: inclusion would have changed the "sum of statistics" to -10.5, 955.0, 0.1 & 1% ± 3

significant overall analysis, or when there is strong indirect evidence from other sources) when statistical significance is of little concern, and when the existence of a real treatment benefit may be quite firmly accepted despite the apparent difference not being conventionally significant (or even, perhaps, being slightly unfavorable). In summary, **over-emphasis on formalistic questions of which differences are moderately significant and which are not is a serious statistical mistake**, and it may have serious medical consequences. Elsewhere,[30,31] there is more extensive discussion of the common sense medico-statistical principles that underly appropriate interpretation of P-values in trials.

5.7 RESULTS OF RADIOTHERAPY TRIALS as an example of the summation of (O-E) values from different trials to provide an overall test of the "null hypothesis" of no treatment effect

As an example of the use of (O-E) for the combination of information from related studies, consider the question of whether radiotherapy after mastectomy affects survival in early breast cancer. There have been about 30 trials of radiotherapy after mastectomy, information from 19 of which was obtained in 1985. To describe the statistical methods, an analysis of the results from just these 19 trials is given in Table 3M.*

In Table 3M, the Grand Total (GT) of all the separate O-E values from the radiotherapy trials is -9.0, suggesting that there may have been about 18 deaths avoided in the radiotherapy groups. The standard deviation (sd) of this grand total is 27.8, however, so that the grand total is less than half a standard deviation away from zero (z=GT/sd = 0.3; NS). Even if radiotherapy had no net effect whatever on survival in the circumstances of these trials, such a small difference from zero could well have arisen just by the play of chance. This overview is not as complete as might be wished because data from about a dozen other trials were still not available in 1985 (Table 3M footnote); for future overviews, however, those will be sought.

---

* A previous review[3] of some of the mature trials of radiotherapy after mastectomy differs in two ways from Table 3M. First, Table 3M includes many trials that are not yet mature and which, therefore, contribute data only on early survival. Second, Table 3M excludes the early Manchester trials from the main analysis because treatment allocation in those trials was based on odd/even birth dates and may, at least in principle, have been subject to some bias. (For the sake of comparison with the previous review, overall results with the Manchester trials included are given in footnotes to these tables.) Nevertheless, the two reviews have much in common, and neither provides good evidence of any favorable net effect of radiotherapy on medium-term (e.g. 10-15 years) survival after mastectomy.

5.8 Use of (O-E) values to provide a description of the typical reduction in the odds of treatment failure (i.e. to describe the "alternative hypothesis")

For practical purposes, what is required in the assessment of an effective treatment is not only evidence that the treatment does do something (i.e. a "test of the null hypothesis") but also an estimate of how big, and hence how medically worthwhile, the effect of treatment is likely to be (i.e. a description of the "alternative hypothesis"). Fortunately, the quantities already calculated (i.e. the standard deviation of the grand total, and z, the number of standard deviations by which the grand total differs from zero) can provide not only a statistically sensitive test of whether treatment has any effect but also a useful description of the size of the treatment effect that is "typical" of the set of trials being reviewed. A convenient and appropriately weighted estimate of the typical treatment effect is provided by the "typical odds ratio" (TOR) — that is, the typical ratio of the odds of an unfavorable outcome among treatment-allocated patients to the corresponding odds among controls. (For example, a typical odds ratio of 0.8 would correspond to a reduction of about 20% in the odds of an unfavorable outcome.) The typical odds ratio is estimated in Table 3M and elsewhere by the surprisingly simple formula $\exp(z/sd)$,* with approximate 95% confidence limits $\exp(z/sd \pm 1.96/sd)$. The typical percent reduction in the odds of an unfavorable outcome is similarly estimated by $r = 100 - 100\exp(z/sd)$, with the approximate standard deviation for this estimated reduction being $-r/z$.

The typical odds ratio suggested by the partial overview of radiotherapy trials in Table 3M is 0.99, with 95% confidence limits of 0.92 to 1.06. Inclusion of an odds ratio of 1.00 in the confidence interval indicates that the result is not conventionally significant, while the extremes of the interval indicate that the data from these particular 19 trials are readily compatible both with a small mortality reduction and with a small mortality increase.

5.9 Graphical display methods for separate trial results, and for an overview

Data for each of the radiotherapy trials in Table 3M are provided separately for women aged under 50 and 50 or older at randomization. Those trials (e.g. 64B1 and 64B2; see Section 6) that consist of more than one part are further subdivided into their constituent parts for separate analysis. In principle this is appropriate and necessary, but in practice it can produce a mass of numbers that is difficult to grasp without juxtaposing graphical and numerical presentations of the data (Figure 1M). In the tabular parts of Figure 1M, for each separate trial the results of the analyses for women aged under 50 and 50 or over — and, for trials with more than one part, the results for each part — are added together, to give just one overall result for that one trial. In the graphical part of Figure 1M (and of all other such figures), a **solid square** indicates the apparent effect of treatment in that trial (i.e. the ratio of the annual odds of death among treatment-allocated patients to that among controls), and a **horizontal line** indicates the 99% confidence limits for this odds ratio.

The visual impression given by the horizontal lines may be slightly misleading, for a small, uninformative trial gives a long, visually striking confidence interval, whereas a large, reliable trial gives a small, unobtrusive confidence interval. To reverse this visual impression, therefore, the sizes of the solid squares have been chosen to be directly proportional to the amount of information each trial contains. So, a large informative trial yields a large black square and a small trial that is much less informative yields a small black square.* The areas of the solid squares indicate that nearly half of the available information comes from one particularly large trial (70B) which has a slightly unpromising result, and that about half the remaining information comes from three moderately large trials (64B, 70A, 71B), none with particularly promising results. Formal summation of the 19 separate O-E values, one per trial, confirms that these studies provide little evidence overall of any real effect on survival. The "typical odds ratio" suggested by this overview of just 19 of the radiotherapy trials is 0.99 ± 0.04, which is nowhere near statistically significant ($z = -0.3$). In Figure 1M (and in subsequent figures), the statistical reliability of the overview result is depicted by a black square in the left-hand margin (with size proportional to the "information content"** of the overview), and a 95% confidence interval for the "typical odds reduction" suggested by the overview is depicted by a diamond-shaped symbol. It can be seen from the diamond-shaped 95% confidence interval in Figure 1M how reliable the overall result is. Of course, since there are so many individual trials it was quite likely that just by the play of chance one or two of them might have yielded results that are 2 to 2.5 sd away from the overall average. Because of this multiplicity of comparisons, fairly extreme (99%) confidence limits have been plotted for the many individual trials. It can be seen from the 99% confidence intervals for the separate trial results that each individual trial result is statistically compatible with the overall result. An overview is not subject to such multiple comparison problems, so for each overview result 95% limits have been plotted. **This convention (99% limits for individual trials, 95% limits for overviews) is used throughout the present report.** An additional general convention is that the symbol $\pm$ will be used to denote **one** standard deviation (as, for example, in "a mortality reduction of 16% $\pm$ 3").

5.10 Similarities between risk ratios, death rate ratios and odds ratios when event rates are low

In practice, when comparing mortality (or recurrence) among treated and control patients in trial analyses when substantial proportions have not yet suffered the endpoint of interest, "risk ratios", "death rate ratios" and "odds ratios" are often not importantly different from each other as methods for describing

---

* It can be shown that z/sd is the "one-step" estimator of the log odds ratio, i.e. the first step from a log odds ratio of zero towards the "maximum-likelihood" estimator in a standard (Newton-Raphson) iterative search for the maximum of the log-likelihood function. Hence, exp(z/sd) is called the one-step estimator of the odds ratio.[32] Any bias in this one-step estimator will be negligible in overviews of randomized trials involving **small** treatment effects and **reasonably large** numbers of outcome events.[32] In practical analyses of substantial trial results, it appears that (as long as there is less than a twofold difference in odds and at least several dozen endpoints) the one-step and the maximum-likelihood are about as accurate as each other as estimators of the true odds ratio: for example, from the ISIS-2 data in Table 1 they yield 0.772 and 0.771 respectively.

---

* Formally, the area of each solid square has been made proportional to the variance of O-E, since when trying to compare two treatments the size of the variance of O-E can be used in a statistical sense as an estimate of the "information content" of the data (i.e. as the local curvature of the log-likelihood[5]).

** The area of the black square describes the amount of information in the overview. It is simply the sum of the areas of the solid squares plotted for the individual trials that contribute to the overview, since the sizes of all the squares throughout this report involve the same scale factor. (The scale factor that was chosen makes the length of the base of each square equal 0.01 times the square root of the corresponding variance.)

Fig. 1M. **Example of graphic display of an overview of 19** trials: MORTALITY in all available unconfounded randomized post-mastectomy radiotherapy trials**

MORTALITY ANALYSES ONLY, FOR RADIOTHERAPY

Study No.	Study Name	No. Events / No. Entered Treatment	Control	O–E	Variance	Odds Ratio ♣ (Treatment : Control)	Odds Redn. (± S.D.)
64B	Oslo	253/563	236/552	3.3	89.0		
69A	Heidelberg XRT	63/84	40/58	4.7	17.9		
70A₁	Manchester RBS1	169/355	183/359	−7.2	67.5		
70B	Kings/Cambridge	676/1376	677/1424	14.6	269.2		
71B	Stockholm A	119/323	132/321	−6.7	53.4		
71C₁	NSABP B-04 *	145/352	170/365	−5.4	66.5		
72A	WSSA Glasgow	48/94	67/123	−3.9	21.3		
73A	Wessex	27/71	40/75	−6.9	13.9		
73C	Mayo Clinic	56/121	53/120	−0.9	22.3		
74B	Edinburgh I	50/173	50/175	1.8	22.2		
75K	Piedmont OA	71/145	56/136	4.9	27.3		
76A₂₋₃	SECSG 1	39/127	52/129	−4.0	19.4		
76C	Glasgow	44/112	42/102	0.0	18.3		
78A	S Swedish BCG	69/386	78/382	−6.3	34.6		
78D₄	Scottish D	13/47	5/46	3.8	4.3		
78G	CCABC Canada	25/154	24/137	−2.1	11.0		
82B₁	Danish BCG 82b	16/224	15/212	0.7	7.3		
82C	Danish BCG 82c	19/232	18/233	0.5	8.8		
84A₂	BMFT 03 Germany	0/22	0/23				
	Total	**1902/4961**	**1938/4972**	**−9.0**	**774.2**		**1% ± 4**

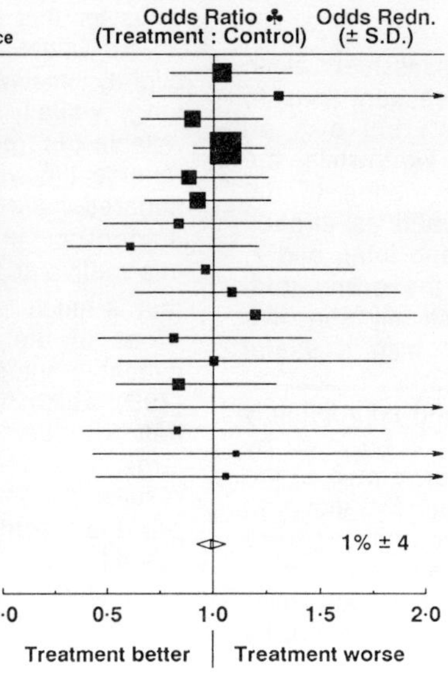

0.0   0.5   1.0   1.5   2.0
Treatment better | Treatment worse

Test for heterogeneity: $X^2_{17} = 15.1$; $2P > 0.1$; NS

Treatment effect $2P > 0.1$; NS

* Published results (3), since individual patient data not available.
** Data from about 10 randomized radiotherapy trials that began before 1.1.1985 were not available in 1985 and are not included here. Data from one large trial (Manchester Christie 49B) have been excluded because of non-standard randomization.
♣ 95% confidence intervals for overview and 99% for individual trials.

---

trial results. Consider, for example, a hypothetical group of 100 patients suffering a steady death rate of 1% per month. After 3 years, only about 30 of the original group would be dead (and not 36, since the death rate of 1% per month refers to those alive at the beginning of each month, and their number decreases as time goes by). Thus, the **risk** of death at 3 years would be 30/100 while the **odds** of death would be 30/70. A 20% reduction in the monthly death rate would reduce the death rate from 1.0% to 0.8% per month. So, by 3 years, only 25 would then be dead. Thus, the risk of death would be 25/100 and the odds of death would be 25/75. Comparison of these with the previous figures (30/100 and 30/70 respectively) shows that in this particular example a 20% reduction in the monthly death rate corresponds after 3 years to a 17% reduction in the risk of death, and a 23% reduction in the odds of death. The percent reduction in odds will always be the largest, but the similarity of these three figures suggests that it often does not matter much whether percentage reductions in the risk of death, in the death rate or in the odds of death are used to describe trial results. For reasons of arithmetic simplicity, use will chiefly be made of percentage reductions in the odds of death (or, for trials where data are available separately for each year of follow-up, percentage reductions in the **annual** odds of death: see below).

5.11 Logrank "year of death" analyses for statistical significance tests, and for estimation of reductions in annual odds of death

In most of the trials to be reviewed, results are available separately in year 1 (i.e. during the year following randomization), year 2, year 3, year 4 and year 5+ (where "5+" includes all events recorded during or after the fifth year). The numbers of deaths in the treatment-allocated and control-allocated groups in a particular period were related to the numbers of patients in each group that were still alive and being followed up at the start of that period (using methods described in Table 2). For each trial this yields up to five separate (O-E) values and their corresponding variances, one per time period (with fewer than five in those trials with fewer than five years of follow-up). The sum of these five (or fewer) values for one trial yields the logrank test statistic for a year-of-death analysis of that trial, with the logrank variance given by the sum of the five variances. The logrank statistic was first recommended in 1966 for the analysis of single cancer trials,[29] and it was first recommended in 1976 for the unbiased, statistically efficient combination of information from an overview of the results of several different cancer trials.[30,31]

The chief advantage of the availability of information from each separate year is not that logrank analyses are more sensitive than crude analyses, for the improvement in sensitivity is only small in trials where most patients survive. It is rather that logrank analyses readily permit separate analyses of the effects of treatment in each separate year, which may improve medical understanding (as long as the important statistical uncertainties in the apparent effects in each separate year are taken properly into account). Hence, when they are available, annual logrank O-E values will be used in preference to overall crude O-E values. In providing a description of the alternate hypothesis, the calculation of exp(z/sd) now yields an estimate of the typical ratio of (and, from this, the percentage reduction in) the **annual** odds of death. The estimate for a given year is, of course, based only on data derived from trials that have follow-up and deaths reported in that year. (Similarly, for analyses of disease-free survival exactly the same statistical methods can be applied to the numbers of patients suffering a first recurrence or prior death.)

5.12 Life-table estimation for descriptive purposes

Suppose that appropriate analysis of a particular set of trials yielded, in years 1, 2, 3, 4 and 5+, a particular set of five odds ratios. In principle, the meaning of these could be illustrated by applying them either to the failure rates of good-prognosis women (yielding a pair of estimated survival curves that are both "good-prognosis") or to the failure rates of poor-prognosis women (yielding a pair of estimated survival curves that are both "poor-prognosis"). Arbitrarily, we shall generally choose to illustrate them by applying them to a hypothetical category of women whose prognosis in years 1, 2, 3, 4 and 5+ is the average of that for all patients in these particular trials.* This yields a pair of survival curves that will in practice be rather similar to those that would have been obtained by crudely mixing all the trials together. The survival curves actually obtained in this way, however, are not subject to any of the objections that might be raised by analyses of a mixture of many trials. In particular, they can differ systematically only if **within** trials the treatment-allocated patients differ systematically from those allocated control.

5.13 Test of heterogeneity between several different trial results

In the case of a treatment (such as radiotherapy, perhaps) that appears to have little or no overall effect on mortality it may be plausible that in each separate trial it will have little or no overall effect (except for the known side-effects of radiotherapy, which should be slight during the first decade or so[28]), in which case a test for heterogeneity would be expected to yield a null result. In confirmation of this expectation, the formal chi-squared test of heterogeneity for the 19 radiotherapy results in Figure 1 does yield a completely non-significant result (chi-square = 14.9 on 17 degrees of freedom, NS). In general, however, when different trials address a particular type of treatment some degree of heterogeneity of the **real** effects of the treatment regimens in those trials must be expected.

Standard statistical tests for heterogeneity between many different trials (or patient categories) are, however, of limited value, partly because they are statistically insensitive, and partly because some heterogeneity of the real effects of treatment in the different trials (or categories) is likely to exist no matter what a formal test for heterogeneity may indicate. In general, therefore, even if a standard "chi-squared" test does not provide conventionally significant evidence of any heterogeneity at all, some important heterogeneity may well still exist that failed to be detected because the heterogeneity test is so crude. Sometimes, when the trials or categories can be arranged in some meaningful order, a test for trend can be used instead, and is then likely to be more informative than a test for heterogeneity. But, whether or not there is some real heterogeneity (and whether or not some test for trend or for heterogeneity happens to yield a conventionally significant result), this does not invalidate the standard overview techniques used to analyze the trials.

5.14 Arithmetic details of tests for trend, for heterogeneity and for interaction

If treatment effects are evaluated in various different circumstances (e.g. in each of 4 different age groups, or in each of 40 different trials) then the following approximate procedures will be used to test for a trend between the separate results where there is a natural ordering (e.g. from younger to older), or for heterogeneity among them where no natural ordering exists.

(a) **Test for trend:** The circumstances are numbered in their natural order (e.g. 1 = age <40, 2 = age 40-49, 3 = age 50-59, 4 = age 60+). O-E and its variance, V, are calculated separately for the treatment effect in each circumstance (e.g. O_1-E_1 and V_1 for circumstance 1), omitting any circumstances for which the variance is zero, i.e. in which there is no useful information. Let k denote the number of circumstances remaining (i.e with non-zero V). Next, the following values are calculated:

$A = V_1+V_2+V_3+...$

$B = 1.V_1+2.V_2+3.V_3+...$

$C = 1.1.V_1+2.2.V_2+3.3.V_3+...$

$D = (O_1-E_1)+(O_2-E_2)+(O_3-E_3)+...$

$E = 1.(O_1-E_1)+2.(O_2-E_2)+3.(O_3-E_3)+...$

$F = (O_1-E_1)^2/V_1+(O_2-E_2)^2/V_2+(O_3-E_3)^2/V_3+...$

A test for a trend between the odds ratios produced by treatment in these different circumstances may be based on calculation of the quantity (E−DB/A). If there is no real heterogeneity between the odds ratios, then it can be shown that this quantity will differ only randomly from zero, and that its standard deviation (sd) will be approximately $\sqrt{(C-BB/A)}$. Values more extreme than ±1.96 sd would therefore correspond approximately to 2P<0.05, etc. Provided the effects of treatment are not large, the statistical properties[5,31] of O-E and V imply that this trend test is asymptotically efficient at detecting a steady multiplicative trend in the odds ratios that are produced by treatment on going from one circumstance to the next.

(b) **Test for heterogeneity:** A test for heterogeneity may be obtained by calculating the quantity (F-DD/A). If there is no real heterogeneity between the odds ratios in the k different circumstances being considered, then this quantity will be distributed approximately as a standard chi-squared distribution with k-1 degrees of freedom. Such tests for heterogeneity among several circumstances can, however, be very crude (see above). Hence, if a test for trend makes medical sense (as, for example, when there is a natural ordering between the circumstances being considered) then a test for trend should generally be used rather than a test for heterogeneity, for a trend test is likely to be much more sensitive to any real differences that may exist between the sizes of the treatment effects in different circumstances.

(c) **Test for "interaction" between the treatment effects in just two different circumstances:** In this case, the tests for trend and for heterogeneity (with k=2) can be shown to yield identical significance levels.

5.15 Practical meaning of a clear effect of treatment in a single large trial

Suppose, hypothetically, that only one trial had ever addressed a particular therapeutic question, that the trial was extraordinarily large, and that it yielded a very definite 45% mortality reduction with very tight 95% confidence limits (e.g. 40%-50%). What would the practical implications be? Although this strong result would still not imply that another large trial of some approximately similar treatment for some approximately similar patients must also yield a 40-50% effect, it might well imply that such a trial should be expected to yield an effect in the same direction and of **approximately** similar size (e.g. a 30% reduction, or a 60% reduction, perhaps). Likewise, although the original trial result would not guarantee that attempts to use that treatment in the future would produce a 40-50% mortality reduction outside trials (since future patients could well differ in various ways from the trial patients and there could well be important differences between the care of patients in trials and out of trials), an **approximately** similar effect on mortality might be expected. In making sensible use of really clear results from a single clinical trial, the key assumption is merely that, in somewhat different medical circumstances, therapeutic effects in the same direction but of only approximately similar magnitude are still

---

* In one particular time period, if r, the estimated annual failure rate, equals [no. of failures/no. of woman-years], and b, the estimated log odds ratio (treatment : control), equals [total of (O-E) values/total of their variances] then p, the estimated probability of avoiding failure for one year, equals exp(-r), and the separate probability estimates would be p+0.5p(p-1)b for treated patients and p-0.5p(p-1)b for the control patients.

likely to exist. Extrapolation too far, of course, may lead to mistaken decisions about treatment, but so too may failure to extrapolate far enough. Thus, even for a single trial result, an estimated risk reduction with tight confidence limits implies only that similar, but not necessarily identical, treatment effects will be achieved in other circumstances.

5.16 Practical meaning of a clear effect of treatment in an overview of many trials

Exactly the same is true of the "typical mortality reductions" and associated confidence limits derived from an overview of many trials. As with a single trial, the statistical calculations address the question "Given the studies that were undertaken, what range of results are statistically compatible with the actual data?". They do **not** involve saying "If different trials had been performed, what would have been seen?".

After calculation of the "typical mortality reduction" and its associated confidence limits, medical judgement — with its attendant uncertainties and disputes — is needed to help determine the circumstances to which that result is likely to be approximately relevant, just as was the case with a large single trial result.

It should be noted that just as a positive trial result does not guarantee that all patients will benefit from the treatment being tested, so too a null result in a trial or an overview does not guarantee that no patient will benefit. It does set limits on what the average difference in medium-term mortality is likely to be but, for example, the null result in Figure 1M is easily compatible with the suggestion that several dozen of the 4000 treated patients might have been protected from death within the first decade or so by radiotherapy, but that the play of chance has obscured this. (It is also compatible, at least in principle, with the opposite possibility of several dozen deaths having been caused by radiotherapy, although any serious adverse effects on causes of death other than breast cancer are likely to be revealed more reliably by the cause-specific mortality overviews planned for future reports than by the present all-causes mortality analyses.) Thus, although it is important to be aware of the clinical trial results, it is also important to be aware of their statistical limitations.

5.17 Fixed-effect "assumption-free" methods, and random-effect "assumed representativeness" methods

The general approach that has been described in the present report for analyzing and interpreting overviews is sometimes called the "fixed effects" method, because the overall result that it gets (by comparing like with like within each separate trial) is not directly influenced by any heterogeneity among the true effects of treatment in different trials. This terminology is, however, unsatisfactory, for it misleadingly suggests that any heterogeneity between the true effects of treatment in different trials is assumed to be zero in this general approach, whereas in fact no such unjustified assumptions are involved — indeed, the "assumption-free" method might be a better name.

When several trials have addressed similar questions it might appear that formal statistical analyses of the heterogeneity of their findings (perhaps assessing it by some "random effects" statistical method) are needed to augment the use of medical judgement in determining how far an overview of their results can be trusted. But, the statistical assumptions needed for such statistical methods to be of direct medical relevance are unlikely to be met. In particular, the different trial designs that were adopted would have to have been randomly selected from some underlying set of possibilities that includes the populations about which predictions are to be made. This is unlikely to be the case, since trial designs are adopted for a variety of reasons, many of which depend in a complex way on the apparent results of earlier trials. Moreover, selective factors that are difficult to define may affect the types of patients in trials, and therapeutic factors that are also difficult to define may differ between trials, or between past trials and future medical practice. Finally, tests of the heterogeneity of the results of many trials may be biased by a tendency for trials with extreme results in either direction to stop recruitment early. Thus, whereas various "fixed effects" methods may actually be assumption-free, various "random effects" methods may unjustifiably assume representativeness.[*] In view of these difficulties, such "assumed representativeness" (random-effects) methods are not used in the present report.

Even if formal statistical estimates of the degree of heterogeneity that exists may not help much in judging how far the overall results should be trusted, it is obviously sensible to scrutinize thoughtfully any "outliers" in an overview of many trial results. There are, however, many different medical questions that may reasonably be asked of such a large body of data, and different readers may prefer different statistical methods. Hence, in the accompanying reports, the data are presented in sufficient detail to allow a variety of different analyses of them.

---

[*] Moreover, such methods may be of limited statistical sensitivity, particularly when just a few major studies provide most of the evidence. Hence, the loss of statistical power may be illustrated in a particularly extreme form by an important practical example involving only two major trials, where the "assumption-free" (fixed-effects) methods yield a statistically definite ($P<0.0001$) answer that is strongly supported by a wide range of indirect evidence, but where various "assumed-representativeness" methods might inappropriately fail to do so. There have been only two major randomized trials of the primary prevention of non-fatal myocardial infarction by long-term antiplatelet therapy.[33] The results of one (129/11037 (1.2%) aspirin versus 213/11034 (1.9%) control, $P<0.0001$) indicated a highly significant 39% ± 9 reduction in the odds of suffering a non-fatal myocardial infarction, but the results of the other (80/3429 (2.3%) aspirin versus 41/1710 (2.4%) control, NS) indicated a reduction of only 3% ± 19, i.e. virtually no difference. (Although the discrepancy between their results appears striking it is in fact only a 2.1 standard deviation discrepancy, so it could well be largely or wholly attributed to the play of chance and/or the data-dependent early closure of the trial with the significant result.) As is generally the case, the overview of the two results may be more reliable than either considered in isolation, and the "assumption-free" overview methods used in the present report indicate an overall reduction of 32% ± 8 ($P<0.0001$) in the odds of non-fatal myocardial infarction.[33] A reduction of about this size is rendered extremely plausible by its similarity to the significant reductions in non-fatal myocardial infarction (or reinfarction) that have been shown for antiplatelet therapy in other circumstances by randomized trial overviews (e.g. 31% ± 5 among patients with previous myocardial infarction, 35% ± 12 among patients with previous stroke, 35% ± 17 among patients with unstable angina, or 49% ± 9 among patients who were in hospital because of a suspected acute heart attack).[33] Hence, the use of "assumption-free" (i.e. fixed-effects) methods in a standard overview of these two primary prevention trials yields an extremely definite answer that is almost certainly qualitatively correct. In contrast, since only one of the trials yields a significantly favorable result while the other yields a completely null result, some "assumed-representativeness" (i.e. random-effects) methods might misleadingly have concluded that the two studies together showed no clear evidence of benefit.

6. MATERIALS AVAILABLE FOR REVIEW

Appendix Table 1 lists (i) all the relevant randomized trials from which data are available for the overview, together with (ii) a few randomized studies that would be relevant if data were available from them, and (iii) a few studies that might well be mistaken for randomized studies, but that cannot contribute to the overview because in fact they used non-random methods (e.g. alternate allocation, or allocation based on odd/even dates of birth or record numbers). The trials are sorted with respect to the calendar year in which they first entered patients. Trials from which data are available are listed to help describe the contents of the various overviews, and the other studies are listed to help describe the completeness of the various overviews. (In the latter context, it is obviously helpful to list those randomized studies from which data were not available, and it may also be helpful to list any non-random studies that might well be mistakenly supposed to have been randomized studies that had been overlooked.)

If these trials are examined in sequence from the 1950s to the 1980s, then some trends may be noticed in the types of trials that were undertaken in different periods. Each study is given a unique trial number (e.g. 64A, 64B1, 64B2, 64C, 64D) which is used to identify it elsewhere. The first two digits indicate the year patient entry began, and the subsequent characters complete the unique identification. Thus, for example, 64A, 64C and 64D are three different trials that began in 1964, while 64B1 and 64B2 are two parts of one trial, in at least one part of which patient entry began in 1964. Trials are split (for analysis) into distinct "parts" whenever the treatment allocation differed. For example, some trials have different randomizations for pre- and post-menopausal patients, while other trials may be divided into earlier and later parts because there were changes during the trial in the randomization proportions (e.g. from a ratio of 1:1 to 2:1) or the treatment options (e.g. from Treatment A vs Treatment B vs Treatment C to Treatment A vs Treatment B only). Whenever a trial is divided into parts, O-E and its variance are calculated separately for each part, and an overall O-E for that trial is obtained by appropriate summation of one O-E from each relevant part.

Appendix Table 1 also includes some basic data from each trial (or from the different parts of each trial), and further details may be found in the accompanying short reports. Specifically, for each trial it gives:

(a) "No. entered" — the number of women known to have been entered into that trial;
(b) "NK 1984" — the number of women for whom there was no information on vital status since January 1, 1984;
(c) "Dead" — the number known to have died before September 1, 1985;
(d) "Recur" — the number known to have relapsed but not known to have died before September 1, 1985.

The numbers of trials recorded as having started in 1983 and 1984 are smaller than the annual numbers started in the preceding few years. This may represent some under-ascertainment of recent trials. Even if it does, however, it is unlikely to make any material difference to the overview, since few deaths would have been reported by mid-1985 for trials started in 1983. Hence, the inadvertent omission of any such trials would probably involve the loss of very little information.

7. STRUCTURE OF RESULTS AND OF DISCUSSION

7.1 Overall analyses followed by subgroup analyses

The material available for analysis is so extensive that complete separation of RESULTS and DISCUSSION is not convenient. Some commentary will therefore accompany many of the results, partly to explain the reasons for particular statistical analyses and partly to discuss the implications of some particular results. The sequence of presentation will involve first the tamoxifen trials, then the chemotherapy trials and finally, more briefly, the ovarian ablation trials.

To be of relevance to medical practice, trials have to give guidance not just on the average effects of all treatments on all patients, but on the expected effects of specific types of treatment for specific types of patient. There are two quite different ways of seeking guidance from trial results (see the last four sections in the Introduction) about one specific type of patient (e.g. node-negative, aged over 50). The more obvious way is to consider the apparent effects of treatment just among "such" patients. This is unbiased, but may be subject to substantial random error even in an overview of many trials. An alternative approach is to be guided, at least in general terms, by the overall analysis (which, although less specific, is less liable to be distorted by the play of chance or by unduly selective emphasis on particular subgroup analyses). Difficulties, however, arise when these two approaches to assessment of the evidence for a particular subgroup of women yield very different answers. These difficulties cannot be bypassed, but they can be limited if specific subgroup analyses are generally considered in conjunction with other indirectly relevant evidence, and the structure of the RESULTS section is designed to facilitate this.

For the tamoxifen trials and for the chemotherapy trials, an overall mortality analysis is given first, including all women in all available trials. This fundamental analysis inevitably involves an extremely heterogeneous range of comparisons. But, it does provide the most statistically reliable test of whether some such types of treatment really can delay death among some types of women, and is largely or wholly unaffected by the important selective biases that can be introduced by a "data-dependent" choice of which analyses to emphasize. Next, various more detailed subgroup analyses of mortality are given to begin the exploration of any possible differences between the effects on mortality of different sub-categories of treatment, or of the effects of treatments in different sub-categories of women.

Some of the first subgroup analyses compare the responses to treatment of women aged under 50 and of women aged 50 or over (i.e. roughly pre- and post-menopausal), and many later sub-analyses are further subdivided by age. To indicate the age-group being analyzed, many Figure and Table numbers have "all ages", "<50" or "50+" appended to them. (Trials 78J and 79A, which included only postmenopausal women but did not provide a subdivision of results by age, are categorized as "50+".) The results of these subgroup analyses and sub-subgroup analyses need to be interpreted extremely cautiously, however, especially where they appear discrepant with the overall analyses, for the play of chance alone is likely to yield several false positive or false negative subgroup effects (see section 3.8 above: "Selection biases from subgroup analyses").

7.2 Mortality analyses followed by recurrence analyses

Following each mortality analysis, the corresponding analysis of recurrence-free survival will be given. (To indicate whether Mortality or Recurrence is being analyzed, many Figure and Table numbers have M or R appended to them.) A few trials were able to supply information only on mortality, so the recurrence data are only just over 90% complete (94% tamoxifen, 90% chemotherapy). Moreover, the dates of recurrence that are available are likely to be somewhat less reliable than dates of death. Neither of these limitations, however, need introduce any appreciable bias. Perhaps more importantly, any effects of treatment on recurrence are of only indirect relevance to effects on death. Despite these limitations, analyses of recurrence data can still serve at least two purposes. First, they may be of intrinsic biological interest. Second, for purely statistical reasons (partly because the numbers of recurrences are often somewhat greater than the numbers of deaths, but chiefly because the proportional effects of treatment are often much larger on recurrence than on mortality), subgroup analyses based on recurrence may be less likely to yield exaggeratedly positive or false negative results. Even, therefore, when interpreting the subgroup analyses just of mortality it may help to consider them in the light of the corresponding subgroup analyses of recurrence (see section 3.10 above: "Use of recurrence data, as well as mortality data, to study interactions unbiasedly").

8. RESULTS OF TAMOXIFEN TRIALS

Table 4 summarizes the 28 randomized trials available for review in which at least two of the adjuvant treatment regimens being compared differed only in that one involved tamoxifen and the other did not (with any other adjuvant treatments in those two particular regimens being the same). Information from those 28 trials is available from a total of 16,513 women, 23% (3782) of whom were reported to have died. Recurrence data were available on 94%, among whom 35% (5379/15541) were reported to have relapsed (or, in a few cases, to have died before relapse). These events were approximately evenly distributed over years 1, 2, 3, 4 and 5+ of follow-up, providing useful information for up to about five years but not beyond. About four-fifths of all patients were at least 50 years old when randomized.

Because of the traditional emphasis on age as a possible modifier of response to breast cancer therapy, most analyses are first presented for "all patients" (which avoids data-dependent emphasis on particular age groups) and are then presented separately for women aged under 50 and for women aged 50 or older. Only one-fifth of the patients in the tamoxifen trials were under 50. Of these younger women, most were studied in trials comparing tamoxifen plus chemotherapy versus chemotherapy alone, and fewer than one-third were studied in trials comparing tamoxifen alone versus no other adjuvant therapy. In contrast, two-thirds of the women aged 50 or older were studied in trials of tamoxifen alone.

8.1 Overall analysis of MORTALITY in tamoxifen trials

A summary of the mortality results in the 28 tamoxifen trials included in the overview, along with an estimate of the effect of tamoxifen in all of the trials combined, is shown in the first section (all ages) of Figure 2M.* For each trial, the quantity O-E and its variance are listed separately, and beside these values the ratio of the annual odds of death (with 99% confidence intervals) in the tamoxifen-allocated group as compared with the control group in that trial is shown graphically (see Section 5.9). There is great variability in the apparent effects of treatment (solid squares) in the individual trials, but the wide confidence intervals for individual trial results (horizontal lines) reflect the unreliability of these separate estimates.

If tamoxifen had no effect on mortality among the patients in any of these 28 trials, each non-zero value for O-E could equally well have been positive or negative, and the grand total of these 28 values would have differed only randomly from zero. However, most of the O-E values were negative, and their grand total was -148.3. This suggests that at least 300 deaths (148.3 x 2) were avoided or substantially delayed by tamoxifen (or that a larger number of deaths were moderately delayed: survival curves, which provide a more informative description of the medical meaning of this mortality difference, follow later). Summation of the corresponding variances for each O-E value yielded a total variance of 851.3, the square root of which (29.2) provides an estimate of the standard deviation of the grand total of the individual O-E values. Dividing the grand total by its standard deviation (z = -148.3/29.2 = -5.1; 2P<0.00001) shows that it is more than 5 sd away from zero. This represents a reduction of 16% ± 3 in the odds of death among women of all ages assigned to tamoxifen treatment and is far too large to be plausibly attributed merely to the play of chance. Unless there is some serious undetected error or bias in favour of treatment, it must be accepted that adjuvant tamoxifen can at least delay death.

The first question is whether any appreciable part of the 16% mortality reduction can be accounted for by initial imbalances between the treatment groups in some importantly prognostic factors. This did not appear to be so. Age itself was not an important prognostic factor in these studies — and, in any case, all the analyses are already "retrospectively stratified" for age (see Statistical Methods). The only importantly prognostic factors were nodal status, some information on which was available for almost all the subjects, and estrogen receptor status, some information on which was available for almost half of the patients in the tamoxifen trials. Analyses that were additionally stratified for nodal status (see Figure 8M below) or for estrogen receptor status (see Figure 11M below) did not reduce the apparent effect of tamoxifen — indeed, stratification for nodal status actually increased it very slightly, to 18% ± 3.

Another check against serious error involves seeing whether almost all the benefit comes from just one or two trials that are unsupported by the other studies. This does not appear to be the case. Four large trials (77H=NATO, 77K=NSABP B-09, 78D=Scottish and 77C=Danish BCG77c) collectively contributed about half of all the deaths; the other half of the deaths occurred in the 24 smaller studies (Figure 2M [All ages]). The reduction in the odds of death observed in the four larger trials (14% ± 4; 2P=0.001) is similar to the reduction observed in the smaller trials (18% ± 5; 2P<0.0001), and to that observed overall (16% ± 3; 2P<0.00001). The largest single contribution to the overall result comes from one of these four large trials (77H=NATO: O-E = -27.1), but this trial contributes less than one-fifth of the total O-E. Even if the NATO trial had never been undertaken, the overall difference in mortality in all the other trials would still be 4.4 standard deviations away from zero, which is highly significant (2P=0.00001).

There is some indirect evidence suggesting that patients in one of the smaller trials, 80E, may not have been properly randomized, for in that trial the poor-prognosis patients appeared more likely to have been given tamoxifen. Perhaps because of this, the results of that trial (Figure 2M [All ages]

---

* Figures are plotted in three sections, the first section being for patients of all ages (e.g. Figure 2M [All ages]), the second being for patients aged less than 50 (Figure 2M [<50]), and the final section being for those aged 50 or over (Figure 2M [≥50]). An M in the figure number denotes Mortality data, and an R denotes Recurrence data. More detailed tabulations of the tamoxifen mortality data are given in Appendix Table 2M. In that table, results for each trial (or trial part, where changes in the protocol necessitate greater subdivision) are listed chronologically in two groups, based on the scheduled duration of tamoxifen treatment (1 year or less, and 2+ years). The data from each study have been analyzed separately for women aged under 50 and for those aged 50 or older at the time of entry into the study. For each age subset and for all patients regardless of age, an O-E and variance is given for each trial (or trial part). When chemotherapy was given to patients in both arms of a trial, the chemotherapy regimen used is indicated in the fourth column of the table.

Table 4M. **MORTALITY data available for evaluation of adjuvant tamoxifen* in early breast cancer** (including some trials with an identical chemotherapy regimen for the different treatment groups)

Randomized adjuvant tamoxifen comparison	Mortality (all causes)	
	No. of trials (or parts)	Deaths/ patients
(a) 5 years vs NIL	2	253/1518
(b) 3 years vs NIL	1	56/179
(c) 2 years vs NIL	16	2014/9810
2 or more years (a+b+c)	**19**	**2323/11507**
(d) 1 year vs NIL	8	1384/4742
(e) 6 months vs NIL	1	75/264
1 year or less (d+e)	**9**	**1459/5006**
ALL TAMOXIFEN TRIALS (a+b+c+d+e)	**28**	**3782/16513** 23% die

MORTALITY ANALYSES M[all ages]

* In each trial about half the patients were allocated adjuvant TAMOXIFEN, the other half being allocated CONTROL (i.e. similar primary and adjuvant treatment, but without adjuvant tamoxifen).

and Appendix Table 2M) suggest a slight disadvantage[*] for patients treated with tamoxifen. The contribution of that trial is so small (22 tamoxifen deaths and 20 control deaths), however, that it makes little difference whether it is included or not, and since no direct evidence of improper randomization was found on exhaustive review of the trial procedures, it remains in the present overview.

The only randomized studies deliberately omitted are all trials from Russia and Japan (from which countries only incomplete details were available: see section 4.3). The exact effects of their unavailability are, of course, not known, but there is no reason to expect their findings to be systematically different. The incomplete preliminary mortality information thus far obtained from the Russian and Japanese trials is also favorable to tamoxifen, so inclusion of that preliminary information would have strengthened the main mortality result.

8.2 Overall analysis of RECURRENCE in tamoxifen trials

Data on recurrence are available for 15,541 (94%) of the randomized patients and are summarized in Figure 2R.[**] The reduction in recurrence produced by tamoxifen in many of the individual trials is so large that there is really no need for an overview to demonstrate it, and the main use of the recurrence data may be to help in the assessment of the effects of treatment in different subgroups of trials or of patients. Overall, the reduction in the annual odds of recurrence suggested by this overview of all the available recurrence data is 30% ± 2 (Figure 2R [All ages]), which is approximately twice as big as the reduction in mortality. Again, this analysis is already stratified for age, and additional stratification for nodal status (see Figure 8R below) or for estrogen receptor status (see Figure 11R below) did not reduce the apparent effect of tamoxifen — indeed, stratification for nodal status actually increased it very slightly, to 33% ± 2.

8.3 Heterogeneity between tamoxifen TRIALS, and heterogeneity between WOMEN

It can generally be assumed that when many trial results are being reviewed some real differences between the **sizes** of the effects of different treatments may well exist (although perhaps not differences in the **directions** of these effects). Even twofold differences in size may, however, be surprisingly difficult to demonstrate reliably. Suppose, for example, that half the tamoxifen trials had studied one particular regimen and half the trials had studied another that is really twice as effective. Even if the two mortality reductions actually observed in these two groups of trials happened to reflect the twofold differences in effectiveness of these two regimens precisely, this twofold difference between the sizes of the two observed mortality reductions might not be statistically significant. Furthermore, the apparent mortality reductions in the two groups of trials might instead — just by chance — have been nearly the same, which would misleadingly suggest that there was no difference between the reductions in the risk of death that are really produced by the two different tamoxifen regimens. Alternatively, the apparent effect in the trials of the less active regimen might be somewhat worse than the truth and might then misleadingly suggest that that regimen had no effect whatsoever.

---

[*] When trial 80E was reanalyzed with a retrospective stratification for nodal status, however (see Figure 7 below), it then indicated a slight advantage for tamoxifen-allocated patients in both survival (O-E = -3.9, variance = 8.1) and recurrence-free survival (-6.5, 11.1).

[**] More detailed tabulations of the recurrence data in the tamoxifen trials are available in Appendix Table 2R.

8 Results of tamoxifen trials

RECURRENCE
ANALYSES
R[all ages]

Table 4R. **RECURRENCE data available for evaluation of adjuvant tamoxifen* in early breast cancer** (including some trials with an identical chemotherapy regimen for the different treatment groups)

Randomized adjuvant tamoxifen comparison	Recurrence (or prior death)	
	No. of trials (or parts)	Events/ patients
(a) 5 years vs NIL	2	424/1518
(b) 3 years vs NIL	1	85/179
(c) 2 years vs NIL	13	3115/9448
2 or more years (a+b+c)	**16**	**3624/11145**
(d) 1 year vs NIL	7	1625/4132
(e) 6 months vs NIL	1	130/264
1 year or less (d+e)	**8**	**1755/4396**
ALL TAMOXIFEN TRIALS (a+b+c+d+e)	**24**	**5379/15541** 35% recur or die

* In each trial about half the patients were allocated adjuvant TAMOXIFEN, the other half being allocated CONTROL (i.e. similar primary and adjuvant treatment, but without adjuvant tamoxifen).

The same problems apply if, instead of two categories of treatment, two categories of women are considered separately (e.g. those aged under 50 years and those 50 or over: see below). Even if a highly significant mortality reduction in one particular category can be accepted as real, the lack of any apparent mortality reduction in the other category might be either a true negative or a false negative.

In general, the main ways of protecting against misinterpretation of such chance fluctuations are: (i) to emphasize the overall mortality analyses and not just the subgroup analyses (and, in particular, not to worry too much about whether the apparent effects of treatment in subgroup analyses are conventionally "significant"), and (ii) to examine subgroup analyses not only of survival but also of recurrence-free survival. In general, the least informative significance tests are likely to be those seeking evidence of non-specific heterogeneity. These will be disposed of first, before the more specific questions are addressed.

8.4 Heterogeneity of MORTALITY results in tamoxifen trials

Although the overall analyses demonstrate that tamoxifen significantly reduces mortality, standard statistical tests of heterogeneity (on several "degrees of freedom") provide no reason to suppose that the magnitude of this effect varies from trial to trial. A test for heterogeneity of the apparent effects in the different trials yields a completely non-significant result (21.5 on 27 degrees of freedom: Figure 2M [All ages]). Such calculations are not particularly informative, however, because they are dominated by unstable contributions from the smaller studies that could obscure any real heterogeneity between the larger trials. A more appropriate way of seeking evidence of heterogeneity might be first to group the smaller trials together into a few meaningful subcategories and then to compare the combined results of these groups of smaller trials with the individual results from the larger trials (Table 5M). The trials were, therefore, divided first by duration of tamoxifen treatment (i.e. two years or more, or one year or less) and then by treatment comparison (i.e. tamoxifen vs no adjuvant therapy at all, or tamoxifen plus chemotherapy vs the same chemotherapy without tamoxifen). This produces four categories, and Table 5M shows the combined results in each of these four categories for the smaller trials; in addition, the separate results are given for the four larger trials. No group of smaller trials or single larger trial yields a result significantly better than the overall average, and a conventional chi-square test of heterogeneity between the 8 lines of Table 5M [all ages] yields a non-significant result (8.7 on 7 degrees of freedom). These additional heterogeneity analyses again provide no reason to suppose that the highly significant overall mortality reduction is due to one or two isolated positive results. Indeed, in Table 5M [all ages] the only result that appears slightly atypical is an almost null mortality result from one of the large trials (77K: NSABP B-09), though even that is not convincingly discrepant with the other trial results.

8.5 Heterogeneity of RECURRENCE results in tamoxifen trials

When, however, the recurrence results in the same 8 trials or groups of trials were compared (Table 5R), there was significant heterogeneity between the 8 recurrence results (chi-square = 22.2 on 7 degrees of freedom, P=0.002). But, although the effect of tamoxifen on recurrence appears larger in some of the trials (for example, 78D: Scottish study of five years of tamoxifen) than in others, none of the 8 results in Table 5R [all ages] is, taken separately, clearly significantly better or worse than the overall estimated reduction in recurrence of 30% for all the trials together. This illustrates the general principle that, when considering a set of trials (as in Table 5), any overall heterogeneity is difficult to interpret reliably without analysis

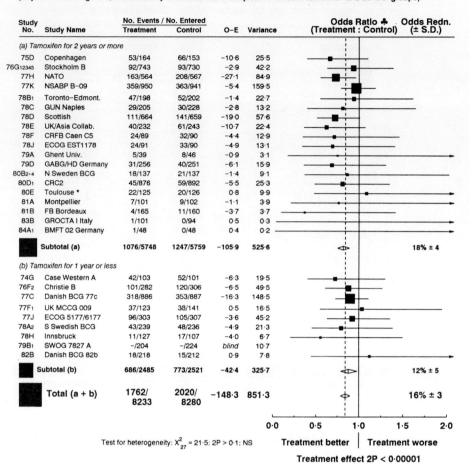

8 Results of tamoxifen trials

RECURRENCE ANALYSES R[all ages]

BELOW, LEFT:
R[<50]
RECURRENCE in women UNDER 50 years old at entry

BELOW, RIGHT:
R[≥50]
RECURRENCE in women 50 OR MORE years old at entry

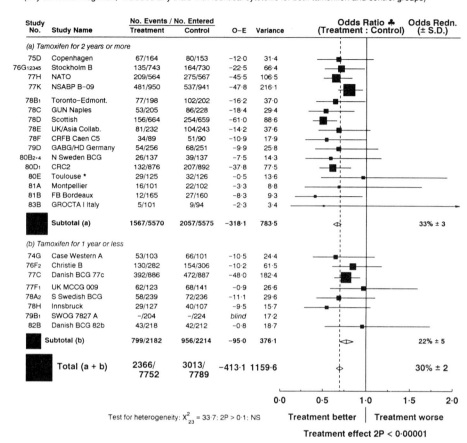

Fig. 2R. RECURRENCE in all available trials of tamoxifen vs no tamoxifen
(any tamoxifen regimen; includes any trials with identical cytotoxic for both tamoxifen and control groups)

Fig. 2R[<50]. RECURRENCE among women aged UNDER 50 at entry

Fig. 2R[≥50]. RECURRENCE among women aged 50 OR OVER at entry

* Significant imbalance in initial nodal status
♣ 95% confidence intervals for overview and 99% for individual trials.

Table 5M. MORTALITY in all available adjuvant tamoxifen trials: four largest trials, and summated results from four categories of smaller* trial

Trial (or category of trials)	Numbers of deaths Tamox	Numbers of deaths Control	Statistical calculations for TAMOXIFEN groups only Total O-E	Var O-E	Odds reduction
TRIALS OF 2+ YEARS OF TAMOXIFEN:					
1. Large trial of tamoxifen vs no adjuvant (NATO)	163	208	-27.1	84.9	27% ± 9
2. Large trial of tamoxifen vs no adjuvant (Scottish)	111	141	-19.0	57.6	28% ± 11
3. Smaller trials of tamoxifen vs no adjuvant	321	381	-36.0	161.2	20% ± 7
4. Largest trial of tamoxifen + chemo vs same chemo (NSABP B-09)	359	363	-5.4	159.5	3% ± 8
5. Smaller trials of tamoxifen + chemo vs same chemo	122	154	-18.5	62.4	26% ± 11
TRIALS OF ≤1 YEAR OF TAMOXIFEN:					
6. Largest trial of tamoxifen vs no adjuvant (Danish BCG 77c)	318	353	-16.3	148.5	10% ± 8
7. Smaller trials of tamoxifen vs no adjuvant	155	185	-15.4	77.5	18% ± 10
8. All trials of tamoxifen + chemo vs same chemo	213	235	-10.7	99.7	10% ± 9
ALL TRIALS	1762	2020	-148.3	851.3	16% ± 3

Test for heterogeneity: chi-square on 7 degrees of freedom = 8.7; NS.

* "Small" means the variance of (O-E) in the all-ages mortality analysis was less than fifty.

MORTALITY ANALYSES M[all ages]

BELOW, LEFT: M[<50] MORTALITY in women UNDER 50 years old at entry

BELOW, RIGHT: M[≥50] MORTALITY in women 50 OR MORE years old at entry

Table 5M[<50]. MORTALITY among women aged less than 50 at entry

Trial (or category of trials)	Numbers of deaths Tamox	Numbers of deaths Control	Statistical calculations for TAMOXIFEN groups only Total O-E	Var O-E	Odds reduction
TRIALS OF 2+ YEARS OF TAMOXIFEN:					
1. Large trial of tamoxifen vs no adjuvant (NATO)	28	29	-3.6	12.5	25% ± 25
2. Large trial of tamoxifen vs no adjuvant (Scottish)	11	13	-1.6	5.1	27% ± 38
3. Smaller trials of tamoxifen vs no adjuvant	38	46	-3.8	18.9	18% ± 21
4. Largest trial of tamoxifen + chemo vs same chemo (NSABP B-09)	156	148	6.3	68.0	-10% ± 13
5. Smaller trials of tamoxifen + chemo vs same chemo	49	43	1.5	21.2	-7% ± 23
TRIALS OF ≤1 YEAR OF TAMOXIFEN:					
6. Largest trial of tamoxifen vs no adjuvant (Danish BCG 77c)	2	0	0.1	0.1	—
7. Smaller trials of tamoxifen vs no adjuvant	8	8	-0.4	3.7	9% ± 49
8. All trials of tamoxifen + chemo vs same chemo	106	100	3.6	46.1	-8% ± 15
ALL TRIALS	398	387	2.1	175.6	-1% ± 8

Test for heterogeneity: chi-square on 7 degrees of freedom = 3.5; NS.

Table 5M[≥50]. MORTALITY among women aged 50 years or more at entry

Trial (or category of trials)	Numbers of deaths Tamox	Numbers of deaths Control	Statistical calculations for TAMOXIFEN groups only Total O-E	Var O-E	Odds reduction
TRIALS OF 2+ YEARS OF TAMOXIFEN:					
1. Large trial of tamoxifen vs no adjuvant (NATO)	135	179	-23.4	72.4	28% ± 10
2. Large trial of tamoxifen vs no adjuvant (Scottish)	100	128	-17.3	52.5	28% ± 12
3. Smaller trials of tamoxifen vs no adjuvant	283	335	-32.1	142.3	20% ± 8
4. Largest trial of tamoxifen + chemo vs same chemo (NSABP B-09)	203	215	-11.7	91.5	12% ± 10
5. Smaller trials of tamoxifen + chemo vs same chemo	73	111	-20.0	41.2	38% ± 12
TRIALS OF ≤1 YEAR OF TAMOXIFEN:					
6. Largest trial of tamoxifen vs no adjuvant (Danish BCG 77c)	316	353	-16.4	148.4	10% ± 8
7. Smaller trials of tamoxifen vs no adjuvant	147	177	-15.1	73.8	18% ± 11
8. All trials of tamoxifen + chemo vs same chemo	107	135	-14.3	53.6	23% ± 12
ALL TRIALS	1364	1633	-150.4	675.7	20% ± 3

Test for heterogeneity: chi-square on 7 degrees of freedom = 7.0; NS.

8 Results of tamoxifen trials

> **RECURRENCE ANALYSES**
> **R[all ages]**
>
> BELOW, LEFT:
> **R[<50]**
> RECURRENCE
> in women
> UNDER 50
> years old
> at entry
>
> BELOW, RIGHT:
> **R[≥50]**
> RECURRENCE
> in women
> 50 OR MORE
> years old
> at entry

Table 5R. **RECURRENCE in all available adjuvant tamoxifen trials: four largest trials, and summated results from four categories of smaller* trial**

Trial (or category of trials)	Numbers of recurrences Tamox	Control	Total O-E	Var O-E	Odds reduction
TRIALS OF 2+ YEARS OF TAMOXIFEN:					
1. Large trial of tamoxifen vs no adjuvant (NATO)	209	275	-45.5	106.5	35% ± 8
2. Large trial of tamoxifen vs no adjuvant (Scottish)	156	254	-61.0	88.6	50% ± 8
3. Smaller trials of tamoxifen vs no adjuvant	483	648	-100.5	246.9	33% ± 5
4. Largest trial of tamoxifen + chemo vs same chemo (NSABP B-09)	481	537	-47.8	216.1	20% ± 6
5. Smaller trials of tamoxifen + chemo vs same chemo	238	343	-63.3	125.4	40% ± 7
TRIALS OF ≤1 YEAR OF TAMOXIFEN:					
6. Largest trial of tamoxifen vs no adjuvant (Danish BCG 77c)	392	472	-48.0	182.4	23% ± 7
7. Smaller trials of tamoxifen vs no adjuvant	217	266	-30.7	106.9	25% ± 8
8. All trials of tamoxifen + chemo vs same chemo	190	218	-16.3	86.8	17% ± 10
ALL TRIALS	2366	3013	-413.1	1159.6	30% ± 2

Test for heterogeneity: chi-square on 7 degrees of freedom = 22.2; P=0.002

* See Table 5M footnote.

Table 5R[<50]. RECURRENCE among women aged less than 50 at entry

Trial (or category of trials)	Numbers of recurrences Tamox	Control	Total O-E	Var O-E	Odds reduction
TRIALS OF 2+ YEARS OF TAMOXIFEN:					
1. Large trial of tamoxifen vs no adjuvant (NATO)	35	38	-8.8	14.4	46% ± 20
2. Large trial of tamoxifen vs no adjuvant (Scottish)	20	36	-9.2	12.0	54% ± 20
3. Smaller trials of tamoxifen vs no adjuvant	80	95	-7.9	41.7	17% ± 14
4. Largest trial of tamoxifen + chemo vs same chemo (NSABP B-09)	204	213	-2.3	89.4	3% ± 10
5. Smaller trials of tamoxifen + chemo vs same chemo	91	106	-10.9	39.2	24% ± 14
TRIALS OF ≤1 YEAR OF TAMOXIFEN:					
6. Largest trial of tamoxifen vs no adjuvant (Danish BCG 77c)	3	0	0.3	0.2	—
7. Smaller trials of tamoxifen vs no adjuvant	13	14	-1.1	6.0	17% ± 37
8. All trials of tamoxifen + chemo vs same chemo	79	80	-0.5	34.0	1% ± 17
ALL TRIALS	525	582	-40.4	237.0	16% ± 6

Test for heterogeneity: chi-square on 7 degrees of freedom = 10.8; NS.

Table 5R[≥50]. RECURRENCE, women aged 50 years or more at entry

Trial (or category of trials)	Numbers of recurrences Tamox	Control	Total O-E	Var O-E	Odds reduction
TRIALS OF 2+ YEARS OF TAMOXIFEN:					
1. Large trial of tamoxifen vs no adjuvant (NATO)	174	237	-36.7	92.1	33% ± 9
2. Large trial of tamoxifen vs no adjuvant (Scottish)	136	218	-51.8	76.6	49% ± 8
3. Smaller trials of tamoxifen vs no adjuvant	403	553	-92.6	209.2	36% ± 6
4. Largest trial of tamoxifen + chemo vs same chemo (NSABP B-09)	277	324	-45.5	126.6	30% ± 7
5. Smaller trials of tamoxifen + chemo vs same chemo	147	237	-52.4	82.3	47% ± 8
TRIALS OF ≤1 YEAR OF TAMOXIFEN:					
6. Largest trial of tamoxifen vs no adjuvant (Danish BCG 77c)	389	472	-48.3	182.2	23% ± 7
7. Smaller trials of tamoxifen vs no adjuvant	204	252	-29.5	100.8	25% ± 9
8. All trials of tamoxifen + chemo vs same chemo	111	138	-15.9	52.8	26% ± 12
ALL TRIALS	1841	2431	-372.7	922.7	33% ± 3

Test for heterogeneity: chi-square on 7 degrees of freedom = 22.8; 2P=0.002.

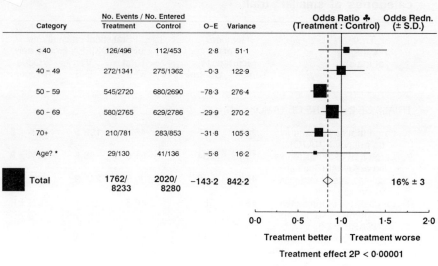

3M. Effects of tamoxifen on MORTALITY among women of DIFFERENT AGES
(tamoxifen regimen; includes any trials with identical cytotoxic for both tamoxifen and control groups)

MORTALITY ANALYSES

♣ 95% confidence intervals for overview and 99% for individual categories.
N.B: The test for trend is between 5 specific categories.
* 79A, which required age > 65, and 78J, which required menopause (and is elsewhere in the "≥ 50" analyses).

of more specific comparisons (e.g. of different durations of tamoxifen: see sections 8.11 and 8.12 below).

8.6 Influence of age on the effect of tamoxifen: MORTALITY data

The data from the tamoxifen trials are given separately for women aged under 50 in Figure 2M [<50], and for women aged 50 or over in Figure 2M [≥50].

Women aged 50 or over (Figure 2M [≥50]): Patients in the tamoxifen group fared somewhat better than those in the control group in all but two of the trials (i.e. all but two of the black squares are to the left of the solid vertical line). When the trials are considered separately, however, only one (77H: NATO) has a mortality result with a 99 percent confidence interval that does not cross the solid vertical line, and hence yields a P-value below 0.01. All of the other trials yield less extreme P-values and thus, considered separately, would not provide clearly significant evidence of benefit among older women. Taken together, however, their generally favorable results reinforce each other. The reduction in the odds of death among older women is significant when all trials in the overview are considered together (12,861 women over 50: 20% ± 3; 2P<0.00001). The dashed vertical line indicates this overall result — i.e. a 20 percent mortality reduction among older women — and comparison of the width of the confidence intervals for the individual trials with the narrowness of the separation between the dashed and the solid vertical lines shows that none of the individual trials was large enough on its own to detect a 20 percent mortality reduction reliably.

Women aged under 50 (Figure 2M [<50]): Overall, among the women aged under 50 there was no apparent effect whatever of tamoxifen on mortality — indeed, the data actually suggested a 1% increase in mortality, although the standard deviation of this apparent increase was large (± 8). This lack of significant benefit is in marked contrast with the results among older women, and the estimated difference between the effects of tamoxifen in these two broad age groups is statistically significant (test for interaction = 2.8 sd, 2P<0.01). But, even though an appreciable effect of tamoxifen on the mortality of younger women has not been directly demonstrated by the trials, neither can it be ruled out by the available evidence. The number of patients aged under 50 was smaller (only 3652 in total), and hence the mortality results are less informative than among the older women. Thus, the available mortality data are, for example, compatible with tamoxifen in women under 50 having an effect on the risk of death that is about half the size of the effect in women over 50.

The results of a finer subdivision with respect to age are given in Figure 3M. The wide confidence intervals in Figure 3M indicate, however, that mortality analyses in such small subgroups are not statistically reliable, so inferences based closely on them cannot be trusted.

8.7 Influence of age on the effect of tamoxifen: RECURRENCE data

The recurrence data also suggest that tamoxifen may have a somewhat smaller effect among younger women (Figure 2R [<50]) than among older women (Figure 2R [≥50]). But, although the difference between the sizes of the recurrence reductions in the two age groups is statistically significant (test for interaction = 3.2 sd; 2P=0.001), this interaction appears to be not "qualitative" but "quantitative" (i.e. a difference only in the size of the benefit). For, although the recurrence reduction produced by tamoxifen is smaller among younger women (16% ± 6; 2P<0.01) than among older women (33% ± 3; 2P<0.00001), both are significantly favorable. Recurrence data are available on 94% of the patients enrolled in all trials, and so the inclusion of additional recurrence data from the remaining studies would probably not have altered to any material extent the apparent relative effectiveness of tamoxifen in the two age groups. In particular, there is no reason to suppose that the additional data would have abolished the apparent delay of recurrence by tamoxifen among younger women.

The results of a finer subdivision with respect to age are given in Figure 3R. There is a significant trend towards an increased tamoxifen effect among older women in this analysis, but no indication of an unfavorable effect among younger women. The recurrence results in the five age groups in Figure 3R are statistically more stable than the corresponding age-specific mortality results were in Figure 3M, but still the confidence limits in the younger age groups are so wide that although

RECURRENCE ANALYSES

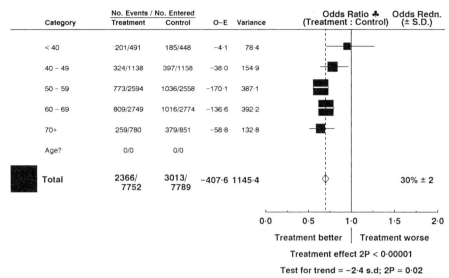

Fig. 3R. Effects of tamoxifen on RECURRENCE among women of DIFFERENT AGES
(any tamoxifen regimen; includes any trials with identical cytotoxic for both tamoxifen and control groups)

♣ 95% confidence intervals for overview and 99% for individual categories.

significant benefit has been demonstrated in the age groups 70+, 60-69, 50-59 and 40-49, even a benefit of, for example, about 20% has not been excluded in the age group under 40 (Figure 3R).

8.8 Description of the SIZE of the effect of tamoxifen: MORTALITY data

Overall in all the women in all the trials, allocation to tamoxifen was associated with a typical reduction of 16% ± 3 in the annual odds of death, at least during the first few years. For a number of reasons, however, this may be an underestimate of the potential benefit of adjuvant tamoxifen. First, the tamoxifen schedules may not have been optimal in terms of the duration or dose of treatment. Second, the effect of actually **using** tamoxifen may be somewhat greater than the effect of being **allocated** to use it. Because of imperfect compliance (i.e. failure to give the full dose or duration of tamoxifen to some of those allocated active treatment and/or the use of tamoxifen in some of those allocated control), the apparent effect of allocated treatment will tend to underestimate slightly the effect of actual use of tamoxifen. Third, retrospective stratification of the analysis not just for age but also for other important risk factors would not weaken the estimated reduction in risk in the tamoxifen group — indeed, in the case of nodal status it would very slightly strengthen it (to 18% ± 3).

Finally, the average reduction in odds of death, 16% ± 3, is a weighted average of the mortality reductions in each separate year after randomization, with greater weight in the earlier years where there is at present more information. However, if the effect of tamoxifen on mortality is less in year 1 than in years 3-5 then the estimated effect of tamoxifen on the 5-year mortality risk may, at present, be a little less than it will be when data are available from all 5 years of all trials. To help determine whether there are any interesting differences between the sizes of the mortality reductions in years 1, 2, 3, 4 and 5+ after randomization (where 5+ includes events in year 5 plus those few events thus far reported that occurred later), Table 6M provides mortality data subdivided by year of follow-up. Overall, the chief mortality difference emerges in years 3 and 4; indeed, there was no apparent difference in survival between the tamoxifen group and the control group during the first year after randomization. There is at present insufficient evidence after year 5 for any conclusion about longer-term effects on mortality to be possible. But, although there is no clear evidence of further divergence after year 4, neither is there indication of any convergence after year 4.

From the annual mortality rates and O-E values in Tables 6M, the approximate effects of tamoxifen in each successive year of follow-up can be illustrated by survival curves (Figure 4M). These calculations are somewhat crude (see Statistical Methods), for the mixture of good and poor prognosis women varies in different years as the contribution from different trials changes, while the estimation of woman-years is only approximate since some investigators provided only the year of death and not the precise date of death. These difficulties should, however, affect both tamoxifen-allocated and control-allocated women similarly, so the comparison between the treatment and control patients remains unbiased and informative. On average, therefore, allocation to the tamoxifen regimens studied in these trials improved the estimated probability of survival five years after randomization from 68% to 73% (Figure 4M [all ages]), an absolute difference of about 5 deaths per 100 women. It cannot be determined from such evidence, however, whether this represents delay of death by about a year in most women, long-term avoidance of death in about five per cent of women, or something in between.

8.9 Description of the SIZE of the effect of tamoxifen: RECURRENCE data

The corresponding analyses of the effects of tamoxifen on disease recurrence rates in each separate year are given in Table 6R and are illustrated in Figure 4R. In contrast with the mortality data, where there was no apparent effect in year 1, a highly significant recurrence reduction was already apparent by the end of the first year. This grew significantly bigger during year 2, and again during year 3, and there was a further non-significant improvement during year 4. Although there is no significant evidence of further divergence after the end of the fourth year, neither is there evidence that the recurrence-free survival curves begin to converge during the first five years (Table 6R).

Estimated five-year recurrence-free survival curves for tamoxifen-allocated and control-allocated women are shown in

Table 6M. Annual MORTALITY rates in trials of adjuvant tamoxifen

TRIAL YEAR	ALLOCATED TAMOXIFEN Dead/Woman-yrs		ALLOCATED CONTROL Dead/Woman-yrs		STATISTICAL CALCULATIONS FOR TREATED GROUP ONLY O-E*	Var(O-E)*	Odds Redn.
(a) All trials of tamoxifen							
1	271/ 7790	3.4%	258/ 7831	3.2%	8.2	125.6	yrs 1-2
2	426/ 6636	6.2%	486/ 6637	7.1%	-29.8	210.1	6% ±5
3	366/ 5252	6.7%	493/ 5177	9.1%	-64.6	195.3	
4	278/ 3873	6.9%	362/ 3699	9.3%	-46.8	146.2	yrs 3+
5+	421/ 5493	7.3%	421/ 5153	7.9%	-15.4	174.2	22% ± 4
All	1762/ 29044	(—)	2020/ 28498	(—)	-148.3	851.3	16% ± 3
4-year risk	21.3% ± 0.6		25.9% ± 0.6				
5-year risk**	27.1% ± 0.6		31.7% ± 0.6				

Difference in 4-year mortality risk from ALLOCATION to control or to tamoxifen = 4.5% ± 0.8
Difference in 5-year mortality risk from ALLOCATION to control or to tamoxifen = 4.7% ± 0.9

(b) Trials of 2 (or, in a few trials, more) years of tamoxifen							
1	155/ 5445	2.8%	146/ 5458	2.6%	5.1	72.1	yrs 1-2
2	249/ 4608	5.2%	293/ 4616	6.2%	-23.2	126.2	9% ± 7
3	215/ 3597	5.7%	296/ 3543	8.1%	-42.6	117.2	
4	171/ 2640	6.2%	242/ 2502	9.2%	-39.8	94.4	yrs 3+
5+	286/ 3854	7.1%	270/ 3575	7.4%	-5.5	115.6	24% ± 5
All	1076/ 20145	(—)	1247/ 19694	(—)	-105.9	525.6	18% ± 4
4-year risk	18.6% ± 0.6		23.8% ± 0.7				
5-year risk**	24.3% ± 0.7		29.4% ± 0.7				

Difference in 4-year mortality risk from ALLOCATION to control or to tamoxifen = 5.2% ± 1.0
Difference in 5-year mortality risk from ALLOCATION to control or to tamoxifen = 5.1% ± 1.0

MORTALITY ANALYSES M[all ages]

BELOW, LEFT: M[<50] MORTALITY in women UNDER 50 years old at entry

BELOW, RIGHT: M[≥50] MORTALITY in women 50 OR MORE years old at entry

---

* The tabulated (O-E) values and their variances are appropriately derived from summation of separate contributions from each trial, so they avoid direct comparisons between patients in different trials.
** Taking the event rate in year 5 to be approximately that in years 5 or over.

WARNING — ABOVE COMPARISON OF (a) WITH (b) IS NOT AGE-STANDARDIZED:
SEE BELOW FOR SEPARATE AGE-SPECIFIC COMPARISONS (<50, ≥50), AND SEE TABLE 7.

Table 6M[<50]. MORTALITY, women aged less than 50 at entry

TRIAL YEAR	ALLOCATED TAMOXIFEN Dead/Woman-yrs		ALLOCATED CONTROL Dead/Woman-yrs		STATISTICAL CALCULATIONS FOR TREATED GROUP ONLY O-E*	Var(O-E)*	Odds Redn.
(a) All trials of tamoxifen							
1	53/ 1720	3.1%	30/ 1698	1.7%	11.1	19.8	yrs 1-2
2	103/ 1411	7.0%	96/ 1397	6.7%	2.5	45.7	-23% ± 14
3	84/ 1074	7.5%	85/ 1064	7.7%	-0.6	38.1	
4	69/ 797	8.3%	87/ 781	10.5%	-9.2	35.0	yrs 3+
5+	89/ 1229	6.7%	89/ 1274	7.0%	-1.7	37.0	10% ± 9
All	398/ 6231	(—)	387/ 6215	(—)	2.1	175.6	-1% ± 8
4-year risk	23.6% ± 1.3		24.2% ± 1.3				
5-year risk**	28.7% ± 1.3		29.5% ± 1.3				

Difference in 4-year mortality risk from ALLOCATION to control or to tamoxifen = 0.7% ± 1.8
Difference in 5-year mortality risk from ALLOCATION to control or to tamoxifen = 0.8% ± 1.9

(b) Trials of 2 (or, in a few trials, more) years of tamoxifen							
1	39/ 1258	3.1%	22/ 1245	1.7%	8.2	14.5	yrs 1-2
2	72/ 1049	6.6%	67/ 1044	6.2%	2.0	32.1	-24% ± 16
3	59/ 801	7.1%	60/ 798	7.2%	-0.6	26.8	
4	46/ 584	7.6%	66/ 569	10.9%	-10.1	25.2	yrs 3+
5+	66/ 936	6.6%	64/ 950	6.8%	-0.8	27.1	14% ± 10
All	282/ 4627	(—)	279/ 4606	(—)	-1.3	125.7	1% ± 9
4-year risk	22.3% ± 1.4		23.9% ± 1.5				
5-year risk**	27.4% ± 1.5		29.0% ± 1.6				

Difference in 4-year mortality risk from ALLOCATION to control or to tamoxifen = 1.6% ± 2.1
Difference in 5-year mortality risk from ALLOCATION to control or to tamoxifen = 1.7% ± 2.2

Table 6M[≥50]. MORTALITY, women aged 50 years or more at entry

TRIAL YEAR	ALLOCATED TAMOXIFEN Dead/Woman-yrs		ALLOCATED CONTROL Dead/Woman-yrs		STATISTICAL CALCULATIONS FOR TREATED GROUP ONLY O-E*	Var(O-E)*	Odds Redn.
(a) All trials of tamoxifen							
1	218/ 6071	3.5%	228/ 6134	3.6%	-2.9	105.8	yrs 1-2
2	323/ 5225	6.0%	390/ 5240	7.2%	-32.3	164.4	12% ± 6
3	282/ 4177	6.5%	408/ 4113	9.5%	-64.0	157.2	
4	209/ 3076	6.5%	275/ 2918	9.0%	-37.5	111.2	yrs 3+
5+	332/ 4264	7.5%	332/ 3878	8.2%	-13.6	137.2	25% ± 4
All	1364/ 22813	(—)	1633/ 22282	(—)	-150.4	675.7	20% ± 3
4-year risk	20.8% ± 0.6		26.3% ± 0.7				
5-year risk**	26.7% ± 0.7		32.3% ± 0.7				

Difference in 4-year mortality risk from ALLOCATION to control or to tamoxifen = 5.6% ± 0.9
Difference in 5-year mortality risk from ALLOCATION to control or to tamoxifen = 5.7% ± 1.0

(b) Trials of 2 (or, in a few trials, more) years of tamoxifen							
1	116/ 4188	2.7%	124/ 4213	2.9%	-3.1	57.6	yrs 1-2
2	177/ 3560	4.8%	226/ 3571	6.2%	-25.2	94.1	17% ± 7
3	156/ 2796	5.3%	236/ 2745	8.3%	-42.0	90.4	
4	125/ 2056	5.8%	176/ 1933	8.7%	-29.7	69.3	yrs 3+
5+	220/ 2918	7.2%	206/ 2625	7.6%	-4.7	88.5	26% ± 5
All	794/ 15517	(—)	968/ 15088	(—)	-104.6	399.9	23% ± 4
4-year risk	17.5% ± 0.7		23.7% ± 0.8				
5-year risk**	23.4% ± 0.8		29.5% ± 0.8				

Difference in 4-year mortality risk from ALLOCATION to control or to tamoxifen = 6.3% ± 1.1
Difference in 5-year mortality risk from ALLOCATION to control or to tamoxifen = 6.1% ± 1.1

8   Results of tamoxifen trials

Table 6R. Annual RECURRENCE rates in trials of adjuvant tamoxifen

RECURRENCE ANALYSES R[all ages]
BELOW, LEFT: R[<50] RECURRENCE in women UNDER 50 years old at entry
BELOW, RIGHT: R[≥50] RECURRENCE in women 50 OR MORE years old at entry

TRIAL YEAR	ALLOCATED TAMOXIFEN Recur/Woman-yrs		ALLOCATED CONTROL Recur/Woman-yrs		STATISTICAL CALCULATIONS FOR TREATED GROUP ONLY		
					O-E*	Var(O-E)*	Odds Redn.
(a) All trials of tamoxifen							
1	657/ 7150	8.8%	868/ 7095	11.5%	-102.1	335.7	yrs 1-2
2	649/ 5730	10.6%	948/ 5415	16.1%	-163.2	341.3	32% ± 3
3	392/ 4349	8.4%	579/ 3867	13.9%	-115.9	212.3	
4	310/ 3089	9.4%	296/ 2639	10.7%	-18.3	133.8	yrs 3+
5+	358/ 4209	8.0%	322/ 3578	8.7%	-13.6	136.6	26% ± 4
All	2366/ 24527	(—)	3013/ 22594	(—)	-413.1	1159.6	30% ± 2
4-year risk	32.3% ± 0.7		42.9% ± 0.7				
5-year risk**	37.7% ± 0.7		47.9% ± 0.7				

Difference in 4-year recurrence risk from ALLOCATION to control or to tamoxifen = 10.5% ± 1.0
Difference in 5-year recurrence risk from ALLOCATION to control or to tamoxifen = 10.1% ± 1.0

(b) Trials of 2 (or, in a few trials, more) years of tamoxifen							
1	426/ 5166	7.9%	595/ 5102	11.0%	-83.1	225.6	yrs 1-2
2	424/ 4136	9.5%	662/ 3899	15.8%	-131.6	233.4	37% ± 4
3	259/ 3099	7.8%	398/ 2737	13.5%	-87.1	143.8	
4	212/ 2197	9.1%	190/ 1846	9.9%	-8.2	88.6	yrs 3+
5+	246/ 3084	7.4%	212/ 2590	8.1%	-8.1	92.2	27% ± 5
All	1567/ 17683	(—)	2057/ 16175	(—)	-318.1	783.6	33% ± 3
4-year risk	30.1% ± 0.8		41.6% ± 0.8				
5-year risk**	35.3% ± 0.8		46.3% ± 0.8				

Difference in 4-year recurrence risk from ALLOCATION to control or to tamoxifen = 11.5% ± 1.1
Difference in 5-year recurrence risk from ALLOCATION to control or to tamoxifen = 11.0% ± 1.1

* The tabulated (O-E) values and their variances are appropriately derived from summation of separate contributions from each trial, so they avoid direct comparisons between patients in different trials.
** Taking the event rate in year 5 to be approximately that in years 5 or over.

WARNING — ABOVE COMPARISON OF (a) WITH (b) IS NOT AGE-STANDARDIZED:
SEE BELOW FOR SEPARATE AGE-SPECIFIC COMPARISONS (<50, ≥50), AND SEE TABLE 7.

Table 6R[<50]. RECURRENCE, women aged less than 50 at entry

TRIAL YEAR	ALLOCATED TAMOXIFEN Recur/Woman-yrs		ALLOCATED CONTROL Recur/Woman-yrs		STATISTICAL CALCULATIONS FOR TREATED GROUP ONLY		
					O-E*	Var(O-E)*	Odds Redn.
(a) All trials of tamoxifen							
1	155/ 1477	10.0%	172/ 1445	11.2%	-9.0	71.7	yrs 1-2
2	164/ 1105	13.7%	180/ 1055	15.8%	-11.9	72.5	13% ± 8
3	80/ 776	9.8%	124/ 712	15.8%	-23.3	43.6	
4	71/ 546	12.3%	42/ 491	8.4%	10.5	24.6	yrs 3+
5+	55/ 851	5.9%	64/ 848	7.6%	-6.6	24.5	19% ± 9
All	525/ 4756	(—)	582/ 4550	(—)	-40.4	237.0	16% ± 6
4-year risk	38.6% ± 1.6		42.3% ± 1.6				
5-year risk**	42.2% ± 1.6		46.7% ± 1.6				

Difference in 4-year recurrence risk from ALLOCATION to control or to tamoxifen = 3.7% ± 2.3
Difference in 5-year recurrence risk from ALLOCATION to control or to tamoxifen = 4.5% ± 2.2

(b) Trials of 2 (or, in a few trials, more) years of tamoxifen							
1	125/ 1213	9.8%	142/ 1187	11.2%	-8.7	58.4	yrs 1-2
2	128/ 938	12.6%	151/ 901	15.6%	-14.7	59.1	18% ± 8
3	67/ 681	9.4%	109/ 624	15.8%	-21.9	37.5	
4	61/ 477	12.2%	34/ 425	7.8%	10.1	20.9	yrs 3+
5+	49/ 745	6.0%	52/ 734	7.2%	-3.9	20.9	18% ± 10
All	430/ 4053	(—)	488/ 3871	(—)	-39.1	196.7	18% ± 6
4-year risk	37.3% ± 1.7		41.9% ± 1.7				
5-year risk**	41.1% ± 1.7		46.0% ± 1.7				

Difference in 4-year recurrence risk from ALLOCATION to control or to tamoxifen = 4.6% ± 2.4
Difference in 5-year recurrence risk from ALLOCATION to control or to tamoxifen = 5.0% ± 2.4

Table 6R[≥50]. RECURRENCE, women aged 50 years or more at entry

TRIAL YEAR	ALLOCATED TAMOXIFEN Recur/Woman-yrs		ALLOCATED CONTROL Recur/Woman-yrs		STATISTICAL CALCULATIONS FOR TREATED GROUP ONLY		
					O-E*	Var(O-E)*	Odds Redn.
(a) All trials of tamoxifen							
1	502/ 5673	8.4%	696/ 5650	11.6%	-93.1	263.9	yrs 1-2
2	485/ 4624	9.8%	768/ 4361	16.2%	-151.3	268.8	37% ± 3
3	312/ 3572	8.1%	455/ 3155	13.4%	-92.6	168.7	
4	239/ 2543	8.8%	254/ 2148	11.2%	-28.8	109.2	yrs 3+
5+	303/ 3358	8.6%	258/ 2730	9.1%	-6.9	112.0	28% ± 4
All	1841/ 19771	(—)	2431/ 18044	(—)	-372.7	922.7	33% ± 3
4-year risk	30.8% ± 0.7		43.0% ± 0.8				
5-year risk**	36.7% ± 0.7		48.2% ± 0.8				

Difference in 4-year recurrence risk from ALLOCATION to control or to tamoxifen = 12.2% ± 1.1
Difference in 5-year recurrence risk from ALLOCATION to control or to tamoxifen = 11.4% ± 1.1

(b) Trials of 2 (or, in a few trials, more) years of tamoxifen							
1	301/ 3954	7.3%	453/ 3915	11.0%	-74.4	167.2	yrs 1-2
2	296/ 3198	8.6%	511/ 2998	15.8%	-116.9	174.3	43% ± 4
3	192/ 2419	7.3%	289/ 2113	12.9%	-65.3	106.3	
4	151/ 1721	8.2%	156/ 1421	10.4%	-18.3	67.8	yrs 3+
5+	197/ 2339	7.4%	160/ 1856	8.4%	-4.2	71.3	30% ± 5
All	1137/ 13630	(—)	1569/ 12304	(—)	-279.0	586.8	38% ± 3
4-year risk	27.9% ± 0.9		41.5% ± 0.9				
5-year risk**	33.6% ± 0.9		46.4% ± 1.0				

Difference in 4-year recurrence risk from ALLOCATION to control or to tamoxifen = 13.6% ± 1.3
Difference in 5-year recurrence risk from ALLOCATION to control or to tamoxifen = 12.8% ± 1.3

Fig. 4M. MORTALITY analysis: survival by year in all available trials of ANY DURATION of tamoxifen (includes any trials with identical cytotoxic for both groups)

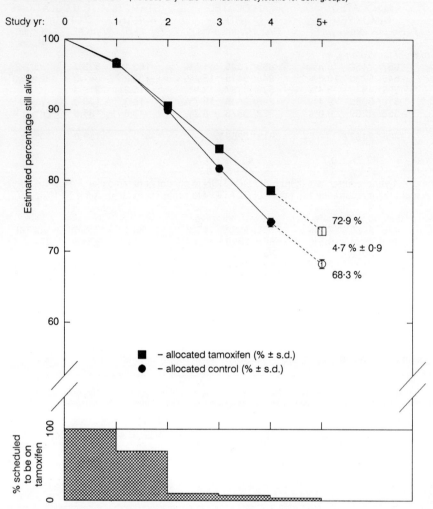

MORTALITY ANALYSES M[all ages]

BELOW, LEFT: M[<50] MORTALITY in women UNDER 50 years old at entry

BELOW, RIGHT: M[≥50] MORTALITY in women 50 OR MORE years old at entry

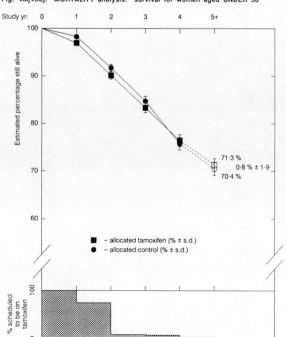

Fig. 4M[<50]. MORTALITY analysis: survival for women aged UNDER 50

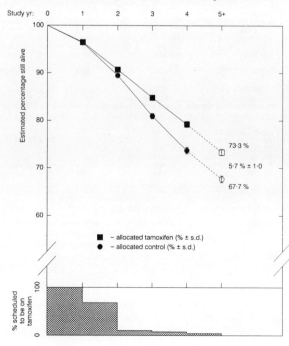

Fig. 4M[≥50]. MORTALITY analysis: survival for women aged 50 OR OVER

8 Results of tamoxifen trials

RECURRENCE ANALYSES R[all ages]

BELOW, LEFT: R[<50] RECURRENCE in women UNDER 50 years old at entry

BELOW, RIGHT: R[≥50] RECURRENCE in women 50 OR MORE years old at entry

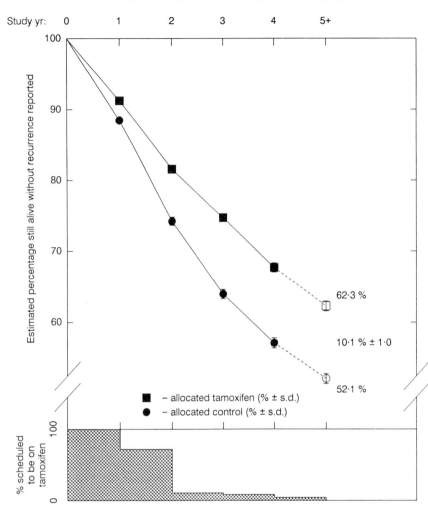

Fig. 4R. RECURRENCE-FREE SURVIVAL by year in all available trials of ANY DURATION of tamoxifen (includes any trials with identical cytotoxic for both groups)

62·3 %
10·1 % ± 1·0
52·1 %

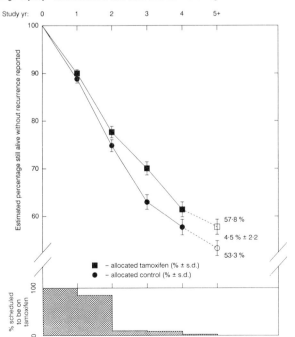

Fig. 4R[<50]. RECURRENCE-FREE SURVIVAL for women aged UNDER 50

57·8 %
4·5 % ± 2·2
53·3 %

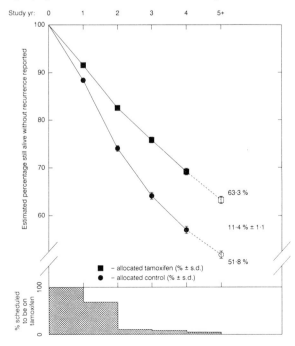

Fig. 4R[≥50]. RECURRENCE-FREE SURVIVAL for women aged 50 OR OVER

63·3 %
11·4 % ± 1·1
51·8 %

Table 7M. **Indirect comparisons between the effects of different tamoxifen regimens on the odds of MORTALITY**

Type of comparison	Reduction (± sd) in MORTALITY in tamoxifen groups				
	Age <50	Age ≥50	Both ages, crude	Both ages, standardized*	
(a) Different duration of tamoxifen					
At least 2 yrs vs no tamoxifen	1% ± 9	23% ± 4	18% ± 4	19% ± 4)	
1 yr or less vs no tamoxifen	-7% ± 15	15% ± 6	12% ± 5	11% ± 5)	NS**
(b) Interaction of tamoxifen with chemotherapy					
Tamoxifen vs nil	21% ± 14	19% ± 4	19% ± 4	19% ± 4)	
Tam+chemo vs same chemo	-9% ± 9	22% ± 6	10% ± 5	17% ± 6)	NS
(c) Different doses of tamoxifen					
20 mg/day vs no tamoxifen	-5% ± 9	22% ± 5	15% ± 4	17% ± 4)	
30-40 mg/day vs no tamoxifen	8% ± 14	18% ± 5	17% ± 5	16% ± 5)	NS
ANY TAMOXIFEN vs NO ADJUVANT TAM (and 95% confidence limits)	-1% ± 8 (-17% to 13%)	20% ± 3 (14% to 26%)	16% ± 3 (10% to 21%)	16% ± 3 (10% to 22%)	

MORTALITY ANALYSES

* The age-standardized reduction in the log of the odds ratio (treatment : control) was defined as one-fifth of that among women aged under 50 plus four-fifths of that among older women. (This ensures that, for the overall result, the age-standardized and crude reductions are the same.)

** Exclusion of deaths in the first year after randomization (during which the 1-year treatment schedules and the longer treatment schedules are the same) would alter these age-standardized percentages to 22% ± 4 and 15% ± 6, the difference between which is still not conventionally significant (2P>0.1).

Figure 4R. Allocation to the tamoxifen regimens studied in these trials improved the estimated probability of recurrence-free survival five years after randomization from 52% to 62% (Figure 4R [all ages]), an absolute difference of about 10 recurrences or deaths per 100 women. It is, of course, impossible to determine just from the recurrence-free survival graphs in Figure 4R whether this represented delay of recurrence by a year or two in most women, complete avoidance of recurrence in about one woman in ten, or something in between these two extremes.

8.10 Further indirect comparisons: factors other than age

Table 7 shows the effects of allocation to adjuvant tamoxifen according to (a) the scheduled duration of tamoxifen administration (one year or less, or two years or longer), (b) the scheduled use of adjuvant chemotherapy (tamoxifen vs no adjuvant therapy, or tamoxifen plus chemotherapy vs the same adjuvant chemotherapy alone), and (c) the dose of tamoxifen to be tested (20 mg per day, or 30-40 mg per day). Further details are provided in the subtotals of Appendix Table 2. Because the effects of tamoxifen appear to be less in younger than in older women, it would be preferable for these other indirect comparisons to be standardized for age, and in Table 7 both the crude all-ages effects and the age-standardized effects are given. (In principle, the standardized comparisons are preferable, but in practice the standardization makes no qualitative difference to the other indirect comparisons.)

8.11 Effects of different durations of tamoxifen treatment: MORTALITY data

Since tamoxifen does have some effect on mortality, it is possible that two or more years of tamoxifen might be more beneficial than one year of tamoxifen. Almost no results from trials directly comparing different durations of tamoxifen therapy were available for an overview of this question. In this context, therefore, the indirect comparison of results from trials of two years or longer and from trials of one year or less of tamoxifen may provide some insight. The age-standardized reduction in the annual odds of death in all trials that tested two or more years of tamoxifen is 19% ± 4 compared with 11% ± 5 in trials of only one year or less of tamoxifen (Table 7M). Exclusion of deaths during the first year after randomization (during which the 1-year schedules and the longer schedules involve exactly the same treatment) produces similar changes in both these percentages, but has little effect on the difference between them (see footnote to Table 7M). Comparison of the survival curves derived from trials of two or more years of tamoxifen (Figure 5M [All ages]) indicates that, among women of all ages, such regimens improved 5-year survival from 70.6% to 75.7%. But, although more prolonged use of tamoxifen does appear to be somewhat more effective at delaying death, this "interaction" is not statistically significant in the mortality analyses.

8.12 Effects of different durations of tamoxifen treatment: RECURRENCE data

Similar analyses of the recurrence data (Table 7R, Figure 5R) suggest rather more strongly that two or more years of tamoxifen may be more effective than one year or less of treatment. These comparisons, however, include recurrences in the first year after randomization, during which the treatment regimens in the trials of one year of tamoxifen and in the trials of more prolonged tamoxifen had not yet begun to differ. If attention is restricted to the period after year 1, then the

RECURRENCE ANALYSES

Table 7R. **Indirect comparisons between the effects of different tamoxifen regimens on the odds of RECURRENCE (or prior death)**

Type of comparison	Reduction (± sd) in RECURRENCE in tamoxifen groups			
	Age <50	Age ≥50	Both ages, crude	Both ages, standardized*
(a) Different duration of tamoxifen				
At least 2 yrs vs no tamoxifen	18% ± 6	38% ± 3	33% ± 3	34% ± 3) **
1 yr or less vs no tamoxifen	3% ± 16	24% ± 5	22% ± 5	21% ± 5)
(b) Interaction of tamoxifen with chemotherapy				
Tamoxifen vs nil	32% ± 10	32% ± 3	32% ± 3	32% ± 3) NS
Tam+chemo vs same chemo	8% ± 7	35% ± 5	26% ± 4	31% ± 4)
(c) Different doses of tamoxifen				
20 mg/day vs no tamoxifen	16% ± 7	34% ± 4	30% ± 3	31% ± 3) NS
30-40 mg/day vs no tamoxifen	15% ± 11	32% ± 4	30% ± 4	29% ± 4)
ANY TAMOXIFEN vs NO ADJUVANT TAM (and 95% confidence limits)	16% ± 6 (4% to 26%)	33% ± 3 (29% to 37%)	30% ± 2 (26% to 34%)	30% ± 2 (26% to 34%)

* See footnote to Table 7M.

** 2P<0.01: but, exclusion of recurrences in the first year after randomization (during which the 1-year treatment schedules and the longer treatment schedules are the same) would alter these age-standardized percentages to 35% ± 3 and 23% ± 6, the difference between which is still conventionally significant (2P=0.02).

difference between the age-standardized treatment effects becomes somewhat smaller (see footnote to Table 7R), but it still remains conventionally significant (2P=0.02).

8.13 Interactions of tamoxifen with chemotherapy: MORTALITY data

Tamoxifen added to chemotherapy produces a significant (17% ± 6; 2P=0.003) age-standardized reduction in the annual odds of death compared with the same chemotherapy alone (Table 7M). This is almost the same as the apparent size of the age-standardized effect on mortality of tamoxifen alone compared with no adjuvant therapy (19% ± 4; 2P<0.00001). These two highly significant treatment effects indicate that tamoxifen reduced mortality to a similar extent whether or not chemotherapy was to be given. This suggests that for many women such chemotherapy alone is not enough without tamoxifen, though it provides no direct information at all on whether chemotherapy itself was helpful (but, see RESULTS OF CYTOTOXIC TRIALS, below). It may, however, help simplify assessment of the effects of chemotherapy, for the similarity of the sizes of the apparent effects of tamoxifen in these two circumstances suggests that any proportional effects of adjuvant tamoxifen on 5-year mortality and any proportional effects of cytotoxic chemotherapy on 5-year mortality may be approximately independent of each other.

8.14 Interactions of tamoxifen with chemotherapy: RECURRENCE data

The age-standardized recurrence data also show a highly significant effect of tamoxifen (32% ± 3; 2P<0.00001) in the absence of chemotherapy, a highly significant effect of tamoxifen (31% ± 4; 2P<0.00001) in the presence of chemotherapy, and no apparent difference between the effects of tamoxifen in these two circumstances. Again, however, although this does suggest that for many women chemotherapy alone is not enough without tamoxifen, it does not give any direct information as to whether chemotherapy itself was helpful. As was the case for mortality, the similarity of the sizes of the apparent effects of tamoxifen in these two circumstances suggests that any effects of adjuvant tamoxifen and of adjuvant cytotoxic chemotherapy on the recurrence rates during the first five years may be largely independent of each other.

8.15 Effects of different tamoxifen doses: MORTALITY data

Regardless of possible differences in side-effects associated with different doses of tamoxifen, there are certainly differences in cost that could be of practical importance in many countries. Most of the studies tested 20 mg (11 trials: 8503 patients) or 30 mg (12 trials: 5405 patients) of tamoxifen daily, and the remaining few studies tested 40 mg per day (5 trials: 2605 patients). No direct comparisons of different tamoxifen doses are available, but the indirect comparison of the age-standardized effects of 20 mg/day and of 30-40 mg/day on mortality does not suggest any difference between the effects of different doses (Table 7M).

8.16 Effects of different tamoxifen doses: RECURRENCE data

The available data on recurrence in trials of 20 mg/day and in the trials of 30-40 mg/day are summarized in Table 7R. As was the case for mortality, this indirect comparison also fails to suggest any greater effect with doses over 20 mg/day.

Fig. 5M. **MORTALITY analysis: survival by year in all available trials of 2 OR MORE YEARS of tamoxifen** (includes any trials with identical cytotoxic for both groups)

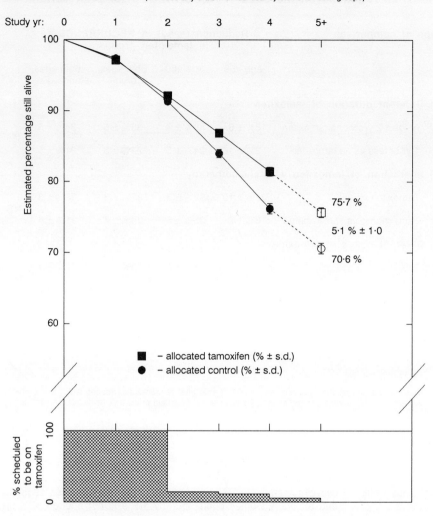

MORTALITY ANALYSES M[all ages]

BELOW, LEFT: M[<50] MORTALITY in women UNDER 50 years old at entry

BELOW, RIGHT: M[≥50] MORTALITY in women 50 OR MORE years old at entry

Fig. 5M[<50]. MORTALITY analysis: survival for women aged UNDER 50

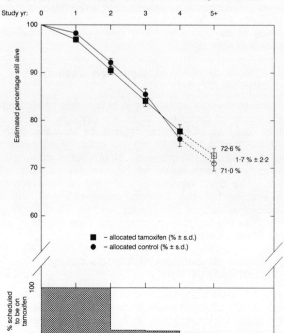

Fig. 5M[≥50]. MORTALITY analysis: survival for women aged 50 OR OVER

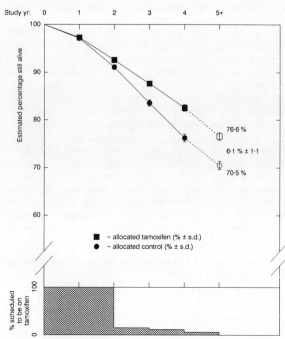

8 Results of tamoxifen trials

RECURRENCE ANALYSES R[all ages]

BELOW, LEFT: R[<50] RECURRENCE in women UNDER 50 years old at entry

BELOW, RIGHT: R[≥50] RECURRENCE in women 50 OR MORE years old at entry

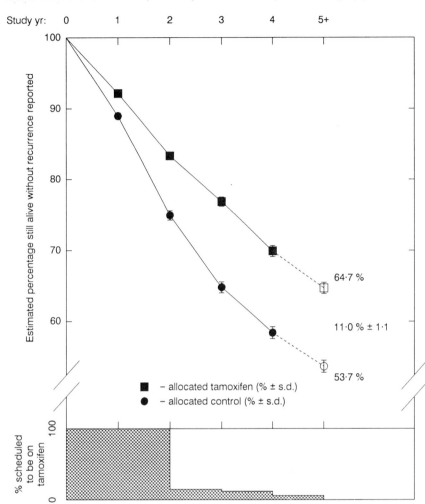

Fig. 5R. RECURRENCE-FREE SURVIVAL by year in all available trials of 2 OR MORE YEARS of tamoxifen (includes any trials with identical cytotoxic for both groups)

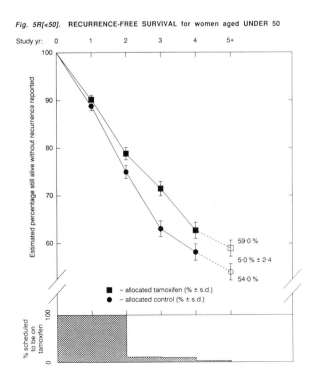

Fig. 5R[<50]. RECURRENCE-FREE SURVIVAL for women aged UNDER 50

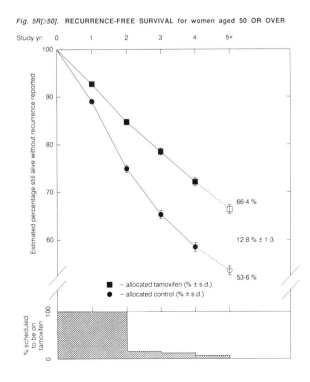

Fig. 5R[≥50]. RECURRENCE-FREE SURVIVAL for women aged 50 OR OVER

Fig. 6M. MORTALITY analysis: survival by year, subdivided by NODAL STATUS
(all ages; women in any tamoxifen trial, whether allocated tamoxifen or control)

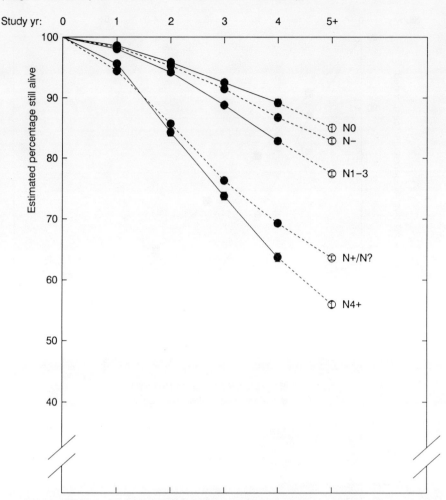

MORTALITY ANALYSES
M[all ages]

8.17 Effects of tamoxifen in different nodal categories: MORTALITY data

Five nodal status categories were defined, but the first two will be merged in the principal analyses (see below):

"N0": No positive axillary lymph nodes were found after the axilla had been "cleared". (The definition of clearance, and the pathological procedures needed to define negativity, were left to the discretion of the investigator.)

"N-": No positive nodes were found either clinically or after the nodes were "sampled". (The definition of sampling was, however, left to the discretion of the investigator.)

"N1-3": 1-3 positive nodes were found after axillary "clearance".

"N4+": At least 4 positive nodes were found after axillary sampling or clearance.

"N+/N?": Status of nodes uncertain, sometimes because no information on nodal status was reported, sometimes because the only information reported was that the nodes were "positive", and sometimes because nodal involvement was found only by sampling (and not by clearance), and the number of involved nodes was not reported to exceed 3.

Categories N0 and N1-3 included only patients whose axillae were "cleared". The thoroughness or reliability of the axillary dissections and the pathological investigations used to define N0, N1-3 and N4+ presumably varied considerably from one trial to another. This does not, however, invalidate the tamoxifen analyses that follow, for these are based on overviews of statistical analyses that each involved only comparison of patients in one nodal category status in one trial with other patients in that same category in that same trial.

The 5-year survival curves for all patients (irrespective of treatment allocation) are plotted in Figure 6M, subdivided by nodal status. It can be seen that, despite the different criteria used in these trials, "nodal status" does effectively divide patients into high, medium and low risk groups, and this division could well have been even clearer if deaths from breast cancer could have been analyzed separately from other deaths. The estimated five-year survival was similar in the category N0 and in the category N- (85% and 83%, respectively). Since the prognosis of the women in both these categories was relatively favorable, these two nodal status categories contain only limited numbers of deaths and so, taken separately, neither can be expected to yield statistically reliable trial results. The two categories (N0/N-) were, therefore, combined in all subsequent analyses of the effects of tamoxifen.

The four parts of Figure 7M display the individual trial results and the overall estimates (stratified for age <50, 50+) for each of the four nodal status categories (N0/N-, N1-3, N4+, and N+/N?), and the summary estimates for these four categories are brought together for graphic display in Figure 8M.

8 Results of tamoxifen trials

RECURRENCE ANALYSES R[all ages]

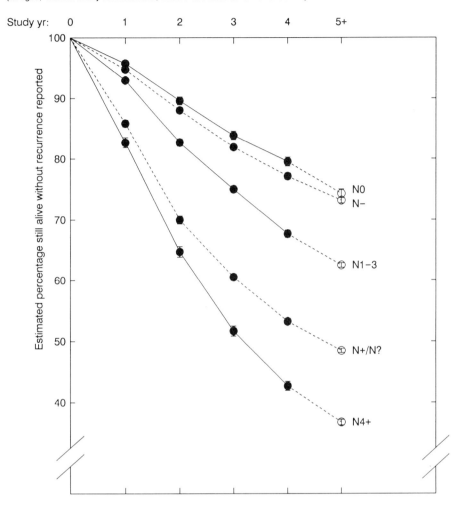

Fig. 6R. **RECURRENCE-FREE SURVIVAL by year, subdivided by NODAL STATUS**
(all ages; women in any tamoxifen trial, whether allocated tamoxifen or control)

The sum of the different contributions from one particular trial in Figure 7M provides an analysis of that trial that is "retrospectively stratified"[31] for what information is available on initial nodal status and for age. Likewise, in the bottom line of Figure 8M the sum of the four separate overviews in the different nodal status categories provides an analysis of the effects of tamoxifen that has been retrospectively stratified for the available information on nodal status and for age. Since comparison of like with like is slightly more sensitive, and since in addition there was a slight tendency for poor-prognosis patients to have been allocated tamoxifen (which was not significant except in trial 80E), this stratified analysis is a little more definitively favorable (18% ± 3: Figure 8M) than was the original unstratified analysis (16% ± 3: Figure 2M). Overall, there was no evidence of any differences between the sizes of the proportional mortality reductions among patients in the four different nodal status categories (e.g. N0/N-: 15% ± 8 reduction; N4+: 17% ± 6 reduction), and the test for trend in Figure 8M between the first three of them (N0/N-, N1-3 and N4+) was completely non-significant. Hence, since N4+ patients are at higher risk of death than those with fewer involved nodes, the **absolute** reductions in 5-year mortality produced by tamoxifen (Table 8M) are greater for women with 4+ nodes recorded (absolute reduction in estimated 5-year mortality = 6.1 deaths per 100 women) than among women without any nodal involvement recorded (absolute reduction = 2.5 deaths per 100 women). There is, however, no direct evidence yet as to whether or not the same would be true for the 10-year mortality, or for the 15-year mortality.

8.18 Effects of tamoxifen in different nodal categories: RECURRENCE data

As was the case for mortality, the nodal categorization does successfully divide the patients into those at higher risk of recurrence within 5 years and those at lower risk (Figure 6R). The individual trial results and the overall estimates for each of the four nodal status categories are displayed in Figure 7R, and the summary estimates for these four categories are graphically displayed together in Figure 8R. The final line of Figure 8R provides an estimate (33% ± 2) of the effect of tamoxifen allocation on the odds of recurrence-free survival that is "retrospectively stratified"[31] for what is known of initial nodal status and for age. In each separate nodal category in Figure 8R the proportional reductions are highly significantly different from zero, but they are not significantly different from each other. Thus, the recurrence data are compatible with the hypothesis that the proportional reductions in the annual odds of recurrence (or prior death) are reasonably similar for node negative, 1-3 node and 4+ node patients. As for the mortality analysis, however, the **absolute** reductions in the 5-year recurrence rate produced by tamoxifen are greater for women with 4+ nodes than for women with 1-3 nodes or without nodal involvement (Table 8R).

Fig. 7M. **Effects of tamoxifen on MORTALITY in various NODAL categories**
(all ages; content and format as in Fig. 2M; nodal status categories as in text)

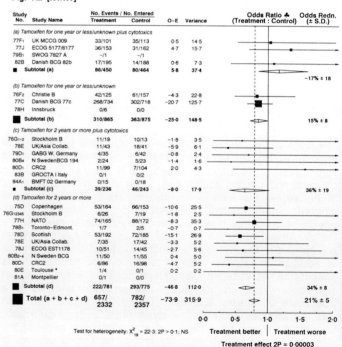

* Significant imbalance in initial nodal status
♣ 95% confidence intervals for overview and 99% for individual trials.

8 Results of tamoxifen trials

Fig. 7R. **Effects of tamoxifen on RECURRENCE in various NODAL categories**
(all ages; content and format as in Fig. 2R; nodal status categories as in text)

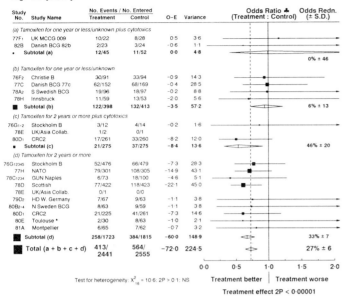

* Significant imbalance in initial nodal status
♣ 95% confidence intervals for overview and 99% for individual trials.

Fig. 8M. **Effects of tamoxifen on MORTALITY in various NODAL categories**
(from Fig. 7M: all ages; all tamoxifen durations; includes any trials with identical cytotoxic for both groups)

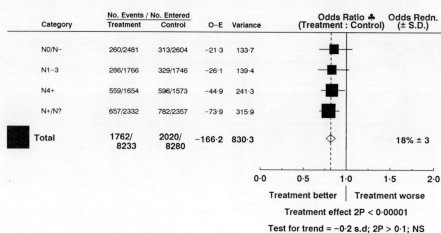

♣ 95% confidence intervals for overview and 99% for individual categories.
N.B: The test for trend is between 3 specific categories.

MORTALITY ANALYSES M[all ages]

Table 8M. **Tamoxifen and MORTALITY in categories of nodal status**
(including trials with identical chemotherapy for both treatment and control groups)

Nodal status at entry	"Life-table" estimates of 5-year mortality risks			Statistical calculations for TAMOXIFEN groups only		
	Allocated CONTROL	Allocated TAMOXIFEN	Difference (± sd)	O-E	Var. of O-E	Mortality redn ± sd
N0/N−	17.6%	15.1%	2.5% ± 1.3	−21.3	133.7	15% ± 8
N1-3	24.7%	20.4%	4.3% ± 1.7	−26.1	139.4	17% ± 8
N4+	47.0%	41.0%	6.1% ± 2.1	−44.9	241.3	17% ± 6
(N+/N?)	(40.0%)	(32.6%)	(7.4% ± 1.6)	(−73.9)	(315.9)	(21% ± 5)
All women	31.8%	27.0%	4.7% ± 0.9	−166.2	830.3	18% ± 3

Note: Any age, any duration of tamoxifen. The statistical calculations for "All women" are retrospectively stratified for status at entry, but the life-table estimates are not.

MORTALITY ANALYSES M[all ages]

8.19 Effects of tamoxifen in different Estrogen Receptor categories: MORTALITY data

Because the main therapeutic effects of tamoxifen may be mediated through the cytoplasmic Estrogen Receptor (ER) protein, ER measurements of the excised primary tumor might perhaps identify a group of early breast cancer patients whose micrometastatic disease is largely or wholly unresponsive to tamoxifen.

Measurements of ER levels were available for nearly half of the patients in the tamoxifen trials. Some investigators provided quantitative information while others submitted only qualitative descriptions of the assay results. Patients were classified as "poor" if the level was less than 10 fmols/mg cytosol protein (or if the tumor was described by the principal investigator as "ER−"). Patients were classified as "ER+" if the level was 10-99 fmols/mg (or was described as ER+) and as "ER++" if the level was greater than 100 fmols/mg (or was described as ER++). Patients without ER measurements available were classified as "ER?". Among the women with measurements of ER levels available, 31% were ER poor, 44% ER+, and 26% ER++.

There are, however, several potential sources of error in the measurement of the receptor proteins. There may, for example, have been a heterogeneous admixture of receptor positive and receptor negative cells within the original primary tumor. Moreover, the active receptor protein may be relatively unstable following surgical excision of the tumor, so low values may arise artefactually. The overview, therefore, assesses only the question of whether the ER **assays** of the primary tumor that were actually conducted in these trials can define a category of "ER negative" women whose micrometastatic disease will be completely unaffected by adjuvant tamoxifen. As with the nodal status classifications, variation in the reliability of the receptor assays used in different trials does not bias the comparisons of treatment within each category of **measured** ER status, because the statistical analyses assume merely that the assay methods used to distribute patients into ER categories were consistent for tamoxifen patients and for controls within each trial. But, if there are many false negative ER assay results in the data then

8 Results of tamoxifen trials

RECURRENCE ANALYSES R[all ages]

Fig. 8R. **Effects of tamoxifen on RECURRENCE in various NODAL categories**
(from Fig. 7R: all ages; all tamoxifen durations; includes any trials with identical cytotoxic for both groups)

Category	No. Events / No. Entered Treatment	No. Events / No. Entered Control	O-E	Variance	Odds Ratio ♣ (Treatment : Control)	Odds Redn. (± S.D.)
N0/N–	413/2441	564/2555	–72.0	224.5		
N1–3	453/1749	597/1728	–93.7	229.9		
N4+	678/1449	790/1374	–119.1	279.3		
N+/N?	822/2113	1062/2132	–159.0	389.5		
Total	2366/7752	3013/7789	–443.8	1123.2		33% ± 2

Treatment better | Treatment worse
Treatment effect 2P < 0.00001
Test for trend = –1.2 s.d; 2P > 0.1; NS

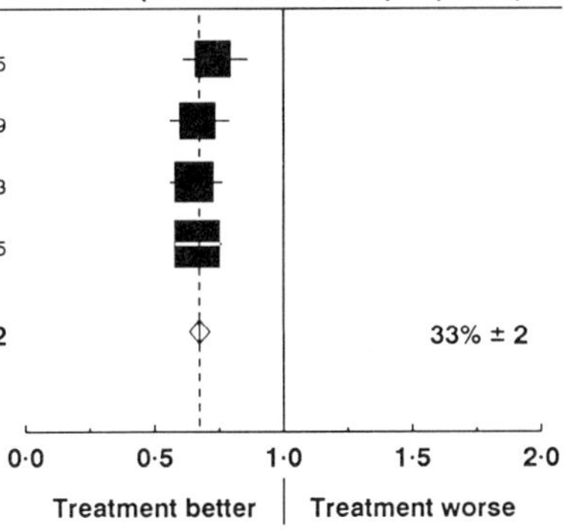

♣ 95% confidence intervals for overview and 99% for individual categories.
N.B: The test for trend is between 3 specific categories.

RECURRENCE ANALYSES R[all ages]

Table 8R. **Tamoxifen and RECURRENCE in categories of nodal status**
(including trials with identical chemotherapy for both treatment and control groups)

Nodal status at entry	"Life-table" estimates of 5-year recurrence risks			Statistical calculations for TAMOXIFEN groups only		
	Allocated CONTROL	Allocated TAMOXIFEN	Difference (± sd)	O-E	Var. of O-E	Recurrence redn ± sd
N0/N–	29.4%	23.5%	5.9% ± 1.6	–72.0	224.5	27% ± 6
N1-3	42.6%	32.1%	10.6% ± 2.0	–93.7	229.9	33% ± 5
N4+	70.0%	55.7%	14.4% ± 2.3	–119.1	279.3	35% ± 5
(N+/N?)	(58.4%)	(43.8%)	(14.7% ± 1.8)	(–159.0)	(389.5)	(34% ± 4)
All women	48.0%	37.7%	10.3% ± 1.0	–443.8	1123.2	33% ± 2

Note: As in Table 8M.

inability of the currently available assay results to identify a group of wholly unresponsive women would not necessarily imply inability of some better ER assay to identify unresponsive women.

Some indication of the biological validity of the available ER assay results can be obtained by seeing how well they predict survival, for a wholly inaccurate ER assay would not be of any prognostic significance, while a reasonably accurate assay should be. The 5-year survival curves for all patients (irrespective of treatment allocation) are plotted in Figure 9M, subdivided by ER status. This indicates that the available ER assay results are at least somewhat informative, since the ER categorization was strongly predictive of mortality, with the 31% of patients who were measured to be ER poor being at particularly high risk.

The four parts of Figure 10M display the individual tamoxifen trial results and the overall estimates for each of the four ER status categories (ER poor, ER+, ER++, ER?), and the summary estimates for these four categories are displayed together graphically in Figure 11M. The prognosis of women classified as ER poor was substantially worse than that of other women (Figure 9M), suggesting that this ER classification did have some biologic meaning. Despite this, the ER measurements performed in these studies did not identify a group of women wholly unresponsive to tamoxifen: in Figures 10M and 11M the estimated mortality reduction produced by tamoxifen allocation among "ER poor" women is 15% ± 7, which is conventionally significant (2P=0.03), though not extremely so. Indeed, since ER poor patients were at highest risk, the **absolute** improvements in 5-year survival produced by tamoxifen (Table 9M) appeared, paradoxically, to be slightly greater for ER poor women than for other women (Table 9M). But, although the mortality improvements produced by tamoxifen among ER poor women and among ER positive (+ or ++) women both differ significantly from zero, they do not differ significantly from each other.

8.20 Effects of tamoxifen in different Estrogen Receptor categories: RECURRENCE data

Data from the tamoxifen trials on both the results of receptor measurements and the time to recurrence are available on only

Fig. 9M. MORTALITY analysis: survival by year, subdivided by ER STATUS
(all ages; women in any tamoxifen trial, whether allocated tamoxifen or control)

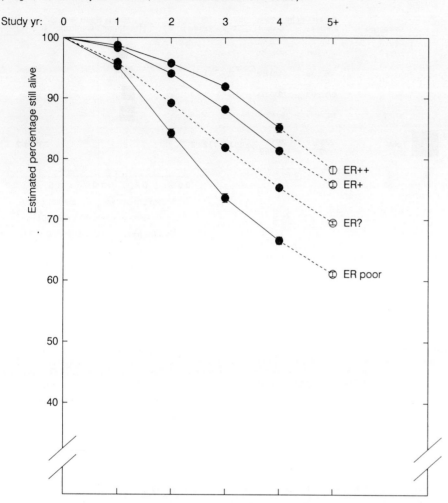

MORTALITY
ANALYSES
M[all ages]

36% of the patients,* which limits the statistical sensitivity of the recurrence analyses within ER categories. As was the case for mortality, however, the ER measurements are significantly related to prognosis (Figure 9R), but do not identify any group of women who are wholly unresponsive to tamoxifen. The detailed results available from each study are shown in Figure 10R for each of the four ER status categories, and the summary estimates for these four categories are displayed together in Figure 11R. The delay in recurrence from tamoxifen treatment is 31% ± 7 (2P<0.00003) for ER poor patients, 40% ± 7 (2P<0.00001) for ER+ patients, and 50% ± 8 (2P<0.00001) for ER++ patients. Although this trend towards a greater relative effect of tamoxifen with a greater ER level is statistically significant (2P<0.02), the difference between the ER poor and the other patients in their response to tamoxifen again appears to be merely a quantitative difference, not a qualitative one. The absolute improvements in 5-year recurrence-free survival were slightly smaller for ER poor women than for other women (Table 9R). But again, although the improvements in recurrence-free survival produced by tamoxifen among ER poor women and among other women both differ very significantly from zero, they do not differ significantly from each other.

The hypothesis that tamoxifen might delay recurrence among women whose primary tumor was measured to be ER poor was generated by the results of the NATO trial (77H), but even if the results from that hypothesis-generating study are removed from the overview estimate, the delay of recurrence among the "ER poor" patients in the remaining trials is still significant (24% ± 8; 2P=0.004). Hence, for recurrence as for mortality, the ER assays used in at least some of these trials did not succeed in picking out a group of wholly unresponsive women.

8.21 Age-specific effects of ER assay results

The ER assay is more likely to be positive in older women than it is in younger women, so the fact that the analyses in Figures 10 and 11 were standardized for age will have tended, if anything, to have decreased the predictive power of the available ER assay results (unstandardized data not shown).

---

* Information from study 77K was available only in tabular form, giving mortality by year, recurrence by year and ER versus mortality, but not ER versus recurrence.

8 Results of tamoxifen trials

RECURRENCE ANALYSES R[all ages]

Fig. 9R. **RECURRENCE-FREE SURVIVAL by year, subdivided by ER STATUS**
(all ages; women in any tamoxifen trial, whether allocated tamoxifen or control)

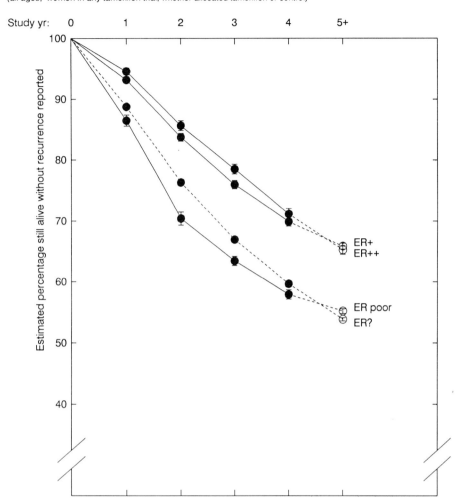

Fig. 10M. Effects of tamoxifen on MORTALITY in various ER categories

(all ages; content and format as in Fig. 2M; ER status categories as in text)

Fig. 10M [ER-poor]

Study No.	Study Name	No. Events / No. Entered: Treatment	Control	O–E	Variance	Odds Redn. (± S.D.)
	(a) Tamoxifen for 2 years or more					
75D	Copenhagen	39/102	43/88	−6·8	17·1	
76G12345	Stockholm B	20/111	17/118	2·8	8·3	
77H	NATO	33/116	46/95	−12·4	17·6	
77K	NSABP B-09	162/318	147/308	5·1	66·0	
78C	GUN Naples	17/67	15/64	−1·6	6·8	
78D	Scottish	30/127	36/129	−2·7	12·7	
78F	CRFB Caen C5	10/24	11/19	−2·3	4·3	
78J	ECOG EST1178	3/3	3/5	0·8	1·0	
79D	GABG/HD Germany	19/123	25/121	−3·0	9·6	
81B	FB Bordeaux	0/16	2/9	−1·1	0·5	
83B	GROCTA I Italy	0/14	0/8			
84A1	BMFT 02 Germany	1/9	0/6	0·3	0·2	
	Subtotal (a)	**334/1030**	**345/970**	**−20·9**	**144·1**	**13% ± 8**
	(b) Tamoxifen for 1 year or less					
74G	Case Western A	19/39	23/35	−4·0	8·7	
77C	Danish BCG 77c	12/26	19/32	−3·1	6·3	
77J	ECOG 5177/6177	51/131	56/133	−2·4	23·4	
78A2	S Swedish BCG	12/47	17/59	−0·8	6·6	
78H	Innsbruck	1/2	0/0			
79B1	SWOG 7827 A	−/10	−/19	*blind*	1·4	
82B	Danish BCG 82b	6/23	3/24	0·8	1·8	
	Subtotal (b)	**103/278**	**122/302**	**−9·9**	**48·0**	**19% ± 13**
	Total (a + b)	**437/ 1308**	**467/ 1272**	**−30·7**	**192·1**	**15% ± 7**

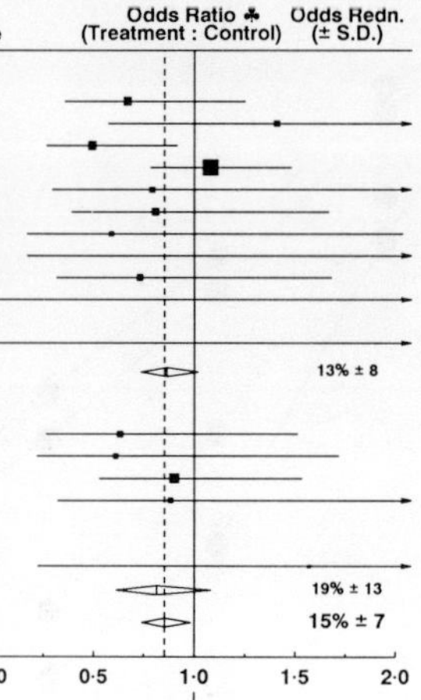

Test for heterogeneity: X^2_{16} = 18·6; 2P > 0·1; NS

Treatment better | Treatment worse

Treatment effect 2P = 0·03

Fig. 10M [ER+]

Study No.	Study Name	No. Events / No. Entered: Treatment	Control	O–E	Variance	Odds Redn. (± S.D.)
	(a) Tamoxifen for 2 years or more					
75D	Copenhagen	8/43	16/49	−2·7	5·0	
76G12345	Stockholm B	29/292	36/302	−4·6	14·8	
77H	NATO	18/72	35/99	−5·4	11·6	
77K	NSABP B-09	109/327	109/326	0·1	48·8	
78C	GUN Naples	6/61	13/77	−3·3	4·2	
78D	Scottish	19/135	16/134	1·4	7·8	
78F	CRFB Caen C5	5/21	9/28	−1·0	3·2	
78J	ECOG EST1178	14/48	12/42	0·7	6·0	
79D	GABG/HD Germany	9/81	13/83	−2·7	5·0	
81B	FB Bordeaux	3/106	4/107	−0·4	1·7	
83B	GROCTA I Italy	1/62	0/59	0·5	0·3	
84A1	BMFT 02 Germany	0/14	0/15			
	Subtotal (a)	**221/1262**	**263/1321**	**−17·5**	**108·3**	**15% ± 9**
	(b) Tamoxifen for 1 year or less					
74G	Case Western A	13/40	21/47	−3·0	6·8	
77C	Danish BCG 77c	20/40	12/44	5·7	7·1	
77J	ECOG 5177/6177	39/142	36/140	1·2	16·8	
78A2	S Swedish BCG	6/51	9/50	−2·9	3·4	
78H	Innsbruck	8/79	16/74	−4·4	5·7	
79B1	SWOG 7827 A	−/110	−/119	*blind*	3·6	
82B	Danish BCG 82b	1/40	2/45	−0·3	0·7	
	Subtotal (b)	**92/502**	**106/519**	**−5·7**	**44·1**	**12% ± 14**
	Total (a + b)	**313/ 1764**	**369/ 1840**	**−23·1**	**152·4**	**14% ± 8**

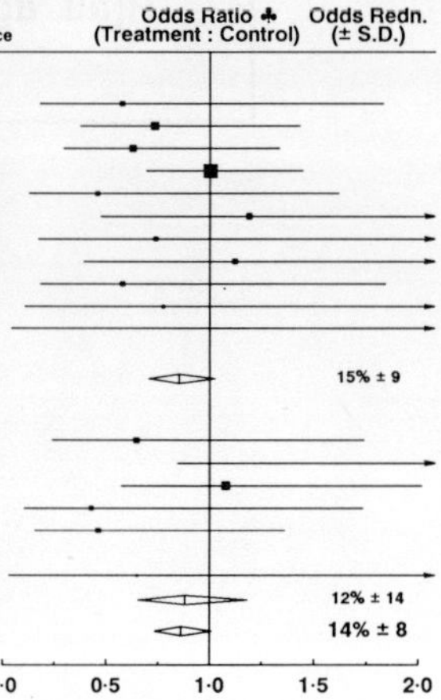

Test for heterogeneity: X^2_{17} = 20·7; 2P > 0·1; NS

Treatment better | Treatment worse

Treatment effect 2P = 0·06

Fig. 10M [ER++]

Study No.	Study Name	No. Events / No. Entered: Treatment	Control	O–E	Variance	Odds Redn. (± S.D.)
	(a) Tamoxifen for 2 years or more					
75D	Copenhagen	6/19	7/16	−1·1	2·6	
76G12345	Stockholm B	20/205	21/194	−0·5	9·2	
77H	NATO	20/73	24/70	−3·8	9·9	
77K	NSABP B-09	43/133	41/141	1·8	18·1	
78C	GUN Naples	3/33	0/38	0·7	0·4	
78D	Scottish	10/122	13/98	−2·9	5·3	
78F	CRFB Caen C5	1/21	6/20	−2·7	1·7	
78J	ECOG EST1178	6/29	15/34	−4·5	4·8	
79D	GABG/HD Germany	3/52	2/47	−0·2	0·8	
81B	FB Bordeaux	1/43	5/44	−1·7	1·5	
83B	GROCTA I Italy	0/24	0/22			
84A1	BMFT 02 Germany	0/9	0/9			
	Subtotal (a)	**113/763**	**134/733**	**−14·9**	**54·2**	**24% ± 12**
	(b) Tamoxifen for 1 year or less					
74G	Case Western A	10/24	8/19	0·0	3·5	
77C	Danish BCG 77c	15/71	11/70	1·3	5·9	
77J	ECOG 5177/6177	6/26	12/31	−2·2	4·0	
78A2	S Swedish BCG	13/70	10/53	−0·4	5·3	
78H	Innsbruck	2/46	1/33	0·0	0·7	
79B1	SWOG 7827 A	−/69	−/76	*blind*	3·6	
82B	Danish BC3 82b	0/11	0/7			
	Subtotal (b)	**53/317**	**50/289**	**−1·8**	**23·0**	**8% ± 20**
	Total (a + b)	**166/ 1080**	**184/ 1022**	**−16·7**	**77·2**	**19% ± 10**

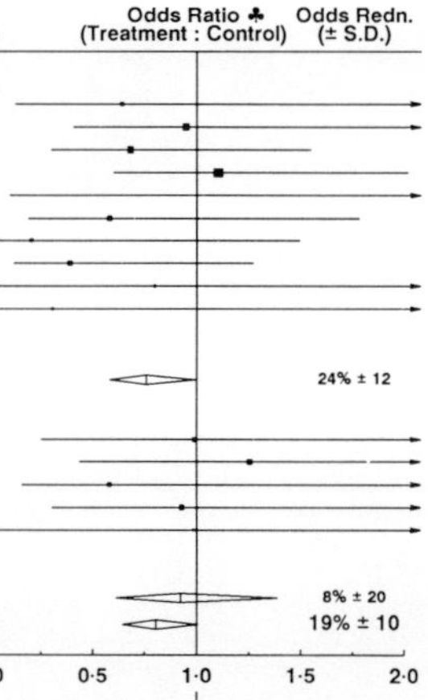

Test for heterogeneity: X^2_{15} = 13·3; 2P > 0·1; NS

Treatment better | Treatment worse

Treatment effect 2P = 0·06

Fig. 10M [ER?]

Study No.	Study Name	No. Events / No. Entered: Treatment	Control	O–E	Variance	Odds Redn. (± S.D.)
	(a) Tamoxifen for 2 years or more					
76G12345	Stockholm B	23/135	19/116	−0·5	9·1	
77H	NATO	92/303	103/303	−6·2	44·5	
77K	NSABP B-09	45/172	66/166	−13·0	24·8	
78B1	Toronto-Edmont.	47/198	52/202	−1·4	22·7	
78C	GUN Naples	3/44	2/49	0·1	1·1	
78D	Scottish	52/280	76/298	−13·2	29·0	
78E	UK/Asia Collab.	40/232	61/243	−10·7	22·4	
78F	CRFB Caen C5	8/23	6/23	0·6	3·3	
78J	ECOG EST1178	1/11	3/9	−1·2	1·0	
79A	Ghent Univ.	5/39	8/46	−0·9	3·1	
80B2-4	N Sweden BCG	18/137	21/137	−1·4	9·1	
80D1	CRC2	45/876	59/892	−5·5	25·3	
80E	Toulouse *	22/125	20/126	0·8	9·9	
81A	Montpellier	7/101	9/102	−1·1	3·9	
83B	GROCTA I Italy	0/1	0/5			
84A1	BMFT 02 Germany	0/16	0/18			
	Subtotal (a)	**408/2693**	**505/2735**	**−53·5**	**209·3**	**23% ± 6**
	(b) Tamoxifen for 1 year or less					
76F2	Christie B	101/282	120/306	−6·5	49·5	
77C	Danish BCG 77c	271/749	311/741	−20·5	128·1	
77F1	UK MCCG 009	37/123	38/141	0·5	16·5	
77J	ECOG 5177/6177	0/4	1/3	−0·5	0·3	
78A2	S Swedish BCG	12/71	12/74	−0·1	5·7	
79B1	SWOG 7827 A	−/15	−/10	*blind*	1·8	
82B	Danish BCG 82b	11/144	10/136	0·1	5·0	
	Subtotal (b)	**438/1388**	**495/1411**	**−26·6**	**206·9**	**12% ± 7**
	Total (a + b)	**846/ 4081**	**1000/ 4146**	**−80·1**	**416·2**	**18% ± 4**

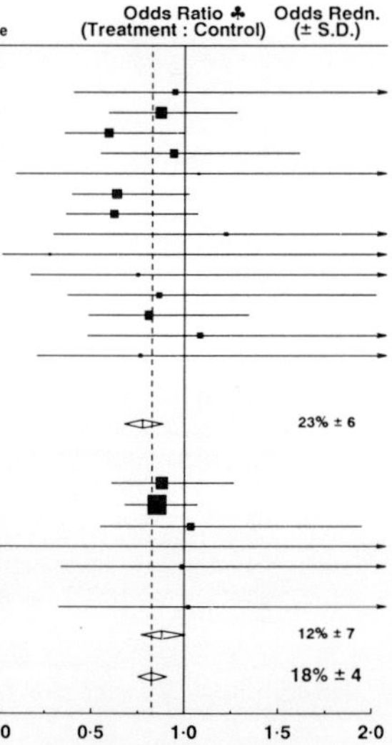

Test for heterogeneity: X^2_{20} = 12·4; 2P > 0·1; NS

Treatment better | Treatment worse

Treatment effect 2P = 0·00009

♣ 95% confidence intervals for overview and 99% for individual trials.

8 Results of tamoxifen trials

Fig. 10R. **Effects of tamoxifen on RECURRENCE in various ER categories**
(all ages; content and format as in Fig. 2R; ER status categories as in text)

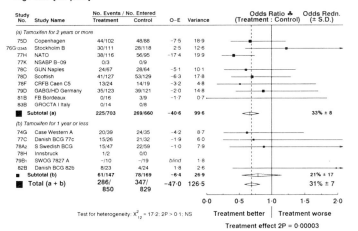

♣ 95% confidence intervals for overview and 99% for individual trials.

Fig. 11M. **Effects of tamoxifen on MORTALITY in various ER categories**
(from Fig. 10M: all ages; all tamoxifen durations; includes any trials with identical cytotoxic for both groups)

Category	No. Events / No. Entered Treatment	Control	O-E	Variance	Odds Ratio ♣ (Treatment : Control)	Odds Redn. (± S.D.)
ER poor	437/1308	467/1272	-30.7	192.1		
ER+	313/1764	369/1840	-23.1	152.4		
ER++	166/1080	184/1022	-16.7	77.2		
ER?	846/4081	1000/4146	-80.1	416.2		
Total	1762/8233	2020/8280	-150.7	837.9		16% ± 3

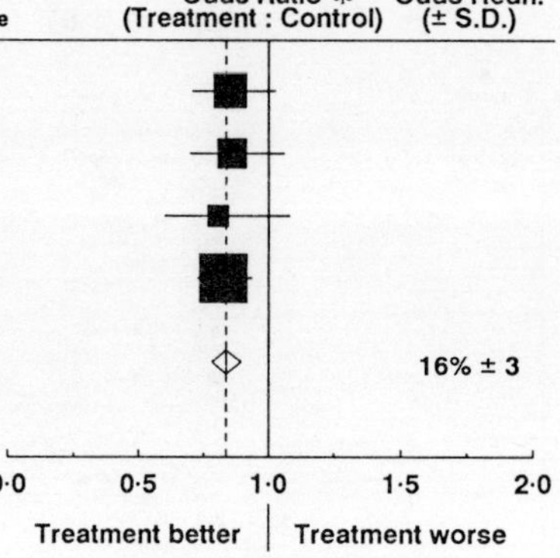

Treatment effect 2P < 0·00001

Test for trend = -0·3 s.d; 2P > 0·1; NS

♣ 95% confidence intervals for overview and 99% for individual categories.
N.B: The test for trend is between 3 specific categories.

MORTALITY ANALYSES M[all ages]

Table 9M. **Tamoxifen and MORTALITY in categories of estrogen receptor status** (including trials with identical chemotherapy for both treatment and control)

Estrogen receptor status at entry	"Life-table" estimates of 5-year mortality risks			Statistical calculations for TAMOXIFEN groups only		
	Allocated CONTROL	Allocated TAMOXIFEN	Difference (± sd)	O-E	Var. of O-E	Mortality redn ± sd
ER poor	41.7%	36.0%	5.6% ± 2.2	-30.7	192.1	15% ± 7
ER+	25.7%	22.5%	3.1% ± 1.7	-23.1	152.4	14% ± 8
ER++	23.8%	19.6%	4.2% ± 2.2	-16.7	77.2	19% ± 10
(ER?)	(33.2%)	(27.7%)	(5.5% ± 1.3)	(-80.1)	(416.2)	(18% ± 4)
All women	31.8%	27.0%	4.7% ± 0.9	-150.7	837.9	16% ± 3

Note: Any age, any duration of tamoxifen. The statistical calculations for "All women" are retrospectively stratified for status at entry, but the life-table estimates are not.

MORTALITY ANALYSES M[all ages]

8 Results of tamoxifen trials

RECURRENCE ANALYSES R[all ages]

Fig. 11R. Effects of tamoxifen on RECURRENCE in various ER categories
(from Fig. 10R: all ages; all tamoxifen durations; includes any trials with identical cytotoxic for both groups)

Category	No. Events / No. Entered Treatment	No. Events / No. Entered Control	O−E	Variance	Odds Ratio ♣ (Treatment : Control)	Odds Redn. (± S.D.)
ER poor	286/850	347/829	−47.0	126.5		
ER+	241/1236	381/1321	−68.8	134.4		
ER++	173/884	254/809	−63.8	91.6		
ER?	1666/4782	2031/4830	−230.6	790.3		
Total	2366/7752	3013/7789	−410.2	1142.8		30% ± 2

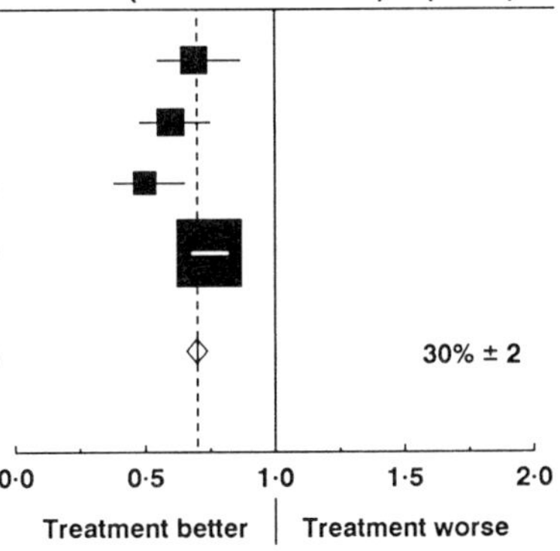

Treatment better | Treatment worse

Treatment effect 2P < 0.00001

Test for trend = −2.4 s.d; 2P = 0.02

♣ 95% confidence intervals for overview and 99% for individual categories.
N.B: The test for trend is between 3 specific categories.

RECURRENCE ANALYSES R[all ages]

Table 9R. **Tamoxifen and RECURRENCE in categories of estrogen receptor status** (including trials with identical chemotherapy for both treatment and control)

Estrogen receptor status at entry	"Life-table" estimates of 5-year recurrence risks			Statistical calculations for TAMOXIFEN groups only		
	Allocated CONTROL	Allocated TAMOXIFEN	Difference (± sd)	O-E	Var. of O-E	Recurrence redn ± sd
ER poor	50.4%	38.7%	11.7% ± 3.0	−47.0	126.5	31% ± 7
ER+	40.4%	27.5%	12.9% ± 2.4	−68.8	134.4	40% ± 7
ER++	42.2%	26.6%	15.6% ± 3.0	−63.8	91.6	50% ± 8
(ER?)	(50.3%)	(41.7%)	(8.6% ± 1.2)	(−230.6)	(790.3)	(25% ± 3)
All women	48.0%	37.6%	10.3% ± 1.0	−410.2	1142.8	30% ± 2

Note: As in Table 9M.

9. RESULTS OF CYTOTOXIC TRIALS

The first part of Table 10 summarizes the trials available for review in which at least two of the adjuvant treatment regimens being compared differed only in that one involved prolonged[*] chemotherapy and the other did not (with any other adjuvant treatments in those two particular regimens being the same). Information from those 32 trials is available from a total of 9069 women, 32% (2872) of whom were reported to have died. Recurrence data were available on 90%, among whom 44% (3564/8172) had either relapsed (or, in a few cases, died before relapse). These events were approximately evenly distributed over years 1, 2, 3, 4 and 5+ of follow-up, providing useful information for up to about five years after randomization but not beyond. About one-third of all patients were less than 50 years old, and two-thirds were aged 50 or over when randomized.

The second part of Table 10 summarizes the trials in which at least two arms compare more with less chemotherapy. Information from these 16 comparisons is available from a total of 5089 women, 37% (1879) of whom were reported to have died. Recurrence data were available on 84%, among whom 48% (2049/4252) had relapsed (or died before relapse). About half were less than 50 years old.

Cytotoxic trials are far more heterogeneous than tamoxifen trials, and this complicates the interpretation of an overview of the randomized trials of adjuvant cytotoxic therapy. The regimens differ not only in duration and in dose intensity but also in the number of drugs used and in the specific drugs employed. Hence, the results from a few fairly effective regimens might be misleadingly diluted by those of a number of rather ineffective regimens, yielding an overview that is reliably informative about neither. Since the most extensively studied drug combinations in these trials were those involving cyclophosphamide, methotrexate and fluorouracil (CMF), analysis of trials evaluating regimens that included this combination (with or without other cytotoxic agents) are presented separately. Even these "CMF-based trials" are unlikely to be homogeneous, however. But, the more finely the trials or patients are subdivided, the more the results are subject to statistical fluctuations. Even if the trials of CMF and the trials of CMF with other cytotoxics are combined, the total number of women enrolled in randomized trials comparing such "CMF-based" regimens with no chemotherapy is still under

---

[*] Trials of chemotherapy used only during the perioperative period (i.e. only in the first few days or weeks after surgery) were not included since the data available for the overview were too limited.

Table 10M. **MORTALITY data available for evaluation of adjuvant cytotoxic in early breast cancer** (including some trials with an identical tamoxifen regimen for the different treatment groups)

Randomized cytotoxic comparison	Mortality (all causes)	
	No. of trials (or parts)	Deaths/patients
(a) CMF vs NIL (incl CMFPr vs NIL)	11	920/3380
(b) CMF with extra cytotoxics vs NIL	5	367/1467
(c) Other polychemotherapy vs NIL	11	743/2315
(d) Single-agent chemotherapy vs NIL	8	971/2257
1. Any cytotoxic therapy vs NIL (a+b+c+d, excluding overlap*)	32	2872/9069
2. Polychemotherapy vs single-agent chemotherapy	10	1144/3005
3. More prolonged vs less prolonged polychemotherapy	6	752/2111
ALL CYTOTOXIC TRIALS (1+2+3, excluding overlap*)	40	4503/13442 34% die

MORTALITY ANALYSES M[all ages]

Note: Prolonged regimes only: excludes "perioperative" trials.

* The totals are less than the sum of the parts, because trials 74D1, 74D2, 75E2, 77B and 77E had 3-way randomizations, and so contribute to more than one analysis. (74D1: more prolonged poly vs less prolonged poly vs nil. 74D2: more prolonged poly vs less prolonged poly vs single. Others: poly vs single vs nil.)

5000 (Table 10). This may be contrasted with the tamoxifen trials, in which tamoxifen was compared with no tamoxifen in 16,500 women, 11,500 of whom received 2 or more years of tamoxifen. Because of the smaller size of the chemotherapy trials, and because of the heterogeneity of the regimens used, the estimated sizes of the effects of particular types of chemotherapy are likely to be less certain, especially in subgroup analyses, than were the estimated sizes of the effects of tamoxifen.

9.1 Overall analysis of MORTALITY in chemotherapy trials

Results from each of the 32 trials comparing adjuvant chemotherapy with none (i.e. with no adjuvant chemotherapy) are shown in Figure 12M [All ages]*. The results are subdivided into four categories according to the characteristics of the drug regimen. Later, the subtotals of these four categories (a,b,c,d) will be used to help compare the effects of those different categories of drug regimen. First, however, what is wanted is an overview of the effects of any adjuvant chemotherapy versus none in these 32 trials. As a first step, the results of each of the three trials (75E, 77B and 77E) that compare polychemotherapy **versus** single-agent **versus** nil, which in Figure 12 contribute two lines of results per trial (comparing poly versus nil, and comparing single versus nil), get combined into one line of results per trial (comparing the two-thirds who got **any** chemotherapy in that trial versus the one-third who did not). This reduces the total number of lines of results from 35 down to 32, i.e. one per trial.

If chemotherapy had no effect on survival, each O-E value would differ only randomly from zero, and so too would the grand total of these 32 values. But most of the O-E values are negative, and the grand total is -71.7. This overall total suggests that about 140 deaths were avoided or substantially delayed by chemotherapy (or that a larger number of deaths were moderately delayed: see later survival curve analyses). Its variance is 592.8 (calculated by summation of the 32 separate variances, one per trial: see Appendix Table 3M). Its standard deviation is therefore 24.3 (the square root of 592.8), so the grand total of the 32 individual O-E values is 2.9 standard deviations below zero (i.e. z = -2.9 = -71.7/24.3; 2P=0.003). This represents an estimated reduction in the odds of death among women of all ages who were assigned to chemotherapy of 11% ± 4.* This analysis is already stratified by age and additional

---

* More detailed tabulations of the mortality data are available in Appendix Table 3M, in which results for each trial (or trial part, where changes in the protocol necessitate greater subdivision) are listed chronologically. The data from each study have been recorded separately for women aged under 50 and for those aged 50 or older at the time of entry into the study. For each age subset and for all patients regardless of age, an O-E and variance is given for comparison of any adjuvant chemotherapy versus none in each trial (or trial part). When tamoxifen was given to patients in both arms of a trial, this is indicated in the fourth column of Appendix Table 3M.

* This 11% reduction estimates the average benefit in the entire population entered into these trials, of whom only about 3/8 were under 50 years of age. Because the effects of treatment appear greater for younger than for older women (see below), the estimated benefit in a population where half are under 50 would be about 14%, as in an earlier report of these data.[2]

RECURRENCE ANALYSES R[all ages]

Table 10R. **RECURRENCE data available for evaluation of adjuvant cytotoxic in early breast cancer** (including some trials with an identical tamoxifen regimen for the different treatment groups)

Randomized cytotoxic comparison	Recurrence (or prior death)	
	No. of trials (or parts)	Deaths/ patients
(a) CMF vs NIL (incl CMFPr vs NIL)	9	1158/2863
(b) CMF with extra cytotoxics vs NIL	5	563/1467
(c) Other polychemotherapy vs NIL	11	1055/2315
(d) Single-agent chemotherapy vs NIL	7	977/1877
1. Any cytotoxic therapy vs NIL (a+b+c+d, excluding overlap*)	29	3564/8172
2. Polychemotherapy vs single-agent chemotherapy	9	1044/2264
3. More prolonged vs less prolonged polychemotherapy	5	1027/2015
ALL CYTOTOXIC TRIALS (1+2+3, excluding overlap*)	**38**	**5307/11708** 45% recur or die

* The totals are less than the sum of the parts, because trials 74D1, 74D2, 75E2, 77B and 77E had 3-way randomizations, and so contribute to more than one analysis. (74D1: more prolonged poly vs less prolonged poly vs nil. 74D2: more prolonged poly vs less prolonged poly vs single. Others: poly vs single vs nil.)

Treatment of Early Breast Cancer

Fig. 12M. **MORTALITY in all available trials of cytotoxic vs no cytotoxic**
(any prolonged cytotoxic; includes any trials with identical tamoxifen for both cytotoxic and control groups)

Study No.	Study Name	Treatment	No. Events / No. Entered Treatment	Control	O−E	Variance	Odds Redn. (± S.D.)
(a) CMF trials (includes CMFPr, but not CMF + other cytotoxics)							
73B	INT Milan 7205	CMF	104/210	105/181	−10·9	38·7	
75E₂	Manchester I *	CMF	18/54	18/54	0·0	8·0	
76C	Glasgow	CMF	44/112	49/100	−5·9	19·6	
76E	EORTC 09771	CMF	40/228	54/223	−9·5	21·3	
77B₁₋₂	Danish BCG 77b *	CMF	62/201	68/196	−4·9	29·5	
77J₂	ECOG EST6177	CMFPr	37/87	32/95	3·0	15·2	
78E	UK/Asia Collab.	CMF	53/241	48/234	2·2	22·3	
78K₃	Ludwig III	CMF	39/171	53/164	−6·3	20·4	
79E	Guy's/Manch. II	CMF	22/136	32/132	−6·7	12·4	
80F	INT Milan 8004	CMF	0/47	9/47	−4·5	2·1	
82C	Danish BCG 82c	CMF	15/234	18/233	−1·5	7·9	
	Subtotal (a)		434/1721	486/1659	−45·1	197·4	20% ± 6
(b) Trials of CMF with extra cytotoxics							
76H₁	West Midlands	CMFVALvor	128/294	134/275	−10·9	56·1	
77G	Vienna	CMFV	22/81	27/82	−2·8	11·3	
79B₁	SWOG 7827 A	CMFVPr	−/204	−/211	blind	8·8	
79C	Case Western B	CMFVPr	9/62	8/71	0·7	4·0	
83B	GROCTA I Italy	CMF then E	1/101	1/86	0·0	0·5	
	Subtotal (b)		180/742	187/725	−11·2	80·7	13% ± 10
(c) Trials of regimes without some or all of C, M, F							
73C₁	Mayo Clinic	CFPr	13/21	5/13	2·7	3·1	
74A	Berlin-Buch	various	55/115	53/85	1·0	19·6	
74D₁	DFCI 74063	AC	2/4	3/4	−1·2	0·7	
74E	UK MCCG 003	CVF/CVM	74/148	74/141	−1·8	29·4	
75G	Northwick Park	MelV	19/65	24/75	−0·9	9·7	
75J	King's CRC I	MelM	97/222	97/212	−0·6	40·8	
76H₂	West Midlands	LeuMF	41/285	44/289	−1·2	20·1	
77E	Oxford *	MelMF	40/106	43/100	−3·3	18·3	
80A	MD Anderson8026	MV	6/113	11/122	−2·2	4·1	
80B₃₋₄	N Sweden BCG	AC	11/76	19/76	−3·7	6·9	
80C₂	SE Sweden BCG B	AC	4/21	8/22	−1·3	2·6	
	Subtotal (c)		362/1176	381/1139	−12·5	155·2	8% ± 8
(d) Single agents							
67A	Birmingham	C/F	69/157	30/83	4·8	18·6	
72B	NSABP B-05	Mel	97/198	101/182	−6·8	38·4	
74C	Edinburgh II	F	108/175	116/170	−8·8	43·2	
75E	Guy's/Manch. I *	Mel	89/187	80/184	5·9	35·1	
76D	Dublin	F	12/20	11/21	1·4	4·8	
77B₁₋₂	Danish BCG 77b *	C	58/194	68/196	−4·8	28·9	
77E	Oxford *	Mel	42/100	43/100	−1·9	18·3	
78A₁	S Swedish BCG	C	26/147	21/143	3·9	11·2	
	Subtotal (d)		501/1178	470/1079	−6·3	198·3	3% ± 7
	Total (a + b + c + d) *		**1477/ 4817**	**1395/ 4252**	**−71·7**	**592·8**	**11% ± 4**

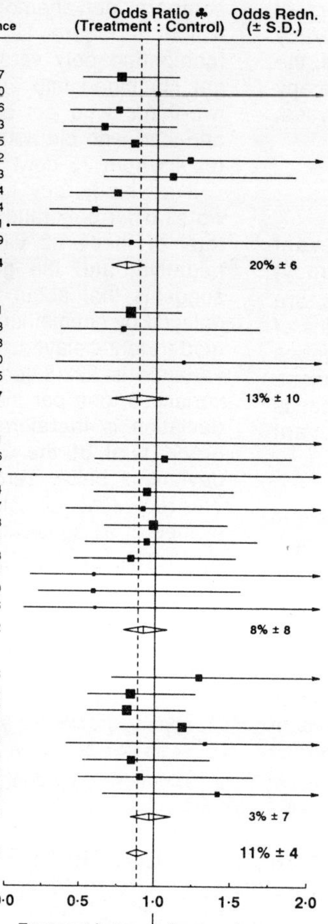

Test for heterogeneity: $X^2_{31} = 36·5$; $2P > 0·1$; NS
Treatment effect $2P = 0·003$

Treatment better | Treatment worse

MORTALITY ANALYSES M[all ages]

BELOW, LEFT:
M[<50]
MORTALITY in women UNDER 50 years old at entry

BELOW, RIGHT:
M[≥50]
MORTALITY in women 50 OR MORE years old at entry

Fig. 12M[<50]. **MORTALITY among women aged UNDER 50 at entry**

Study No.	Study Name	Treatment	Treatment	Control	O−E	Variance	Odds Redn. (± S.D.)
(a) CMF trials (includes CMFPr, but not CMF + other cytotoxics)							
73B	INT Milan 7205	CMF	41/95	46/75	−8·3	15·4	
75E₂	Manchester I *	CMF	8/21	9/24	−0·2	3·8	
76C	Glasgow	CMF	16/47	18/34	−4·8	6·9	
76E	EORTC 09771	CMF	15/96	23/90	−5·2	8·6	
77B₁₋₂	Danish BCG 77b *	CMF	46/149	44/136	−2·3	20·5	
77J₂	ECOG EST6177	CMFPr	0/1	1/8	−0·1	0·1	
78E	UK/Asia Collab.	CMF	21/131	20/97	−2·9	9·7	
78K₃	Ludwig III	CMF	1/2	4/4	−0·4	0·5	
79E	Guy's/Manch. II	CMF	9/63	19/60	−7·0	6·3	
80F	INT Milan 8004	CMF	0/26	5/29	−2·2	1·2	
82C	Danish BCG 82c	CMF	0/4	0/14			
	Subtotal (a)		157/635	189/554	−33·3	72·4	37% ± 9
(b) Trials of CMF with extra cytotoxics							
76H₁	West Midlands	CMFVALvor	48/119	55/119	−6·9	23·0	
77G	Vienna	CMFV	9/32	9/27	−0·8	4·1	
79B₁	SWOG 7827 A	CMFVPr	−/14	−/13	blind	0·2	
79C	Case Western B	CMFVPr	2/19	0/20	1·0	0·5	
83B	GROCTA I Italy	CMF then E	1/38	1/39	0·0	0·5	
	Subtotal (b)		61/222	69/218	−6·3	28·3	20% ± 17
(c) Trials of regimes without some or all of C, M, F							
73C₁	Mayo Clinic	CFPr	3/5	1/3	0·6	0·7	
74A	Berlin-Buch	various	22/59	16/28	−1·0	6·9	
74D₁	DFCI 74063	AC	1/2	1/1	−0·3	0·2	
74E	UK MCCG 003	CVF/CVM	32/61	25/47	2·1	11·4	
75G	Northwick Park	MelV	7/21	8/24	0·0	3·4	
75J	King's CRC I	MelM	35/85	29/62	−1·8	13·6	
76H₂	West Midlands	LeuMF	19/122	20/131	−0·3	9·2	
77E	Oxford *	MelMF	13/45	19/46	−3·2	6·3	
80A	MD Anderson8026	MV	1/51	6/63	−2·1	1·7	
80B₃₋₄	N Sweden BCG	AC	1/16	7/22	−2·1	1·8	
80C₂	SE Sweden BCG B	AC	0/0	1/1			
	Subtotal (c)		134/467	133/430	−7·2	56·2	12% ± 13
(d) Single agents							
67A	Birmingham	C/F	38/77	13/28	1·7	8·7	
72B	NSABP B-05	Mel	24/68	38/61	−7·2	11·8	
74C	Edinburgh II	F	39/59	42/56	−4·3	15·2	
75E	Guy's/Manch. I *	Mel	26/69	34/74	−2·6	13·2	
76D	Dublin	F	5/8	5/10	0·9	2·0	
77B₁₋₂	Danish BCG 77b *	C	39/128	42/130	−2·1	19·0	
77E	Oxford *	Mel	17/44	19/48	−0·3	7·8	
78A₁	S Swedish BCG	C	17/98	10/90	3·4	6·4	
	Subtotal (d)		205/551	205/497	−10·3	84·2	11% ± 10
	Total (a + b + c + d) *		**557/ 1875**	**524/ 1497**	**−54·6**	**219·7**	**22% ± 6**

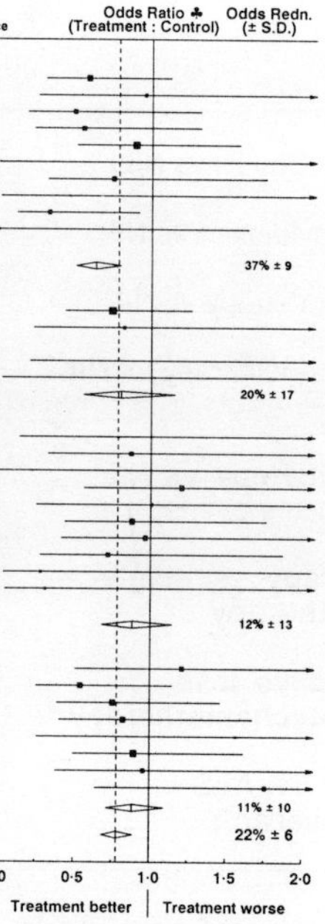

Test for heterogeneity: $X^2_{29} = 32·6$; $2P > 0·1$; NS
Treatment effect $2P = 0·0002$

Fig. 12M[≥50]. **MORTALITY among women aged 50 OR OVER at entry**

Study No.	Study Name	Treatment	Treatment	Control	O−E	Variance	Odds Redn. (± S.D.)
(a) CMF trials (includes CMFPr, but not CMF + other cytotoxics)							
73B	INT Milan 7205	CMF	63/115	59/106	−2·6	23·3	
75E₂	Manchester I *	CMF	10/33	9/30	0·2	4·2	
76C	Glasgow	CMF	28/65	31/66	−1·0	12·7	
76E	EORTC 09771	CMF	25/132	31/133	−4·3	12·7	
77B₁₋₂	Danish BCG 77b *	CMF	16/52	24/66	−2·6	9·1	
77J₂	ECOG EST6177	CMFPr	37/86	31/87	3·1	15·1	
78E	UK/Asia Collab.	CMF	32/110	28/135	4·8	13·2	
78K₃	Ludwig III	CMF	38/169	49/160	−5·9	19·9	
79E	Guy's/Manch. II	CMF	13/73	13/72	0·3	6·0	
80F	INT Milan 8004	CMF	0/21	4/18	−2·3	0·9	
82C	Danish BCG 82c	CMF	15/230	18/232	−1·5	7·9	
	Subtotal (a)		277/1086	297/1105	−11·9	125·0	9% ± 9
(b) Trials of CMF with extra cytotoxics							
76H₁	West Midlands	CMFVALvor	80/175	79/156	−4·0	33·1	
77G	Vienna	CMFV	13/49	18/55	−2·0	7·2	
79B₁	SWOG 7827 A	CMFVPr	−/190	−/198	blind	8·6	
79C	Case Western B	CMFVPr	7/43	8/51	−0·4	3·5	
83B	GROCTA I Italy	CMF then E	0/63	0/47			
	Subtotal (b)		119/520	118/507	−4·9	52·4	9% ± 13
(c) Trials of regimes without some or all of C, M, F							
73C₁	Mayo Clinic	CFPr	10/16	4/10	1·9	2·4	
74A	Berlin-Buch	various	33/56	37/57	2·0	12·8	
74D₁	DFCI 74063	AC	1/2	2/3	−0·8	0·5	
74E	UK MCCG 003	CVF/CVM	42/87	49/94	−4·0	18·0	
75G	Northwick Park	MelV	12/44	16/51	−0·9	6·2	
75J	King's CRC I	MelM	62/137	68/150	1·2	27·2	
76H₂	West Midlands	LeuMF	22/163	24/158	−0·9	10·9	
77E	Oxford *	MelMF	27/61	24/52	−0·9	11·0	
80A	MD Anderson8026	MV	5/62	5/59	0·0	2·4	
80B₃₋₄	N Sweden BCG	AC	10/60	12/54	−1·6	5·1	
80C₂	SE Sweden BCG B	AC	4/21	7/21	−1·3	2·6	
	Subtotal (c)		228/709	248/709	−5·3	99·0	5% ± 10
(d) Single agents							
67A	Birmingham	C/F	31/80	17/55	3·1	9·9	
72B	NSABP B-05	Mel	73/130	63/121	0·4	26·6	
74C	Edinburgh II	F	69/116	74/114	−4·5	28·0	
75E	Guy's/Manch. I *	Mel	63/118	46/110	8·5	21·9	
76D	Dublin	F	7/12	6/11	0·5	2·7	
77B₁₋₂	Danish BCG 77b *	C	19/66	24/66	−2·7	9·9	
77E	Oxford *	Mel	25/56	24/52	−1·5	10·5	
78A₁	S Swedish BCG	C	9/49	11/53	0·2	4·7	
	Subtotal (d)		296/627	265/582	3·9	114·1	−4% ± 10
	Total (a + b + c + d) *		**920/ 2942**	**871/ 2755**	**−17·1**	**373·1**	**4% ± 5**

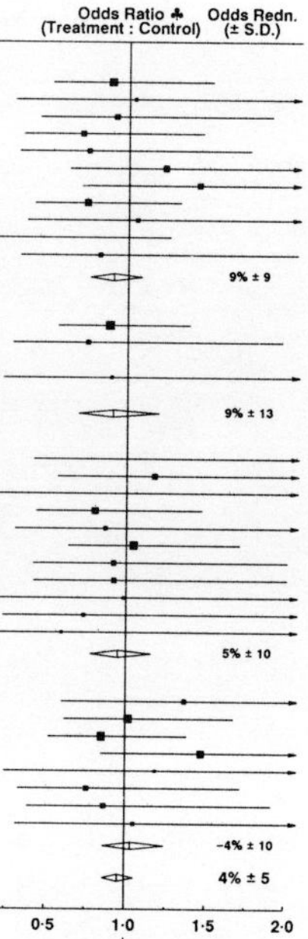

Test for heterogeneity: $X^2_{30} = 23·3$; $2P > 0·1$; NS
Treatment effect $2P > 0·1$; NS

♣ 95% confidence intervals for overview and 99% for individual trials.
* Control patients in 3-way trials (polychem. vs. single agent vs. no cytotoxic) contribute to 2 subtotals but only once to the grand total and X^2.

9 Results of cytotoxic trials

RECURRENCE ANALYSES R[all ages]

BELOW, LEFT:
R[<50]
RECURRENCE in women UNDER 50 years old at entry

BELOW, RIGHT:
R[≥50]
RECURRENCE in women 50 OR MORE years old at entry

Fig. 12R. **RECURRENCE in all available trials of cytotoxic vs no cytotoxic**
(any prolonged cytotoxic; includes any trials with identical tamoxifen for both cytotoxic and control groups)

Study No.	Study Name	Treatment	No. Events / No. Entered Treatment	Control	O−E	Variance	Odds Ratio ♣ (Treatment : Control)	Odds Redn. (± S.D.)
(a) CMF trials (includes CMFPr, but not CMF + other cytotoxics)								
73B	INT Milan 7205	CMF	129/210	131/181	−21.6	49.3		
75E2	Manchester I *	CMF	26/54	28/54	−3.0	11.2		
76C	Glasgow	CMF	59/112	60/100	−6.9	23.4		
76E	EORTC 09771	CMF	76/228	101/223	−20.9	37.3		
77B1+2	Danish BCG 77b *	CMF	76/201	101/196	−19.6	38.3		
78E	UK/Asia Collab.	CMF	88/241	97/234	−8.5	36.7		
79E	Guy's/Manch. II	CMF	35/136	52/132	−12.3	18.4		
80F	INT Milan 8004	CMF	3/47	19/47	−8.4	4.9		
82C	Danish BCG 82c	CMF	34/234	43/233	−5.1	17.2		
	Subtotal (a)		526/1463	632/1400	−106.4	236.6		36% ± 5
(b) Trials of CMF with extra cytotoxics								
76H1	West Midlands	CMFVALvor	182/294	200/275	−28.7	71.6		
77G	Vienna	CMFV	31/81	37/82	−4.2	15.1		
79B1	SWOG 7827 A	CMFVPr	−/204	−/211	blind	16.6		
79C	Case Western B	CMFVPr	12/62	21/71	−4.2	7.5		
83B	GROCTA I Italy	CMF then E	5/101	3/86	0.7	1.9		
	Subtotal (b)		262/742	301/725	−40.2	112.7		30% ± 8
(c) Trials of regimes without some or all of C, M, F								
73C1	Mayo Clinic	CFPr	14/21	10/13	−1.2	4.1		
74A	Berlin-Buch	various	63/115	65/85	−9.2	23.2		
74D1	DFCI 74063	AC	2/4	3/4	−1.2	0.9		
74E	UK MCCG 003	CVF/CVM	91/148	94/141	−8.4	36.0		
75G	Northwick Park	MelV	25/65	30/75	−0.7	12.5		
75J	King's CRC I	MelM	127/222	136/212	−10.9	51.8		
76H2	West Midlands	LeuMF	80/285	83/289	−2.1	37.8		
77E	Oxford *	MelMF	52/106	60/100	−7.8	23.4		
80A	MD Anderson8026	MV	21/113	30/122	−4.3	11.5		
80B3+4	N Sweden BCG	AC	25/76	30/76	−4.9	10.9		
80C2	SE Sweden BCG B	AC	5/21	9/22	−1.5	2.9		
	Subtotal (c)		505/1176	550/1139	−52.1	214.9		22% ± 6
(d) Single agents								
67A	Birmingham	C/F	76/157	44/83	−3.0	22.3		
74C	Edinburgh II	F	129/175	133/170	−10.2	48.9		
75E	Guy's/Manch. I *	Mel	109/187	103/184	−0.6	43.7		
76D	Dublin	F	14/20	12/21	1.7	5.2		
77B1+2	Danish BCG 77b *	C	71/194	101/196	−18.6	37.6		
77E	Oxford *	Mel	48/100	60/100	−7.8	22.8		
78A1	S Swedish BCG	C	38/147	39/143	1.2	17.7		
	Subtotal (d)		485/980	492/897	−37.2	198.2		17% ± 6
	Total (a + b + c + d) *		1778/4361	1786/3811	−216.4	707.9		26% ± 3

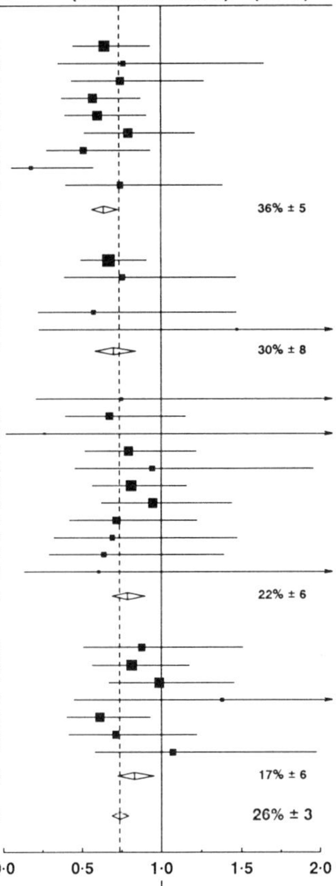

Test for heterogeneity: $X^2_{28} = 34.1; 2P > 0.1; NS$

Treatment effect $2P < 0.00001$

Treatment better | Treatment worse

Fig. 12R[<50]. RECURRENCE among women aged UNDER 50 at entry

Study No.	Study Name	Treatment	Treatment	Control	O−E	Variance	Odds Ratio ♣	Odds Redn. (± S.D.)
(a) CMF trials (includes CMFPr, but not CMF + other cytotoxics)								
73B	INT Milan 7205	CMF	52/95	55/75	−14.2	20.2		
75E2	Manchester I *	CMF	12/21	15/24	−2.1	5.2		
76C	Glasgow	CMF	20/47	19/34	−4.3	7.3		
76E	EORTC 09771	CMF	30/96	42/90	−9.2	15.4		
77B1+2	Danish BCG 77b *	CMF	57/149	67/130	−12.8	26.8		
78E	UK/Asia Collab.	CMF	38/131	40/99	−8.7	16.0		
79E	Guy's/Manch. II	CMF	13/63	29/60	−11.5	8.7		
80F	INT Milan 8004	CMF	2/26	11/29	−4.2	2.9		
82C	Danish BCG 82c	CMF	0/4	0/1				
	Subtotal (a)		224/632	278/542	−67.1	102.4		48% ± 7
(b) Trials of CMF with extra cytotoxics								
76H1	West Midlands	CMFVALvor	70/119	88/119	−15.4	29.9		
77G	Vienna	CMFV	13/32	14/27	−2.2	5.8		
79B1	SWOG 7827 A	CMFVPr	−/14	−/13	blind	0.9		
79C	Case Western B	CMFVPr	4/19	5/20	0.0	2.0		
83B	GROCTA I Italy	CMF then E	2/38	2/39	0.1	1.0		
	Subtotal (b)		91/222	111/218	−17.7	39.7		36% ± 13
(c) Trials of regimes without some or all of C, M, F								
73C1	Mayo Clinic	CFPr	3/5	2/3	−0.1	0.6		
74A	Berlin-Buch	various	26/59	20/28	−3.9	7.8		
74D1	DFCI 74063	AC	1/2	1/1	−0.2	0.5		
74E	UK MCCG 003	CVF/CVM	38/61	30/47	−0.9	13.1		
75G	Northwick Park	MelV	10/21	9/24	1.4	4.3		
75J	King's CRC I	MelM	42/85	41/62	−9.0	16.3		
76H2	West Midlands	LeuMF	35/122	36/131	0.4	16.5		
77E	Oxford *	MelMF	19/45	27/48	−4.2	9.8		
80A	MD Anderson8026	MV	8/51	14/63	−1.8	4.7		
80B3+4	N Sweden BCG	AC	3/16	11/22	−3.1	2.9		
80C2	SE Sweden BCG B	AC	0/0	1/1				
	Subtotal (c)		185/467	191/430	−21.6	76.5		25% ± 10
(d) Single agents								
67A	Birmingham	C/F	41/77	16/28	−0.6	9.4		
74C	Edinburgh II	F	46/59	46/56	4.6	16.3		
75E	Guy's/Manch. I *	Mel	30/69	43/74	−7.5	15.3		
76D	Dublin	F	5/8	6/10	0.3	2.2		
77B1+2	Danish BCG 77b *	C	45/128	67/130	−14.0	24.3		
77E	Oxford *	Mel	20/44	27/48	−3.3	9.9		
78A1	S Swedish BCG	C	23/98	24/90	−0.7	10.9		
	Subtotal (d)		210/483	229/436	−30.6	88.3		29% ± 9
	Total (a + b + c + d) *		710/1804	700/1424	−123.5	275.9		36% ± 5

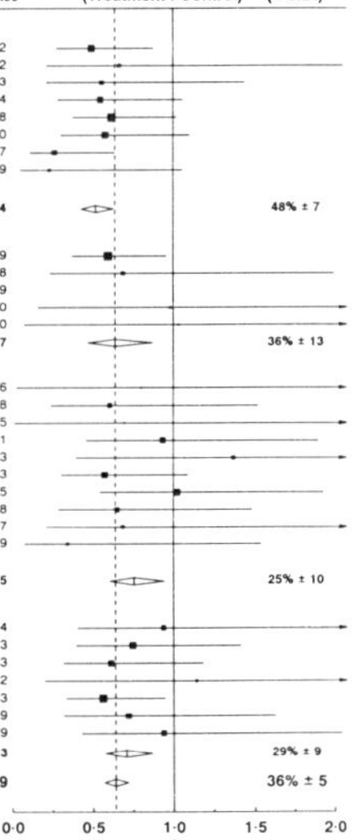

Test for heterogeneity: $X^2_{26} = 26.5; 2P > 0.1; NS$

Treatment effect $2P < 0.00001$

Fig. 12R[≥50]. RECURRENCE among women aged 50 OR OVER at entry

Study No.	Study Name	Treatment	Treatment	Control	O−E	Variance	Odds Ratio ♣	Odds Redn. (± S.D.)
(a) CMF trials (includes CMFPr, but not CMF + other cytotoxics)								
73B	INT Milan 7205	CMF	77/115	76/106	−7.4	29.1		
75E2	Manchester I *	CMF	14/33	13/30	−0.9	6.0		
76C	Glasgow	CMF	39/65	41/66	−2.6	16.0		
76E	EORTC 09771	CMF	46/132	59/133	−11.7	21.9		
77B1+2	Danish BCG 77b *	CMF	19/52	34/66	−6.7	11.6		
78E	UK/Asia Collab.	CMF	50/110	57/135	0.2	20.7		
79E	Guy's/Manch. II	CMF	22/73	23/72	−0.8	9.7		
80F	INT Milan 8004	CMF	1/21	8/18	−4.2	2.0		
82C	Danish BCG 82c	CMF	34/230	43/232	−5.1	17.2		
	Subtotal (a)		302/831	354/858	−39.2	134.2		25% ± 7
(b) Trials of CMF with extra cytotoxics								
76H1	West Midlands	CMFVALvor	112/175	112/156	−13.3	41.7		
77G	Vienna	CMFV	18/49	23/55	−2.1	9.2		
79B1	SWOG 7827 A	CMFVPr	−/190	−/198	blind	15.7		
79C	Case Western B	CMFVPr	8/43	16/51	−4.1	5.5		
83B	GROCTA I Italy	CMF then E	3/63	1/47	0.7	0.9		
	Subtotal (b)		171/520	190/507	−22.4	73.0		26% ± 10
(c) Trials of regimes without some or all of C, M, F								
73C1	Mayo Clinic	CFPr	11/16	8/10	−1.0	3.5		
74A	Berlin-Buch	various	37/56	45/57	−5.3	15.4		
74D1	DFCI 74063	AC	1/2	2/3	−1.0	0.5		
74E	UK MCCG 003	CVF/CVM	53/87	64/94	−7.4	22.9		
75G	Northwick Park	MelV	15/44	21/51	−2.1	8.2		
75J	King's CRC I	MelM	85/137	95/150	−1.9	35.5		
76H2	West Midlands	LeuMF	45/163	47/158	−2.4	21.3		
77E	Oxford *	MelMF	33/61	33/52	−3.6	13.6		
80A	MD Anderson8026	MV	13/62	17/59	−2.5	6.8		
80B3+4	N Sweden BCG	AC	22/60	19/54	−1.8	7.9		
80C2	SE Sweden BCG B	AC	5/21	8/21	−1.5	2.9		
	Subtotal (c)		320/709	359/709	−30.5	138.4		20% ± 8
(d) Single agents								
67A	Birmingham	C/F	35/80	28/55	−2.3	12.9		
74C	Edinburgh II	F	83/116	87/114	−5.4	32.5		
75E	Guy's/Manch. I *	Mel	79/118	60/110	6.9	28.4		
76D	Dublin	F	9/12	6/11	1.4	3.0		
77B1+2	Danish BCG 77b *	C	26/66	34/66	−4.6	13.3		
77E	Oxford *	Mel	28/56	33/52	−4.5	12.8		
78A1	S Swedish BCG	C	15/49	15/53	1.9	6.8		
	Subtotal (d)		275/497	263/461	−6.6	109.8		6% ± 9
	Total (a + b + c + d) *		1068/2557	1086/2387	−93.0	432.0		19% ± 4

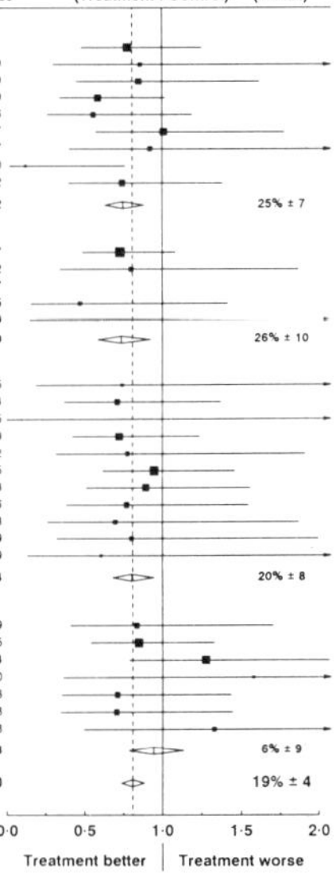

Test for heterogeneity: $X^2_{26} = 26.7; 2P > 0.1; NS$

Treatment effect $2P < 0.00001$

♣ 95% confidence intervals for overview and 99% for individual trials.

* Control patients in 3-way trials (polychem. vs. single agent vs. no cytotoxic) contribute to 2 subtotals but only once to the grand total and X^2.

Fig. 13M. **Effects of cytotoxic versus no cytotoxic on MORTALITY among women of DIFFERENT AGES at entry**
(any prolonged cytotoxic; includes any trials with identical tamoxifen for both groups)

MORTALITY ANALYSES

Category	No. Events / No. Entered Treatment	Control	O-E	Variance	Odds Ratio ♣ (Treatment : Control)	Odds Redn. (± S.D.)
< 40	180/519	153/388	−20.5	63.0		
40 – 49	377/1356	371/1109	−37.4	152.1		
50 – 59	531/1683	509/1529	−9.7	217.0		
60 – 69	350/1156	332/1132	−12.3	138.7		
70+	39/103	30/94	3.0	11.5		
Total	1477/4817	1395/4252	−76.8	582.3		12% ± 4

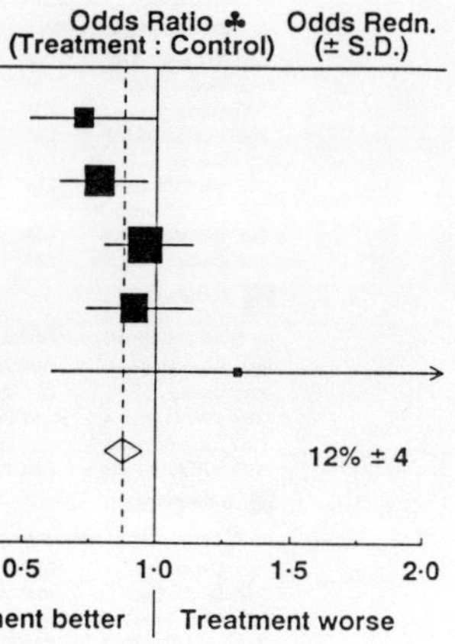

Treatment better | Treatment worse

Treatment effect 2P = 0.001

Test for trend = 2.4 s.d; 2P = 0.02

♣ 95% confidence intervals for overview and 99% for individual categories.

stratification of all the separate trial results for nodal status, which is the only important prognostic factor available, merely alters the overall result to 10% ± 4, which is still statistically significant (2P=0.01: see Figure 19M below). No material selection biases exist in this all-ages, all-trials analysis. Hence, this analysis provides strong evidence that for some types of women adjuvant chemotherapy can at least delay death.

9.2 Overall analysis of RECURRENCE in chemotherapy trials

Data on recurrence are available for 8172 (90%) of the randomized patients from 29 of the 32 trials, and are summarized in Figure 12R.* In general, the recurrence reductions are larger, and hence more easily demonstrable, than were the corresponding mortality reductions, and the typical odds reduction suggested by the overview of all 29 of these trials is 26% ± 3 (a difference of 8 standard deviations; 2P<0.00001). Again, the analysis is already stratified for age, and additional stratification for nodal status (see Figure 19R below) leaves this recurrence result exactly unchanged.

9.3 Heterogeneity between chemotherapy TRIALS, and heterogeneity between WOMEN

It can generally be assumed that when many chemotherapy trial results are being reviewed some real differences between the **sizes** of the effects of different treatments may well exist (although perhaps not differences in the **directions** of these effects). Even twofold differences in size may, however, be surprisingly difficult to demonstrate reliably. (For an example of this, see section 8.3 above: "Heterogeneity between tamoxifen trials".) The same problems apply if, instead of two categories of treatment, two categories of women are considered separately (e.g. those aged under 50 years and those 50 or over: see below). Even if a highly significant mortality reduction in one particular category can be accepted as real, the lack of any apparent mortality reduction in the other category might be either a true negative or a false negative.

In general, the main ways of protecting against misinterpretation of such chance fluctuations are: (i) to emphasize the overall mortality analyses, and not just the subgroup analyses (and, in particular, not to worry too much about whether the apparent effects of treatment in subgroup analyses are conventionally "significant"), and (ii) to examine subgroup analyses not only of survival but also of recurrence-free survival. In general, the least informative significance tests are likely to be those seeking evidence of non-specific heterogeneity. These will be disposed of first, before the more specific questions are addressed.

9.4 Heterogeneity of MORTALITY results in chemotherapy trials

The 11% ± 4 mortality reduction in the overview of all trials (Figure 12M) is strong evidence of some overall effect of chemotherapy on survival, but the separate mortality results are subject to substantial random error. A test of heterogeneity of the apparent effects in the different analyses yields a completely non-significant result (chi-square 36.5 with 31 degrees of freedom: Figure 12M [All ages]). This lack of any **statistically** significant heterogeneity between these trials cannot, however, guarantee that there is no **medically** significant heterogeneity between them, because such overall heterogeneity tests are statistically unreliable. A more appropriate way of seeking evidence of heterogeneity might be first to group the trials or women into a few meaningful subcategories and then to compare the combined results of these subcategories. In later analyses of heterogeneity, therefore, the data from the individual chemotherapy trials have been grouped according to specific patient variables (e.g. age) and according to various aspects of the treatment regimens (e.g. the number of drugs employed or the type of polychemotherapy regimen).

9.5 Heterogeneity of RECURRENCE results in chemotherapy trials

Because the proportional effect of treatment on recurrence-free survival is more than twice as great as that on mortality, it can be demonstrated reliably not only in the overview but also in some of the individual trials. Thus, for example, the effect of

---

* As for the mortality data, more detailed tabulations of the recurrence data are available in the Appendix (see Appendix Table 3R).

RECURRENCE ANALYSES

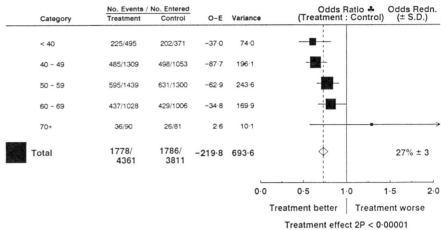

Fig. 13R. Effects of cytotoxic versus no cytotoxic on RECURRENCE among women of DIFFERENT AGES at entry
(any prolonged cytotoxic; includes any trials with identical tamoxifen for both groups)

♣ 95% confidence intervals for overview and 99% for individual categories.

chemotherapy on recurrence reaches a statistical significance level of 2P<0.01 in 7 of the individual trials shown in Figure 12R [All ages]. Despite this, however, there is still no statistically significant heterogeneity between the 29 trials from which recurrence information is available (chi-square = 34.1 on 28 degrees of freedom: Figure 12R [All ages]). But again, such non-specific heterogeneity tests are likely to be less informative than more specific analyses of heterogeneity might be (see below).

9.6 Influence of age on the effect of chemotherapy: MORTALITY data

The data from the chemotherapy trials are given separately for women aged under 50 in Figure 12M [<50], and for women aged 50 or over in Figure 12M [≥50].

Women aged under 50 (Figure 12M [<50]): In contrast with tamoxifen, which had its clearest effect among older women, the chemotherapy regimens tested in these trials had their clearest effects in younger women. The graphic display of the odds ratio determined for each study illustrates the variability in the apparent size of the effect of chemotherapy and even in the apparent direction of the effect (beneficial or detrimental). Patients in the chemotherapy group fared somewhat better than those in the control group in most of the trials (i.e. most of the black squares are to the left of the solid vertical line). When the trials are considered separately, however, only one (79E: Guy's/Manchester II) has a mortality result with a 99% confidence interval that does not cross the solid vertical line, and hence yields a P-value below 0.01. For all the other trials, the 99% confidence intervals for the mortality reductions among women aged under 50 overlap unity, and thus, considered separately, would not provide clearly significant evidence of benefit among younger women. But, when all the trials were combined their generally favorable results reinforce each other so that the overview estimate was substantially to the left of unity (Figure 12M [<50]), indicating a highly significant (22% ± 6; 2P=0.0002) reduction in mortality. The dashed vertical line indicates this overall result — i.e. a 22% mortality reduction among younger women — and the solid vertical line indicates no difference between treatment and control (i.e. an odds ratio of 1.0). Comparison of the width of the confidence intervals for the individual trials with the narrowness of the separation between the dashed and solid vertical lines shows that none of the individual trials was large enough on its own to detect reliably a 22% reduction in the odds of death.

Women aged 50 or over (Figure 12M [≥50]): Overall, among the women aged 50 or over there was little apparent effect of chemotherapy on mortality — indeed, the data suggested only a 4% reduction, but the standard deviation of this difference was large (± 5). This lack of statistical significance is in marked contrast with the definiteness of the results among younger women, and the estimated difference between the effects of chemotherapy in these two age groups is statistically significant (test for interaction = 2.4 sd, 2P=0.02).

But, as noted in the Introduction, the danger when trial results are subdivided with respect to age (or with respect to any other factor) to identify one group where treatment appears particularly promising is that the size of the effect seen in that group may be **misleadingly** large, while the size of the effect seen in those excluded from it may be **misleadingly** small. (For a particularly striking example of this, see Table 1.) Moreover it is particularly difficult to "prove a negative": lack of statistical proof of effect is not the same as proof of lack of effect. The lack of a clearly significant mortality reduction among older women does not, therefore, provide good evidence that the average effect on mortality of the treatments tested was negligible among women aged 50 or more (and still less does it mean that future cytotoxic regimens will have little effect among women in this age group). Thus, for example, the available mortality data are compatible with chemotherapy having an effect among women aged 50 or over that is about half the size of the effect among women under 50.

The results of a finer subdivision with respect to age are given in Figure 13M. The observed mortality reduction is statistically significant (2P<0.01) both among women aged under 40 and among women aged 40-49, but not among women aged 50-59, 60-69 or over 70. (Of course, the number of women over 70 who were randomized is far too small for meaningful interpretation of the effects of chemotherapy in this group). This trend towards a smaller proportional effect on mortality among older women than among younger women is

Table 11M. Indirect and direct comparisons between the effects of different cytotoxic chemotherapy regimens on the odds of **MORTALITY**

MORTALITY ANALYSES

Type of comparison	Reduction (± sd) in MORTALITY in groups with more chemotherapy			
	Age <50	Age ≥50	Both ages, crude	Both ages, standardized*
(a) Single-agent and poly- chemotherapy				
Single-agent vs no chemo**	11% ± 10	-4% ± 10	3% ± 7	2% ± 7)
Polychemo vs no chemo	26% ± 7	8% ± 6	15% ± 4	15% ± 4) NS
Polychemo vs single-agent	21% ± 9	17% ± 8	18% ± 6	18% ± 6
P-value for direct comparison				2P=0.002
(b) Interaction of chemotherapy with tamoxifen				
Chemo*** (-tam) vs nil	22% ± 6	3% ± 5	11% ± 4	11% ± 4)
Chemo*** (+tam) vs same tam	31% ± 30	12% ± 13	15% ± 12	20% ± 14) NS
(No direct comparison possible)				
(c) Duration of chemotherapy				
<1 yr vs nil (mean ≈ 6 mths†)***	18% ± 13	16% ± 10	17% ± 8	17% ± 8)
1 year chemo vs nil	19% ± 9	7% ± 8	12% ± 6	12% ± 6) NS
2 years chemo vs nil**	25% ± 10	-6% ± 9	7% ± 6	7% ± 6)
More vs less prolonged ch.** ***	-8% ± 12	-12% ± 12	-10% ± 8	-11% ± 9
P-value for direct comparison				NS
ANY CHEMOTHERAPY vs NO ADJUVANT CHEMO (and 95% confidence limits)	22% ± 6 (11% to 32%)	4% ± 5 (-6% to 14%)	11% ± 4 (4% to 18%)	11% ± 4 (4% to 18%)

* The age-standardized reduction in the log of the odds ratio (treatment : control) was defined as three-eighths of that among women aged under 50 plus five-eighths of that among older women. (This ensures that, for the overall result, the age-standardized and crude reductions are the same.)

** **Negative** values indicate **worse** outcome among those allocated more chemotherapy.

*** All such trials involved polychemotherapy.

† Range 4-10 months; but, most such trials were of exactly 6 months of chemotherapy.

statistically significant (2P<0.02: Figure 13M), but this significant trend does **not** prove that treatment has been clearly shown to have no effect on mortality among older women.

9.7 Influence of age on the effect of chemotherapy: RECURRENCE data

The recurrence data also suggest that chemotherapy may have a somewhat greater effect among younger women than among older women but, in contrast with the mortality data, the reduction in recurrence is highly significant both among women aged under 50 (36% ± 5; 2P<0.00001: Figure 12R [<50]) and among women aged 50 or over (19% ± 4; 2P<0.00001: Figure 12R [≥50]).

The results of a finer subdivision with respect to age are given in Figure 13R. Chemotherapy appears to have a greater proportional effect among younger women (trend of size of effect with respect to age = 3.2 sd, 2P=0.001), and the apparent risk reduction for women under 40 (39% ± 9) is twice as large as that for women aged 60-69 (19% ± 7). But, even on its own the recurrence reduction among women aged 60-69 is still statistically significant (2P<0.01). (Too few women aged over 70 were treated for any reliable estimate of the effect of treatment in this age group.) It is noteworthy that, comparing women slightly under and slightly over 50 years of age, there is no qualitative difference between the recurrence reductions achievable (36% ± 6 at ages 40-49, and 23% ± 6 at ages 50-59: each 2P<0.0001). These particular age-specific recurrence analyses, together with the others in Figure 13R, suggest that any difference between the effects of chemotherapy among premenopausal and among postmenopausal women is likely to involve merely a moderate difference between the **sizes** of the effects rather than a difference as to whether or not any substantial effect exists. In other words, such differences appear to be "quantitative" rather than "qualitative" (which is, of course, not what would have been expected if the therapeutic effect of chemotherapy were due entirely to indirect effects mediated through suppression of ovarian function).

9.8 Further comparisons: factors other than age

Table 11 shows the results of adjuvant chemotherapy trials

RECURRENCE ANALYSES

Table 11R. **Indirect and direct comparisons between the effects of different cytotoxic chemotherapy regimens on the odds of RECURRENCE (or prior death)**

Type of comparison	Reduction (± sd) in RECURRENCE in groups with more chemotherapy				
	Age <50	Age ≥50	Both ages, crude	Both ages, standardized*	
(a) Single-agent and poly- chemotherapy					
Single-agent vs no chemo**	29% ± 9	6% ± 9	17% ± 6	15% ± 7	) 2P=
Polychemo vs no chemo	39% ± 5	23% ± 5	30% ± 4	29% ± 4	) 0.03
Polychemo vs single-agent P-value for direct comparison	18% ± 9	20% ± 8	19% ± 6	19% ± 6 2P=0.002	
(b) Interaction of chemotherapy with tamoxifen					
Chemo*** (-tam) vs nil	36% ± 5	19% ± 5	27% ± 3	26% ± 3	) NS
Chemo*** (+tam) vs same tam	36% ± 19	19% ± 12	24% ± 10	26% ± 10	)
(No direct comparison possible)					
(c) Duration of chemotherapy					
<1 yr vs nil (mean ≈ 6 mths†)***	26% ± 10	24% ± 8	25% ± 6	25% ± 6	)
1 year chemo vs nil	38% ± 7	22% ± 7	29% ± 5	28% ± 5	) NS
2 years chemo vs nil	39% ± 8	13% ± 7	24% ± 5	24% ± 5	)
More vs less prolonged ch.** *** P-value for direct comparison	9% ± 10	-3% ± 10	3% ± 7	2% ± 7 NS	
ANY CHEMOTHERAPY vs NO ADJUVANT CHEMO (and 95% confidence limits)	36% ± 5 (28% to 43%)	19% ± 4 (11% to 27%)	26% ± 3 (21% to 32%)	26% ± 3 (20% to 31%)	

*, **, ***, † See notes to table 11M.

according to (a) the number of cytotoxic drugs involved, (b) the scheduled use of adjuvant tamoxifen (chemotherapy vs no adjuvant therapy, or chemotherapy plus tamoxifen vs the same tamoxifen alone), and (c) the scheduled duration of the chemotherapy. Further details are provided in the subtotals of Appendix Table 3. Because the effects of chemotherapy appear to be less in older women, it might be preferable for these comparisons to be standardized for age, and in Table 11 both the crude all-ages effects and the age-standardized effects are given. (In principle, the standardized comparisons are preferable, but in practice the standardization makes no qualitative difference to the comparisons.*)

In contrast with the tamoxifen trials (in which no direct randomized comparisons of different regimens were available), some "second-generation" chemotherapy trials have been undertaken that have involved direct comparisons — which generally provide a better basis for inference — of combination regimens with single-agent regimens, as well as direct comparisons of regimens of different duration. The overall results of these comparisons are among the findings summarized in Table 11, while the individual trials that contribute to them are described in Figure 14 (prolonged multiple-agent vs single-agent chemotherapy; see below) and in Figure 16 (more prolonged vs less prolonged chemotherapy; see below).

9.9 DIRECT and INDIRECT comparisons of the effects of multiple-agent and of single-agent chemotherapeutic regimens: MORTALITY data

Overviews of the results of the randomized trials comparing single-agent chemotherapy with no chemotherapy and of those comparing combination chemotherapy with no chemotherapy are shown in Table 11M. Indirect comparison of these overall results suggested that **poly**chemotherapy may be more effective than **single**-agent chemotherapy, and the direct comparisons reinforce this (Table 11M and Figure 14M).

The age-standardized reduction in the annual odds of death suggested by the trials comparing the various types of **poly**chemotherapy with no chemotherapy was 15% ± 4, a difference of 3.4 standard deviations from zero (2P<0.001), whereas that observed in trials comparing the various types of **single**-agent chemotherapy with no chemotherapy was only 2% ± 7. But, although this does not provide clear evidence that the single-agent chemotherapeutic regimens delayed death (Figure

---

* For comparability with the overall results, standardization is to a population with 3/8 under 50 years of age. Standardization to a population in which half were under 50, as in a previous report of these results,[2] would have given slightly greater weight to the promising results among younger women, but this would have increased the estimated treatment effect only slightly.

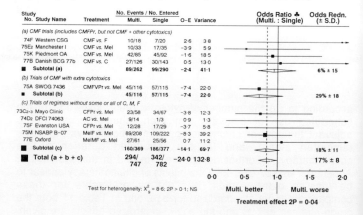

Fig. 14M. MORTALITY: MULTIPLE-AGENT cytotoxic versus SINGLE-AGENT cytotoxic (all available trials, including any with identical tamoxifen for both groups)

Fig. 14M[<50]. MORTALITY among women aged UNDER 50 at entry

Fig. 14M[≥50]. MORTALITY among women aged 50 OR OVER at entry

♣ 95% confidence intervals for overview and 99% for individual trials.

12M), neither does it suffice on its own to establish that such single-agent regimens really are less effective than the polychemotherapy regimens that were tested. For, these comparisons between the apparent effects on mortality of polychemotherapy and of single-agent chemotherapy are indirect, and are not even statistically significant.

An overview of the mortality results from the direct comparisons of polychemotherapy with single-agent chemotherapy (generally melphalan) was significantly in favour of polychemotherapy, both among women aged under 50 and among those aged 50 or over (Table 11M and Figure 14M). The age-standardized mortality difference in these direct comparisons is 18% ± 6, which provides highly significant (2P=0.002) evidence that at least some types of polychemotherapy can delay death more effectively than some of the single agents with which they were compared. The largest single contribution to this result came from one trial (75A: SWOG 7436) of CMFVPr versus melphalan (Figure 14M), but even if that particular study were deleted, a significant mortality difference would remain (sum of remaining O-E values = -31.4 with variance 203.3; 2P=0.03).

So, although polychemotherapy has been demonstrated to reduce mortality both when compared with no chemotherapy and when compared with single-agent chemotherapy, there is no

9 Results of cytotoxic trials

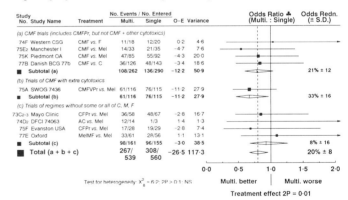

RECURRENCE ANALYSES R[all ages]

BELOW, LEFT: R[<50] RECURRENCE in women UNDER 50 years old at entry

BELOW, RIGHT: R[≥50] RECURRENCE in women 50 OR MORE years old at entry

Fig. 14R. RECURRENCE: MULTIPLE-AGENT cytotoxic versus SINGLE-AGENT cytotoxic (all available trials, including any with identical tamoxifen for both groups)

Fig. 14R[<50]. RECURRENCE among women aged UNDER 50 at entry

Fig. 14R[≥50]. RECURRENCE among women aged 50 OR OVER at entry

♣ 95% confidence intervals for overview and 99% for individual trials.

direct evidence in the mortality analyses of any effect of single-agent chemotherapy (Table 11M: age-standardized reduction in mortality 2% ± 7; but, with a wide 95% confidence interval: 12% adverse to 15% benefit). In comparing the effects of polychemotherapy in various subcategories, therefore, combination of the results of trials of polychemotherapy vs nil with the results of trials of polychemotherapy versus single-agent chemotherapy may be statistically more reliable than combination of just the results of those trials comparing polychemotherapy vs nil since it is based on some 10,000 women, more than 3000 of whom have died. This is, of course, a data-derived choice of which subgroups of the chemotherapy regimens to combine together, and in view of this it must be regarded as somewhat liable to just the type of "selective bias" that overviews should avoid. So, if larger numbers of women had been available in finer subcategories then this particular type of "combination" analysis could and would have been avoided.

Figure 15M[*] includes mortality data on all available randomized trials of polychemotherapy versus nil/ single-agent therapy, and therefore summarizes almost all of the direct

---

[*] This combines data from Figures 12M and 14M, and one trial (77A) that did not fit in either of these two previous analyses.

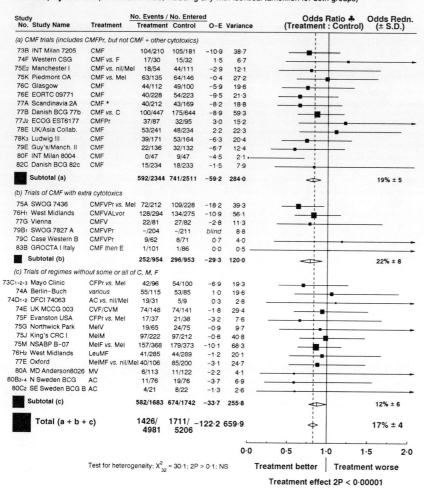

Fig. 15M. MORTALITY: MULTIPLE-AGENT cytotoxic versus SINGLE-AGENT, OR NO, cytotoxic (all available trials, including any with identical tamoxifen for both groups)

Fig. 15M[<50]. MORTALITY among women aged UNDER 50 at entry

Fig. 15M[≥50]. MORTALITY among women aged 50 OR OVER at entry

* Perioperative therapy ± 1 yr CMF
♣ 95% confidence intervals for overview and 99% for individual trials.

9 Results of cytotoxic trials

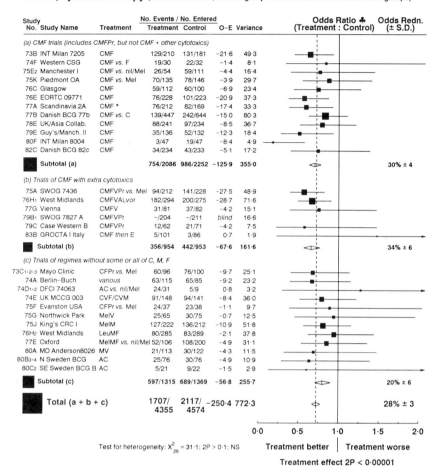

Fig. 15R. **RECURRENCE: MULTIPLE-AGENT cytotoxic versus SINGLE-AGENT, OR NO, cytotoxic therapy** (all available trials, including any with identical tamoxifen for both groups)

RECURRENCE ANALYSES R[all ages]

BELOW, LEFT: R[<50] RECURRENCE in women UNDER 50 years old at entry

BELOW, RIGHT: R[≥50] RECURRENCE in women 50 OR MORE years old at entry

Fig. 15R[<50]. RECURRENCE among women aged UNDER 50 at entry

Fig. 15R[≥50]. RECURRENCE among women aged 50 OR OVER at entry

* Perioperative therapy ± 1 yr CMF
♣ 95% confidence intervals for overview and 99% for individual trials.

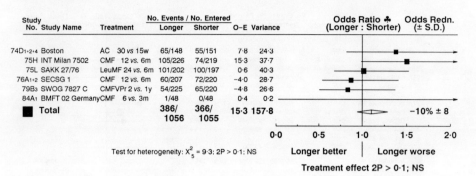
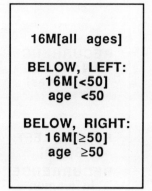

Fig. 16M. **MORTALITY: direct comparison of DIFFERENT DURATIONS of the same cytotoxic** (any cytotoxic regimen: includes any trials with identical tamoxifen for both groups)

Fig. 16M[<50]. MORTALITY among women aged UNDER 50 at entry

Fig. 16M[≥50]. MORTALITY among women aged 50 OR OVER at entry

♣ 95% confidence intervals for overview and 99% for individual trials.

evidence from these trials that polychemotherapy can delay death. The reduction in the annual odds of death is 17% ± 4, which is about the same as the 15% ± 4 age-standardized reduction seen in the comparison of polychemotherapy versus no chemotherapy (Table 11M). Estimation of this 17% ± 4 mortality reduction has, of course, been achieved at the expense of (a) being completely non-specific about which polychemotherapeutic regimens were tested, and (b) having one-quarter of the control patients treated with single-agent chemotherapy (so, for example, if single-agent chemotherapy is about half as effective as polychemotherapy, then the expected overall result in Figure 15M would correspond to only about seven-eighths of the full effect of polychemotherapy versus nil). Even though combination of the trials in Figure 15M does provide a reliable test of whether polychemotherapy reduces mortality, it is still difficult to use such evidence to help decide which types of patients can best expect to benefit from therapy. For, the play of chance can still produce differences between particular subgroups that are at least as great as any real differences between treatments that are likely to be discovered. In the case of age, for example, although the reduction in mortality observed with polychemotherapy versus nil/single-agent chemotherapy is highly significant among women aged under 50 (Figure 15M [<50]: 25% ± 5; 2P<0.00001) and less than half as big among those aged 50 or over (Figure 15M [≥50]: 11% ± 5; 2P<0.02), the apparent difference between the sizes of these two reductions is only just conventionally significant (2P=0.04).

9.10 DIRECT and INDIRECT comparisons of the effects of multiple-agent and single-agent chemotherapeutic regimens: RECURRENCE data

In contrast with the results for mortality, the age-standardized results in Table 11R indicate that significant reductions in recurrence were observed both in trials comparing polychemotherapy with no chemotherapy (29% ± 4; 2P<0.00001) and in trials comparing single-agent chemotherapy with no chemotherapy (15% ± 7; 2P=0.02). As was the case for mortality, however, both the indirect and direct comparisons (Table 11R and Figure 14R) indicate that, even though single-agent chemotherapy does reduce the annual odds of recurrence to some extent, polychemotherapy may be more effective. Figure 15R therefore includes recurrence data from all available randomized trials of polychemotherapy versus nil/single-agent therapy. This wide combination of trial results minimizes purely random errors, particularly within specific subgroups (e.g. by age; see Figures 15R [<50] and 15R [≥50]), but again it does so at the expense of being a data-derived comparison.

9.11 Interactions of chemotherapy with tamoxifen: MORTALITY data

The question as to whether chemotherapy has a substantially different effect in the presence or absence of tamoxifen cannot be answered reliably by the currently available data, for only some 2000 women have been studied in the presence of tamoxifen (i.e. in trials of cytotoxic chemotherapy plus tamoxifen versus tamoxifen alone). Table 11M compares the mortality data available from these trials with the results from trials that studied polychemotherapy in the absence of tamoxifen, and further details are given in Appendix Table 3M. No significant differences were apparent between the effects of polychemotherapy in the absence (11% ± 4; 2P<0.01) and in the presence (20% ± 14; NS) of tamoxifen, which suggests that the questions of whether tamoxifen should be used and of whether chemotherapy should be used may be largely independent of each other. Again, however, the random errors are far too large for such comparisons to be statistically stable (but at least the results in Table 11M do not provide any reason for eliminating

9 Results of cytotoxic trials

Fig. 16R. RECURRENCE: direct comparison of DIFFERENT DURATIONS of the same cytotoxic (any cytotoxic regimen: includes any trials with identical tamoxifen for both groups)

```
16R[all ages]

BELOW, LEFT:
16R[<50]
age <50

BELOW, RIGHT:
16R[≥50]
age ≥50
```

Study No.	Study Name	Treatment	No. Events / No. Entered Longer	Shorter	O-E	Variance	Odds Redn. (± S.D.)
74D1,2,4	Boston	AC 30 vs 15w	83/148	85/151	1.4	32.6	
75H	INT Milan 7502	CMF 12 vs. 6m	125/226	105/219	10.4	47.7	
75L	SAKK 27/76	LeuMF 24 vs. 6m	133/202	136/197	-2.7	52.1	
76A1,2	SECSG 1	CMF 12 vs. 6m	92/207	109/220	-7.9	42.2	
79B3	SWOG 7827 C	CMFVPr 2 vs. 1y	73/225	86/220	-7.4	34.5	
	Total		506/1008	521/1007	-6.3	209.1	3% ± 7

Test for heterogeneity: $\chi^2_4 = 5.3$; 2P > 0.1; NS
Treatment effect 2P > 0.1; NS

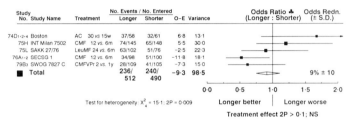

Fig. 16R[<50]. RECURRENCE among women aged UNDER 50 at entry

Test for heterogeneity: $\chi^2_4 = 15.1$; 2P = 0.009
Treatment effect 2P > 0.1; NS

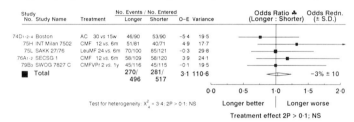

Fig. 16R[≥50]. RECURRENCE among women aged 50 OR OVER at entry

Test for heterogeneity: $\chi^2_4 = 3.4$; 2P > 0.1; NS
Treatment effect 2P > 0.1; NS

♣ 95% confidence intervals for overview and 99% for individual trials.

---

from the overall assessment of the effects of chemotherapy those trials in which tamoxifen was present in both arms).

9.12 Interactions of chemotherapy with tamoxifen: RECURRENCE data

The age-standardized effects of chemotherapy in the absence of tamoxifen on recurrence rates (26% ± 3 reduction; 2P<0.00001) were about the same size as those in trials where polychemotherapy plus tamoxifen was compared with tamoxifen alone (26% ± 10 reduction; 2P<0.01). As was the case for the mortality results, therefore, this similarity of the two recurrence reductions is compatible with any real effects of tamoxifen and of polychemotherapy being approximately independent of each other.

9.13 DIRECT comparisons of the effects of different duration of therapy: MORTALITY data

Just over 2000 women have been randomized into the 6 trials that compared more prolonged chemotherapeutic regimens (e.g. 6 to 24 months) with less prolonged regimens (e.g. 3 to 6 months). The mortality results from these trials, which are based on a total of over 700 deaths, are displayed in Figure 16M: review of these direct comparisons did not indicate that, for these particular chemotherapeutic regimens, more prolonged treatment conferred any survival advantage over less prolonged treatment. Indeed, if anything, it slightly suggested the opposite, since combination of the results in older and younger women indicated a non-significant difference of -11% ± 9 which, being negative, is in favour of **less** prolonged therapy (Table 11M, Figure 16M). In the two trials of 12 months versus 6 months of CMF, the results were significantly in favour of shorter treatment in one and non-significantly against it in the other, so that when they are considered together there is no significant excess mortality among those allocated 12 courses of treatment.

9.14 DIRECT comparisons of the effects of different duration of therapy: RECURRENCE data

Data on recurrence are available from 5 of the 6 trials comparing two durations of therapy (Table 11R: Figure 16R). No significant difference (2% ± 7; NS) was apparent between the effects of therapy on recurrence rates observed in direct randomized comparisons of shorter and longer periods of treatment. Thus, for these particular polychemotherapeutic regimens, the available recurrence data do not indicate that more prolonged treatment confers any material advantage (Table 11R, Figure 16R). The 95% confidence interval is wide, however, indicating that more randomized evidence is still needed to address this important question really reliably.

9.15 Effects of chemotherapy in different nodal categories: MORTALITY data

As with the analyses of the tamoxifen trials, patients were placed in one of five nodal categories based on information available at randomization: N-, N0, N1-3, N4+ and "N+/N?" (see above for definitions).

The 5-year survival curves for all patients (irrespective of treatment allocation) are plotted in Figure 17M, illustrating the extent to which these "nodal status" categories do divide patients into high, medium and low risk groups. Mortality was similar in the N0 and N- categories, and these were combined in all subsequent analyses into a single "N0/N-" category, as was done in the tamoxifen trials.

The four parts of Figure 18M display the individual results from all trials of chemotherapy versus untreated control for each of the four nodal status categories (N0/N-, N1-3, N4+, and N+/N?), and the summary estimates for these four categories are graphically displayed together in Figure 19M. The bottom line of Figure 19M provides an analysis of mortality in the chemotherapy trials that is retrospectively stratified for the

Fig. 17M. MORTALITY analysis: survival by year, subdivided by NODAL STATUS
(all ages; all women in any cytotoxic trial, whether allocated cytotoxic or control)

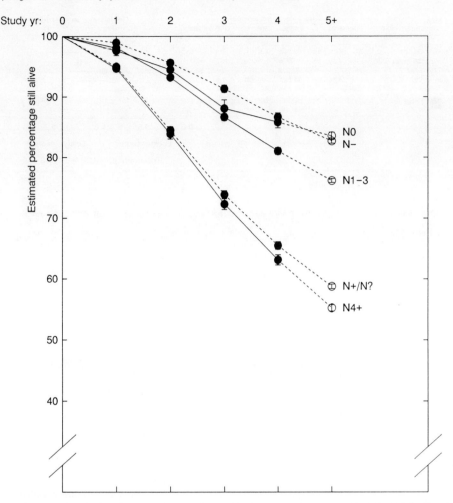

MORTALITY
ANALYSES
M[all ages]

available data on both age and nodal status, and overall it indicates a reduction of 10% ± 4 in the annual odds of death.

The effect of chemotherapy just among node-negative women may have to be estimated indirectly from the overall results, since the number of deaths among women with N0/N- disease is small. (This is partly because such women have a good prognosis and partly because many chemotherapy trials excluded such women.) Hence, the statistical uncertainty as to the effects of chemotherapy on mortality among such women is larger than for patients in the other three nodal categories. These fluctuations become still worse when individual trials are considered, so although some of the results among node-negative women appear remarkably promising (e.g. trial 80F INT Milan: 0 vs 9 deaths), others appear rather unpromising (e.g. trial 74E MCCG: 10 vs 6 deaths).

When all available studies are considered together, the proportional reductions in mortality in the four nodal status categories are reasonably similar to each other (Figure 19M), and a test for a trend among them is not statistically significant. Taken separately, however, the four results are each subject to substantial random error. Paradoxically, therefore, the overall result may provide a more reliable guide to the effects of treatment in each separate nodal status category than is provided by the apparent effects of treatment in that one category alone.

The absolute differences in estimated 5-year survival that are produced by chemotherapy in the four nodal status categories of Figure 19M are given in Table 12M; as expected, however, they are subject to such large random errors that they cannot separately be trusted.

9.16 Effects of chemotherapy in different nodal categories: RECURRENCE data

As expected, the nodal categories do divide the patients quite effectively into those at high, medium and low risk of recurrence (Figure 17R). The individual trial results and the overall estimates for each of the nodal status categories are displayed in Figure 18R, and the summary estimates for these four categories are displayed graphically in Figure 19R. Again, the bottom line of Figure 19R provides an analysis of recurrence-free survival that is retrospectively stratified for age and for the available information on nodal status, and overall it indicates a reduction of 26% ± 3 in the annual odds of first recurrence (or prior death). Although the proportional reduction in recurrence among the N0/N- patients appears somewhat smaller than the reductions observed in the other three nodal categories, this heterogeneity of effect is not significant (either by a test for trend, or by a test comparing this one category with the remaining three).

Thus, these results do not provide good evidence of any real differences between the four **proportional** reductions in the annual odds of recurrence. But, even if the proportional reductions produced by chemotherapy are similar then, at least during the first few years, the **absolute** reductions in risk will tend to be greater for poor-prognosis than for good-prognosis

9 Results of cytotoxic trials

RECURRENCE
ANALYSES
R[all ages]

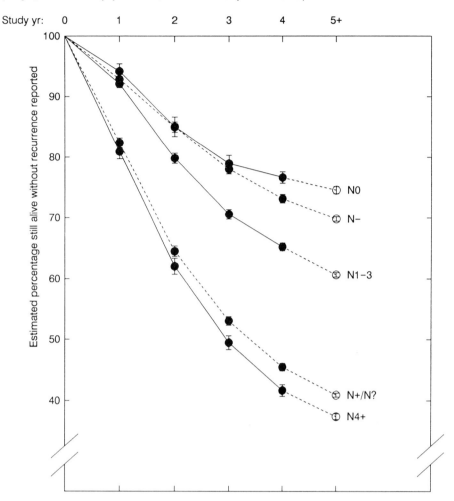

Fig. 17R. **RECURRENCE-FREE SURVIVAL by year, subdivided by NODAL STATUS**
(all ages; all women in any cytotoxic trial, whether allocated cytotoxic or control)

patients (see Table 12R). The absolute differences in estimated 5-year recurrence-free survival in the four nodal status categories are given in Table 12R. As for mortality, however, the random errors involved are too large for reliable direct estimation of any differences between one nodal status category and another in either the relative or the absolute effects of such treatments.

9.17 Description of the SIZE of the effect of chemotherapy: MORTALITY data

It is difficult to choose how best to describe the apparent size of the effects of chemotherapy on mortality, because it depends substantially on which age group is selected for emphasis (e.g. 1: all ages, 2: age under 50, 3: age 50 or over; see Figure 12M), and also on which types of chemotherapy are grouped together (e.g. 1: any prolonged chemotherapy versus no chemotherapy, 2: any prolonged polychemotherapy versus no chemotherapy, 3: any prolonged CMF-based chemotherapy versus no chemotherapy, or 4: any prolonged polychemotherapy versus either no, or single-agent, chemotherapy). This already amounts to a dozen different descriptive analyses and, as there is no really satisfactory way of deciding which to emphasize, survival curves for all twelve are given separately (Figures 20M-23M and Tables 13M-16M).

The analysis of all trials of prolonged chemotherapy versus no chemotherapy among all women (Figure 12M) provided the most natural test of whether chemotherapy was having any effect on survival. The small but significant result that it produced (an 11% ± 4 reduction in the odds of death, 2P=0.003) is illustrated in Figure 20M, and the underlying calculations for year 1, year 2, year 3, year 4 and year 5+ are summarized in Table 13M. Table 13M gives the event rates among treated and control patients, together with the corresponding O-E values for each follow-up year,* for patients in trials comparing any chemotherapy with no chemotherapy. But, although the comparison of any chemotherapy with no chemotherapy does provide a natural test of whether chemotherapy does anything, it does not provide a natural description of the effects of any particular type of chemotherapy. For, it includes trials of various types of polychemotherapy and of various types of single-agent chemotherapy, and there is independent evidence that polychemotherapy is better than single-agent therapy (Table 11M). Hence, subsequent figures and tables make more specific comparisons: polychemotherapy versus no chemotherapy (Figure 21M and Table 14M), CMF-based

---

* The O-E values are calculated by proper combination of one O-E value from each trial. In contrast, the death rates involve mixing together patients from many different trials. In principle, this is less proper than the use of O-E values. In practice, however, the percentage reductions suggested by the ratios of the death rates in each year are almost the same as the percentage reductions calculated from the O-E values in that year, so the death rates are used for descriptive purposes.

***Fig. 18M.* Effects of cytotoxic on MORTALITY in various NODAL categories**
(all ages; content and format as in Fig. 12M; nodal status categories as in text)

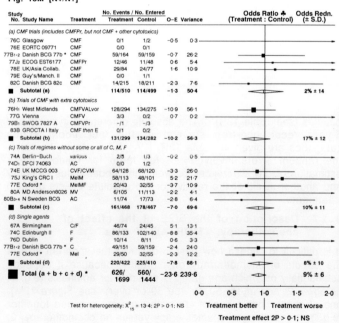

Fig. 18R. **Effects of cytotoxic on RECURRENCE in various NODAL categories**
(all ages; content and format as in Fig. 12R; nodal status categories as in text)

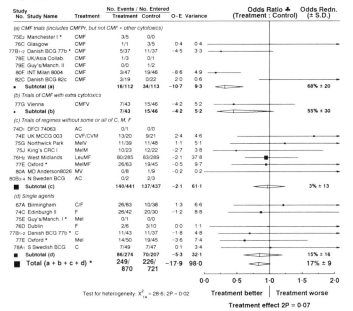

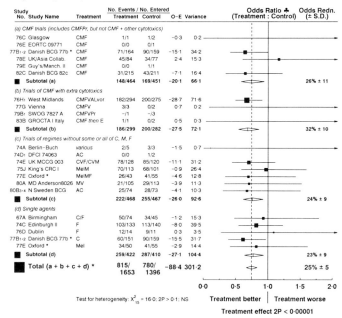

♣ 95% confidence intervals for overview and 99% for individual trials.
* Control patients in 3-way trials (polychem. vs. single agent vs. no cytotoxic) contribute to 2 subtotals but only once to the grand total and X^2.

Fig. 19M. **Effects of cytotoxic on MORTALITY in various NODAL categories**
(from Fig. 18M: all ages; any prolonged cytotoxic; includes any trial with identical tamoxifen for both groups)

Category	No. Events / No. Entered Treatment	No. Events / No. Entered Control	O-E	Variance	Odds Ratio ♣ (Treatment : Control)	Odds Redn. (± S.D.)
N0/N-	169/870	140/722	-4.3	64.3		
N1-3	283/1338	316/1222	-30.2	127.5		
N4+	399/910	379/864	-2.7	143.5		
N+/N?	626/1699	560/1444	-23.6	239.6		
Total	1477/4817	1395/4252	-60.7	574.9		10% ± 4

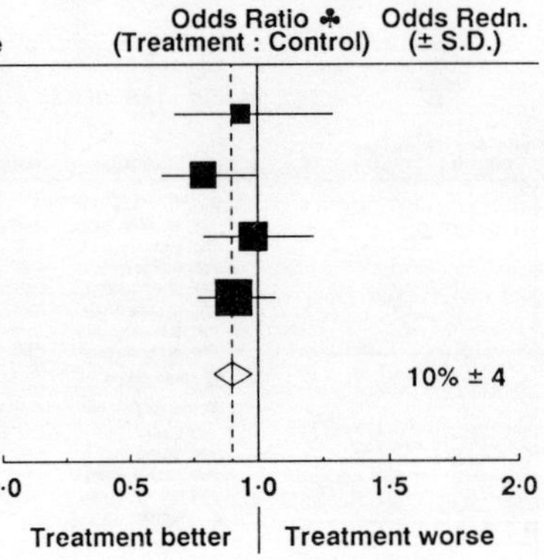

Treatment better | Treatment worse

Treatment effect 2P = 0.01

Test for trend = 0.8 s.d; 2P > 0.1; NS

♣ 95% confidence intervals for overview and 99% for individual categories.
N.B: The test for trend is between 3 specific categories.

MORTALITY ANALYSES M[all ages]

Table 12M. **Cytotoxics and MORTALITY in categories of nodal status**
(including trials with identical tamoxifen regimens for both treatment and control groups)

Nodal status at entry	"Life-table" estimates of 5-year mortality risks			Statistical calculations for CYTOTOXIC groups only		
	Allocated CONTROL	Allocated CYTOTOXIC	Difference (± sd)	O-E	Var. of O-E	Mortality redn ± sd
N0/N-	17.7%	16.8%	0.9% ± 1.9	-4.3	64.3	6% ± 12
N1-3	26.3%	21.3%	5.0% ± 1.9	-30.2	127.5	21% ± 8
N4+	45.0%	44.6%	0.4% ± 2.6	-2.7	143.5	2% ± 8
(N+/N?)	(42.9%)	(39.6%)	(3.3% ± 2.0)	(-23.6)	(239.6)	(9% ± 6)
All women	33.8%	31.1%	2.7% ± 1.1	-60.7	574.9	10% ± 4

Note: Any age, any prolonged cytotoxic regimen. The statistical calculations for "All women" are retrospectively stratified for status at entry, but the life-table estimates are not.

MORTALITY ANALYSES M[all ages]

regimens versus no chemotherapy (Figure 22M and Table 15M), and polychemotherapy versus no/single agent (Figure 23M and Table 16M). Among women aged less than 50 there was in Figures 20-22 (and in the data-derived Figure 23) a significant improvement in survival with no suggestion of convergence of the curves during the first 5 years of follow-up. In contrast, in Figures 20-22 no significant reduction in absolute mortality is seen from the use of chemotherapy in women aged 50 or more. But, as noted in relation to Figures 12M and 13M, the numbers of older women are too small for **negative** conclusions to be reliable.

It would be of interest to be able to compare the effects on death rates in different follow-up years. Overall, there is no statistically significant heterogeneity between the percentage reductions in the five separate annual mortality rates, but such comparisons lack power. Even if the follow-up years are grouped into years 1-2 and years 3+ (Tables 13-16) then no consistent pattern is seen. At the very least, however, the year-by-year analyses show that there is no evidence that the early mortality gains begin to be lost at any time within the first five years.

9.18 Description of the SIZE of the effect of chemotherapy: RECURRENCE data

The effects of treatment on recurrence are larger, and hence the problems of statistical stability are somewhat more tractable. Results corresponding to those given for mortality are presented in Tables 13R-16R and are plotted in Figures 20R-23R. As already noted, for recurrence the reductions produced by chemotherapy among women aged under 50 and among those aged 50 or over are both highly significant (absolute difference produced in Figures 20R [<50] and 20R [≥50] by any chemotherapy: 12.8% ± 2.0 per 100 women and 6.3% ± 1.7 per 100 women respectively; both 2P<0.00001).

For recurrence, the greatest differences by far arose in the first year, which is one reason why so many trials have been able to demonstrate quickly an effect of chemotherapy on recurrence rates. But, even after the first year the additional annual event rates continue to differ in favour of treatment, so there is no evidence that the early gains begin to be lost at any time during the first five years — indeed, even if the first two years are omitted entirely the further divergence in years 3+ is statistically significant (Table 13R).

9 Results of cytotoxic trials

RECURRENCE ANALYSES R[all ages]

Fig. 19R. **Effects of cytotoxic on RECURRENCE in various NODAL categories**
(from Fig. 18R: all ages; any prolonged cytotoxic; includes any trial with identical tamoxifen for both groups)

Category	No. Events / No. Entered Treatment	Control	O−E	Variance	Odds Ratio ♣ (Treatment : Control)	Odds Redn. (± S.D.)
N0/N−	249/870	226/721	−17.9	98.0		
N1−3	342/1125	405/1038	−60.8	154.2		
N4+	372/713	375/656	−43.2	131.2		
N+/N?	815/1653	780/1396	−88.4	301.2		
Total	1778/4361	1786/3811	−210.2	684.7		26% ± 3

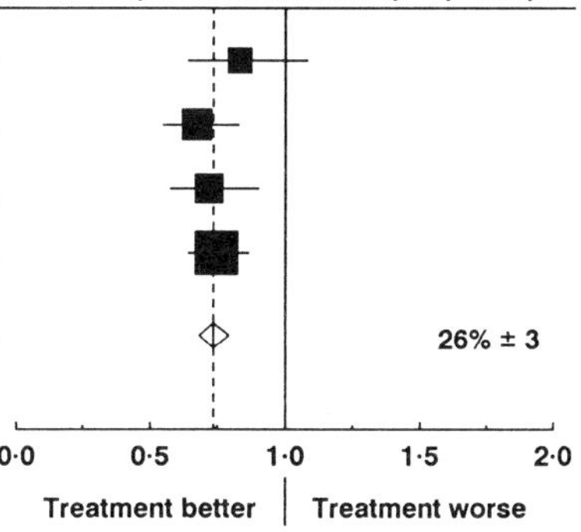

0.0   0.5   1.0   1.5   2.0
Treatment better | Treatment worse

Treatment effect 2P < 0·00001

Test for trend = −1·0 s.d; 2P > 0·1; NS

♣ 95% confidence intervals for overview and 99% for individual categories.
N.B: The test for trend is between 3 specific categories.

RECURRENCE ANALYSES R[all ages]

Table 12R. **Cytotoxics and RECURRENCE in categories of nodal status**
(including trials with identical tamoxifen regimens for both treatment and control groups)

Nodal status at entry	"Life-table" estimates of 5-year recurrence risks			Statistical calculations for CYTOTOXIC groups only		
	Allocated CONTROL	Allocated CYTOTOXIC	Difference (± sd)	O-E	Var. of O-E	Recurrence redn ± sd
N0/N−	31.4%	27.0%	4.4% ± 2.4	−17.9	98.0	17% ± 9
N1-3	45.4%	32.8%	12.6% ± 2.5	−60.8	154.2	33% ± 7
N4+	67.4%	57.4%	10.1% ± 3.3	−43.2	131.2	28% ± 7
(N+/N?)	(63.8%)	(54.0%)	(9.7% ± 2.2)	(−88.4)	(301.2)	(25% ± 5)
All women	52.6%	43.5%	9.1% ± 1.3	−210.2	684.7	26% ± 3

Note: The statistical calculations for "All women" are retrospectively stratified for status at entry, but the life-table estimates are not.

Fig. 20M. MORTALITY analysis: survival by year in all available trials of ANY TYPE of prolonged cytotoxic (incl. any trials with identical tamoxifen for both groups)

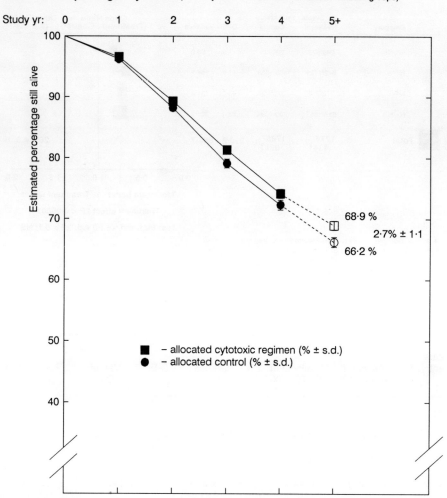

MORTALITY ANALYSES M[all ages]

BELOW, LEFT: M[<50] MORTALITY in women UNDER 50 years old at entry

BELOW, RIGHT: M[≥50] MORTALITY in women 50 OR MORE years old at entry

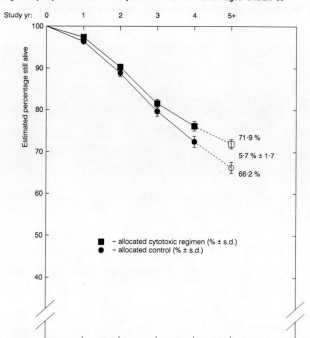

Fig. 20M[<50]. MORTALITY analysis: survival for women aged UNDER 50

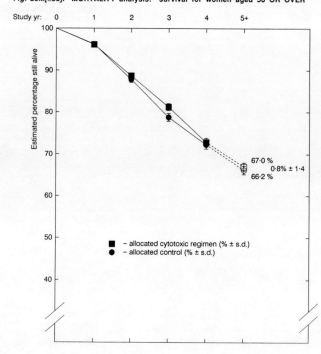

Fig. 20M[≥50]. MORTALITY analysis: survival for women aged 50 OR OVER

9 Results of cytotoxic trials

RECURRENCE ANALYSES R[all ages]

BELOW, LEFT: R[<50] RECURRENCE in women UNDER 50 years old at entry

BELOW, RIGHT: R[≥50] RECURRENCE in women 50 OR MORE years old at entry

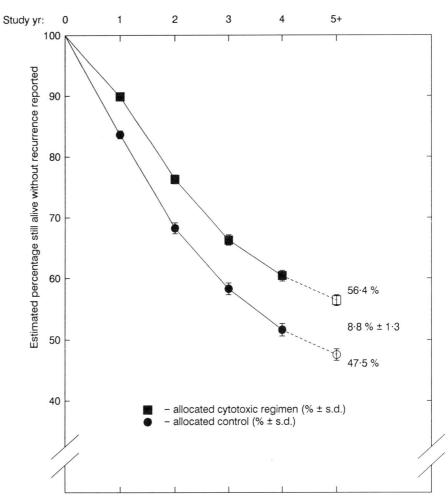

Fig. 20R. RECURRENCE-FREE SURVIVAL by year in all available trials of ANY TYPE of prolonged cytotoxic (incl. any trials with identical tamoxifen for both groups)

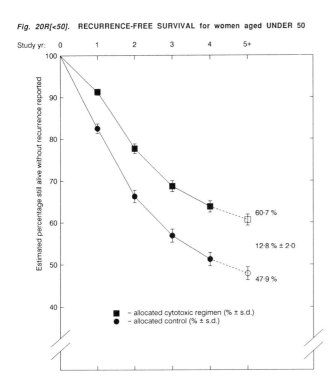

Fig. 20R[<50]. RECURRENCE-FREE SURVIVAL for women aged UNDER 50

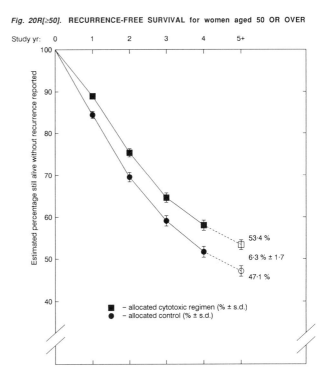

Fig. 20R[≥50]. RECURRENCE-FREE SURVIVAL for women aged 50 OR OVER

Fig. 21M. **MORTALITY analysis: survival by year in all available trials of MULTI-AGENT prolonged cytotoxic** (incl. any trials with identical tamoxifen for both groups)

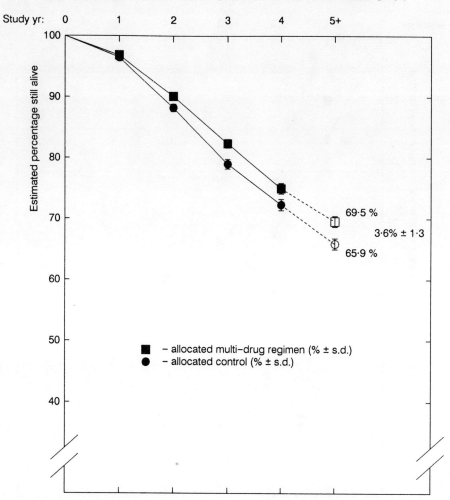

MORTALITY ANALYSES M[all ages]

BELOW, LEFT: M[<50]
MORTALITY in women UNDER 50 years old at entry

BELOW, RIGHT: M[≥50]
MORTALITY in women 50 OR MORE years old at entry

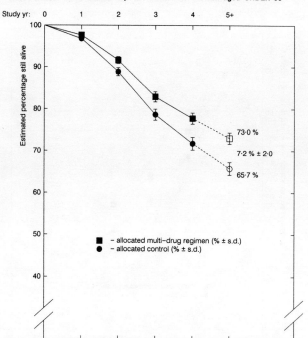

Fig. 21M[<50]. MORTALITY analysis: survival for women aged UNDER 50

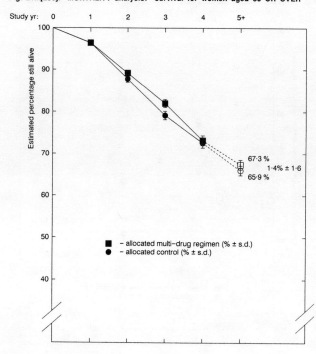

Fig. 21M[≥50]. MORTALITY analysis: survival for women aged 50 OR OVER

9 Results of cytotoxic trials

RECURRENCE ANALYSES R[all ages]

BELOW, LEFT:
R[<50]
RECURRENCE in women UNDER 50 years old at entry

BELOW, RIGHT:
R[≥50]
RECURRENCE in women 50 OR MORE years old at entry

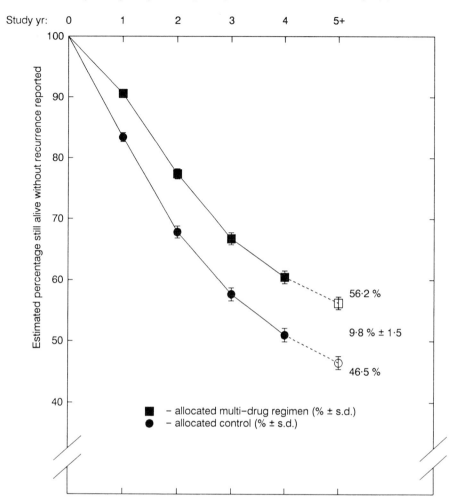

Fig. 21R. RECURRENCE-FREE SURVIVAL by year in all available trials of MULTI-AGENT prolonged cytotoxic (incl. any with identical tamoxifen for both groups)

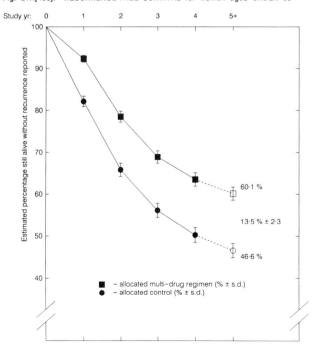

Fig. 21R[<50]. RECURRENCE-FREE SURVIVAL for women aged UNDER 50

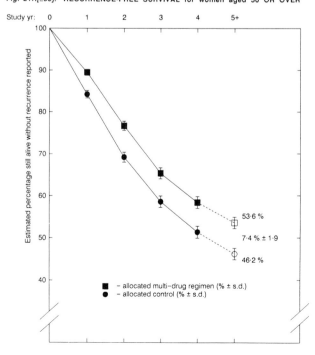

Fig. 21R[≥50]. RECURRENCE-FREE SURVIVAL for women aged 50 OR OVER

Fig. 22M. MORTALITY analysis: survival by year in all available trials of CMF-BASED prolonged cytotoxic (incl. any trials with identical tamoxifen for both groups)

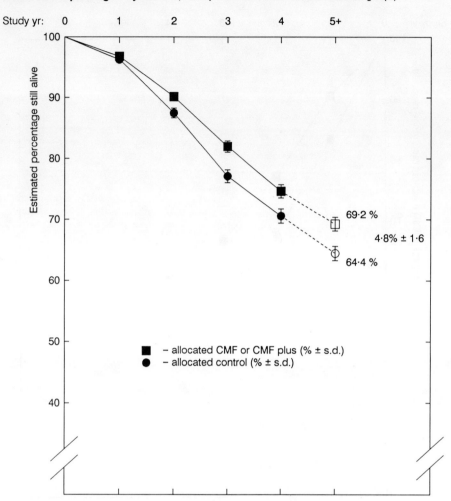

Fig. 22M[<50]. MORTALITY analysis: survival for women aged UNDER 50

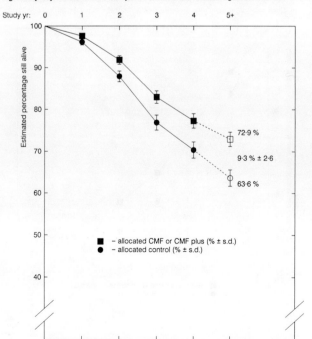

Fig. 22M[≥50]. MORTALITY analysis: survival for women aged 50 OR OVER

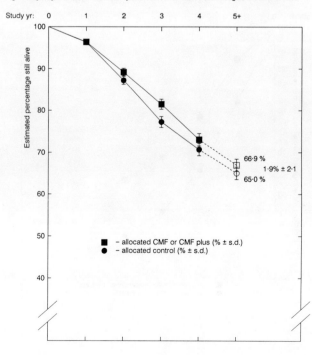

9 Results of cytotoxic trials

RECURRENCE ANALYSES R[all ages]

BELOW, LEFT: R[<50] RECURRENCE in women UNDER 50 years old at entry

BELOW, RIGHT: R[≥50] RECURRENCE in women 50 OR MORE years old at entry

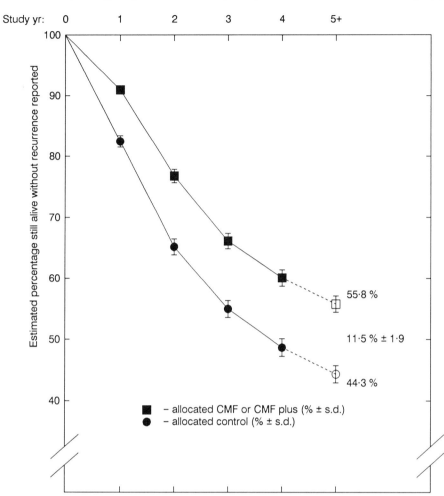

Fig. 22R. RECURRENCE-FREE SURVIVAL by year in all available trials of CMF-BASED prolonged cytotoxic vs none (incl. any trials with identical tamoxifen for both groups)

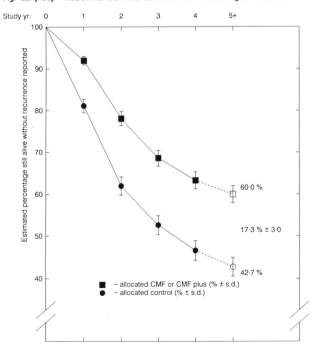

Fig. 22R[<50]. RECURRENCE-FREE SURVIVAL for women aged UNDER 50

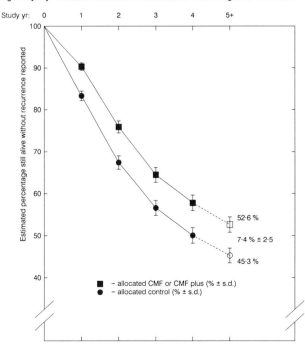

Fig. 22R[≥50]. RECURRENCE-FREE SURVIVAL for women aged 50 OR OVER

Fig. 23M. MORTALITY analysis: survival by year in all available trials of MULTIPLE versus NO/SINGLE-AGENT cytotoxic (incl. any with identical tam. for both groups)

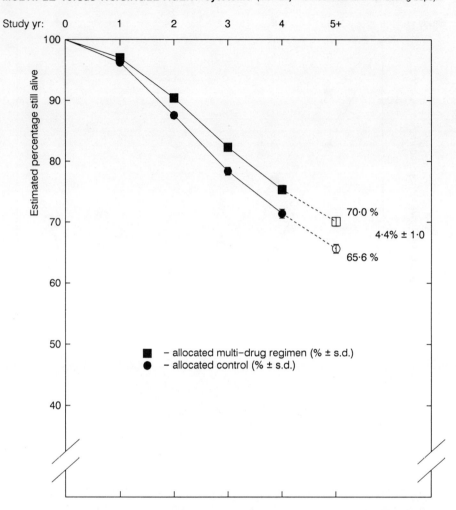

MORTALITY ANALYSES M[all ages]

BELOW, LEFT: M[<50] MORTALITY in women UNDER 50 years old at entry

BELOW, RIGHT: M[≥50] MORTALITY in women 50 OR MORE years old at entry

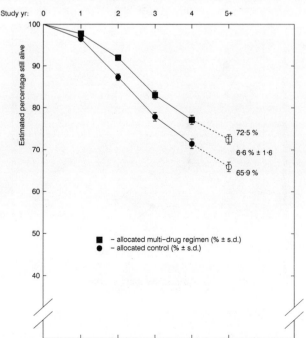

Fig. 23M[<50]. MORTALITY analysis: survival for women aged UNDER 50

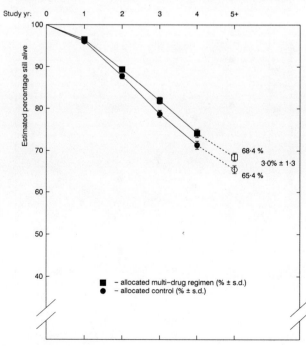

Fig. 23M[≥50]. MORTALITY analysis: survival for women aged 50 OR OVER

9 Results of cytotoxic trials

RECURRENCE ANALYSES R[all ages]

BELOW, LEFT:
R[<50]
RECURRENCE in women UNDER 50 years old at entry

BELOW, RIGHT:
R[≥50]
RECURRENCE in women 50 OR MORE years old at entry

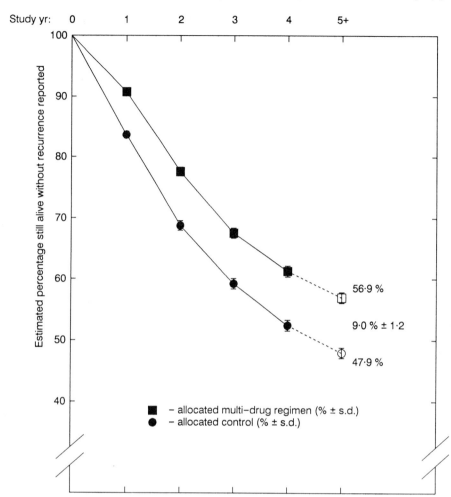

Fig. 23R. RECURRENCE-FREE SURVIVAL by year in all available trials of MULTIPLE versus NO/SINGLE-AGENT cytotoxic (incl. any with identical tam. for both groups)

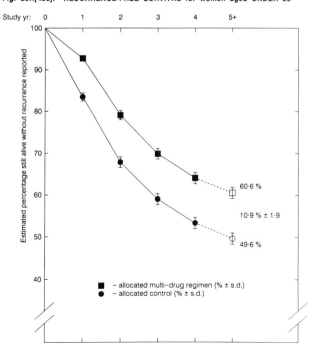

Fig. 23R[<50]. RECURRENCE-FREE SURVIVAL for women aged UNDER 50

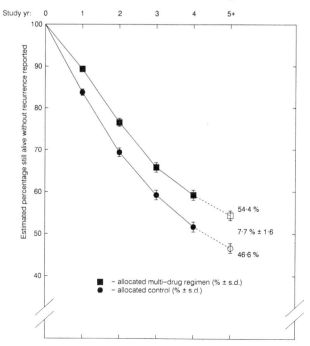

Fig. 23R[≥50]. RECURRENCE-FREE SURVIVAL for women aged 50 OR OVER

Table 13M. Annual MORTALITY rates in trials of prolonged cytotoxic

TRIAL YEAR	ALLOCATED ANY CHEMOTHERAPY Dead/Woman yrs		ALLOCATED CONTROL Dead/Woman yrs		STATISTICAL CALCULATIONS FOR TREATED GROUP ONLY		
					O-E*	Var(O-E)	Odds Redn.
(a) Women aged less than 50 years at entry							
1	49/ 1782	2.6%	53/ 1425	3.6%	-8.0	23.4	yrs 1-2
2	126/ 1599	7.3%	100/ 1272	7.8%	-3.2	48.8	14% ± 11
3	139/ 1351	9.7%	116/ 1067	10.3%	-4.0	53.5	
4	79/ 1132	6.6%	83/ 852	9.1%	-11.9	35.3	yrs 3+
5+	164/ 2718	5.5%	172/ 1875	8.6%	-27.4	58.7	25% ± 7
All	557/ 8583	(—)	524/ 6490	(—)	-54.6	219.8	22% ± 6

4-year risk   23.9% ± 1.1   27.6% ± 1.3
5-year risk**   28.1% ± 1.1   33.8% ± 1.3
Difference in 4-year mortality from ALLOCATION to control or chemotherapy = 3.7% ± 1.7
Difference in 5-year mortality from ALLOCATION to control or chemotherapy = 5.7% ± 1.7

(b) Women aged 50 years or more at entry							
1	107/ 2714	3.9%	99/ 2533	3.8%	0.9	48.3	yrs 1-2
2	184/ 2314	7.7%	196/ 2124	8.7%	-11.2	85.1	7% ± 8
3	170/ 1898	8.4%	185/ 1713	10.3%	-17.2	78.4	
4	165/ 1520	10.3%	116/ 1367	8.3%	15.0	61.8	yrs 3+
5+	294/ 3491	8.1%	275/ 3099	8.5%	-4.7	99.5	3% ± 6
All	920/ 11937	(—)	871/ 10835	(—)	-17.1	373.1	4% ± 5

4-year risk   27.1% ± 1.0   27.7% ± 1.0
5-year risk**   33.0% ± 1.0   33.8% ± 1.0
Difference in 4-year mortality from ALLOCATION to control or chemotherapy = 0.6% ± 1.4
Difference in 5-year mortality from ALLOCATION to control or chemotherapy = 0.8% ± 1.4

* The tabulated (O-E) values and their variances are appropriately derived from summation of separate contributions from each trial, so they avoid direct comparisons between patients in different trials.
** Taking the event rate in year 5 to be approximately that in years 5 or over.

MORTALITY ANALYSES

Table 14M. Annual MORTALITY: multi-drug cytotoxic *v.* no cytotoxic

TRIAL YEAR	ALLOCATED POLY-CHEMOTHERAPY Dead/Woman yrs		ALLOCATED CONTROL Dead/Woman yrs		STATISTICAL CALCULATIONS FOR TREATED GROUP ONLY		
					O-E*	Var(O-E)	Odds Redn.
(a) Women aged less than 50 years at entry							
1	31/ 1249	2.4%	37/ 1136	3.2%	-5.0	16.2	yrs 1-2
2	73/ 1107	6.2%	84/ 1005	8.2%	-10.6	35.0	26% ± 12
3	95/ 927	9.5%	99/ 824	11.5%	-8.8	41.3	
4	49/ 759	6.2%	61/ 648	8.8%	-9.1	24.7	yrs 3+
5+	104/ 1596	6.1%	110/ 1250	8.4%	-13.2	39.8	26% ± 8
All	352/ 5637	(—)	391/ 4864	(—)	-46.7	156.9	26% ± 7

4-year risk   22.3% ± 1.3   28.3% ± 1.5
5-year risk**   27.0% ± 1.3   34.3% ± 1.5
Difference in 4-year mortality from ALLOCATION to control or polychemotherapy = 6.0% ± 2.0
Difference in 5-year mortality from ALLOCATION to control or polychemotherapy = 7.2% ± 2.0

(b) Women aged 50 years or more at entry							
1	77/ 2109	3.6%	79/ 2120	3.6%	-0.4	37.0	yrs 1-2
2	136/ 1755	7.5%	163/ 1739	8.9%	-12.6	67.6	12% ± 9
3	119/ 1398	8.0%	140/ 1369	9.8%	-12.7	58.2	
4	122/ 1075	10.7%	92/ 1068	8.3%	13.5	47.6	yrs 3+
5+	170/ 2016	8.0%	189/ 1987	9.2%	-9.7	65.9	5% ± 7
All	624/ 8353	(—)	663/ 8283	(—)	-22.0	276.4	8% ± 6

4-year risk   26.8% ± 1.2   27.4% ± 1.1
5-year risk**   32.7% ± 1.2   34.1% ± 1.1
Difference in 4-year mortality from ALLOCATION to control or polychemotherapy = 0.6% ± 1.6
Difference in 5-year mortality from ALLOCATION to control or polychemotherapy = 1.4% ± 1.6

MORTALITY ANALYSES

9 Results of cytotoxic trials

RECURRENCE ANALYSES

Table 13R. Annual RECURRENCE rates in trials of prolonged cytotoxic

TRIAL YEAR	ALLOCATED ANY CHEMOTHERAPY Fail/Woman yrs		ALLOCATED CONTROL Fail/Woman yrs		STATISTICAL CALCULATIONS FOR TREATED GROUP ONLY O-E*	Var(O-E)	Odds Redn.
(a) Women aged less than 50 years at entry							
1	165/ 1671	8.7%	248/ 1271	17.5%	-64.9	84.2	yrs 1-2
2	231/ 1362	14.9%	206/ 945	19.6%	-27.6	82.6	43% ± 6
3	139/ 1085	11.6%	108/ 709	14.1%	-11.2	48.8	
4	68/ 881	7.1%	59/ 548	9.9%	-9.6	26.6	yrs 3+
5+	107/ 1986	5.0%	79/ 1128	6.6%	-10.1	33.7	25% ± 8
All	710/ 6985	(—)	700/ 4603	(—)	-123.4	275.9	36% ± 5

4-year risk	36.2% ± 1.3		48.7% ± 1.6	
5-year risk**	39.3% ± 1.3		52.1% ± 1.5	

Difference in 4-year recurrence from ALLOCATION to control or chemotherapy = 12.6% ± 2.1
Difference in 5-year recurrence from ALLOCATION to control or chemotherapy = 12.8% ± 2.0

(b) Women aged 50 years or more at entry

TRIAL YEAR	Fail/Woman yrs		Fail/Woman yrs		O-E	Var(O-E)	Odds Redn.
1	271/ 2275	11.1%	350/ 2068	15.6%	-51.0	129.7	yrs 1-2
2	289/ 1748	15.2%	288/ 1471	17.6%	-19.9	116.3	25% ± 6
3	205/ 1313	14.2%	175/ 1089	15.0%	-4.8	78.1	
4	112/ 1005	10.2%	109/ 822	12.5%	-10.7	46.8	yrs 3+
5+	191/ 2284	8.0%	164/ 1734	8.9%	-6.4	61.1	11% ± 7
All	1068/ 8626	(—)	1086/ 7184	(—)	-93.0	432.0	19% ± 4

4-year risk	42.0% ± 1.2		48.3% ± 1.3	
5-year risk**	46.6% ± 1.2		52.9% ± 1.2	

Difference in 4-year recurrence from ALLOCATION to control or chemotherapy = 6.3% ± 1.8
Difference in 5-year recurrence from ALLOCATION to control or chemotherapy = 6.3% ± 1.7

RECURRENCE ANALYSES

Table 14R. Annual RECURRENCE: multi-drug cytotoxic v. no cytotoxic

TRIAL YEAR	ALLOCATED POLY-CHEMOTHERAPY Fail/Woman yrs		ALLOCATED CONTROL Fail/Woman yrs		STATISTICAL CALCULATIONS FOR TREATED GROUP ONLY O-E*	Var(O-E)	Odds Redn.
(a) Women aged less than 50 years at entry							
1	104/ 1215	7.7%	206/ 1056	17.8%	-59.8	65.4	yrs 1-2
2	167/ 986	15.0%	171/ 775	20.0%	-22.8	65.9	47% ± 6
3	105/ 771	12.3%	90/ 577	14.7%	-8.0	39.6	
4	50/ 606	7.8%	50/ 441	10.4%	-7.0	21.9	yrs 3+
5+	74/ 1272	5.3%	63/ 839	7.3%	-8.8	25.8	24% ± 9
All	500/ 4850	(—)	580/ 3687	(—)	-106.4	218.6	39% ± 5

4-year risk	36.5% ± 1.6		49.7% ± 1.7	
5-year risk**	39.9% ± 1.6		53.4% ± 1.7	

Difference in 4-year recurrence from ALLOCATION to control or polychemotherapy = 13.2% ± 2.4
Difference in 5-year recurrence from ALLOCATION to control or polychemotherapy = 13.5% ± 2.3

(b) Women aged 50 years or more at entry

TRIAL YEAR	Fail/Woman yrs		Fail/Woman yrs		O-E	Var(O-E)	Odds Redn.
1	204/ 1808	10.5%	302/ 1775	15.8%	-49.7	107.4	yrs 1-2
2	212/ 1364	14.3%	243/ 1232	17.8%	-24.0	93.2	31% ± 6
3	161/ 993	14.7%	145/ 889	15.3%	-2.7	63.5	
4	84/ 732	10.6%	86/ 660	12.4%	-6.4	36.8	yrs 3+
5+	132/ 1461	8.3%	127/ 1227	10.1%	-9.4	44.8	12% ± 8
All	793/ 6359	(—)	903/ 5783	(—)	-92.2	345.6	23% ± 5

4-year risk	41.6% ± 1.4		48.6% ± 1.4	
5-year risk**	46.4% ± 1.4		53.8% ± 1.4	

Difference in 4-year recurrence from ALLOCATION to control or polychemotherapy = 7.1% ± 2.0
Difference in 5-year recurrence from ALLOCATION to control or polychemotherapy = 7.4% ± 1.9

Table 15M. Annual MORTALITY: CMF-containing cytotoxic v. none

TRIAL YEAR	ALLOCATED CMF-BASED CHEM. Dead/Woman yrs		ALLOCATED CONTROL Dead/Woman yrs		STATISTICAL CALCULATIONS FOR TREATED GROUP ONLY		
					O-E*	Var(O-E)	Odds Redn.
(a) Women aged less than 50 years at entry							
1	20/ 796	2.4%	28/ 713	3.9%	-5.4	11.4	yrs 1-2
2	43/ 693	5.9%	55/ 613	8.6%	-8.9	22.0	35% ± 14
3	60/ 582	9.7%	67/ 496	12.5%	-7.7	27.2	
4	34/ 476	6.8%	35/ 392	8.5%	-3.8	15.6	yrs 3+
5+	61/ 990	5.7%	73/ 702	9.6%	-13.7	24.5	31% ± 10
All	218/ 3537	(—)	258/ 2916	(—)	-39.5	100.7	32% ± 8

4-year risk   22.7% ± 1.7   29.6% ± 2.0
5-year risk**   27.1% ± 1.7   36.4% ± 2.0
Difference in 4-year mortality from ALLOCATION to control or CMF-based chem. = 6.9% ± 2.6
Difference in 5-year mortality from ALLOCATION to control or CMF-based chem. = 9.3% ± 2.6

(b) Women aged 50 years or more at entry							
1	53/ 1428	3.6%	54/ 1442	3.7%	-0.2	25.4	yrs 1-2
2	88/ 1143	7.4%	112/ 1126	9.4%	-11.9	45.5	16% ± 11
3	81/ 881	8.6%	102/ 849	11.4%	-12.6	41.0	
4	73/ 662	10.3%	56/ 643	8.5%	6.3	28.9	yrs 3+
5+	101/ 1140	8.4%	91/ 1099	8.1%	1.6	36.7	4% ± 9
All	396/ 5254	(—)	415/ 5159	(—)	-16.8	177.4	9% ± 7

4-year risk   26.9% ± 1.5   29.3% ± 1.5
5-year risk**   33.1% ± 1.5   35.0% ± 1.5
Difference in 4-year mortality from ALLOCATION to control or CMF-based chem. = 2.3% ± 2.1
Difference in 5-year mortality from ALLOCATION to control or CMF-based chem. = 1.9% ± 2.1

MORTALITY ANALYSES

Table 16M. Annual MORTALITY: multi-drug v. single, or no, cytotoxic

TRIAL YEAR	ALLOCATED POLY-CHEMOTHERAPY Dead/Woman yrs		ALLOCATED SINGLE/NONE Dead/Woman yrs		STATISTICAL CALCULATIONS FOR TREATED GROUP ONLY		
					O-E*	Var(O-E)	Odds Redn.
(a) Women aged less than 50 years at entry							
1	42/ 1825	2.3%	67/ 1893	3.5%	-11.5	25.7	yrs 1-2
2	101/ 1654	5.9%	167/ 1692	9.5%	-30.1	59.0	39% ± 9
3	147/ 1394	9.7%	157/ 1408	10.8%	-7.5	64.8	
4	80/ 1096	7.1%	96/ 1102	8.3%	-6.6	39.2	yrs 3+
5+	163/ 2564	6.0%	189/ 2374	7.7%	-17.2	64.5	17% ± 7
All	533/ 8532	(—)	676/ 8468	(—)	-73.0	253.2	25% ± 5

4-year risk   22.9% ± 1.1   28.6% ± 1.2
5-year risk**   27.5% ± 1.1   34.1% ± 1.1
Difference in 4-year mortality from ALLOCATION to single/none or to poly = 5.7% ± 1.6
Difference in 5-year mortality from ALLOCATION to single/none or to poly = 6.6% ± 1.6

(b) Women aged 50 years or more at entry							
1	100/ 2799	3.5%	120/ 2967	3.9%	-5.6	51.4	yrs 1-2
2	184/ 2396	7.4%	228/ 2525	8.7%	-16.5	92.7	14% ± 8
3	173/ 1968	8.4%	220/ 2058	10.2%	-19.6	87.3	
4	154/ 1555	9.5%	161/ 1630	9.4%	0.6	69.3	yrs 3+
5+	282/ 3519	7.7%	306/ 3549	8.3%	-8.1	106.0	10% ± 6
All	893/ 12237	(—)	1035/ 12729	(—)	-49.3	406.7	11% ± 5

4-year risk   25.9% ± 1.0   28.7% ± 1.0
5-year risk**   31.6% ± 0.9   34.6% ± 0.9
Difference in 4-year mortality from ALLOCATION to single/none or to poly = 2.8% ± 1.4
Difference in 5-year mortality from ALLOCATION to single/none or to poly = 3.0% ± 1.3

MORTALITY ANALYSES

9 Results of cytotoxic trials

RECURRENCE ANALYSES

Table 15R. Annual RECURRENCE: CMF-containing cytotoxic v. none

TRIAL YEAR	ALLOCATED CMF-BASED CHEM. Fail/Woman yrs		ALLOCATED CONTROL Fail/Woman yrs		STATISTICAL CALCULATIONS FOR TREATED GROUP ONLY		
					O-E*	Var(O-E)	Odds Redn.
(a) Women aged less than 50 years at entry							
1	69/ 774	8.1%	138/ 657	18.9%	-40.6	43.8	yrs 1-2
2	104/ 615	15.1%	126/ 454	23.7%	-24.9	45.3	52% ± 8
3	64/ 477	12.2%	54/ 326	15.1%	-5.8	23.8	
4	32/ 382	7.7%	32/ 252	11.5%	-5.9	13.8	yrs 3+
5+	46/ 784	5.2%	39/ 441	8.2%	-7.6	15.5	31% ± 12
All	315/ 3030	(—)	389/ 2130	(—)	-84.9	142.1	45% ± 6

4-year risk   36.8% ± 2.0   53.5% ± 2.3
5-year risk** 40.0% ± 2.0   57.3% ± 2.2
Difference in 4-year recurrence from ALLOCATION to control or CMF-based chem. = 16.7% ± 3.1
Difference in 5-year recurrence from ALLOCATION to control or CMF-based chem. = 17.3% ± 3.0

(b) Women aged 50 years or more at entry							
1	120/ 1149	9.7%	202/ 1129	16.7%	-41.8	68.2	yrs 1-2
2	142/ 814	15.9%	155/ 726	19.1%	-13.4	60.2	35% ± 7
3	94/ 557	15.2%	83/ 488	16.0%	-2.4	36.6	
4	45/ 405	10.3%	43/ 356	11.5%	-2.4	19.0	yrs 3+
5+	72/ 744	8.9%	61/ 631	9.5%	-1.7	23.2	8% ± 11
All	473/ 3669	(—)	544/ 3329	(—)	-61.6	207.2	26% ± 6

4-year risk   42.2% ± 1.9   49.9% ± 1.9
5-year risk** 47.4% ± 1.8   54.7% ± 1.8
Difference in 4-year recurrence from ALLOCATION to control or CMF-based chem. = 7.7% ± 2.6
Difference in 5-year recurrence from ALLOCATION to control or CMF-based chem. = 7.4% ± 2.5

RECURRENCE ANALYSES

Table 16R. Annual RECURRENCE: multi-drug v. single, or no, cytotoxic

TRIAL YEAR	ALLOCATED POLY-CHEMOTHERAPY Fail/Woman yrs		ALLOCATED SINGLE/NONE Fail/Woman yrs		STATISTICAL CALCULATIONS FOR TREATED GROUP ONLY		
					O-E*	Var(O-E)	Odds Redn.
(a) Women aged less than 50 years at entry							
1	128/ 1641	7.3%	289/ 1644	16.6%	-77.4	87.3	yrs 1-2
2	223/ 1364	14.7%	256/ 1261	18.6%	-26.6	93.8	44% ± 6
3	135/ 1069	11.7%	133/ 973	12.9%	-6.5	55.2	
4	69/ 800	8.2%	75/ 729	9.7%	-5.7	31.2	yrs 3+
5+	106/ 1713	5.6%	95/ 1408	6.9%	-8.7	37.1	16% ± 8
All	661/ 6587	(—)	848/ 6015	(—)	-124.9	304.6	34% ± 5

4-year risk   35.9% ± 1.4   46.7% ± 1.4
5-year risk** 39.4% ± 1.3   50.4% ± 1.3
Difference in 4-year recurrence from ALLOCATION to single/none or to poly = 10.8% ± 1.9
Difference in 5-year recurrence from ALLOCATION to single/none or to poly = 10.9% ± 1.9

(b) Women aged 50 years or more at entry							
1	262/ 2304	10.6%	414/ 2388	16.2%	-68.2	141.7	yrs 1-2
2	277/ 1783	14.3%	326/ 1734	17.2%	-26.3	123.0	30% ± 5
3	205/ 1341	14.0%	203/ 1301	14.6%	-4.4	84.6	
4	108/ 1004	10.1%	133/ 981	12.8%	-13.9	51.4	yrs 3+
5+	194/ 2162	8.2%	193/ 1942	9.8%	-12.6	66.8	14% ± 7
All	1046/ 8594	(—)	1269/ 8347	(—)	-125.5	467.6	24% ± 4

4-year risk   40.8% ± 1.2   48.3% ± 1.2
5-year risk** 45.6% ± 1.2   53.4% ± 1.1
Difference in 4-year recurrence from ALLOCATION to single/none or to poly = 7.6% ± 1.7
Difference in 5-year recurrence from ALLOCATION to single/none or to poly = 7.7% ± 1.6

10. RESULTS OF THE OVARIAN ABLATION TRIALS

Table 17M. **MORTALITY results in all available unconfounded randomized* trials of ovarian ablation**

Study No.	Name of trial	Age <50 at entry Basic data		Statistics		Age 50+ at entry Basic data		Statistics		All ages Basic data		Statistics	
		Treated	Control	O-E	Var	Treated	Control	O-E	Var	Treated	Control	O-E	Var
57A1	Norwegian RH *pre-*	12/64	15/80	-0.1	6.0	2/12	3/13	-0.2	1.2	14/76	18/93	-0.3	7.2
57A2	Norwegian RH *post-*	2/4	3/3	-1.4	0.8	59/86	57/84	-3.3	18.8	61/90	60/87	-4.8	19.7
60A	Ontario CTRF	5/6	2/3	1.6	0.9	110/163	106/160	0.4	37.4	115/169	108/163	2.0	38.3
64A	Saskatchewan CF**	44/139	45/103	-6.8	18.5	22/44	44/69	-3.2	10.7	66/183	89/172	-10.0	29.1
65B1	PMH Toronto *pre-*	42/68	46/62	-6.9	16.3	0/0	0/0	0.0	0.0	42/68	46/62	-6.9	16.3
65B2	PMH Toronto *post-*	35/71	36/71	-0.9	13.8	83/145	98/146	-8.1	32.5	118/216	134/217	-9.0	46.3
71A	CRFB Caen A	0/1	0/0	0.0	0.0	19/26	15/25	2.9	6.5	19/27	15/25	2.9	6.5
78K2	Ludwig II	47/177	48/179	-0.2	17.5	0/0	0/0	0.0	0.0	47/177	48/179	-0.2	17.5
79B2	SWOG 7827B***	-/74	-/75	0.0	2.4	-/12	-/15	0.0	0.5	-/86	-/90	—	2.8
ALL TRIALS (Odds reduction ± sd)		190/604	202/576	-17.2 (20% ± 10)	76.1	296/488	324/512	-11.4 (10% ± 9)	107.6	486/1092	526/1088	-28.5 (14% ± 7)	183.6

* The Christie A (48A) and Malmö Sweden (62A) trials are excluded because treatment allocation was by date of birth. The CCABC (78G) ablation comparison is confounded by prednisone. The Boston (61A), NSABP B-03 (63B), and Bradford (74J2) trials are temporarily excluded for lack of data.
** The Saskatchewan CF (64A) trial has been included without adjustment despite a significant imbalance in entry ages.
*** Results blinded, but contribute to totals.

One of the first reasonably controlled trials of cancer therapy ever undertaken was an early study of ovarian irradiation by Paterson and co-workers at the Christie Hospital in Manchester, England (48A).[34] It started in 1948, allocated some 600 premenopausal women between ovarian ablation and control on the basis of their dates of birth, and reported moderately promising results. Around the early 1960s, a second group of ovarian ablation trials was initiated (including one of the first NSABP studies, 63B); in these, patients were randomized in a manner that precluded prior knowledge by physician or patient of the treatment arm to which the patient would be allocated. In the late 1970s, a third group of randomized ovarian ablation trials among premenopausal women began, but by 1985 these had not yet yielded mature results.

10.1 MORTALITY in ovarian ablation trials

The available mortality information from the randomized trials is summarized in Table 17M. In total it involves only about 1000 women aged less than 50 and 1000 women aged 50 or over, so random errors are large (and, because of the current unavailability of information from some of the early trials, slight biases may exist). Thus, although the available mortality results are moderately promising (an apparent reduction in the odds of death of 22% ± 10 among women aged less than 50, and 11% ± 9 among the older women: overall 16% ± 7), more follow-up of the existing trials is needed, and more patients need to be randomized in the future with respect to some surgical, radiotherapeutic or hormonal form of ablation of ovarian function. Because of the epidemiological observation that artificial menopause may increase the subsequent incidence of myocardial infarction, a satisfactory interpretation of the trials of ablation may have to involve analyses not just of total mortality but also of cause-specific mortality.

10.2 RECURRENCE in ovarian ablation trials

The available recurrence information from the randomized trials is summarized in Table 17R. Information from two more trials (60A and 78K2) is missing, yet both overall and, particularly, among younger women there is a clearly significant (P<0.001) delay of recurrence.

10 Results of ovarian ablation trials

TABLE 17R. **RECURRENCE results in all available unconfounded randomized* trials of ovarian ablation**

Study No.	Name of trial	Age <50 at entry Basic data Treated	Control	Statistics O-E	Var	Age 50+ at entry Basic data Treated	Control	Statistics O-E	Var	All ages Basic data Treated	Control	Statistics O-E	Var
57A1	Norwegian RH *pre-*	12/64	21/80	-3.2	7.6	3/12	3/13	0.3	1.4	15/76	24/93	-2.9	9.0
57A2	Norwegian RH *post-*	2/4	3/3	-1.4	0.9	62/86	60/84	-3.6	21.1	64/90	63/87	-5.0	22.0
64A	Saskatchewan CF**	46/139	49/103	-10.3	20.7	22/44	44/69	-3.1	11.4	68/183	93/172	-13.4	32.0
65B1	PMH Toronto *pre-*	47/68	47/62	-5.0	17.1	0/0	0/0	0.0	0.0	47/68	47/62	-5.0	17.1
65B2	PMH Toronto *post-*	38/71	40/71	-2.7	16.0	95/145	111/146	-8.0	36.3	133/216	151/217	-10.7	52.3
71A	CRFB Caen A	1/1	0/0	0.0	0.0	22/26	17/25	3.9	7.4	23/27	17/25	3.9	7.4
79B2	SWOG 7827B***	-/74	-/75	0.0	6.8	-/12	-/15	0.0	0.7	-/86	-/90	0.0	7.5
ALL TRIALS (Odds reduction ± sd)		159/421	177/394	-25.8 (31% ± 10)	68.9	205/325	237/352	-10.5 (13% ± 11)	78.3	364/746	414/746	-36.3 (22% ± 7)	147.3

* The Christie A (48A) and Malmö Sweden (62A) trials are excluded because treatment allocation was by date of birth. The CCABC (78G) ablation comparison is confounded by prednisone. The Ontario CTRF (60A), Boston (61A), NSABP B-03 (63B), Bradford (74J2) and Ludwig II (78K2) trials are temporarily excluded for lack of data.

** *** See Table 17M footnotes.

11. SUMMARY OF RESULTS

11.1 General principles

The present overview has some implications for patient management, for the conduct of future trials, and for future collaboration among trialists. These will be discussed separately; in addition, many relevant topics have already been discussed in the Introduction and the Results. The reviews of radiotherapy and of ovarian ablation are less detailed than those of tamoxifen and of chemotherapy, and will not be discussed further save to note that, at least for ovarian ablation (whether by surgery, by ovarian irradiation or by newer hormonal methods), what is chiefly needed is randomization of substantially larger numbers of women between active treatment and control.

Both for tamoxifen and for chemotherapy, the **proportional** risk reductions did not appear to differ systematically between high and low risk women. In order to illustrate the **absolute** risk reductions that might typically be achieved with such treatments, the proportional risk reductions that were observed in the trials of them will be applied to women who, in the absence of treatment, would have had a 5-year risk of death of 35% and a 5-year risk of death or of recurrence of 50%. (The absolute difference in 5-year survival suggested by this method would, of course, be smaller for good-prognosis patients and larger for poor-prognosis patients.) Opinions differ on whether the results among women under 50 and over 50 years of age should chiefly be considered separately (treating information in one age group as being largely irrelevant to the other) or whether they should chiefly be considered together (treating the apparent difference between the actual trial results under and over 50 years of age as being, at least in part, a "data-derived" subgroup finding), and so the results are provided for women under 50, for women over 50, and for women of all ages.

11.2 Subgroup analyses of mortality and of recurrence

The overall results for tamoxifen and for chemotherapy have established beyond reasonable doubt that both types of treatment

Table 18M. **Absolute improvement expected in 5-year survival per 100 tamoxifen-allocated women** (of whom 35 would have been expected to die within 5 years if allocated no adjuvant treatment*)

AGE at start (years)	5-year survivors per 100 patients		
	Allocated CONTROL (65% assumed*)	Allocated TAMOXIFEN (for at least 2 yrs)	Absolute difference (± SD & P-value)
Under 50	(65)	67	2 (±2, NS)
50 or over	(65)	72	7 (±1, P<0.0001)
All ages	(65)	71	6 (±1, P<0.0001)

* Estimates and P-values are from Figure 5M, scaled to a control survival probability of 0.65. Example: In Figure 5M the survival estimates for women over 50 are 0.705 and 0.766, so it is estimated that such treatment would change a 5-year survival of 0.65 into 0.72 (calculated as 0.65 raised to the power [log 0.766/log 0.705]), i.e. an absolute difference of 7 per 100 women.

MORTALITY ANALYSES

Table 19M. **Absolute improvement expected in 5-year survival per 100 CMF-allocated women** (of whom 35 would have been expected to die within 5 years if allocated no adjuvant treatment*)

AGE at start (years)	5-year survivors per 100 patients		
	Allocated CONTROL (65% assumed*)	Allocated TAMOXIFEN (for at least 6 mths)	Absolute difference (± SD & P-value)
Under 50	(65)	74	9 (±3, P<0.0001)
50 or over	(65)	67	2 (±2, NS)
All ages	(65)	70	5 (±2, P<0.001)

* Estimates and P-values are from Figure 22M, scaled to a control survival probability of 0.65 as in Table 18M.

MORTALITY ANALYSES

can, for some women, improve 5-year survival and, still more so, 5-year recurrence-free survival. Information on longer-term differences (e.g. at 10 years or beyond) will be provided by future overviews. Many questions, however, remain as to which treatment schedules appear most promising and which patients are likely to gain most from treatment in terms of 5-year differences. Subgroup analyses, either of mortality or of recurrence, may help answer some of these questions, but the appropriate interpretation of subgroup analyses can be difficult. Trials usually observe fewer deaths than recurrences, so there may be more statistical "noise" in a mortality analysis than in a recurrence analysis. Of much greater relevance for statistical power, however, is the fact that there may be a smaller "signal" to be detected in mortality analyses than in recurrence analysis, since the proportional effects of treatment on the odds of an unfavourable outcome may be smaller for death than for recurrence, especially if breast cancer deaths are diluted by other causes of death that are largely unaffected by treatment or if the treatment of relapse is particularly effective for control patients.

Because the effects of treatment involve a smaller "signal-to-noise" ratio (i.e. a smaller number of standard deviations away from zero) for overall survival than for recurrence-free survival, subgroup analyses may be statistically less reliable for overall survival than for recurrence-free survival. This means that, from a purely statistical viewpoint, the relative effects of treatment in different subcategories of women or of trials may be measured less accurately for death than for recurrence. A delay of recurrence is not, of course, equivalent to a delay of death. But, where treatment is of some real benefit in two different categories of women but the play of chance makes the effects of treatment on mortality appear to be negligible in one of those two categories, recurrence analyses that indicate benefit in both categories may help avoid a mistaken interpretation of the mortality analyses.

11.3 Indirect comparisons: (a) effects of different treatments, (b) effects in different patient categories

In general, indirect comparisons between tamoxifen and chemotherapy have been avoided. There are two reasons why comparisons of the apparent effects of tamoxifen with the apparent effects of chemotherapy may **not** be particularly useful. First, the two effects are both subject to appreciable random error, as indicated by their standard deviations (Tables 18 and 19). Second, the question of whether chemotherapy is better than tamoxifen in younger women (or vice versa in older women) may be of limited practical relevance because the two types of treatment are not necessarily mutually exclusive. If one of these treatments produces a worthwhile benefit then it can be used, if the other produces a worthwhile benefit then it can be used, and if both do so (that is, if the combination provides a worthwhile improvement over either alone) then both can be used concurrently — or, if unfavourable drug-drug interactions are feared, consecutively. There is already highly significant evidence that tamoxifen is of additional benefit whether or not

RECURRENCE ANALYSES

Table 18R. **Absolute improvement expected in 5-year recurrence-free survival per 100 tamoxifen-allocated women**
(of whom 50 would have been expected to suffer recurrence or death within 5 years if allocated no adjuvant treatment*)

AGE at start (years)	5-year recurrence-free survivors per 100 patients		
	Allocated CONTROL (50% assumed*)	Allocated TAMOXIFEN (for at least 2 yrs)	Absolute difference (± SD & P-value)
Under 50	(50)	55	5 (±2, P<0.05)
50 or over	(50)	63	13 (±1, P<0.0001)
All ages	(50)	61	11 (±1, P<0.0001)

* Estimates and P-values are from Figure 5R, scaled to a control recurrence-free survival probability of 0.50 (as in Table 18M, but with 0.50 replacing 0.65).

RECURRENCE ANALYSES

Table 19R. **Absolute improvement expected in 5-year recurrence-free survival per 100 CMF-allocated women**
(of whom 50 would have been expected to suffer recurrence or death within 5 years if allocated no adjuvant treatment*)

AGE at start (years)	5-year recurrence-free survivors per 100 patients		
	Allocated CONTROL (50% assumed*)	Allocated TAMOXIFEN (for at least 6 mths)	Absolute difference (± SD & P-value)
Under 50	(50)	67	17 (±3, P<0.0001)
50 or over	(50)	57	7 (±7, P<0.05)
All ages	(50)	60	10 (±2, P<0.0001)

* Estimates and P-values are from Figure 22R, scaled to a control recurrence-free survival probability of 0.50 (as in Table 18M, but with 0.50 replacing 0.65).

chemotherapy is to be given (Table 7), and there is suggestive evidence that chemotherapy may be of additional benefit even if tamoxifen is to be given (Table 11). Research in this area continues, but in the absence of good evidence that the therapeutic effects of one are much modified by the presence of the other (Tables 7 and 11), the decisions as to whether to use chemotherapy and whether to use tamoxifen may be largely independent of each other.

Some other indirect comparisons between the sizes of the treatment effects are inevitable, especially when variables cannot be subjected to randomization (e.g. years from diagnosis, or age) or when data from randomized comparisons are limited or non-existent (e.g. different durations of tamoxifen). In this overview, indirect comparisons have been made (Table 7) to help assess the appropriate duration of tamoxifen treatment, the optimal dose of tamoxifen, and (Table 11) to compare the effects of CMF with those of other combination chemotherapy regimens. Such indirect comparisons do, however, need to be interpreted cautiously. For, although many of the biases inherent in non-random methods (such as those involving historical controls) are avoided by comparing the effects of treatment observed in randomized trials in one circumstance with the effects of treatment observed in randomized trials in another circumstance, some potential for bias may remain.

11.4 Effects of tamoxifen

The tamoxifen duration that has been most extensively studied in the trials is a 2-year regimen, so for definiteness the particular results that will be used to illustrate the tamoxifen overview will be those of "2+" years of treatment (i.e. of 2 years or, in a few trials, longer). The present trials provided no evidence that tamoxifen doses of 30-40 mg/day were any more effective than doses of 20 mg/day (Table 7), and so information from all trials of 2+ years of tamoxifen will be considered together, irrespective of the daily dose. The overall mortality results for these tamoxifen schedules are summarized in Table 18M. The results are given separately for younger women, for older women, and for all women. The mortality reduction produced by tamoxifen was most certain among women aged 50 or over.

The recurrence data (Table 18R) also suggest that tamoxifen may have a somewhat smaller effect among younger women than among older women. But, although the difference between the sizes of the recurrence reductions in the two age groups is statistically significant, the recurrence reductions are significantly favorable both among women aged less than 50 and among those aged 50 or over.

11.5 Effects of chemotherapy

The only cytotoxic drug combination that has been extensively studied in these trials is CMF (i.e. cyclophosphamide, methotrexate and 5-fluorouracil), so for definiteness the particular results that will be used to illustrate the chemotherapy overview will be those of "CMF-based" treatments (i.e. of CMF or, in a few trials, CMF plus some extra drugs) versus controls with no adjuvant cytotoxics. Direct randomized comparisons did not provide any evidence that chemotherapy durations of a year or more were any more effective than durations of about 6 months, and nor did the indirect comparisons (Table 11). Hence, information from all CMF-based trials will be considered together, irrespective of the treatment duration (or intensity: unfortunately, the present overview does not include any assessment of the possible relevance of treatment intensity). The overall mortality results for these chemotherapy schedules are summarized in Table 19M. In contrast with the results for tamoxifen, the effects of CMF-based chemotherapy were most certain among women under the age of 50.

Chemotherapy significantly reduced recurrence rates among both younger and older women (Table 19R), which — as was the case with tamoxifen — suggests that any differences between the effects of treatment among younger and among older women may be in the sizes of the effects rather than in their direction. (This means that the differences between the treatment effects in women over and under 50 may be only "quantitative" rather than "qualitative", even though chemotherapy may be able to have some effects on women with significant ovarian function that it could not have on other women.)

11.6 Duration of benefits

For both tamoxifen and chemotherapy, of course, a key question is how long any differences in survival are likely to persist. At present, there is no evidence that the survival curves do begin to converge at any time during the first five years — rather the reverse, in fact. But, a difference in survival at 5 years might, at least in principle, disappear at 10 or at 15 years, indicating that death from the disease was merely delayed (though perhaps to a worthwhile extent), but not avoided. Conversely, an early difference might persist, or even increase, in the longer term. The most reliable way to find out what actually happens is by direct observation of large numbers of randomized patients for many years, which can be achieved by extension of the follow-up period in the present overview.

Additional years of follow-up may appreciably alter the interpretation of the results of this overview, especially as they apply to particular subgroups. For example, if proportional reductions in mortality are initially similar among patients with a poor prognosis and among those with a good prognosis, then — at least during the first few years of follow-up — the absolute mortality reductions produced by treatment will be greatest in those with a poor prognosis. With further follow-up, however, this may not continue to be the case, and if any clear differences in mortality are still apparent after 10 or 15 years, the greatest absolute differences might then be among women who initially had a reasonably good prognosis. The interpretation of the long-term effects of treatment may also be helped by analyses of cause-specific mortality occurring before disease recurrence, both to check against any specific adverse effects of treatment and to help avoid dilution of any favorable effects of treatment by random fluctuations in mortality from unrelated causes.

11.7 Effects of full compliance with allocated treatment

The sizes of the differences in outcome between those allocated treatment and those allocated control in Tables 18 and 19 would, in expectation, have been somewhat larger if all those allocated active treatment had actually received it and if all those allocated control had completely avoided active treatment. This is particularly relevant to the mortality analyses, for on relapse many control patients may have been given active treatment.

Instead of addressing the purely theoretical question of "treatment versus no treatment ever", adjuvant trials can be used to address the important practical question of "immediate versus deferred treatment" (i.e. adjuvant treatment versus treatment only on relapse). Some of the trials contributing to this overview (e.g. the Scottish Cancer Trials Office study of five years of tamoxifen, 78D) were designed explicitly as trials of immediate versus deferred therapy, and the active treatment of relapsed control patients was implicit in many of the remaining studies. Hence, all control patients who relapsed in those studies should then have been offered active treatment. Inevitably, however, some relapsed controls in some studies may have failed to get effective treatment. If such treatment had always been given on relapse, then the mortality differences (though not the recurrence differences) in Tables 18 and 19 would presumably have been somewhat smaller: they would not, however, have disappeared.

11.8 Generalizability of trial results to clinical practice

Even if both mortality and recurrence analyses are available, and even with the size of the present overview, it may still be difficult to identify **directly** from subgroup analyses the exact types of adjuvant treatment that are most effective or the exact categories of women most likely to benefit, since separate analyses (particularly of overall mortality) in subgroups of the trials or of the total population are liable to be unduly influenced by the play of chance. This does not, of course, mean that such overviews are uninformative about real clinical situations, which always involve considering a specific treatment for a specific patient (e.g. age 55, poorly differentiated 3 cm medial tumour, 2 axillary nodes involved, active multiple sclerosis, etc). What it does mean, however, is that information about specific situations may best be sought not directly, but indirectly.

Even if, in the trials that are currently available, the expected effect of a specific treatment for a specific type of patient cannot reliably be estimated directly from randomized assessment of exactly that treatment in exactly that type of patient, a semi-quantitative estimate of the effect may still be obtained indirectly, by approximate analogy with the results of an overview of a somewhat wider category of treatments among a much wider category of patients. Such an analogy does not involve the implausible assumption that treatment effects in different circumstances must be exactly the same size, but merely the more limited assumption that they are likely to point substantially in the same direction. The plausibility of this latter assumption depends on medical judgement as to how importantly different the circumstances really are, which is a subjective procedure. But, although excessive extrapolation of particular results to other circumstances may lead to some women being inappropriately treated, so too may insufficient extrapolation.

12. IMPLICATIONS

12.1 Implications for patient management

The overall findings are summarized in Tables 18 and 19 (and findings for particular subgroups are summarized in Tables 7 and 11). In Tables 18 and 19 the absolute differences in 5-year survival from the particular regimens that were tested involved about five or six fewer deaths per 100 women treated, and those in 5-year recurrence-free survival involved about 10 or 11 fewer per 100. (From such data it is, of course, not possible to tell whether a moderate number of women got a large benefit or whether a large number of women got a moderate benefit.)

Patient management depends on a wide range of considerations, of which trial results are only one part. Hence, the present report makes no **recommendations** as to what treatments should be prescribed (nor does it deal with practical aspects of the dosages or side effects of any of the treatments considered, which have been reviewed by others[35]). Trial results (or, better, overviews of them) provide information, and not instruction, to clinicians and patients. Where individual trials or overviews do produce statistically definite answers — as, for example, in Tables 18 and 19 — clinicians who treat early breast cancer should merely make themselves clearly aware of those answers (and of any important limitations of the methods that produced them), in order that they may bear them in mind when deciding what treatments to recommend. Thus, physicians should be aware of the approximate size and of the great statistical stability of the improvements in 5-year survival observed in the trials of tamoxifen in older women and in the trials of polychemotherapy in younger women, of the significant improvements in 5-year recurrence-free survival both under and over 50 years of age produced by tamoxifen and by chemotherapy, and of the uncertain relevance of these differences at 5 years to long-term outcome. (Likewise, physicians should be aware that the evidence thus far reviewed from the randomized trials of radiotherapy has not provided clearly significant evidence that such treatment can delay death.) Such trial results are, however, only one of the many factors to be borne in mind when deciding which treatment(s) to recommend.

12.2 Implications for the conduct of future trials

The present results may influence both the content and the methodology of future trials. They draw attention to many inadequately answered questions, several of which may be answered reliably only by further randomization of large numbers of patients and by continuing the collaborative follow-up of those already randomized. For example, both for tamoxifen and for chemotherapy one key question is what the survival differences now seen at 5 years will look like at 10, 15 or 20 years. For tamoxifen, a key question that now needs to be answered is what the effects of really prolonged treatment will be. After, say, 2 years should treatment with tamoxifen stop,* should it continue for several years, or should it even continue indefinitely? Another key question is whether, for women under 50 years of age, randomization of much larger numbers will reveal any significant mortality reduction from prolonged (e.g. 5 years) adjuvant tamoxifen. (There is already significant evidence among these younger women that recurrence can be delayed by tamoxifen, but opinions differ as to how promising this observation should be regarded as being.) Also, there is the question of whether measurement by some improved method of the Estrogen Receptor protein in the primary tumor (or some other measurements) can identify more reliably than has been possible in some of the major trials a group of women who will be wholly unresponsive to tamoxifen. For ovarian ablation, there is simply the need for randomization of several thousand more women, with long-term follow-up of cause-specific mortality.

Finally, for adjuvant chemotherapy, there is an almost limitless range of unanswered questions for future trials about the choice of drugs, the intensity of dosage (which still needs to be addressed on a really large scale by directly randomized comparisons), the duration of treatment (ditto), the effects on mortality of cytotoxic treatments for older women, etc, etc.

12.3 The "Uncertainty Principle" as the fundamental eligibility criterion for clinical trials

Fears have been expressed that results from an overview might prevent some trials that physicians believe to be important from being successful. This should, however, rarely be the case, unless a situation arises where patients but not physicians become generally convinced that a particular treatment policy is inferior. To be "important", a trial must address a therapeutic question that many physicians consider still unanswered — and, where physicians and patients remain uncertain, a randomized trial should then be both informative and ethical.

Some successful trials[17] have actually incorporated the ethical need for uncertainty as the fundamental principle that determines trial eligibility. Thus, for example, the categories of patient eligible for a trial of tamoxifen versus control might be defined primarily by the physician's uncertainty rather than by the patient's age, stage, pathology or biochemistry. If, after assessment of, and discussion with, a particular patient with early breast cancer, the responsible physician remains **substantially uncertain** whether to recommend adjuvant tamoxifen then that particular patient is automatically eligible for randomization, irrespective of any other factors. If, for any medical, mental or social reason(s), the responsible physician is reasonably certain that treatment should be given (or should not be given) to a particular patient, then that patient is automatically ineligible. Thus, for example, the global question as to whether placebo control remains ethical in Stage I disease would be replaced by a large number of individual doctors considering this question for one patient after another (perhaps with different answers for different individual Stage I patients). If, at present, doctors are often substantially uncertain whether to recommend adjuvant treatment then a placebo-controlled trial in Stage I disease is, at present, both ethical and necessary. (Likewise, a "stop/continue" tamoxifen randomization after, say, 2 years could be ethical if it is open only to patients for whom the responsible physician is substantially uncertain whether

---

*   Trials that involve late randomization between stop and continue may be much more reliable than trials involving early randomization between a few years and several years of treatment, for with early randomization the numbers of failures that happen to take place in each treatment group before the treatments should diverge add a lot of uninformative random variation to a standard analysis of outcome by allocated treatment.

continued tamoxifen treatment is indicated.) Some clinical trials can be greatly simplified, and increased in size, by appropriate use of this "uncertainty principle".

Where the uncertainty principle is used as the basis for the design of certain trials, it is likely that the patients actually randomized will be quite heterogeneous. Such heterogeneity can help trials, or future overviews, to address the important question of who needs treatment. In contrast, if entry to a trial is tightly restricted, producing a very homogeneous group of patients, then that trial will not generate direct evidence as to whether the entry restrictions were appropriate (i.e. as to whether treatment would actually have been of some value among the excluded women). If eligibility criteria for trials are unnecessarily strict then even subsequent overviews may not be able to get good direct evidence as to who really needs treatment, for in this and all other respects overviews can help to answer only the questions that trialists have chosen to ask.

12.4 Numbers needed in future randomized trials

Turning finally from the content to the methodology of future trials, the present overview illustrates yet again the need for extraordinarily large numbers of women to be studied if results are to be generated that are sufficiently reliable to justify having a substantial worldwide influence on the treatment strategies for an important disease. It may not represent a disproportionate effort to randomize several thousand women in order to decide how to treat several hundred thousand women worldwide — indeed, it could be argued that it represents a much greater waste, not just of money but of human lives, to use ineffective treatments, or to fail to use effective treatments. But, if such numbers are to be randomized then not only may more resources have to be assigned to the support of the large collaborative groups, but also those groups may have to make substantial efforts to widen their networks of collaborators (and, perhaps, to simplify certain protocols, so as to reduce the cost per randomized patient). In vascular disease, systematic overviews of many small trials have engendered some trials far larger than existed previously (e.g. ISIS-1,[18] GISSI-1,[36] and ISIS-2[17]) that have in turn demonstrated extremely definite mortality reductions, and the same may ultimately happen for some types of cancer therapy.

12.5 Implications for future overviews in breast cancer

The present overview has been necessitated by the frustrating ability of some treatments to produce only modest therapeutic effects. One possible reaction to this situation is to dismiss such modest effects as being of little real interest, and to pin most hope on the future emergence of some different and vastly more effective type of therapy. Opinions differ as to the likelihood of any such therapy emerging over the next decade or two, but whether or not this happens many questions about moderate differences between one and another variant of treatment are still likely to continue to need to be answered reliably. Any such questions that are widely seen to need an answer may well eventually be addressed by dozens of different trials around the world. If **moderate** treatment effects need to be recognized or refuted then systematic overviews of the results of those trials may well be helpful, both so that moderate selective biases can be avoided and so that random errors can be kept small. Thus, over the next decade or two several important therapeutic questions about moderate differences in major causes of death will need to be answered in the light of systematic overviews of as much as possible of the importantly relevant randomized evidence. Some organizational or statistical aspects of the present collaboration may then provide a useful example to those organizing such overviews.

In the particular case of early breast cancer, the present overview is to be consolidated and extended in various ways during the 1990s so that the best use can be made of the randomized information that already exists around the world. First, there is a need for more detailed collaboration with some groups (particularly in Russia and Japan) with which only partial collaboration has thus far been possible. Second, there are many questions about the effects of these and other adjuvant treatments (on long-term mortality, on cause-specific mortality before first recurrence, on the incidence of contralateral disease and in particular subgroups of patients) that can best be answered by continued collaboration during the 1990s. Reliable assessment of moderate benefits from widely practicable treatments may avoid a few thousand unnecessary deaths a year, and the present results already give a foretaste of what might be achieved in the future.

13. ACKNOWLEDGEMENTS

We are indebted to the thousands of women who took part in the trials, and to the many medical, statistical and administrative trial investigators who carefully checked any queries and provided detailed information on trials.

The trialists, secretariat (*), and working party members (†) in this study were, in alphabetical order: O Abe†, M Abeloff†, D Ahmann†, K Andersen†, G Bastert, M Baum†, AR Bianco†, F Boccardo†, G Bonadonna†, A Bryant, R Buchanan, M Buyse†, A Buzdar, P Carbone†, J Carpenter, R Chlebowski, EA Clarke, M Clarke*, R Collins*, R Cooper, N Corcoran, J Crowley†, F Cummings, J Cuzick†, N Day†, T Delozier†, A de Schryver, L Douville*, J Dubois, K Durrant*, B Fisher†, J Forbes†, P Forrest†, R Gelber†, R Gelman†, J Glick†, J Godwin*, A Goldhirsch†, R Gray*, T Guidi*, C Harwood*, T Hatschek, J Hayward†, C Henderson†, C Hill*, H Høst, J Houghton†, A Howell†, C Hubay, J Ingle, R Jakesz†, F Jungi, M Kaufmann†, K Kelly, M Kissin, L-G Larsson, M Lippman†, A Litton, J Lythgoe, R Margreiter, L Mauriac, C McArdle, G Mead*, J Meakin, M Morrison†, H Mouridsen†, A Naja, R Nissen-Meyer, Y Nomura†, GD Oates, K Osborne†, T Palshof, U Peek, R Peto*, K Pritchard†, J Ragaz, C Redmond†, O Repelaer van Driel, G Ribeiro†, R Rubens, L Rutqvist†, S Ryden†, E Scanlon, H Scheurlen, V Semiglazov†, F Senanayake, HJ Senn†, R Simon†, M Söderberg, M Spittle, H Stewart†, R Swindell†, D Tormey†, P Valagussa†, J van Dongen†, A Wallgren†, R Weiss, K Welvaart, R Wittes†, N Wolmark†, and W Wood†.

The scientific stimuli for this collaboration were the Clinical Trial Service Unit of the Imperial Cancer Research Fund and the Medical Research Council, the Breast Trials Unit of the Cancer Research Campaign, the UK Breast Cancer Trials Co-ordinating Subcommittee, the Project on Controlled Therapeutic Trials of the Union Internationale Contre le Cancer, and the Cancer Office of the World Health Organization.

This study was chiefly supported by the Imperial Cancer Research Fund and the Cancer Research Campaign, and also by the Medical Research Council (UK), the National Cancer Institute (US), the General Motors Cancer Research Foundation, the World Health Organization (Cancer Unit), and the Wendy Will Case Memorial Cancer Research Fund.

Address correspondence to: EBCTCG Secretariat, ICRF-MRC Clinical Trial Service Unit, Nuffield Department of Clinical Medicine, Radcliffe Infirmary, University of Oxford, Oxford, United Kingdom.

14. REFERENCES

(1) ANON. Review of mortality results in randomized trials in early breast cancer. Lancet 1984; ii: 1205.

(2) EARLY BREAST CANCER TRIALISTS' COLLABORATIVE GROUP. Effects of adjuvant tamoxifen and of cytotoxic therapy on mortality in early breast cancer. New Engl J Med 1988; 319: 1681-1692.

(3) CUZICK J, STEWART H, PETO R, ET AL. Overview of randomized trials of post-operative adjuvant radiotherapy in breast cancer. Cancer Treatment Reports 1987; 71: 15-29.

(4) PETO R. Why do we need systematic overviews of randomized trials? Stat Med 1987; 6: 233-240.

(5) YUSUF S, PETO R, LEWIS J, COLLINS R, SLEIGHT P. Beta-blockade during and after myocardial infarction: an overview of the randomized trials. Prog Cardiovasc Dis 1985; 27: 335-371.

(6) ANTIPLATELET TRIALISTS' COLLABORATION. Secondary prevention of vascular disease by prolonged antiplatelet treatment. Brit Med J 1988; 296: 320-331.

(7) CARTER SK. Acute lymphocytic leukemia. In Randomized trials in cancer: a critical review by sites (Staquet M, ed). New York: Raven Press, 1978, pp.1-24.

(8) ROZENCWEIG M, VON HOFF DD, DAVIS HL, JACOBS EM, MUGGIA FM, DEVITA VT. Hodgkin's Disease. In Randomized trials in cancer: a critical review by sites (Staquet M, ed). New York: Raven Press, 1978, pp.103-130.

(9) YUSUF S, COLLINS R, PETO R. Why do we need some large, simple randomized trials? Statistics in Medicine 1984; 3: 409-420.

(10) SILVERBERG E. Cancer Statistics, 1988. "Ca — A cancer journal for clinicians". American Cancer Society, New York, 1988; 38(1): 5-22.

(11) DICKERSIN K, CHAN SS, CHALMERS TC, SACKS HS, SMITH H. Publication bias in randomized control trials. Controlled Clinical Trials 1987; 8: 343-353.

(12) SIMES RJ. Publication bias: the case for an international registry of clinical trials. J Clin Oncology 1986; 4: 1529-1541.

(13) REES JKH, GRAY RG, SWIRSKY D, HAYHOE FGJ. Principal results of the Medical Research Council's 8th Acute Myeloid Leukaemia Trial. Lancet 1986; ii: 1236-41.

(14) BYAR DP. Why data bases should not replace randomized clinical trials. Biometrics 1980; 36: 337-342.

(15) PETO R. Statistical aspects of cancer trials. In: The treatment of cancer (Halnan K, ed). London: Chapman & Hall, 1982, pp.867-871.

(16) GAIL M, SIMON R. Testing for qualitative interactions between treatment effects and patient subsets. Biometrics 1985; 41: 361-372.

(17) ISIS-2 COLLABORATIVE GROUP. Randomised trial of intravenous streptokinase, oral aspirin, both, or neither among 17,187 cases of suspected acute myocardial infarction: ISIS-2. Lancet 1988; ii: 349-360.

(18) ISIS-1 COLLABORATIVE GROUP. Randomised trial of intravenous atenolol among 16 027 cases of suspected acute myocardial infarction: ISIS-1. Lancet 1986; ii: 57-66.

(19) COLLINS R, GRAY R, GODWIN J, PETO R. Avoidance of large biases and large random errors in the assessment of moderate treatment effects: the need for systematic overviews. Stat Med 1987; 6: 245-250.

(20) FEINLEIB M. Breast cancer and artificial menopause: a cohort study. J Natl Cancer Inst 1968; 41: 315-329.

(21) COLDITZ GA, WILLETT WC, STAMPFER MJ, ROSNER B, SPEIZER FE, HENNKENS CH. Menopause and the risk of coronary heart disease in women. New Engl J Med 1987; 316: 1105-1110.

(22) STAMPFER MJ, WILLETT WC, COLDITZ GA, ROSNER B, SPEIZER FE, HENNEKENS CH. A prospective study of postmenopausal estrogen therapy and coronary heart disease. New Engl J Med 1985; 313: 1044-1049.

(23) KEY TJA, PIKE MC. The dose-effect relationship between "unopposed" oestrogens and endometrial mitotic rate: its central role in explaining and predicting endometrial cancer risk. Br J Cancer 1988; 57: 205-212.

(24) VESSEY MP. The Jephcott Lecture 1989: An overview of the benefits and risks of combined oral contraceptives. In Oral contraceptives and breast cancer: the implications of present findings for informed consent and informal choice (Mann RD, ed). Carnforth: Parthenon Publishing Group 1990.

(25) BYAR DP, CORLE DK. Hormone therapy for prostate cancer: results of the Veterans Administration Cooperative Urological Research Group Studies. National Cancer Institute Monographs 1988; 7: 165-170.

(26) INTERNATIONAL AGENCY FOR RESEARCH ON CANCER. IARC monographs on the evaluation of the carcinogenic risk of chemicals to humans, Volume 41: Some antineoplastic and immunosuppressive agents. Lyon: IARC, 1981, pp.1-411.

(27) UNITED NATIONS SCIENTIFIC COMMITTEE ON THE EFFECTS OF ATOMIC RADIATION. Sources, effects and risks of ionizing radiation. New York: United Nations, 1988.

(28) UNITED NATIONS SCIENTIFIC COMMITTEE ON THE EFFECTS OF ATOMIC RADIATION. Sources and effects of ionizing radiation. New York: United Nations, 1977.

(29) MANTEL N. Evaluation of survival data and two new rank order statistics arising in its consideration. Cancer Chemotherapy Reports 1966; 50: 163-170.

(30) PETO R, PIKE MC, ARMITAGE P, ET AL. Design and analysis of randomised clinical trials requiring prolonged observation of each patient. Br J Cancer 1976; 34: 585-612.

(31) PETO R, PIKE MC, ARMITAGE P, ET AL. Design and analysis of randomised clinical trials requiring prolonged observation of each patient. Br J Cancer 1977; 35: 1-39.

(32) GREENLAND S, SALVAN A. Bias in the one-step method for pooling study results. Stat Med 1990; 9: 247-252

(33) HENNEKENS CH, BURING JE, SANDERCOCK P, COLLINS R, PETO R. Aspirin and other antiplatelet agents in the secondary and primary prevention of cardiovascular disease. Circulation 1989; 80: 749-756.

(34) PATERSON R. Breast cancer: a report of two clinical trials. J Roy Coll Surgeons Edin 1962; 7: 243-254

(35) HENDERSON IC. Adjuvant systemic therapy of early breast cancer and endocrine therapy of metastatic breast cancer. In: Harris JR, Hellman S, Henderson IC, Kinne DW, eds. Breast diseases. Philadelphia: J B Lippincott, 1987: 324-353.

(36) GRUPPO ITALIANO PER LO STUDIO DELLA STREPTOCHINASI NELL' INFARTO MIOCARDICO (GISSI). Effectiveness of intravenous thrombolytic treatment in acute myocardial infarction. Lancet 1986; i: 397-402.

APPENDIX TABLE 1. List of all relevant randomized trials included in the overview (plus a few others not properly randomized and some without data yet available)

Study No.	Study Name	Treatment Options §	Age at Entry	Menopausal status Pre-	Peri-	Post-	Nodal status N?	Neg	Pos	ER status Poor	Pos	Number Entered	Outcome NK1984	Dead	Recur
48A +	Christie A	Ovlrr ¦ –	25-54				55	182	361			598	260	338	?
49A +	Manchester Q	R ¦ –	28-77					all				720*	?	522	?
49B +	Manchester P	R ¦ –	29-77					all				741	?	489	?
57A1	Norwegian RH	Ovlrr ¦ –	25-54	all				158	11			169	136	32	7
57A2	Norwegian RH	Ovlrr ¦ –	46-77		43	134	1	80	96			177	56	121	6
57B	Illinois	Nm ¦ –	data from this trial were not available in 1985												
58A	NSABP B-01	Thio ¦ –	data from this trial were not available in 1985												
60A	Ontario CTRF	Ovlrr ¦ –	43-83			all	8	128	196			332	58	223	?
61A	Boston	Ooph ¦ –	data from this trial were not available in 1985												
61B	NSABP B-02	R ¦ Thio ¦ F ¦ –	data from this trial were not available in 1985												
62A +	Malmö	Ooph ¦ –	21-65	135		145		129	151			280	?	149	9
62B1	Berlin-Buch ABC	C ¦ R ¦ –	data from this trial were not available in 1985												
62B2	Berlin-Buch ab	C ¦ –	data from this trial were not available in 1985												
63A ++	EFSCH Columbia	Thio ¦ –	29-88					72	110			182	21	129	5
63B	NSABP B-03	Ooph ¦ –	data from this trial were not available in 1985												
64A	Saskatchewan CF	Ooph ¦ –	26-73**	10		1		189	166	1		355	9 *	155	6
64B1	Oslo X-ray	R ¦ –	26-72	232	2	318		349	203			552	277	275	12
64B2	Oslo Co-60	R ¦ –	31-70	212	2	332		366	197			563	347	214	23
65B1	PMH Toronto	Ovlrr ¦ –	35-44	all			8	13	109			130	42	88	6
65B2	PMH Toronto	Ovlrr ¦ OvlrrPr ¦ –	35-76	252	122	275	78	138	433			649	280	369	41
67A	Birmingham	C ¦ F ¦ –	24-69 *	107		137		122	122			244	?	99	25
69A	Heidelberg XRT	R ¦ –	27-84			all						142 *	39	103	?
70A	Manchester RBS1	R ¦ –	21-70	246	49	394	all					714	14	352	63
70B	Kings/Cambridge	R ¦ –	24-73	857	223	1720		2104	696			2800	63	1353	274
71A	CRFB Caen A	Ovlrr ¦ –	48-70			all			all			52	2	34	6
71B	Stockholm A	R(pre) ¦ R(post) ¦ –	27-70	338	132	255	401	304***			4	960	118	373	68
71C+++	NSABP B-04	s2 ¦ s1 ¦ Rs1	?									1079	?	472	?
72A	WSSA Glasgow	R1s1 ¦ R2s1 ¦ R1s2	31-75	110	49	176		252	83***			335	75	184	70
72B	NSABP B-05	Mel ¦ –	?					all				380	?	198	?
72C	Villejuif Paris	R ¦ –	data from this trial were not available in 1985												
73A	Wessex	R ¦ –	30-70	44	19	82		64	82			146	26	67	17
73B	INT Milan 7205	CMF ¦ –	26-75	190		201		all				391		209	51
73C1	Mayo Clinic	CFPr ¦ CFPrR ¦ – ¦ R	22-75	11		23		all				34		18	3
73C2	Mayo Clinic	CFPr ¦ CFPrR ¦ Mel	34-72					all				192		93	25
73C3	Mayo Clinic	CFPr ¦ CFPrR ¦ Mel	27-54	all				all				53	2	25	5
73C4	Mayo Clinic	CFPr ¦ CFPrR	30-55	all				all				49	1	22	7
74A1	Berlin-Buch	Chem ¦ –	31-70	42	6	62		all		39	63**	110	2	72	13
74A2	Berlin-Buch	Chem ¦ –	24-69	48	2	40	1		89	46	42	90	1	36	7
74B	Edinburgh I	R ¦ –	26-69			all						348	1	100	?
74C	Edinburgh II	F ¦ –	22-69	134	37	174		72	273	27	71	345	4	224	38
74D1	DFCI 74-063	AC ¦ AC ¦ –	38-74	3	1	3	1	1	6		1	8	1	5	
74D2	DFCI 74-063	Mel ¦ AC ¦ AC	27-72	12	1	19	1	1	30	2	4	32	4	19	5
74D4	DFCI 75-122B	AC ¦ AC	24-76	105	18	144	1	1	266	81	117	268	15	101	43
74E	UK MCCG 003	CMFV ¦ –	28-77	84	44	125	35	41	214			290	60	148	37
74F	Western CSG	F ¦ CMF	26-78	22	2	38		all				62	11	32	9
74G	Case Western A	CMF ¦ TCMF ¦ TCMFBCG	22-74	97	23	198		all		118	200	318	2	144	42
74J1	Bradford RI	R ¦ Chem	data from this trial were not available in 1985												
74J2	Bradford RI	R ¦ Chem ¦ ROoph ¦ ChemOoph	data from this trial were not available in 1985												
75A	SWOG 7436	Mel ¦ CMFVPr	20-80	181		250	7		433			440	66	(blinded)	
75C +	Alabama BCP	Mel ¦ CMF	31-70	74	27	69		all				171	1	79	19
75D1	Copenhagen	T ¦ –	29-61	130	83		all			137	76	213		75	18
75D2	Copenhagen	T ¦ DES	51-70			all		all		82	72	155		61	14

Nodes (see text for definitions of N?): 'Neg' = N0/N–, 'Pos' = N+/N1-3/N4+.
ER status (see text): 'Poor' = ER– or <10 fmol/mg, 'Pos' = ER+/++ or ≥10 fmol/mg.
Numbers: 'all' = no. randomized, 'NK1984' = not known to be dead but no data since 1-1-1984.
Tests for imbalance: * P < 0.05; ** P < 0.01; *** P < 0.001.
? = unknown.
+ : Results from these trials are not included in the overview analyses because the methods used for randomization (e.g. date of birth) did not preclude the possibility of foreknowledge of treatment allocation.
++ : Results from this trial are not included in the overview analyses because 19 patients, who had been inappropriately excluded because of non-compliance with chemotherapy, have been irrevocably lost to follow-up.
+++ : Published results used since individual patient data were not available for this trial.

APPENDIX TABLE 1 (continued)

Study No.	Study Name	Treatment Options §	Age at Entry	Menopausal status Pre-	Peri-	Post-	Nodal status N?	Neg	Pos	ER status Poor	Pos	Number Entered	Outcome NK1984	Dead	Recur
75E1	Guy's L-Pam	Mel ¦ -	26-76	110	29	121			all	49	161	260	14	125	28
75E2	Manchester I	Mel ¦ - ¦ CMF	25-70	65	6	93	6		159 *	54	61	165	15	62	23
75F	Evanston USA	Mel ¦ CFPr ¦ CFPrBCG	29-80	30		83			all	55	42	113	4	53	13
75G	Northwick Park	MelV ¦ -	31-71	50		90	87	53				140	1	43	12
75H1	INT Milan 7502	CMF ¦ CMF	26-56	all					all	57	138	331		126	34
75H2	INT Milan 7502	CMF ¦ CMF	42-71			all			all	16	32	114		53	17
75J	King's CRC M/M	MelM ¦ -	25-78	138	76	190	1	45	388			434	121	194	69
75K	Piedmont OA	Mel ¦ MelR ¦ CMF ¦ CMFR	20-78		82	16	119	6	12	263		281	6	127	21
75L	SAKK 27/76	LeuMF ¦ LeuMF	29-72 *	205	60	134	10		389	399	10	201	68		
75M	NSABP B-07	Mel ¦ FMel	?						all			741	?	336	?
75N	Arizona Univ.	ACR ¦ AC	data from this trial were not available in 1985												
75P	COG Wisconsin	Mel ¦ CMFV	data from this trial were not available in 1985												
75Q	Fox Chase Phil.	Mel ¦ CMF	data from this trial were not available in 1985												
76A1	SECSG 1	CMF ¦ CMF	24-79	142		165			all	46	113	307	30	72	32
76A2	SECSG 1	CMF ¦ CMF ¦ CMFR	25-79	68		101			all	18	46	169	23	78	20
76A3	SECSG 1	CMF ¦ CMFR	26-76	56		90			all	26	57	146	11	43	26
76C	Glasgow	CMF ¦ CMFR ¦ R	25-74	127		191	9	7	306 *	103	132	322	21	135	41
76D	Dublin	F ¦ -	31-72	18		23		16	25			41		23	3
76E	EORTC 09771	CMF ¦ -	28-72	210		220	1		451			452	176	94	83
76F	Christie B	T ¦ -	38-70			all	173	185	230			588	1	221	63
76G1	Stockholm B	TR ¦ TLeuCMF ¦ R ¦ LeuCMF	47-70	1		75	1	38	283	50	205	322	4	113	42
76G2	Stockholm B	TR ¦ TCMF ¦ R ¦ CMF	45-69	1		22		13	107	25	87	120	1	5	11
76G3	Stockholm B	TR ¦ R	51-70	3		3		13	79	12	71	92	1	6	4
76G4	Stockholm B	T ¦ -	47-77	5		65	15	771	4 *	120	535	790	17 *	55	54
76G5	Stockholm B	TR ¦ R	45-70	1		29	146	3		22	95	149	1	6	3
76H1	West Midlands	CMFVALvor ¦ -	22-73	263	106	200			all	134	240 *	569	80	262	120
76H2	West Midlands	LeuMF ¦ -	22-75	282	104	188			all	138	219	574	74	85	78
76K	HD 1 W. Germany	LeuF ¦ -	data from this trial were not available in 1985												
76M	Helsinki	LevamR ¦ ACLevamV ¦ ACLevamRV	data from this trial were not available in 1985												
77A	Scandinavia 2A	periCMFV ¦ periCMFV,CMF	25-77	139	50	187	5	6	372			383 *	21	83	75
77B1	Danish BCG 77b	- ¦ Levam ¦ C ¦ CMF	22-59	all				102	359	2	4	461		174	52
77B2	Danish BCG 77b	- ¦ C ¦ CMF	19-58 *	all				41	210 *	16	35	251		67	26
77B3	Danish BCG 77b	C ¦ CMF	25-58	all				78	422	30	93	500		87	46
77C1	Danish BCG 77c	- ¦ Levam ¦ T	42-92			all		177	569	3	17	746		362	78
77C2	Danish BCG 77c	- ¦ T	42-88 *			all		192	1078	56	213	1270		427	140
77E	Oxford	Mel ¦ MelMF ¦ -	25-71	131	52	119	10	158	138			306		125	35
77F	UK MCCG 009	LeuMFV ¦ TLeuMFV ¦ CMFV ¦ TCMFV ¦ T	23-74	152	49	139		60	288			348	149	96	67
77G	Vienna Surg.	- ¦ CMFV ¦ AzimCMFV	28-70	83	17	141	7	128	106	58	130	241	1	68	26
77H	NATO	T ¦ -	29-75	129		999	24	606	501	211	314	1131	25	371	117
77J1	ECOG EST5177	CMF ¦ CMFPr ¦ TCMFPr	?	all			6	2	654	310	345	662	?	196	?
77J2	ECOG EST6177	- ¦ CMFPr ¦ TCMFPr	?			all	3	1	261	91	170	265	?	100	?
77K	NSABP B-09	FMel ¦ TFMel	?						626	927	1891	?	722	1018	
78A1	S Swedish BCG	R ¦ CR ¦ C	27-63	all				145	284	124	137	429	14	74	38
78A2	S Swedish BCG	R ¦ TR ¦ T	47-71			all		295	423	174	332	718	9	142	66
78D	Toronto-Edmont.	T ¦ -	45-81			all	6		394			400	3	99	80
78C1	GUN Naples	T ¦ -	34-70	12		48		28	32	20	25	60	4	17	11
78C2	GUN Naples	T ¦ -	42-72	6		40		23	25	14	25	46	2	7	5
78C3	GUN Naples	T ¦ -	26-79	27		109		78	58	41	73	136	13	10	29
78C4	GUN Naples	T ¦ -	32-72	16		50		44	22	11	31	66	1		12
78C5	GUN Naples	CMF ¦ TCMF	26-54	all					all	45	55	125	8	25	23
78D1	Scottish PilotB	T ¦ -	47-76			all		6	101	10	27	107	1	49	10
78D2	Scottish B	T ¦ -	45-79			all			1	364	66	160	365		98	55
78D3	Scottish C	T ¦ -	27-79	215		543	1	757		165	282	758	2	87	82
78D4	Scottish D	- ¦ R ¦ T ¦ TR	35-78	28		65		81	12	15	20	93		18	11
78E1	Hong Kong	CMF ¦ T ¦ TCMF	22-70	76	19	68	28	1	139			168	1	40	25
78E2	Sri Lanka	CMF ¦ T ¦ TCMF	26-68	76	3	58	1	2	136			139	27	11	18
78E3	Bombay	CMF ¦ T ¦ TCMF	22-60	24	5	21			all			50	10	6	19
78E4	Great Britain	CMF ¦ T ¦ TCMF	24-69	34	22	57	27	1	90			118	2	43	23
78F	CRFB Caen C5	T ¦ -	47-79			all			all	43	90	179	3	56	29
78G1	CCABC Canada	CMF ¦ CMFR ¦ CMFOvIrrPr ¦ CMFOvIrrPrR	22-56	133		1	1		133	9	124	134		18	12
78G2	CCABC Canada	CMF ¦ CMFR	29-52	all					all		all	25		1	1
78G3	CCABC Canada	CMF ¦ CMFR	27-55	131			1		131	all		132		30	10
78H	Innsbruck	T ¦ -	31-79 *	91			6	112	116	2	232	234	7	28	41
78J	ECOG EST1178	T ¦ -	?			180	2	179		8	153	181	?	57	?
78K2	Ludwig II	CMFPr ¦ CMFPrOoph	?	all					all			356	?	95	?
78K3	Ludwig III	TPr ¦ TCMFPr ¦ -	?			all			all			503	?	143	?
78K4	Ludwig IV	TPr ¦ -	? *			all			all			349	?	97	?
78L	Glasgow	Mel ¦ CMF	data from this trial were not available in 1985												
78M1	NCCTG-773051	CFPr ¦ TCFPr	data from this trial were not available in 1985												
78M2	NCCTG-773051	CFPr ¦ TCFPr ¦ -	data from this trial were not available in 1985												
78P	Helsinki	R ¦ ACV ¦ ACRV	data from this trial were not available in 1985												
78Q	Düsseldorf U.	CMF ¦ CMFR	data from this trial were not available in 1985												
78R	Heidelberg U.	LeuF ¦ -	data from this trial were not available in 1985												

Nodes (see text for definitions of N?): 'Neg' = N0/N-, 'Pos' = N+/N1-3/N4+.
ER status (see text): 'Poor' = ER- or <10 fmol/mg, 'Pos' = ER+/++ or ≥10 fmol/mg.
Numbers: 'all' = no. randomized, 'NK1984' = not known to be dead but no data since 1-1-1984.
Tests for imbalance: * $P < 0.05$; ** $P < 0.01$; *** $P < 0.001$.
? = unknown.

Appendix tables and figure

APPENDIX TABLE 1 (continued)

Study No.	Study Name	Treatment Options §	Age at Entry	Menopausal status Pre-	Peri-	Post-	Nodal status N?	Neg	Pos	ER status Poor	Pos	Number Entered	Outcome NK1984	Dead	Recur
79A	Ghent Univ.	T ¦ -	?			all		all				85	?	13	?
79B1	SWOG 7827 A	T ¦ CMFVPr ¦ TCMFVPr	32-88	2		634	5		634	39	565	639	19	(blinded)	
79B2	SWOG 7827 B	CMFVPr ¦ CMFVPrOoph	24-57	172		3	1		175	23	142	176	6	(blinded)	
79B3	SWOG 7827 C	CMFVPr ¦ CMFVPr	22-76	197		248	1		444	293	13	445	14	(blinded)	
79C1	Case Western B	TOoph ¦ TCMFVPrOoph	24-53 *	39	2				all	8	33	41		2	6
79C2	Case Western B	T ¦ TCMFVPr	44-74		1	91			all	13	79	92		15	10
79D1	GABG I Germany	AC ¦ TAC	22-70	171		206			all	192	185	377	10	58	48
79D2	HD 2 W. Germany	T ¦ -	45-74	4		116		all		52	78	130	12	13	3
79E1	Guy's CMF	CMF ¦ -	23-67	111	30	40	1		189	11	127	190	4	33	24
79E2	Manchester II	CMF ¦ -	25-69	32	9	36		2	76	37	33	78	2	21	9
80A	MD Anderson8026	TFACVPr ¦ TCMFVAPr	23-74	94	14	81	7	17	211			235	42	17	34
80B2	N SwedenBCG 192	- ¦ T	54-74					119	59			178		20	11
80B3	N SwedenBCG 193	- ¦ TOvIrr ¦ AC ¦ TACOvIrr	30-54					2	54			56		11	10
80B4	N SwedenBCG 194	- ¦ T ¦ AC ¦ TAC	55-72					3	93			96		19	15
80C	SE Sweden BCG B	- ¦ AC	47-65			all			all	6	20	43		12	2
80D	CRC2	T ¦ C ¦ TC ¦ -	26-83	509	284	792	7	1007	754			1768		104	235
80E	Toulouse	T ¦ -	49-75			all	1	93	157***			251		41	20
80F	INT Milan 8004	CMF ¦ -	30-69	53	3	38			93	1	93		94		9	13
80G	EST1180/SW.8294	CMFPr ¦ -	data from this trial were not available in 1985												
80J1	Vienna Gyn.	CMF ¦ T ¦ TCMF	data from this trial were not available in 1985												
80J2	Vienna Gyn.	T ¦ -	data from this trial were not available in 1985												
80L	SWOG 7985	CMFV ¦ -	data from this trial were not available in 1985												
80P1	H. Salpétrière	FMThioVPr ¦ TFMThioVPr	data from this trial were not available in 1985												
80P2	H. Salpétrière	AFMThioVPr ¦ TAFMThioVPr	data from this trial were not available in 1985												
80Q	Coimbra	AC ¦ ACR	data from this trial were not available in 1985												
81A	Montpellier	T ¦ -	50-76 *			all	1	127	75			203	14	16	22
81B	FB Bordeaux	CMF ¦ TCMF	22-75	125	38	162			all	25	300	325	1	15	24
81E	NSABP B-13	FMLvor ¦ -	data from this trial were not available in 1985												
81F	Ludwig V	CMFLvor ¦ -	data from this trial were not available in 1985												
81H	ECOG EST3181	FACRThio ¦ FACThio	data from this trial were not available in 1985												
82A	NSABP B-14	T ¦ -	data from this trial were not available in 1985												
82B	Danish BCG 82b	CMF ¦ CMFR ¦ TCMF	25-59	all				65	589	62	146	654		49	60
82C	Danish BCG 82c	T ¦ TR ¦ TCMF	42-75			all		66	633	41	180	699		52	56
82F	Barcelona	CMF ¦ CMF	data from this trial were not available in 1985												
82G	MD Anderson8227	FACVPr ¦ -	data from this trial were not available in 1985												
82H	Oslo	T ¦ -	data from this trial were not available in 1985												
82K	Berlin-Buch	CMFA ¦ CMFA	data from this trial were not available in 1985												
83B	GROCTA I Italy	T ¦ CMFE ¦ TCMFE	31-65	133		145	3		278	33	239	281	2	2	15
83C	Amsterdam	T ¦ -	data from this trial were not available in 1985												
83E	Toulouse	FAC ¦ FACR ¦ TROoph	data from this trial were not available in 1985												
84A1	BMFT 02 Germany	CMF ¦ CMF ¦ TCMF ¦ TCMF	31-71	21	4	39	33	1	62	15	47	96	1	?	
84A2	BMFT 03 Germany	CMF ¦ CMFR	30-71	12	2	31	8	1	36	17	28	45		?	

Nodes (see text for definitions of N?): 'Neg' = N0/N-, 'Pos' = N+/N1-3/N4+.
ER status (see text): 'Poor' = ER- or <10 fmol/mg, 'Pos' = ER+/++ or ≥10 fmol/mg.
Numbers: 'all' = no. randomized, 'NK1984' = not known to be dead but no data since 1-1-1984.
Tests for imbalance: * $P < 0.05$; ** $P < 0.01$; *** $P < 0.001$.
? = unknown.

§ List of treatment abbreviations:

-	Control
¦	Versus
A	Carminomycin / Daunorubicin (Cerubidine; Daunoblastin; Daunomycin) / Doxorubicin (Adriamycin; Adriblastin)
Azim	Azimexon
BCG	Bacillus Calmette-Guérin / Methanol extract of BCG (MER)
C	Cyclophosphamide (Cytoxan; Endoxana; Sendoxan)
Chem	Chemotherapy (unspecified)
DES	Diethylstilboestrol
E	4-Epi-Doxorubicin (Epirubicin; Pharmorubicin)
F	5-Fluoro-Uracil (Efudix) / FT-207 / Futraful / Tegafur
Leu	Chlorambucil (Leukeran)
Levam	Levamisole
Lvor	Folinic Acid (Citrovorum factor; Leucovorin; Refolinon; Rescufolin)
M	Methotrexate (Emtexate; Maxtrex)
Mel	Melphalan (Alkeran; L-Pam; L-phenylalanine mustard; Sarkolysin)
Nm	Nitrogen Mustard
Ooph	Oophorectomy / Ovarian irradiation (OvIrr)
P/Pr	Prednisolone / Prednisone (Decortisyl; Econosone)
peri	Perioperative
R	Radiotherapy (DXT; X-Ray; Tele-Cobalt)
s1/s2	Surgery options
T	Tamoxifen (Emblon; Kessar; Noltam; Nolvadex; Tamofen)
Thio	Triethylenephosphoramide (Thiotepa)
V	Vinblastine (Velban; Velbe) / Vincristine (Oncovin)

Note. Many of the treatment agents named above are registered Trade Marks.

APPENDIX TABLE 2M. MORTALITY (i.e. all-cause mortality) in all available trials of adjuvant tamoxifen versus control with no tamoxifen.

(Logrank analyses of yearly rates, excluding deaths after 1 Sep 1985)

Study No.	Study Name	Tamoxifen Dose mg/day	Time years	Chemo both arms	Age <50 at entry Deaths/Entered Tamox	Control	Statistics O-E (odds redctn)	Var	Age 50+ at entry Deaths/Entered Tamox	Control	Statistics O-E (odds redctn)	Var	Sum of Statistics all patients O-E (odds redctn)	Var	Chisq.

(a) Trials of 2 or more years of tamoxifen

75D1	Copenhagen	30	2	–	18/ 68	21/ 60	-3.0	8.7	16/ 44	20/ 41	-3.4	7.7	*	*	
75D2	Copenhagen	30	2	–	0/ 0	0/ 0	0.0	0.0	19/ 52	25/ 52	-4.2	9.2	-10.6	25.5	4.4
76G1	Stockholm B	40	2	–	0/ 2	1/ 1	-0.7	0.2	25/ 82	25/ 71	-1.7	11.4	*	*	
76G1	Stockholm B	40	2	CMF	1/ 3	0/ 0	0.0	0.0	29/ 79	32/ 84	-1.5	13.1	*	*	
76G2	Stockholm B	40	2	–	0/ 0	0/ 2	0.0	0.0	1/ 19	2/ 16	-0.6	0.7	*	*	
76G2	Stockholm B	40	2	CMF	0/ 0	0/ 0	0.0	0.0	1/ 45	1/ 38	-0.1	0.5	*	*	
76G3	Stockholm B	40	2	–	0/ 0	0/ 0	0.0	0.0	4/ 44	2/ 48	1.0	1.4	*	*	
76G4	Stockholm B	40	2	–	0/ 3	0/ 5	0.0	0.0	29/393	26/389	1.5	13.3	*	*	
76G5	Stockholm B	40	2	–	0/ 3	0/ 0	0.0	0.0	2/ 70	4/ 76	-0.8	1.5	-2.9	42.2	0.2
77H	NATO	20	2	–	28/ 69	29/ 57	-3.6	12.5	135/495	179/510	-23.4	72.4	-27.1	84.9	8.6
77K	NSABP B-09	20	2	MelF	156/391	148/398	6.3	68.0	203/559	215/543	-11.7	91.5	-5.4	159.5	0.2
78B1	Toronto-Edmont.	30	2	–	2/ 4	2/ 6	0.3	0.9	45/194	50/196	-1.7	21.8	-1.4	22.7	0.1
78C1	GUN Naples	30	2	–	1/ 3	1/ 9	0.5	0.3	7/ 28	8/ 20	-2.4	3.2	*	*	
78C2	GUN Naples	30	2	–	0/ 4	0/ 3	0.0	0.0	4/ 28	3/ 11	-1.0	1.4	*	*	
78C3	GUN Naples	30	2	–	1/ 15	0/ 15	0.4	0.2	4/ 49	5/ 57	-0.2	2.2	*	*	
78C4	GUN Naples	30	2	–	0/ 2	0/ 14	0.0	0.0	0/ 16	0/ 34	0.0	0.0	*	*	
78C5	GUN Naples	30	2	CMF	11/ 52	11/ 55	0.5	5.2	1/ 8	2/ 10	-0.5	0.7	-2.8	13.2	0.6
78D1	Scottish PilotB	20	5	–	1/ 1	2/ 4	0.3	0.4	19/ 54	27/ 48	-8.0	9.7	*	*	
78D2	Scottish B	20	5+	–	3/ 9	1/ 2	-0.3	0.3	39/175	55/179	-7.7	21.7	*	*	
78D3	Scottish C	20	5+	–	4/ 97	10/ 98	-3.0	3.4	35/281	38/282	-1.1	17.6	*	*	
78D4	Scottish D	20	5+	–	2/ 8	0/ 6	0.8	0.5	5/ 16	6/ 17	-0.5	2.5	*	*	
78D4	Scottish D	20	5+	–	1/ 7	0/ 7	0.5	0.2	2/ 16	2/ 16	0.0	1.0	-19.0	57.6	6.3
78E1	Hong Kong	40	2	–	2/ 14	4/ 20	-0.4	1.4	8/ 25	5/ 23	1.4	3.0	*	*	
78E1	Hong Kong	40	2	CMF	6/ 29	5/ 17	-1.4	2.3	9/ 26	9/ 26	-1.9	2.3	*	*	
78E2	Sri Lanka	40	2	–	2/ 19	2/ 17	-0.1	1.0	2/ 13	1/ 19	0.8	0.7	*	*	
78E2	Sri Lanka	40	2	CMF	0/ 25	1/ 21	-0.5	0.2	3/ 10	0/ 15	1.8	0.7	*	*	
78E3	Bombay	40	2	–	0/ 6	1/ 8	-0.8	0.2	1/ 1	0/ 4	-0.4	0.2	*	*	
78E3	Bombay	40	2	CMF	1/ 11	2/ 8	-0.6	0.7	1/ 1	0/ 4	0.8	0.2	*	*	
78E4	Great Britain	40	2	–	3/ 7	6/ 10	-0.7	1.9	1/ 21	10/ 20	-5.0	2.6	*	*	
78E4	Great Britain	40	2	CMF	4/ 11	2/ 9	0.7	1.4	5/ 20	12/ 20	-4.3	3.8	-10.7	22.4	5.1
78F	CRFB Caen C5	40	3	–	1/ 2	0/ 2	0.5	0.3	23/ 87	32/ 88	-4.9	12.7	-4.4	12.9	1.5
78J	ECOG EST1178	20	2	–	0/ 0	0/ 0	0.0	0.0	24/ 91	33/ 90	-4.9	13.1	-4.9	13.1	1.8
79A	Ghent Univ.	20	2	–	0/ 0	0/ 0	0.0	0.0	5/ 39	8/ 46	-0.9	3.1	-0.9	3.1	0.3
79D1	GABG W. Germany	30	2	AC	13/ 75	12/ 89	1.3	5.8	13/114	20/ 99	-5.9	7.1	*	*	
79D2	HD2 W.Germany	30	2	–	0/ 2	1/ 3	-0.3	0.2	5/ 65	7/ 60	-1.2	2.8	-6.1	15.9	2.3
80B2	N SwedenBCG 192	40	2	–	0/ 0	0/ 0	0.0	0.0	9/ 89	11/ 89	-0.9	4.8	*	*	
80B4	N SwedenBCG 194	40	2	–	0/ 0	0/ 0	0.0	0.0	7/ 24	5/ 25	0.9	2.7	*	*	
80B4	N SwedenBCG 194	40	2	AC	0/ 0	0/ 0	0.0	0.0	2/ 24	5/ 23	-1.4	1.6	-1.3	9.1	0.2
80D1	CRC2	20	2	–	8/142	7/131	0.5	3.7	14/275	23/312	-3.1	9.0	*	*	
80D1	CRC2	20	2	periC	9/134	7/134	1.3	3.9	14/325	22/315	-4.1	6.8	-5.5	25.3	1.2
80E	Toulouse **	30	2	–	0/ 0	0/ 4	0.0	0.0	22/125	20/122	0.8	9.9	0.8	9.9	0.1
81A	Montpellier	30	2	–	0/ 0	0/ 0	0.0	0.0	7/101	9/102	-1.1	3.9	-1.1	3.9	0.3
81B	FB Bordeaux	30	2	CMF	3/ 67	3/ 59	-0.3	1.5	1/ 98	8/101	-3.4	2.2	-3.7	3.7	3.7
82A	NSABP B-14	20	2	–				Data from this trial were not available in 1985							
83B	GROCTA Italy	30	5	CMF:E	1/ 38	0/ 35	0.5	0.2	0/ 63	0/ 59	0.0	0.0	0.5	0.2	0.9
84A1	BMFT02 German	30	2	CMF	0/ 9	0/ 11	0.0	0.0	0/ 12	0/ 16	0.0	0.0	*	*	
84A1	BMFT02 German	30	2	CMF	0/ 8	0/ 8	0.0	0.0	1/ 19	0/ 13	0.4	0.2	0.4	0.2	0.7

(b) Trials of 1 year (or less) of tamoxifen

74G	Case Western A	40	1	CMF	12/ 31	21/ 42	-1.8	6.9	30/ 72	31/ 59	-4.6	12.6	-6.3	19.5	2.0
76F2	Christie B	20	1	–	5/ 12	3/ 9	0.6	1.8	96/270	117/297	-7.2	47.7	-6.5	49.5	0.9
77C1	Danish BCG 77c	30	1	–	1/ 3	0/ 0	0.0	0.0	114/249	129/251	-7.1	52.3	*	*	
77C2	Danish BCG 77c	30	1	–	1/ 7	0/ 1	0.1	0.1	202/627	224/635	-9.3	96.1	-16.3	148.5	1.8
77F1	UK MCCG 009	20	6m	LeuMFV	11/ 33	11/ 31	-1.2	4.8	5/ 31	14/ 42	-2.8	4.2	*	*	
77F1	UK MCCG 009	20	6m	CVF/CV	10/ 28	4/ 27	1.7	3.0	11/ 31	9/ 41	2.8	4.5	0.5	16.5	0.0
77J1	ECOG EST5177	20	1	CMFPr	60/186	57/189	2.3	26.6	5/ 34	11/ 31	-3.7	3.6	*	*	
77J2	ECOG EST6177	20	1	CMFPr	1/ 5	0/ 1	0.2	0.1	30/ 78	37/ 86	-2.3	14.8	-3.6	45.2	0.3
78A2	S Swedish BCG	30	1	–	0/ 1	0/ 2	0.0	0.0	43/238	48/234	-4.9	21.3	-4.9	21.3	1.1
78H	Innsbruck	20	1	–	3/ 31	5/ 27	-1.0	1.9	8/ 96	12/ 80	-3.0	4.8	-4.0	6.7	2.3
78M	Buenos Aires	20	1	CMF				Data from this trial were not available in 1985							
79B1	SWOG 7827A ***	20	1	CMFVPr	/ 14	/ 16		0.2	/190	/208		10.5		10.7	
80J1	Vienna Gyn.	20	1	CMF				Data from this trial were not available in 1985							
80J2	Vienna Gyn.	20	1	–				Data from this trial were not available in 1985							
82B	Danish BCG 82b	30	1	CMF	11/146	7/144	2.0	4.3	7/ 72	8/ 68	-1.1	3.5	0.9	7.8	0.1
83C1	Amsterdam	30	1+	–				Data from this trial were not available in 1985							

Trials of 2+ yrs of tam	any	2+	any	282/1340	279/1326	-1.3	125.7	794/4408	968/4433	-104.6	399.9	-105.9	525.6	21.3
						(1 ± 9)				(23 ± 4)		(18 ± 4)		
Trials of ≤1 yr of tam	any	≤1	any	116/ 497	108/ 489	3.4	49.9	570/1988	665/2032	-45.7	275.8	-42.4	325.7	5.5
						(-7 ± 15)				(15 ± 6)		(12 ± 5)		
Trials in absence of chem	any	any	–	87/ 541	96/ 521	-9.3	40.4	981/4497	1172/4564	-104.4	489.3	-113.7	529.7	24.4
						(21 ± 14)				(19 ± 4)		(19 ± 4)		
Trials in presence of chem	any	any	+	311/1296	291/1294	11.4	135.2	383/1899	461/1901	-46.0	186.4	-34.6	321.6	3.7
						(-9 ± 9)				(22 ± 6)		(10 ± 5)		
Trials of 20mg/day	20	any	any	303/1167	284/1137	5.8	131.7	669/3056	833/3143	-84.2	340.4	-78.4	472.1	13.0
						(-5 ± 9)				(22 ± 5)		(15 ± 4)		
Trials of 30 or 40mg/day	30/40	any	any	95/ 670	103/ 678	-3.7	43.9	695/3340	800/3322	-66.2	335.3	-69.9	379.2	12.9
						(8 ± 14)				(18 ± 5)		(17 ± 5)		
TOTAL: 28 TRIALS	any	any	any	398/1837	387/1815	2.1	175.6	1364/6396	1633/6465	-150.4	675.7	-148.3	851.3	25.8
						(-1 ± 8)				(20 ± 3)		(16 ± 3)		

* Items marked with an asterisk contribute to the totals directly below them
** Included without adjustment, despite significant imbalance in nodal status
*** Blinded (but contributes to totals)

Approximate chi-squared test for heterogeneity on 27 degrees of freedom = 21.5 (NS)

Appendix tables and figure

APPENDIX TABLE 2R. RECURRENCE (i.e. recurrence or prior death) in all available trials of adjuvant tamoxifen versus control with no tamoxifen.
(Logrank analyses of yearly rates, excluding events after 1 Sep 1985)

Study No.	Study Name	Tamoxifen Dose mg/day	Time, years	Chemo both arms	Age <50 at entry Events/Entered Tamox	Control	Statistics O-E (odds redctn)	Var	Age 50+ at entry Events/Entered Tamox	Control	Statistics O-E (odds redctn)	Var	Sum of Statistics all patients O-E	Var	Chisq.
(a) Trials of 2 or more years of tamoxifen															
75D1	Copenhagen	30	2	-	25/ 68	25/ 60	-2.1	11.3	20/ 44	23/ 41	-3.3	8.7	*	*	
75D2	Copenhagen	30	2	-	0/ 0	0/ 0	0.0	0.0	22/ 52	32/ 52	-6.6	11.4	-12.0	31.4	4.6
76G1	Stockholm B	40	2	-	0/ 2	1/ 1	-0.7	0.2	32/ 82	26/ 71	-0.1	13.5	*	*	
76G1	Stockholm B	40	2	CMF	2/ 3	0/ 0	0.0	0.0	39/ 79	53/ 84	-10.6	18.3	*	*	
76G2	Stockholm B	40	2	-	0/ 0	0/ 2	0.0	0.0	1/ 19	3/ 16	-1.2	0.9	*	*	
76G2	Stockholm B	40	2	CMF	0/ 0	0/ 0	0.0	0.0	6/ 45	6/ 38	-0.4	2.8	*	*	
76G3	Stockholm B	40	2	-	0/ 0	0/ 0	0.0	0.0	7/ 44	3/ 48	2.2	2.4	*	*	
76G4	Stockholm B	40	2	-	0/ 3	0/ 5	0.0	0.0	46/393	63/389	-9.5	26.1	*	*	
76G5	Stockholm B	40	2	-	0/ 3	0/ 0	0.0	0.0	2/ 70	7/ 76	-2.2	2.2	-22.5	66.4	7.6
77H	NATO	20	2	-	35/ 69	38/ 57	-8.8	14.4	174/495	237/510	-36.7	92.1	-45.5	106.5	19.4
77K	NSABP B-09	20	2	MelF	204/391	213/398	-2.3	89.4	277/559	324/543	-45.5	126.6	-47.8	216.1	10.6
78B1	Toronto-Edmont.	30	2	-	2/ 4	2/ 6	0.4	0.8	75/194	100/196	-16.6	36.2	-16.2	37.0	7.1
78C1	GUN Naples	30	2	-	1/ 3	3/ 9	-0.1	0.8	10/ 28	14/ 20	-5.4	4.9	*	*	
78C2	GUN Naples	30	2	-	1/ 4	0/ 3	0.4	0.2	7/ 28	4/ 11	-1.0	2.1	*	*	
78C3	GUN Naples	30	2	-	3/ 15	5/ 15	-1.2	1.8	12/ 49	19/ 57	-2.6	7.0	*	*	
78C4	GUN Naples	30	2	-	0/ 2	1/ 14	-0.1	0.1	1/ 16	10/ 34	-2.8	2.2	*	*	
78C5	GUN Naples	30	2	CMF	17/ 52	25/ 55	-4.0	9.1	1/ 8	5/ 10	-1.6	1.3	-18.4	29.4	11.5
78D1	Scottish PilotB	20	5	-	1/ 1	2/ 4	0.6	0.2	20/ 54	36/ 48	-15.3	11.0	*	*	
78D2	Scottish B	20	5+	-	3/ 9	2/ 2	-1.0	0.6	54/175	94/179	-22.7	31.0	*	*	
78D3	Scottish C	20	5+	-	13/ 97	28/ 98	-8.1	9.6	53/281	75/282	-11.4	29.8	*	*	
78D4	Scottish D	20	5+	-	2/ 8	1/ 6	0.3	0.7	6/ 16	8/ 17	-1.4	2.9	*	*	
78D4	Scottish D	20	5+	-	1/ 7	3/ 7	-1.0	0.9	3/ 16	5/ 16	-1.0	1.9	-61.0	88.6	42.0
78E1	Hong Kong	40	2	-	6/ 14	6/ 20	1.1	2.7	10/ 25	9/ 23	0.1	4.2	*	*	
78E1	Hong Kong	40	2	CMF	9/ 29	5/ 17	0.0	2.9	4/ 14	16/ 26	-3.6	3.9	*	*	
78E2	Sri Lanka	40	2	-	7/ 19	4/ 17	1.4	2.3	3/ 13	5/ 19	-0.4	1.7	*	*	
78E2	Sri Lanka	40	2	CMF	2/ 25	4/ 21	-1.1	1.4	3/ 10	1/ 15	1.6	0.8	*	*	
78E3	Bombay	40	2	-	3/ 6	3/ 6	-1.5	0.8	2/ 6	7/ 8	-3.3	1.4	*	*	
78E3	Bombay	40	2	CMF	5/ 11	3/ 8	0.6	1.7	1/ 1	1/ 4	0.8	0.2	*	*	
78E4	Great Britain	40	2	-	5/ 7	6/ 10	0.4	2.2	4/ 21	17/ 20	-9.2	4.0	*	*	
78E4	Great Britain	40	2	CMF	6/ 11	4/ 9	0.8	2.2	11/ 20	13/ 20	-1.8	5.0	-14.2	37.6	5.4
78F	CRFB Caen C5	40	3	-	1/ 2	1/ 2	0.2	0.5	33/ 87	50/ 88	-11.1	17.5	-10.9	17.9	6.7
78J	ECOG EST1178	20	2	-	Recurrence data from this trial were not available in 1985										
79A	Ghent Univ.	20	2	-	Recurrence data from this trial were not available in 1985										
79D1	GABG W. Germany	30	2	AC	19/ 75	24/ 89	-1.3	9.3	28/114	35/ 99	-7.5	12.7	*	*	
79D2	HD 2 W. Germany	30	2	-	1/ 2	1/ 3	0.3	0.5	6/ 65	8/ 60	-1.3	3.4	-9.9	25.8	3.8
80B2	N SwedenBCG 192	40	2	-	0/ 0	0/ 0	0.0	0.0	14/ 89	17/ 89	-1.4	7.3	*	*	
80B4	N SwedenBCG 194	40	2	-	0/ 0	0/ 0	0.0	0.0	9/ 24	10/ 25	-1.2	3.7	*	*	
80B4	N SwedenBCG 194	40	2	AC	0/ 0	0/ 0	0.0	0.0	3/ 24	12/ 23	-4.8	3.3	-7.5	14.3	3.9
80D1	CRC2	20	2	-	25/142	35/131	-6.0	13.4	42/275	72/312	-12.1	26.0	*	*	
80D1	CRC2	20	2	periC	24/134	29/135	-2.6	12.1	41/325	71/315	-17.1	25.9	-37.8	77.5	18.4
80E	Toulouse **	30	2	-	0/ 0	2/ 4	0.0	0.0	29/125	30/122	-0.5	13.6	-0.5	13.6	0.0
81A	Montpellier	30	2	-	0/ 0	0/ 0	0.0	0.0	16/101	22/102	-3.3	8.8	-3.3	8.8	1.2
81B	FB Bordeaux	30	2	CMF	5/ 67	9/ 59	-2.8	3.3	7/ 98	18/101	-5.5	6.0	-8.3	9.3	7.4
82A	NSABP B-14	20	2	-	Data from this trial were not available in 1985										
83B	GROCTA Italy	30	5	CMF;E	2/ 38	3/ 35	-0.6	1.2	3/ 63	6/ 59	-1.7	2.1	-2.3	3.4	1.5
84A1	BMFT 02 Germany	30	2	CMF	Recurrence data from this trial were not available in 1985										
84A1	BMFT 02 Germany	30	2	CMF	Recurrence data from this trial were not available in 1985										
(b) Trials of 1 year (or less) of tamoxifen															
74G	Case Western A	40	1	CMF	14/ 31	27/ 42	-3.7	8.5	39/ 72	39/ 59	-6.8	15.8	-10.5	24.4	4.5
76F2	Christie B	20	1	-	6/ 12	5/ 9	0.3	2.3	124/270	149/297	-10.4	59.2	-10.2	61.5	1.7
77C1	Danish BCG 77c	30	1	-	1/ 3	0/ 0	0.0	0.0	138/249	158/251	-14.3	61.6	*	*	
77C2	Danish BCG 77c	30	1	-	0/ 7	0/ 3	0.2	0.2	251/627	314/635	-34.0	120.6	-48.0	182.4	12.6
77F1	UK MCCG 009	20	6m	LeuMFV	18/ 33	15/ 31	0.2	6.9	11/ 31	25/ 42	-4.8	7.2	*	*	
77F1	UK MCCG 009	20	6m	CVF/CV	17/ 28	10/ 27	2.2	5.6	16/ 31	18/ 41	1.5	6.9	-0.9	26.6	0.0
77J1	ECOG EST5177	20	1	CMFPr	Recurrence data from this trial were not available in 1985										
77J2	ECOG EST6177	20	1	CMFPr	Recurrence data from this trial were not available in 1985										
78A2	S Swedish BCG	30	1	-	0/ 1	0/ 2	0.0	0.0	58/238	72/234	-11.1	29.6	-11.1	29.6	4.1
78H	Innsbruck	20	1	-	7/ 31	9/ 27	-1.4	3.7	22/ 96	31/ 80	-8.0	12.0	-9.5	15.7	5.7
78M	Buenos Aires	20	1	CMF	Data from this trial were not available in 1985										
79B1	SWOG 7827 A ***	20	1	CMFVPr	/ 14	/ 16		1.0	/190	/208		16.2		17.2	
80J1	Vienna Gyn.	20	1	CMF	Data from this trial were not available in 1985										
80J2	Vienna Gyn.	20	1	-	Data from this trial were not available in 1985										
82B	Danish BCG 82b	30	1	CMF	28/146	26/144	0.9	12.0	15/ 72	16/ 68	-1.7	6.7	-0.8	18.7	0.0
83C1	Amsterdam	30	1+	-	Data from this trial were not available in 1985										
Trials of 2+ yrs. of tam		any	2+	any	430/1323	488/1307	-39.1	196.7	1137/4247	1569/4268	-279.0	586.8	-318.1	783.6	129.1
							(18 ± 6)				(38 ± 3)		(33 ± 3)		
Trials of ≤1 yr of tam		any	≤1	any	95/ 306	94/ 299	-1.3	40.2	704/1876	862/1915	-93.7	335.8	-95.0	376.1	24.0
							(3 ± 16)				(24 ± 5)		(22 ± 5)		
Trials in absence of chem		any	any	-	151/ 541	183/ 521	-26.7	70.3	1306/4367	1732/4428	-259.0	661.0	-285.6	731.3	111.5
							(32 ± 10)				(32 ± 3)		(32 ± 3)		
Trials in presence of chem		any	any	+	374/1088	399/1085	-13.7	166.6	535/1756	699/1755	-113.8	261.7	-127.5	428.3	37.9
							(8 ± 7)				(35 ± 5)		(26 ± 4)		
Trials of 20mg/day		20	any	any	358/ 976	392/ 947	-27.7	160.9	873/2814	1185/2890	-189.0	448.8	-216.7	609.7	77.0
							(16 ± 7)				(34 ± 4)		(30 ± 3)		
Trials of 30 or 40mg/day		30/40	any	any	167/ 653	190/ 659	-12.6	76.0	968/3309	1246/3293	-183.8	473.9	-196.4	550.0	70.1
							(15 ± 11)				(32 ± 4)		(30 ± 4)		
TOTAL : 24 TRIALS		any	any	any	525/1629	582/1606	-40.4	237.0	1841/6123	2431/6183	-372.7	922.7	-413.1	1159.6	147.1
							(16 ± 6)				(33 ± 3)		(30 ± 2)		

* Items marked with an asterisk contribute to the totals directly below them
** Included without adjustment, despite significant imbalance in nodal status
*** Results blinded (but contribute to totals)

Approximate chi-squared test for heterogeneity on 23 degrees of freedom = 33.7 (NS)

APPENDIX TABLE 3M. MORTALITY (i.e. all-cause mortality) in all available trials of one or more types of adjuvant chemotherapy versus control with no cytotoxics.

(Logrank analyses of yearly rates, excluding deaths after 1 Sep 1985)

Study No.	Study Name	Type(s) of chemotherapy	Tam to all?	Age <50 at entry Deaths/Entered Chemo	Control	Statistics O-E (odds redctn)	Var	Age 50+ at entry Deaths/Entered Chemo	Control	Statistics O-E (odds redctn)	Var	Sum of Statistics for young and old O-E	Var	Chisq.
57B	Illinois	Nm	no	Data from this trial were not available in 1985										
61B	NSABP B-02	TtF	no	Data from this trial were not available in 1985										
62B1	Berlin-Buch ABC	C	no	Data from this trial were not available in 1985										
62B2	Berlin-Buch ab	C	no	Data from this trial were not available in 1985										
63A	EFSCH Columbia	Tt	no	Trial omitted (Data on inappropriately excluded randomised patients lost)										
67A	Birmingham	C/F	no	38/ 77	13/ 28	1.7	8.7	31/ 80	17/ 55	3.1	9.9	4.8	18.6	1.2
72B	NSABP B-05	Mel	no	24/ 68	38/ 61	-7.2	11.8	73/130	63/121	0.4	26.6	-6.8	38.4	1.2
73B	INT Milan 7205	CMF	no	41/ 95	46/ 75	-8.3	15.4	63/115	59/106	-2.6	23.3	-10.9	38.7	3.1
73C1	Mayo Clinic	CFPr	no	2/ 2	1/ 3	0.8	0.7	6/ 8	3/ 5	0.9	1.5	*		
73C1	Mayo Clinic	CFPr	no	1/ 3	0/ 0	0.0	0.0	4/ 8	1/ 5	1.1	0.9	2.7	3.1	2.3
74A1	Berlin-Buch	various	no	14/ 23	12/ 19	0.8	5.2	20/ 32	26/ 36	0.6	7.7	*		
74A2	Berlin-Buch	various	no	8/ 36	4/ 9	-1.8	1.6	13/ 24	11/ 21	1.4	5.1	1.0	19.7	0.1
74C	Edinburgh II	F	no	39/ 59	42/ 56	-4.3	15.2	69/116	74/114	-4.5	28.0	-8.8	43.2	1.8
74D1	DFCI 74063	AC	no	1/ 2	1/ 1	-0.3	0.2	1/ 2	2/ 3	-0.8	0.5	-1.2	0.7	2.0
74E	UK MCCG 003	CVF/CVM	no	32/ 61	25/ 47	2.1	11.4	42/ 87	49/ 94	-4.0	18.0	-1.8	29.4	0.1
75E1	Guy's L-Pam	Mel	no	17/ 47	25/ 50	-2.8	9.1	46/ 83	37/ 80	4.5	16.4	*		
75E2	Manchester I	CMF/Mel	no	17/ 43	9/ 24	0.1	5.2	27/ 68	9/ 30	2.7	6.8	4.4	37.6	0.5
75G	Northwick Park	MelV	no	7/ 21	8/ 24	0.0	3.4	12/ 44	16/ 51	-0.9	6.2	-0.9	9.7	0.1
75J	King's CRC I	MelM	no	35/ 65	29/ 62	-1.8	13.6	62/137	66/150	1.2	27.2	-0.6	40.8	0.0
76C	Glasgow	CMF	no	16/ 47	18/ 34	-4.8	6.9	28/ 65	31/ 66	-1.0	12.7	-5.9	19.5	1.8
76D	Dublin	F	no	5/ 8	5/ 10	0.9	2.1	7/ 12	6/ 11	0.5	2.7	1.4	4.8	0.4
76E	EORTC 09771	CMF	no	15/ 96	23/ 90	-5.1	8.6	25/132	31/133	-4.3	12.7	-9.4	21.3	4.2
76H1	West Midlands	CMFVALvor	no	48/119	59/119	-6.9	23.0	80/175	75/156	-4.0	33.1	-10.9	56.1	2.1
76H2	West Midlands	LeuMF	no	19/122	20/131	-0.3	9.2	22/163	24/158	-0.9	10.9	-1.2	20.1	0.1
76M	Helsinki	VAC		Data from this trial were not available in 1985										
77B1	Danish BCG 77b	CMF/C	no	50/155	30/ 70	-4.9	15.6	23/ 75	18/ 40	-4.2	8.4	*		
77B2	Danish BCG 77b	CMF/C	no	35/122	14/ 60	2.0	10.1	12/ 43	6/ 26	0.4	3.9	-6.7	38.0	1.2
77E	Oxford	MelMF/Mel	no	30/ 89	19/ 48	-1.9	9.9	52/117	24/ 52	-1.5	13.7	-3.4	23.6	0.5
77G	Vienna Surg.	CMFV	no	9/ 32	9/ 27	-0.8	4.1	13/ 49	18/ 55	-2.0	7.2	-2.8	11.3	0.7
77J2	ECOG EST6177	CMFPr	no	0/ 1	1/ 8	-0.1	0.1	37/ 86	31/ 87	3.1	15.1	3.0	15.2	0.6
78A1	S Swedish BCG	C	no	17/ 98	10/ 90	3.7	6.5	9/ 49	11/ 53	0.2	4.7	3.9	11.2	1.4
78E1	Hong Kong	CMF	yes	6/ 29	2/ 14	0.2	1.6	2/ 14	8/ 25	-1.4	2.1	*		
78E1	Hong Kong	CMF	no	5/ 17	4/ 20	0.8	2.1	9/ 26	5/ 23	2.1	3.3	*		
78E2	Sri Lanka	CMF	yes	0/ 25	2/ 19	-1.1	0.5	3/ 10	2/ 13	0.8	1.0	*		
78E2	Sri Lanka	CMF	no	1/ 21	2/ 17	-0.7	0.7	0/ 15	1/ 19	-0.4	0.2	*		
78E3	Bombay	CMF	yes	1/ 11	0/ 6	0.4	0.2	1/ 1	0/ 6	0.8	0.2	*		
78E3	Bombay	CMF	no	2/ 8	1/ 6	-0.2	0.5	0/ 4	1/ 8	-0.3	0.2	*		
78E4	Great Britain	CMF	no	2/ 9	6/ 10	-1.9	1.8	12/ 20	10/ 20	1.2	4.7	*		
78E4	Great Britain	CMF	yes	4/ 11	3/ 7	-0.2	1.6	5/ 20	1/ 21	2.1	1.5	2.2	22.4	0.2
78K3	Ludwig III	CMF	yes	1/ 2	4/ 4	-0.4	0.5	38/169	49/160	-5.9	19.9	-6.3	20.4	2.0
78P	Helsinki	VAC		Data from this trial were not available in 1985										
79B1	SWOG 7827 A **	CMFVPr	yes	/ 14	/ 13		0.2	/190	/198		8.6		8.8	
79C1	Case Western B	CMFVPr	yes	2/ 17	0/ 19	1.0	0.5	0/ 0	0/ 5	0.0	0.0	*		
79C2	Case Western B	CMFVPr	yes	0/ 2	0/ 1	0.0	0.0	7/ 43	8/ 46	-0.4	3.5	0.7	4.0	0.1
79E1	Guy's CMF	CMF	no	5/ 46	14/ 49	-5.4	4.4	7/ 51	7/ 44	-0.5	3.3	*		
79E2	Manchester II	CMF	no	4/ 17	5/ 11	-1.7	1.9	6/ 22	6/ 28	0.9	2.8	-6.7	12.4	3.6
80A	MD Anderson8026	MV	yes	16/ 51	6/ 63	-2.1	1.7	5/ 62	5/ 59	0.0	2.4	-2.2	4.1	1.1
80B3	N SwedenBCG 193	AC	yes	0/ 9	3/ 11	-1.1	0.7	1/ 6	0/ 3	0.3	0.2	*		
80B3	N SwedenBCG 193	AC	no	1/ 7	4/ 11	-1.0	1.1	2/ 7	0/ 2	0.8	0.4	*		
80B4	N SwedenBCG 194	AC	no	0/ 0	0/ 0	0.0	0.0	5/ 23	5/ 25	-0.1	2.4	*		
80B4	N SwedenBCG 194	AC	yes	0/ 0	0/ 0	0.0	0.0	2/ 24	7/ 24	-2.4	2.1	-3.7	6.9	2.0
80C2	SE Sweden BCG B	AC	no	0/ 0	1/ 1	0.0	0.0	4/ 21	7/ 21	-1.3	2.6	-1.3	2.6	0.7
80F	INT Milan 8004	CMF	no	0/ 26	5/ 29	-2.2	1.2	0/ 21	4/ 18	-2.3	0.9	-4.5	2.1	9.6
80G	EST1180/SW.8294	CMFPr	no	Data from this trial were not available in 1985										
80J1	Vienna Gyn.	CMF	yes	Data from this trial were not available in 1985										
80L	SWOG 7985	CMFV	no	Data from this trial were not available in 1985										
81E	NSABP B-13	M/F + Lvor	no	Data from this trial were not available in 1985										
82C	Danish BCG 82c	CMF	yes	0/ 4	0/ 1	0.0	0.0	15/230	18/232	-1.5	7.9	-1.5	7.9	0.3
82F2	NCCTG-773051	CFPr	no	Data from this trial were not available in 1985										
82G	MD Anderson8227	FACVPr	no	Data from this trial were not available in 1985										
83B	GROCTA Italy	CMF then E	yes	1/ 38	1/ 39	0.0	0.5	0/ 63	0/ 47	0.0	0.0	0.0	0.5	0.0
Trials in absence of tam		any	no	540/1662	503/1300	-51.6	211.8	822/2110	756/1916	-10.9	323.8	-62.5	535.5	7.3
						(22 ± 6)				(3 ± 5)		(11 ± 4)		
Trials in presence of tam		polychem***	yes	17/ 213	21/ 197	-3.0	8.0	98/ 832	115/ 839	-6.2	49.3	-9.2	57.3	1.5
						(31 ± 30)				(12 ± 13)		(15 ± 12)		
TOTAL : 32 TRIALS		any	yes/no	557/1875	524/1497	-54.6	219.8	920/2942	871/2755	-17.1	373.1	-71.7	592.9	8.7
						(22 ± 6)				(4 ± 5)		(11 ± 4)		

* Items marked with an asterisk contribute to the totals directly below them
** Results blinded (but contribute towards totals)
*** No trials are available of single-agent chemotherapy in the presence of tamoxifen

Approximate chi-squared test for heterogeneity on 31 degrees of freedom = 36.5 (NS)

APPENDIX TABLE 3R. RECURRENCE (i.e. recurrence or prior death) in all available trials of one or more types of adjuvant chemotherapy versus control with no cytotoxics.

(Logrank analyses of yearly rates, excluding events after 1 Sep 1985)

Study No.	Study Name	Type(s) of chemotherapy	Tam to all?	Age <50 at entry Events/Entered Chemo / Control		Statistics O-E (odds redctn)	Var	Age 50+ at entry Events/Entered Chemo / Control		Statistics O-E (odds redctn)	Var	Sum of Statistics for young and old O-E (odds redctn)	Var	Chisq.
57B	Illinois	Nm	no	Data from this trial were not available in 1985										
61B	NSABP B-02	TtF	no	Data from this trial were not available in 1985										
62B1	Berlin-Buch ABC	C	no	Data from this trial were not available in 1985										
62B2	Berlin-Buch ab	C	no	Data from this trial were not available in 1985										
63A	EFSCH Columbia	Tt		Trial omitted (Data on inappropriately excluded randomised patients lost)										
67A	Birmingham	C/F	no	41/ 77	16/ 28	-0.6	9.4	35/ 80	28/ 55	-2.4	12.9	-3.0	22.3	0.4
72B	NSABP B-05	Mel	no	Recurrence data from this trial were not available in 1985										
73B	INT Milan 7205	CMF	no	52/ 95	55/ 75	-14.2	20.2	77/115	76/106	-7.4	29.1	-21.6	49.3	9.5
73C1	Mayo Clinic	CFPr	no	2/ 2	2/ 3	-0.1	0.6	7/ 8	5/ 5	-0.6	2.0	*	*	
73C1	Mayo Clinic	CFPr	no	1/ 3	0/ 0	0.0	0.0	4/ 8	3/ 5	-0.4	1.5	-1.2	4.1	0.3
74A1	Berlin-Buch	various	no	17/ 23	16/ 19	-2.3	6.0	22/ 32	30/ 36	-4.3	9.7	*	*	
74A2	Berlin-Buch	various	no	9/ 36	4/ 9	-1.6	1.8	15/ 24	15/ 21	-1.0	5.7	-9.2	23.2	3.6
74C	Edinburgh II	F	no	46/ 59	46/ 56	-4.8	16.3	83/116	87/114	-5.4	32.5	-10.2	48.9	2.1
74D1	DFCI 74063	AC	no	1/ 2	1/ 1	-0.2	0.5	1/ 2	2/ 3	-1.1	0.5	-1.2	0.9	1.6
74E	UK MCCG 003	CVF/CVM	no	38/ 61	30/ 47	-0.9	13.1	53/ 87	64/ 94	-7.4	22.9	-8.4	36.0	1.9
75E1	Guy's L-Pam	F	no	20/ 47	28/ 50	-4.7	10.2	58/ 83	47/ 80	3.8	21.2	*	*	
75E2	Manchester I	CMF/Mel	no	22/ 43	15/ 24	-3.3	6.4	35/ 68	13/ 30	1.4	8.8	-2.8	46.6	0.2
75G	Northwick Park	MelV	no	10/ 21	9/ 24	1.3	4.3	15/ 44	21/ 51	-2.1	8.2	-0.7	12.5	0.0
75J	King's CRC I	MelM	no	42/ 85	41/ 62	-9.0	16.3	85/137	95/150	-1.9	35.5	-10.9	51.8	2.3
76C	Glasgow	CMF	no	20/ 47	19/ 34	-4.3	7.3	39/ 65	41/ 66	-2.6	16.0	-6.9	23.4	2.0
76D	Dublin	F	no	5/ 8	6/ 10	0.3	2.2	9/ 12	6/ 11	1.4	3.1	1.7	5.2	0.5
76E	EORTC 09771	CMF	no	30/ 96	42/ 90	-9.2	15.4	46/132	59/133	-11.7	21.9	-20.9	37.3	11.7
76H1	West Midlands	CMFVALvor	no	70/119	88/119	-15.4	29.9	112/175	112/156	-13.3	41.7	-28.7	71.6	11.5
76H2	West Midlands	LeuMF	no	35/122	36/131	0.4	16.5	45/163	47/158	-2.4	21.3	-2.1	37.8	0.1
76M	Helsinki	VAC	no	Data from this trial were not available in 1985										
77B1	Danish BCG 77b	CMF/C	no	60/155	44/ 70	-14.5	18.6	27/ 75	24/ 40	-7.3	10.0	*	*	
77B2	Danish BCG 77b	CMF/C	no	42/122	23/ 60	-2.8	12.4	18/ 43	10/ 26	-0.3	5.7	-24.8	46.8	13.1
77E	Oxford	MelMF/Mel	no	39/ 89	27/ 48	-5.1	12.5	61/117	33/ 52	-5.3	16.6	-10.4	29.1	3.7
77G	Vienna Surg.	CMFV	no	13/ 32	14/ 27	-2.2	5.8	18/ 49	23/ 55	-2.1	9.2	-4.2	15.1	1.2
77J2	ECOG EST6177	CMFPr	no	Recurrence data from this trial were not available in 1985										
78A1	S Swedish BCG	C	no	23/ 98	24/ 90	-0.7	10.9	15/ 49	15/ 53	1.9	6.8	1.2	17.6	0.1
78E1	Hong Kong	CMF	yes	9/ 29	6/ 14	-1.7	2.8	4/ 14	10/ 25	-0.7	2.8	*	*	
78E1	Hong Kong	CMF	no	5/ 17	6/ 20	-0.3	2.5	16/ 26	9/ 23	4.3	5.2	*	*	
78E2	Sri Lanka	CMF	yes	2/ 25	7/ 19	-3.1	1.9	3/ 10	3/ 13	0.6	1.2	*	*	
78E2	Sri Lanka	CMF	no	4/ 21	4/ 17	-0.5	1.8	1/ 15	5/ 19	-1.6	1.3	*	*	
78E3	Bombay	CMF	yes	5/ 11	3/ 6	0.3	1.7	1/ 1	2/ 6	0.9	0.1	*	*	
78E3	Bombay	CMF	no	3/ 8	3/ 6	-1.3	0.9	1/ 4	7/ 8	-3.1	1.3	*	*	
78E4	Great Britain	CMF	no	4/ 9	6/ 10	-1.2	2.1	13/ 20	17/ 20	-4.3	5.4	*	*	
78E4	Great Britain	CMF	yes	6/ 11	5/ 7	-0.9	2.2	11/ 20	4/ 21	4.1	3.4	-8.5	36.7	2.0
78K3	Ludwig III	CMF	yes	Recurrence data from this trial were not available in 1985										
78P	Helsinki	VAC	no	Data from this trial were not available in 1985										
79B1	SWOG 7827 A **	CMFVPr	yes	/ 14	/ 13		0.9	/190	/198		15.7		16.6	
79C1	Case Western B	CMFVPr	yes	4/ 17	4/ 19	0.5	1.8	0/ 0	0/ 5	0.0	0.0	*	*	
79C2	Case Western B	CMFVPr	yes	0/ 2	1/ 1	-0.5	0.3	8/ 43	16/ 46	-4.1	5.5	-4.2	7.5	2.3
79E1	Guy's CMF	CMF	no	8/ 46	23/ 49	-9.4	6.5	13/ 51	13/ 44	-1.1	5.6	*	*	
79E2	Manchester II	CMF	no	5/ 17	6/ 11	-2.1	2.1	9/ 22	10/ 28	0.4	4.0	-12.3	18.4	8.2
80A	MD Anderson8026	MV	yes	8/ 51	13/ 63	-1.8	4.7	13/ 62	17/ 59	-2.5	6.8	-4.3	11.5	1.6
80B3	N SwedenBCG 193	AC	yes	1/ 9	4/ 11	-1.0	1.1	3/ 6	0/ 3	0.8	0.5	*	*	
80B3	N SwedenBCG 193	AC	no	2/ 7	7/ 11	-2.2	1.8	4/ 7	0/ 2	1.3	0.8	*	*	
80B4	N SwedenBCG 194	AC	no	0/ 0	0/ 0	0.0	0.0	12/ 23	10/ 25	-0.8	3.9	*	*	
80B4	N SwedenBCG 194	AC	yes	0/ 0	0/ 0	0.0	0.0	3/ 24	9/ 24	-3.2	2.7	-4.9	10.9	2.2
80C2	SE Sweden BCG B	AC	no	0/ 0	1/ 1	0.0	0.0	5/ 21	8/ 21	-1.5	2.9	-1.5	2.9	0.7
80F	INT Milan 8004	CMF	no	2/ 26	11/ 29	-4.2	2.9	1/ 21	8/ 18	-4.2	2.0	-8.4	4.9	14.5
80G	EST1180/SW.8294	CMFPr	no	Data from this trial were not available in 1985										
80J1	Vienna Gyn.	CMF	yes	Data from this trial were not available in 1985										
80L	SWOG 7985	CMFV	no	Data from this trial were not available in 1985										
81E	NSABP B-13	M/F + Lvor	no	Data from this trial were not available in 1985										
82C	Danish BCG 82c	CMF	yes	0/ 4	0/ 1	0.0	0.0	34/230	43/232	-5.1	17.2	-5.1	17.2	1.5
82F2	NCCTG-773051	CFPr	no	Data from this trial were not available in 1985										
82G	MD Anderson8227	FACVPr	no	Data from this trial were not available in 1985										
83B	GROCTA Italy	CMF then E	yes	2/ 38	2/ 39	0.0	1.0	3/ 63	1/ 47	0.7	1.0	0.7	1.9	0.3
Trials in absence of tam		any	no	671/1593	653/1231	-115.2 (36 ±	257.4 5)	955/1894	943/1708	-80.9 (19 ±	375.3 5)	-196.1 (27 ±	632.7 3)	60.8
Trials in presence of tam		polychem***	yes	39/ 211	47/ 193	-8.2 (36 ±	18.5 19)	113/ 663	143/ 679	-12.1 (19 ±	56.7 12)	-20.4 (24 ±	75.1 10)	5.5
TOTAL : 29 TRIALS		any	yes/no	710/1804	700/1424	-123.4 (36 ±	275.9 5)	1068/2557	1086/2387	-93.0 (19 ±	432.0 4)	-216.4 (26 ±	707.9 3)	66.2

* Items marked with an asterisk contribute to the totals directly below them
** Results blinded (but contribute towards totals)
*** No trials are available of single-agent chemotherapy in the presence of tamoxifen

Approximate chi-squared test for heterogeneity on 28 degrees of freedom = 34.0 (NS)

APPENDIX FIGURE 1F.—Front of 1985 data form. Those who sent individual patient data either on forms or on magnetic tapes were asked to use this format.

BCTCS/NCI/UICC/WHO overview of mortality by randomly allocated treatment in early breast cancer trials: provision of one line of CONFIDENTIAL data for each patient ever randomised (INCLUDING any ineligible, withdrawn, unevaluable, lost or "protocol deviant" patients).

Name of trialist or trial group who owns these data (see overleaf): .... Data Sheet No.. . . . . . .

Name of trial: .... Convention used for dates (mark one box): EITHER |_| Day,Month,Year, OR |_| Month,Day,Year

Convention used for numbering treatment groups (enter suitable abbreviations):

Treatment group 1 = ... , Trt. gp. 2 = ... , Trt. gp. 3 = ... , Trt. gp. 4 = ...

Patient number	ESSENTIAL MINIMUM DATA (coded as overleaf)										OPTIONAL EXTRA DATA (coded as overleaf)							
	Trt. gp.	Date randomised	Entry age	Dead/ other	Date died/ last traced	Axi- lla *	Men- op. *	Ovary ablat. *			Sur- gery	Est Rec *	Com- ply? *	Tt. on rlpse *	Other death *	2nd lry *	Re- cur? *	Approx. date of 1st recur.	Dist. recur.?	Approx. date distant recur.

* Leave these columns BLANK if not applicable or if data not (or not yet) conveniently available

USE OF "ACTION" PAPER: Please PRESS HARD (perhaps using a ballpoint pen) so that when you send off the top copy (to the Clinical Trial Service Unit, Radcliffe Infirmary, Oxford, England) you will be able, if you wish, to keep the second copy for your self. (Also, please insert some thick card under the second copy to protect the lower forms in the pad !)

Appendix tables and figure

APPENDIX FIGURE 1B.—Back of the 1985 data form, suggesting various coding conventions.

GUARANTEE OF CONFIDENTIALITY OF DATA: ANY DATA PROVIDED OVERLEAF TO THE BCTCS/NCI/UICC/WHO SECRETARIAT WILL BE HELD SECURELY AND IN STRICT CONFIDENCE AND WILL REMAIN THE PROPERTY OF THE TRIALISTS SUPPLYING IT UNTIL THEY CHOOSE TO PUBLISH IT.

NOTES ON FORMAT OF DATA REQUESTED OVERLEAF: Please accompany these forms by an explanatory letter about any special coding conventions (e.g. on menopausal status, ER or compliance) you have used, plus notes on any special features of the study(s) you wish to draw particular attention to.

+ Dates that are not (or not yet) known exactly:
  either leave DAY blank, and give (approximate or provisional) month and year,
  or leave DAY AND MONTH blank, and just give approximate year.

* Items marked with an asterisk are to be left blank if not applicable, or if data are not yet conveniently available.

ESSENTIAL MINIMUM DATA:

Patient number: Any convenient convention you wish, in case any correspondence becomes necessary. (If reporting several trials, please try to use a numbering system that implicitly specifies both the trial and the patient.)

Trt. gp. : Treatment group number: 1 or 2 only, for 2-group trials, or a wider range for trials with more arms, as defined by you at the top of the form. NB: Even if, in reality, some quite different (or even opposite !) treatment was inadvertantly given, what is wanted is the originally allocated treatment. (For patients erroneously entered more than once, give only the first allocation.)

Date randomised & Entry age: Please describe ALL patients EVER randomised, including even lost or withdrawn patients, and ignore all non-randomised patients.

Dead/other: 1=known to be dead, 2=alive or lost, but alive when last traced.

Date died/last traced: Date of death, or date last known to be alive, as accurately as possible. (See above note on approximate dates)

Axilla (status on entry) 0 or blank = not (yet) conveniently available; 1, 2, 3, = axillary clearance (1 = N0, 2 = N1-3, 3 = N4+); 4, 5 = axillary sample only (4 = N-, 5 = N+); 6, 7, = no sample (6 = clinically N-, 7 = clinically N+).

Menop. (status on entry): 1=pre-, 2=peri-, 3=post-menopausal. Use your own definitions of menopausal status at entry.

Ovary ablat.: 1=Bilateral ovarian ablation (surgical or otherwise) before, at, or shortly after, entry; 2=other.

OPTIONAL EXTRA DATA

Surgery: Type of mastectomy first attempted: 1=radical, 2=total, 3=simple, 4=partial, 5=none.

Est Rec: Estrogen receptor status: If exact value is not available use 1=negative, 2=marginal, 3=positive, otherwise use 4, 5, 6, 7, 8, 9, for 10, 10-19, 20-29, 30-49, 50-99 and 100+ fmds/mg protein respectively.

Comply? : Compliance before any relapse: 1=poor, 2=mid, 3=good compliance, at least until relapse. Use your own definitions: "mid" should involve some appreciable compliance, but some appreciable non-compliance.

Tt. on rlse: Treatment after relapse: 1=hormonal treatment after relapse, 2=cytotoxic treatment attempted vigorously after relapse, 3=both hormonal and vigorous cytotoxic treatment attempted.

Other death: Underlying cause of death, if before first relapse the patient died of another cause while no overt breast cancer was apparent: 0 or blank=not applicable, 1=acute treatment toxicity, 2=(other) infective, 3=acute leukaemia, 4=other 2nd neoplasm, 5=heart, 6=thrombotic or embolic, 7=other vascular, 8=suicide, 9=all other causes (except, of course, breast cancer).

2nd 1ry (i.e. 2nd primary in opposite breast).

Recur?: 1=no, 2=already apparent at around time of entry, 3=arose later.

Approx. date of 1st recur?: 1=no recurrence yet recorded, 2=some recurrence (distant or local or ipsilateral 2nd 1ry)

Dist. recur.: Give the best estimate you can: see above note on approximate dates

Approx. date dist. recur.: Any distant recurrence? 1=no, 2=yes; ignore local disease and regional nodes.
Give the best estimate you can: see note on approximate dates.

Appendix: tables and figures

Short Reports

TITLE: OVARIAN IRRADIATION AS ADJUVANT THERAPY FOR PRIMARY BREAST CANCER

STUDY NUMBER: 57A1-2
STUDY TITLE: NORWEGIAN RH

ABBREVIATED TRIAL NAME: NORWEGIAN RH
STUDY NUMBER: 57A1-2

INVESTIGATORS' NAMES AND AFFILIATIONS:
R Nissen-Meyer,
Norwegian Radium Hospital, Oslo 2, Norway

TRIAL DESIGN: TREATMENT GROUPS:
1. Ov Irr: Ovarian irradiation to a dose of 1000 rads over a period of 6 days following primary surgery.
2. Nil: Ovarian irradiation at the time of first recurrence.

ELIGIBILITY CRITERIA FOR ENTRY:
1. AGE: No restriction.
2. CLINICAL STAGES: Stages I + II. For premenopausal patients, the size, site and histological grading of the tumor were also assessed, and patients considered to have a good prognosis were randomized. The other premenopausal patients were offered ovarian irradiation without randomization and are not therefore a part of this trial.
3. PRIMARY TREATMENT: Mastectomy and axillary dissection or sampling, postoperative local irradiation to the supraclavicular (3600 rads), axillary (3600+1800 rads) and chest wall (3600 rads) over 5 weeks.
4. NODE DISSECTION: All premenopausal women had axillary clearance. All postmenopausal women had axillary clearance (138) or axillary sampling (38) except one.
5. NUMBER OF NODES REMOVED: Not specified.
6. MENSTRUAL STATUS: No restriction.

METHOD OF RANDOMIZATION:
Telephone to a central office where treatment was allocated using random number lists. Treatment allocation was not released until patient identifiers were recorded in the central office.

RANDOMIZATION STRATIFICATIONS:
Premenopausal, postmenopausal.

YEARS OF ACCRUAL:
57A1 (premenopausal): Nov. 1957 - Dec. 1963
57A2 (postmenopausal): Dec. 1957 - June 1961

TOTAL NUMBER OF PATIENTS RANDOMIZED:
346 (169 premenopausal, 177 postmenopausal) (Premenopausal N0 = 158, N1-3 = 10, N4 = 1; postmenopausal N0 = 80, N1-3 = 58, N4 = 38, unknown = 1)

AIM, RATIONALE, SPECIAL FEATURES:
Because of the known beneficial effects of ovarian irradiation in women with metastatic breast cancer and because several uncontrolled series (Taylor 1939, Horsley 1944) had suggested an improvement with prophylactic castration, and in spite of one historically controlled study suggesting no advantage to prophylactic castration (McWhirter, 1956), this study was designed in 1957 to test the effect of primary ovarian irradiation in operable breast cancer in a randomized controlled setting and to study the differential effect of prophylactic castration in different age groups and prognostic groups of women.

Urinary excretion of oestrone and pregnanediol was studied and surprisingly was found to be present for many years after the menopause. This excretion decreased after ovarian irradiation even in women who were already naturally postmenopausal (Nissen-Meyer and Sanner 1963).

DISCUSSION:
Relapse rates were significantly reduced by ovarian irradiation in the entire group and in both premenopausal (p=0.05) and postmenopausal groups (p=0.01). Overall survival was also significantly prolonged in the postmenopausal group (p=0.01) but not significantly so in premenopausal women, although there was a trend to fewer deaths in premenopausal women who received ovarian irradiation. In the postmenopausal group, a beneficial effect appeared to be present for node negative and node positive cases and in those under and those over 60 years. The finding of a significant reduction in recurrence and prolongation in the survival of postmenopausal women is unique to this trial and may relate to the play of chance or to the interesting findings of persistent urinary excretion of oestrone and pregnanediol in the postmenopausal women in this study and the reduction of that excretion by ovarian irradiation.

PUBLICATIONS:
Nissen-Meyer R, Sanner T. The excretion of oestrone, pregnanediol and pregnanetriol in breast cancer patients. I Excretion after spontaneous menopause. Acta Endocrinologica 1963; 44: 325-333.

Nissen-Meyer R, Sanner T. The excretion of oestrone, pregnanediol and pregnanetriol in breast cancer patients. II. Effects of ovariectomy, ovarian irradiation and corticosteroids. Acta Endocrinologica 1963; 44: 334-345.

Nissen-Meyer R. Castration as part of the primary treatment for operable female breast cancer. A statistical evaluation of clinical results. Acta Radiologica Supplementum 1965; 249.

Nissen-Meyer R. Ovarian suppression and its supplement by additive hormonal treatment. INSERM 1975; 55: 151-158.

Table 1. Outcome by allocated treatment for trial 57A1 (Norwegian RH)
Key: T = total free of relevant event at start of relevant year(s)

(i) Death (D) in each separate year,
subdivided by age when randomized (D <50 & D 50+ / T <50 & T 50+)

Year	1. OvIrr	2. Nil
1	1+ 0/64+ 12	2+ 0/80+ 13
2	1+ 1/63+ 12	1+ 0/78+ 13
3	1+ 0/62+ 11	0+ 0/77+ 13
4	0+ 0/61+ 11	2+ 0/77+ 13
5+	9+ 0/61+ 10	10+ 3/75+ 13

(ii) First recurrence or prior death (F) in each separate year,
subdivided by age when randomized (F <50 & F 50+ / T <50 & T 50+)

Year	1. OvIrr	2. Nil
1	2+ 1/64+ 12	5+ 0/80+ 13
2	1+ 1/62+ 11	1+ 0/75+ 13
3	0+ 0/61+ 10	3+ 0/74+ 13
4	1+ 0/61+ 10	3+ 1/71+ 13
5+	8+ 1/60+ 10	9+ 2/68+ 12

(iii) Recurrence with survival (R) and Death (D) in all years together,
subdivided by various characteristics when randomized (R & D / T)

Age at Entry	1. OvIrr	2. Nil
< 40	0+1/11	1+7/16
40 - 49	0+11/53	5+8/64
50 - 59	1+2/12	0+3/13
Any age (including "not known")	1+14/76	6+18/93

	Nodal Status (N0, N1-3, N4+)		
< 50	N0 (by axillary	0+12/60	6+12/73
50 +	clearance)	1+2/12	0+3/13
< 50	N1-N3 (by axillary clearance)	0+0/3	0+3/7
< 50	N4+ (by axillary sample/clearance)	0+0/1	0+0/0

Table 1. Outcome by allocated treatment for trial 57A2 (Norwegian RH)
Key: T = total free of relevant event at start of relevant year(s)

(i) Death (D) in each separate year,
subdivided by age when randomized (D <50 & D 50+ / T <50 & T 50+)

Year	1. OvIrr	2. Nil
1	0+ 0/4+ 86	0+ 11/3+ 84
2	0+ 10/4+ 86	2+ 8/3+ 73
3	1+ 8/4+ 76	0+ 6/1+ 65
4	0+ 3/3+ 68	0+ 4/1+ 59
5+	1+ 38/3+ 65	1+ 28/1+ 55

(ii) First recurrence or prior death (F) in each separate year,
subdivided by age when randomized (F <50 & F 50+ / T <50 & T 50+)

Year	1. OvIrr	2. Nil
1	0+ 6/4+ 86	1+ 18/3+ 84
2	1+ 9/4+ 80	1+ 9/2+ 66
3	0+ 8/3+ 71	0+ 3/1+ 57
4	0+ 7/3+ 63	0+ 7/1+ 54
5+	1+ 32/3+ 56	1+ 23/1+ 47

(iii) Recurrence with survival (R) and Death (D) in all years together,
subdivided by various characteristics when randomized (R & D / T)

Age at Entry	1. OvIrr	2. Nil
40 - 49	0+2/4	0+3/3
50 - 59	2+29/42	2+23/38
60 - 69	1+28/42	1+28/39
70 +	0+2/2	0+6/7
Any age (including "not known")	3+61/90	3+60/87

	Nodal Status (N0, N1-3, N4+)		
< 50	N0 (by axillary	0+0/1	0+1/1
50 +	clearance)	3+19/39	2+19/39
< 50	N1-N3 (by axillary	0+2/3	0+1/1
50 +	clearance)	0+21/26	1+22/28
< 50	N4+ (by axillary	0+0/0	0+1/1
50 +	sample/clearance)	0+19/20	0+16/17
50 +	Remainder (i.e. other N+ or N?)	0+0/1	0+0/0

Estrogen Receptors (ER, fmol/mg)

Breakdown by ER status was not available.

TITLE: THE ONTARIO CANCER TREATMENT AND RESEARCH FOUNDATION CLINICAL TRIAL ON THE COMPARATIVE EFFECT OF OVARIAN IRRADIATION IN CARCINOMA OF THE BREAST IN THE POSTMENOPAUSAL PATIENT

STUDY NUMBER: 60A
STUDY TITLE: ONTARIO CTRF

ABBREVIATED TRIAL NAME: ONTARIO CTRF
STUDY NUMBER: 60A

INVESTIGATORS' NAMES AND AFFILIATIONS:
EA Clarke (1), JC Fetterly (2), NC Ryan (1).

(1) Division Epidemiology and Statistics, Ontario Cancer Treatment and Research Foundation, Toronto, Ontario, Canada.
(2) Cancer Control Agency of British Columbia, Victoria, British Columbia, Canada.

TRIAL DESIGN: TREATMENT GROUPS:
1. Ov Irr: Ovarian irradiation, 1500 rads to the center of the pelvis, given in 5 treat-ments, over 5-10 days, using a field size 10-12 cm longitudinal by 15-16 cm transverse, by either cobalt 60 or 250 Kv.
2. Nil: No ovarian irradiation.

ELIGIBILITY CRITERIA FOR ENTRY:
1. AGE: < 70.
2. CLINICAL STAGES: I, II and III (TNM Staging).
3. PRIMARY TREATMENT: Local excision, simple or radical mastectomy, plus irradiation of the breast, chest wall and/or draining nodal areas, according to "local center policy".
4. AXILLARY NODE DISSECTION: Optional.
5. NUMBER OF NODES REMOVED: Not specified.
6. MENSTRUAL STATUS: 12 or more months postmenopausal.

METHOD OF RANDOMIZATION:
Telephone to a central office at the Ontario Cancer Foundation Centre in London, Ontario where a clinical trials secretary drew an envelope containing an allocation card. Allocation cards were pre-numbered for each center and for each stage within center. The allocations on the cards had previously been made by the use of a random number table. All patient allocations given out were also recorded in a trial logbook.

RANDOMIZATION STRATIFICATIONS:
By center, and within center by stage of disease.

YEARS OF ACCRUAL: 1968-1977.

TOTAL NUMBER OF PATIENTS RANDOMIZED: 332; 169 to ovarian ablation, 163 to no further treatment.

AIM, RATIONALE, SPECIAL FEATURES:
This multi-center trial was undertaken to determine the effectiveness of ovarian ablation in delaying the recurrence of disease and improving survival in post-menopausal women with breast carcinoma.

ADDITIONAL ELIGIBILITY OR STRATIFICATION CRITERIA:
Patients were identified by the six Regional Treatment Centers of the Ontario Cancer Treatment and Research Foundation situated in Hamilton, Kingston, Ottawa, Thunder Bay, London and Windsor. Patients were excluded from randomization if they presented to the Regional Centre more than three months after primary surgery, if they had had a breast biopsy only, if they had a previous history of bilateral oophorectomy, if they had a hematologic, psychiatric, or other disorder considered to be a contra-indication to pelvic irradiation or if they had a Papanicolaou smear suggesting pelvic carcinoma on admission to the center.

Nodal status was felt to be negative clinically in 128 patients who did not receive axillary clearance or sampling as part of their primary surgery. Nodal status was felt clinically or documented surgically to be positive in 204 patients but the numbers of positive nodes in these women were not routinely recorded at the time this study was being done. Thus, information concerning nodal status is quite incomplete.

Estrogen and progesterone receptors were not being measured in Ontario at the time this study was being done.

Nine women were under age 50; 151 were age 50-59, 166 age 60-69 and 6 age 70 or over. Age and nodal distributions were similar between control and treatment groups.

FOLLOW-UP, COMPLIANCE AND TOXICITY:
Of the 169 assigned to ovarian ablation, 25 did not receive this therapy due to oversight. One control patient received ovarian ablation as a result of a misunderstanding of the treatment code allocated.

When the initial treatment was completed, a clinical abstract was mailed to the central site (London centre). Each year, on the anniversary of the patient's surgery, a follow-up form was completed by each center, listing date and site of recurrence of disease or metastases and the date and cause of death. Patients were thus actively followed from entry to 1977 when study entry ended.

Further follow-up was undertaken in 1985 using the source records of the Ontario Cancer Registry. This, together with the assistance of the centres, resulted in all but one patient being followed to January 1, 1984.

DISCUSSION:
The results of this study suggest that ovarian ablation by radiation is of little value in post-menopausal women.

PUBLICATIONS: None.

Table 1. Outcome by allocated treatment for trial 60A (Ontario CTRF)
Key: T = total free of relevant event at start of relevant year(s)

(i) Death (D) in each separate year,
subdivided by age when randomized (D <50 & D 50+ / T <50 & T 50+)

Year	1. OvIrr	2. Nil
1	2+ 6/ 6+ 163	0+ 12/ 3+ 160
2	1+ 17/ 4+ 157	0+ 18/ 3+ 148
3	0+ 19/ 3+ 140	0+ 13/ 3+ 130
4	0+ 12/ 3+ 121	0+ 9/ 3+ 117
5+	2+ 56/ 3+ 109	2+ 54/ 3+ 108

(ii) First recurrence or prior death (F) in each separate year,
subdivided by age when randomized (F <50 & F 50+ / T <50 & T 50+)

Year	1. OvIrr	2. Nil

(recurrence data not available)

(iii) Recurrence with survival (R) and Death (D) in all years together,
subdivided by various characteristics when randomized (R & D / T)

Age at Entry	1. OvIrr	2. Nil
40-49	0+5/6	0+2/3
50-59	0+45/73	0+46/78
60-69	0+63/86	0+58/80
70+	0+2/4	0+2/2
Any age (including "not known")	0+115/169	0+108/163

Nodal Status (N0, N1-3, N4+)

		1. OvIrr	2. Nil
<50	N0 (by clinical evidence only)	0+1/2	0+1/2
50+		0+45/68	0+32/56
<50	Remainder (i.e. other N+ or N?)	0+4/4	0+1/1
50+		0+65/95	0+74/104

Estrogen Receptors (ER, fmol/mg)

Breakdown by ER status was not available.

Short reports

TITLE: THE NORWEGIAN TRIAL TO ASSESS THE ROLE OF POST-OPERATIVE RADIOTHERAPY IN BREAST CANCER

STUDY NUMBER: 64B1-2
STUDY TITLE: OSLO X-RAY and OSLO CO-60

ABBREVIATED TRIAL NAME: OSLO X-RAY and OSLO CO-60
STUDY NUMBER: 64B1-2

INVESTIGATORS' NAMES AND AFFILIATIONS:
Herman Høst and Ivar O Brennhovd,
The Norwegian Radium Hospital, Oslo, Norway.

TRIAL DESIGN: TREATMENT GROUPS:
1. R (Series I): During the first part of the trial, 36 Gy were delivered to the supraclavicular fossa and 25-31 Gy to the chest wall via orthovoltage beam (200 Kv) with an additional 18 Gy to the axilla. Total treatment time was 4 weeks.

   R (Series II): During the second part of the trial, which took place from 1968-72, Cobalt therapy was used to treat the sternal, supra- and infraclavicular regions and the apex of the axilla. The target dose to each region was 50 Gy, in 20 fractions over 4 weeks.

2. Nil: No postoperative radiotherapy.

ELIGIBILITY CRITERIA FOR ENTRY:
1. AGE: No restriction.
2. CLINICAL STAGES: All operable disease according to the criteria of operability defined by Haagensen (1949).
3. PRIMARY TREATMENT: Halsted radical mastectomy and ovarian ablation with a radiation dose of between 6.5-9 Gy.
4. AXILLARY SURGERY: Surgical clearance to apex of axilla.
5. MENSTRUAL STATUS: No restrictions.

METHOD OF RANDOMIZATION:
By method of random numbers.

RANDOMIZATION STRATIFICATIONS: None.

YEARS OF ACCRUAL: 1964-67 for series I; 1968-72 for series II.

TOTAL NUMBER OF PATIENTS RANDOMIZED: 1115 (Series I = 552; Series II = 563).

PUBLICATIONS:
Høst H, and Brennhovd IO. The effect of postoperative radiotherapy in breast cancer. Int. J. Radiat. Oncol. Biol. Phys. 1977; 2:1061-67.

Høst H. Postoperative radiotherapy in breast cancer. In "Clinical Trials in Early Breast Cancer - 2nd Heidelberg Symposium, 1981." M Baum, R Kay & H Scheurlen, Eds. Experimentia Supplementum 41, pp 580-586. Birkhauser - Verlag. Stuttgart (1981).

Høst H, Brennhovd IO, and Loeb M. Postoperative radiotherapy in breast cancer: longterm results from the Oslo Study. Int. J. Radiat. Oncol. Biol. Phys. 1986; 12:727-732.

Table 1. Outcome by allocated treatment for trial 64B1 (Oslo X-ray)
Key: T = total free of relevant event at start of relevant year(s)

(i) Death (D) in each separate year,
subdivided by age when randomized (D <50 & D 50+ / T <50 & T 50+)

Year	1. R	2. Nil
1	0+ 6/ 111+ 174	0+ 4/ 108+ 159
2	5+ 10/ 111+ 168	3+ 7/ 108+ 155
3	6+ 6/ 106+ 158	6+ 7/ 105+ 148
4	5+ 9/ 100+ 152	5+ 11/ 99+ 141
5+	21+ 74/ 95+ 143	24+ 66/ 94+ 130

(ii) First recurrence or prior death (F) in each separate year,
subdivided by age when randomized (F <50 & F 50+ / T <50 & T 50+)

Year	1. R	2. Nil
1	0+ 13/ 111+ 174	2+ 11/ 108+ 159
2	6+ 10/ 111+ 161	11+ 13/ 106+ 148
3	10+ 15/ 105+ 151	3+ 8/ 95+ 135
4	4+ 11/ 95+ 136	1+ 10/ 92+ 127
5+	18+ 58/ 91+ 125	24+ 59/ 91+ 117

(iii) Recurrence with survival (R) and Death (D) in all years together,
subdivided by various characteristics when randomized (R & D / T)

Age at Entry	1. R	2. Nil
< 40	0+7/20	1+5/16
40 - 49	1+30/91	2+33/91
50 - 59	2+50/89	2+37/73
60 - 69	0+53/80	4+53/78
70 +	0+2/5	0+5/8
Any age (including "not known")	3+142/285	9+133/267

	Nodal Status (N0, N1-3, N4+)	1. R	2. Nil
<50	N0 (by axillary	0+1/1	0+0/0
50+	clearance)	0+1/1	0+1/1
<50	N0 (by clinical	1+18/70	1+19/70
50+	evidence only)	0+49/103	5+51/103
<50	N1-N3 (by axillary	0+0/0	0+1/1
50+	clearance)		
<50	N4+ (by axillary	0+7/7	0+5/5
50+	sample/clearance)	0+20/23	0+15/15
<50	Remainder (i.e.	0+11/33	2+13/31
50+	other N+ or N?)	2+35/47	1+28/41

Table 1. Outcome by allocated treatment for trial 64B2 (Oslo Co-60)
Key: T = total free of relevant event at start of relevant year(s)

(i) Death (D) in each separate year,
subdivided by age when randomized (D <50 & D 50+ / T <50 & T 50+)

Year	1. R	2. Nil
1	0+ 2/ 90+ 188	3+ 2/ 101+ 184
2	4+ 7/ 90+ 186	3+ 6/ 98+ 182
3	3+ 5/ 86+ 179	4+ 9/ 95+ 176
4	1+ 5/ 83+ 174	6+ 6/ 91+ 167
5+	25+ 59/ 82+ 169	16+ 48/ 85+ 161

(ii) First recurrence or prior death (F) in each separate year,
subdivided by age when randomized (F <50 & F 50+ / T <50 & T 50+)

Year	1. R	2. Nil
1	3+ 4/ 90+ 188	5+ 10/ 101+ 184
2	9+ 13/ 86+ 184	6+ 13/ 96+ 174
3	3+ 10/ 77+ 171	7+ 9/ 90+ 161
4	2+ 8/ 74+ 161	3+ 4/ 83+ 152
5+	20+ 48/ 72+ 153	16+ 44/ 80+ 148

(iii) Recurrence with survival (R) and Death (D) in all years together,
subdivided by various characteristics when randomized (R & D / T)

Age at Entry	1. R	2. Nil
< 40	1+3/13	1+3/9
40 - 49	3+30/77	4+29/92
50 - 59	1+32/94	4+41/103
60 - 69	4+43/89	5+26/76
70 +	0+3/5	0+4/5
Any age (including "not known")	9+111/278	14+103/285

	Nodal Status (N0, N1-3, N4+)	1. R	2. Nil
<50	N0 (by axillary	0+2/5	0+0/0
50+	clearance)	0+2/4	0+2/3
<50	N0 (by clinical	3+15/51	2+13/56
50+	evidence only)	4+44/119	5+35/128
<50	N1-N3 (by axillary	0+0/1	0+0/1
50+	clearance)	0+2/2	0+1/2
<50	N4+ (by axillary	0+7/9	0+6/11
50+	sample/clearance)	0+10/16	1+8/9
<50	Remainder (i.e.	1+9/24	3+13/33
50+	other N+ or N?)	1+20/47	3+25/42

Estrogen Receptors (ER, fmol/mg)

Breakdown by ER status was not available.

TITLE: THE SASKATCHEWAN CANCER FOUNDATION RANDOMIZED TRIAL OF PROPHYLACTIC OOPHORECTOMY IN OPERABLE CARCINOMA OF THE BREAST

STUDY NUMBER: 64A
STUDY TITLE: SASKATCHEWAN CF

ABBREVIATED TRIAL NAME: SASKATCHEWAN CF
STUDY NUMBER: 64A

INVESTIGATORS' NAMES AND AFFILIATIONS:
AJS Bryant and JA Weir
The Allan Blair Memorial Clinic, Regina and the Saskatchewan Cancer Foundation, Saskatchewan, Canada.

TRIAL DESIGN: TREATMENT GROUPS:
1. Ooph: Surgical oophorectomy
2. Nil: No further systemic therapy

ELIGIBILITY CRITERIA FOR ENTRY:
1. AGE: Unrestricted
2. CLINICAL STAGE: T1,2;N1;M0
3. PRIMARY TREATMENT: Radical mastectomy, modified radical mastectomy (only 1% before 1971; 42% after January 1972). Radiotherapy to the internal mammary nodes in Stage 1 cancers (node negative) with subareolar or medial half tumours. Radiotherapy to the axillary supraclavicular and parasternal areas for stage II (node positive) cancers.
4. NODE DISSECTION: Done as part of radical or modified radical mastectomy in all cases.
5. NUMBER OF NODES REMOVED: Not specified.
6. MENSTRUAL STATUS: Patients were required to have ovarian activity, shown either by the presence of active menstruation or on the basis of cornification in a lateral vaginal wall smear.

METHOD OF RANDOMIZATION: Randomization was made by withdrawal of a sealed envelope. Patients assigned to oophorectomy, or their family physicians, or surgeons were at liberty to decline oophorectomy. If oophorectomy was declined, the card was replaced.

RANDOMIZATION STRATIFICATIONS: None.

YEARS OF ACCRUAL: 1964-1974.

TOTAL NUMBER OF PATIENTS RANDOMIZED: 369 (but see below).

AIM, RATIONALE, SPECIAL FEATURES:
At the time this study was designed in 1963 the place of prophylactic oophorectomy in operable instances of breast carcinoma was a subject of some dispute. This study was therefore undertaken in an attempt to clarify the role of this surgical procedure.

ADDITIONAL ELIGIBILITY CRITERIA OR STRATIFICATION METHODS:
Patients were excluded if they had tumours fixed to the thoracic cage, tumours with extensive skin involvement or fixed metastatic axillary nodes. Patients with carcinoma of the breast associated with pregnancy or lactation and those who had a medical or surgical contra-indication to operation were also excluded.

COMPLIANCE AND TOXICITY:
Following randomization to surgical oophorectomy and with the informed consent of the patient, family doctor and surgeon, surgical castration was carried out at the same hospital admission as for surgical treatment of the breast. The number of patients declining oophorectomy was few but a measure of bias may exist since the cards of those patients who declined oophorectomy were inappropriately replaced and re-used. The patients who refused were not followed up initially and, although 11 have been traced, a few others were not and are not included in this analysis. Follow-up of these patients is now being attempted however for future analysis and publication. There was no loss to follow-up of patients who accepted their assigned treatment.

DISCUSSION:
The results of this study suggest that both survival and relapse-free survival are improved for the entire group of patients who underwent surgical oophorectomy. This improvement becomes significant for both overall survival and relapse-free survival in those patients with positive nodes, particularly those women under 50 years of age with one to three positive nodes. The magnitude of the difference in overall survival and relapse-free survival at ten years is consistent with the observation that about one third of women with breast cancer have hormone-dependent disease and would therefore be likely to benefit from therapeutic castration. In these women, early prophylactic oophorectomy may destroy micrometastases when the tumor burden is small. The results of this study are in agreement with those of several others but differ from the results of the National Surgical Adjuvant Breast Project. The latter study however, terminated at five years of follow-up, whereas in the present study, the advantages of prophylactic oophorectomy did not become apparent until after that time in the whole series or in the subgroup under 50 years of age with positive axillary nodes. Based on these results, prophylactic oophorectomy would seem justified in patients under 50 with positive axillary nodes, particularly if further identification of hormone dependent tumours by current methods such as estrogen and progesterone receptors could be made.

PUBLICATIONS:
Bryant AJS, Weir JA. Prophylactic oophorectomy in operable instances of carcinoma of the breast. Surg. Gynecol. Obstet. 1981, 153:660-664.

Table 1. Outcome by allocated treatment for trial 64A (Saskatchewan CF)
Key: T = total free of relevant event at start of relevant year(s)

(i) Death (D) in each separate year,
subdivided by age when randomized (D <50 & D 50+ / T <50 & T 50+)

Year	1. Ooph	2. Nil
1	3+ 2/ 139+ 44	2+ 2/ 103+ 69
2	10+ 4/ 136+ 42	4+ 5/ 101+ 67
3	8+ 2/ 126+ 38	7+ 3/ 97+ 62
4	2+ 2/ 118+ 36	4+ 4/ 90+ 59
5+	21+ 12/ 116+ 34	28+ 30/ 86+ 55

(ii) First recurrence or prior death (F) in each separate year,
subdivided by age when randomized (F <50 & F 50+ / T <50 & T 50+)

Year	1. Ooph	2. Nil
1	10+ 5/ 139+ 44	11+ 5/ 103+ 69
2	10+ 6/ 128+ 39	10+ 9/ 92+ 64
3	7+ 0/ 118+ 33	9+ 1/ 82+ 55
4	5+ 2/ 111+ 33	6+ 3/ 73+ 54
5+	14+ 9/ 106+ 31	13+ 26/ 67+ 51

(iii) Recurrence with survival (R) and Death (D) in all years together,
subdivided by various characteristics when randomized (R & D / T)

Age at Entry	1. Ooph	2. Nil
< 40	1+17/37	1+6/21
40 - 49	1+27/102	3+39/82
50 - 59	0+18/38	0+30/51
60 - 69	0+4/6	0+13/17
70 +	0+0/0	0+1/1
Any age (including "not known")	2+66/183	4+89/172

	Nodal Status (N0, N1-3, N4+)	1. Ooph	2. Nil
< 50	N0 (by axillary clearance)	1+12/73	1+14/60
50 +		0+10/25	0+18/31
< 50	N1-N3 (by axillary clearance)	0+12/33	1+17/27
50 +		0+4/9	0+15/25
< 50	N4+ (by axillary sample/clearance)	1+20/32	2+14/16
50 +		0+8/10	0+11/13
< 50	Remainder (i.e. other N+ or N?)	0+0/1	0+0/0

	Estrogen Receptors (ER, fmol/mg)		
< 50	ER poor or ER < 10	1+0/1	0+0/0

Short reports

TITLE: OVARIAN IRRADIATION (Ov Irr) & PREDNISONE (Pr) FOLLOWING SURGERY AND RADIOTHERAPY FOR BREAST CARCINOMA

STUDY NUMBER: 65B1-2
STUDY TITLE: PMH TORONTO

ABBREVIATED TRIAL NAME: PMH TORONTO
STUDY NUMBER: 65B1-2

INVESTIGATORS' NAMES AND AFFILIATIONS:
JW Meakin (1), WEC Allt (1), FA Beale (1), RS Bush (1), RM Clark (1), PJ Fitzpatrick (1), NV Hawkins (1), RDT Jenkin (1), JF Pringle (1), JG Reid (1) WD Rider (1) JL Hayward (2), RD Bulbrook (2)

(1) The Princess Margaret Hospital, Toronto, Canada.
(2) The Imperial Cancer Research Fund, Lond, UK.

TRIAL DESIGN: TREATMENT GROUPS:
1. Ov Irr: Ovarian irradiation to a dose of 2000 rads in 5 days.
2. Ov Irr + Pr: Ovarian irradiation in the same dose plus prednisone 7.5 mg daily for 5 years.
3. Nil: No adjuvant hormone therapy.

ELIGIBILITY CRITERIA FOR ENTRY:
1. Age: 35-70.
2. CLINICAL STAGE: Stages I, II & operable III (TNM Staging).
3. PRIMARY TREATMENT: Any operation that removed local disease was acceptable. All but 3 patients received mastectomy. All received postoperative radiotherapy to the chest wall and regional lymph nodes, including the ipsilateral internal mammary chain to a dose of 4000 rads in 16 sessions over 3 weeks.
4. NODE DISSECTION: 90% of premenopausal and 85% of postmenopausal received axillary dissection.
5. NUMBER OF NODES REMOVED: Not specified.
6. MENSTRUAL STATUS: No restrictions. Patients with previous hysterectomy but not oophorectomy were considered premenopausal up to the age of 50 years.

METHOD OF RANDOMIZATION:
Telephone call to a central office at the Princess Margaret Hospital where allocations were based on random number tables.

RANDOMIZATION STRATIFICATIONS:
Patients <45 years were not randomized to the prednisone-containing arm 2. Stratification was by age (<45 vs ≥45) and menstrual status. Patients were considered premenopausal if their last menses had occurred within 6 months of breast surgery.

YEARS OF ACCRUAL: 1965-1972

TOTAL NUMBER OF PATIENTS RANDOMIZED: 779

AIM, RATIONALE, SPECIAL FEATURES:
Because of the ambiguity of data relating to prophylactic adjuvant castration and because of the known value of Pr in treating metastatic breast cancer, this trial was begun in 1965 to test the value of prophylactic Ov Irr with or without Pr in delaying recurrence or prolonging survival.

ADDITIONAL ELIGIBILITY OR STRATIFICATION CRITERIA:
All patients referred within 3 months of their surgery were considered for study. Patients were excluded for the following reasons: poor health; previous or concurrent neoplasm; previous oophorectomy; prior antineoplastic chemical or hormonal therapy; contraindication to prednisone therapy; probable lack of availability for follow-up; and/or pregnancy within one year of surgery.

COMPLIANCE AND TOXICITY:
Follow-up was performed every 3 months for the first 2 years, every 6 months for the next 3 years and then annually. Of the 779 patients allocated, 23 were ineligible; 51 were protocol violations. Of the 435 patients assigned to Ov Irr, 15 received less or more than the protocol dosage. Menses recurred following Ov Irr in 7% of those <45 years of age and in 1% of those ≥45. All women in whom menses recurred had received the assigned radiation dosage. Of the 73 premenopausal women randomized to receive Pr, 49% received Pr as per protocol while Pr was discontinued early in 51% (median duration of therapy-4 years). Of the 112 postmenopausal women who received Pr, 54% received Pr as per protocol, while Pr was discontinued early in 46% (median duration of therapy-2 years).

DISCUSSION:
This study shows that the addition of prednisone to ovarian irradiation produces significant delay in recurrence and prolongs survival in premenopausal women. Data from this study are in agreement with those from the Manchester and Oslo (study 57A) trials in demonstrating an apparent delay in recurrence and prolongation of survival in pre-menopausal women after adjuvant Ov Irr alone, even though the differences were not statistically significant in this study. The lack of agreement with the results of the NSABP trial may relate to the effect of chance, or an irradiated ovary may result in a different psychologic state than ovarian removal.

PUBLICATIONS:
Meakin JW, Allt WEC, Beale FA, Brown TC, Bush RS, Clark RM, Fitzpatrick PJ, Hawkins NV, Jenkin RDT, Reid JG,, Rider WD, Hayward JL, Bulbrook RD. Ovarian irradiation and prednisone therapy following surgery and radiotherapy for carcinoma of the breast. Can. Med. Assoc. J. 1979, 120: 1221-1239.

Meakin JW, Allt WEC, Beale FA, Bush RS, Clark RM, Fitzpatrick PJ, Hawkins NV, Jenkin RDT, Pringle JF, Reid JG, Rider WD, Hayward JL, Bulbrook RD. Ovarian irradiation and prednisone following surgery and radiotherapy for carcinoma of the breast. Breast Cancer Res. Treat. 1983; 3(Suppl.1):45-48.

Table 1. Outcome by allocated treatment for trial 65B1 (PMH Toronto)
Key: T = total free of relevant event at start of relevant year(s)

(i) Death (D) in each separate year, subdivided by age when randomized (D <50 & D 50+ / T <50 & T 50+)

Year	1. OvIrr	3. Nil
1	6+ 0/ 68+ 0	6+ 0/ 62+ 0
2	5+ 0/ 62+ 0	10+ 0/ 56+ 0
3	9+ 0/ 57+ 0	10+ 0/ 46+ 0
4	4+ 0/ 48+ 0	4+ 0/ 36+ 0
5+	18+ 0/ 44+ 0	16+ 0/ 32+ 0

(ii) First recurrence or prior death (F) in each separate year, subdivided by age when randomized (F <50 & F 50+ / T <50 & T 50+)

Year	1. OvIrr	3. Nil
1	13+ 0/ 68+ 0	20+ 0/ 62+ 0
2	10+ 0/ 55+ 0	10+ 0/ 42+ 0
3	8+ 0/ 45+ 0	4+ 0/ 32+ 0
4	2+ 0/ 37+ 0	1+ 0/ 28+ 0
5+	14+ 0/ 35+ 0	12+ 0/ 27+ 0

(iii) Recurrence with survival (R) and Death (D) in all years together, subdivided by various characteristics when randomized (R & D / T)

Age at Entry	1. OvIrr	3. Nil
< 40	1+19/25	0+16/18
40 - 49	4+23/43	1+30/44
Any age (including "not known")	5+42/68	1+46/62

	Nodal Status (N0, N1-3, N4+)		
< 50	N0 (by clinical evidence only)	0+3/4	1+5/9
< 50	Remainder (i.e. other N+ or N?)	5+39/64	0+41/53

Table 1. Outcome by allocated treatment for trial 65B2 (PMH Toronto)
Key: T = total free of relevant event at start of relevant year(s)

(i) Death (D) in each separate year, subdivided by age when randomized (D <50 & D 50+ / T <50 & T 50+)

Year	1. OvIrr	2. OvIrrPr	3. Nil
1	6+ 5/ 71+ 145	6+ 6/ 77+ 139	2+ 6/ 71+ 146
2	3+ 15/ 65+ 140	8+ 13/ 71+ 133	7+ 14/ 69+ 140
3	5+ 11/ 62+ 125	4+ 10/ 63+ 120	4+ 17/ 62+ 126
4	0+ 11/ 57+ 114	5+ 12/ 59+ 110	4+ 9/ 58+ 109
5+	21+ 41/ 57+ 103	9+ 44/ 54+ 98	19+ 52/ 54+ 100

(ii) First recurrence or prior death (F) in each separate year, subdivided by age when randomized (F <50 & F 50+ / T <50 & T 50+)

Year	1. OvIrr	2. OvIrrPr	3. Nil
1	9+ 26/ 71+ 145	14+ 16/ 77+ 139	7+ 23/ 71+ 146
2	4+ 14/ 62+ 119	5+ 9/ 63+ 123	9+ 18/ 64+ 123
3	5+ 7/ 58+ 105	6+ 15/ 58+ 114	10+ 15/ 55+ 105
4	2+ 11/ 53+ 98	2+ 9/ 52+ 99	3+ 4/ 45+ 90
5+	18+ 37/ 51+ 87	8+ 42/ 50+ 90	11+ 51/ 42+ 86

(iii) Recurrence with survival (R) and Death (D) in all years together, subdivided by various characteristics when randomized (R & D / T)

Age at Entry	1. OvIrr	2. OvIrrPr	3. Nil
< 40	0+0/0	0+0/1	0+0/0
40 - 49	3+35/71	3+32/76	4+36/71
50 - 59	9+55/95	5+47/89	9+61/95
60 - 69	3+27/49	1+36/48	4+35/48
70 +	0+1/1	0+2/2	0+2/3
Any age (including "not known")	15+118/216	9+117/216	17+134/217

	Nodal Status (N0, N1-3, N4+)			
< 50	N0 (by clinical evidence only)	0+3/9	0+3/16	1+6/17
50 +		5+11/31	2+16/34	3+18/31
< 50	Remainder (i.e.	3+32/62	3+29/61	3+30/54
50 +	oths. N+ or N?)	7+72/114	4+69/105	10+80/115

Estrogen Receptors (ER, fmol/mg)

Breakdown by ER status was not available.

TITLE: LONG-TERM ADJUVANT CHEMOTHERAPY TRIAL

ABBREVIATED TRIAL NAME: BIRMINGHAM
STUDY NUMBER: 67A

INVESTIGATORS' NAMES AND AFFILIATIONS:
GD Oates (1), RM Baddeley (1), KAH Waterhouse (2), J Powell (2).
  (1) Dept. of Surgery, General Hospital, Birmingham, UK.
  (2) Birmingham and West Midlands Regional Cancer Registry, Queen Elizabeth Medical Centre, Birmingham, UK.

TRIAL DESIGN: TREATMENT GROUPS:
1. C: Cyclophosphamide 4-5 days post operatively - 200 mg iv daily until hospital discharge or WBC $<4.0 \times 10^9$/l then cyclophosphamide 100-200 mg p.o. daily for 12 months.
2. F: 12 months 5-fluorouracil, 12 mg/kg daily x 5 days, starting 7 days post operatively, then 6 mg/kg on alternate days until WBC $<3.0 \times 10^9$/l, repeated after 30 day intervals for 12 months.
3. Nil: No adjuvant chemotherapy.

ELIGIBILITY CRITERIA FOR ENTRY:
1. AGE: <70.
2. CLINICAL STAGES: T1-2, N0-1, M0 (stages I and II).
3. PRIMARY TREATMENT: Radical mastectomy (Patey), with excision of pectoralis minor.
4. NODE DISSECTION: Mandatory.
5. NUMBER OF NODES REMOVED: Not specified.
6. MENSTRUAL STATUS: No restriction.
7. RECEPTOR STATUS: No restriction.

METHOD OF RANDOMIZATION:
Randomization was performed using sealed envelopes and was confined to one office and one investigator. Treatment allocations were based on random number tables, but were not balanced in blocks. All randomized patients, including 6 ineligible patients, have been followed up for mortality data.

RANDOMIZATION STRATIFICATIONS:
Node negative, node positive.

YEARS OF ACCRUAL: 1967-1977.

TOTAL NUMBER OF PATIENTS RANDOMIZED: 244.

PUBLICATIONS: None.

Table 1. Outcome by allocated treatment for trial 67A (Birmingham)
Key: T = total free of relevant event at start of relevant year(s)

(i) Death (D) in each separate year,
    subdivided by age when randomized (D <50 & D 50+ / T <50 & T 50+)

Year	1. C	2. F	3. Nil
1	1+ 1/ 43+ 39	0+ 0/ 34+ 41	1+ 0/ 28+ 55
2	5+ 3/ 42+ 38	2+ 1/ 34+ 41	0+ 0/ 27+ 55
3	2+ 3/ 37+ 35	8+ 1/ 32+ 40	1+ 3/ 27+ 55
4	2+ 2/ 35+ 32	4+ 2/ 24+ 39	4+ 4/ 26+ 52
5+	11+ 6/ 33+ 30	3+ 12/ 20+ 37	7+ 10/ 22+ 48

(ii) First recurrence or prior death (F) in each separate year,
     subdivided by age when randomized (F <50 & F 50+ / T <50 & T 50+)

Year	1. C	2. F	3. Nil
1	6+ 3/ 43+ 39	3+ 0/ 34+ 41	2+ 2/ 28+ 55
2	4+ 6/ 37+ 36	9+ 5/ 31+ 41	5+ 4/ 26+ 53
3	3+ 3/ 33+ 30	2+ 3/ 22+ 36	2+ 8/ 21+ 49
4	1+ 2/ 30+ 27	1+ 0/ 20+ 33	2+ 3/ 19+ 41
5+	8+ 3/ 29+ 25	4+ 10/ 19+ 33	5+ 11/ 17+ 38

(iii) Recurrence with survival (R) and Death (D) in all years together,
      subdivided by various characteristics when randomized (R & D / T)

Age at Entry	1. C	2. F	3. Nil
< 40	0+5/13	0+4/7	1+3/6
40 - 49	2+16/31	3+13/28	2+10/22
50 - 59	2+9/22	1+7/20	5+7/29
60 - 69	0+6/17	3+9/23	6+10/26
Any age (including "not known")	4+36/83	7+33/78	14+30/83

Nodal Status (N0, N1-3, N4+)

		1. C	2. F	3. Nil
< 50	N0 (by clinical	0+4/20	1+8/20	1+1/10
50 +	evidence only)	1+3/20	2+8/24	3+5/28
< 50	Remainder (i.e.	2+17/24	2+9/15	2+12/18
50 +	other N+ or N?)	1+12/19	2+8/19	8+12/27

Estrogen Receptors (ER, fmol/mg)

Breakdown by ER status was not available.

TITLE: HEIDELBERG RADIOTHERAPY TRIAL

ABBREVIATED TRIAL NAME: HEIDELBERG XRT
STUDY NUMBER: 69A

STUDY NUMBER: 69A
STUDY TITLE: HEIDELBERG XRT

INVESTIGATORS' NAMES AND AFFILIATIONS:
HR Scheurlen (1), H Kuttig (2), B Henningsen (3).

(1) Institute for Medical Statistics, Heidelberg, W. Germany.
(2) University Department of Radiotherapy, Heidelberg.
(3) University Department of Surgery, Heidelberg.

TRIAL DESIGN: TREATMENT GROUPS:
1. R: Therapy to residual node sites only: apex of the axilla, supraclavicular fossa and the ipsilateral internal mammary chain. 65 Gy given by cobalt source in daily fractions over 6 wks. A ^{60}Co source was used.
2. Nil: No postoperative radiotherapy.

ELIGIBILITY CRITERIA FOR ENTRY:
1. AGE: All ages.
2. CLINICAL STAGES: T_{1-3}, N_{0-2}, and M_0 lesions.
3. PRIMARY TREATMENT: Radical mastectomy.
4. AXILLARY SURGERY: Clearance of axilla with histological examination of enlarged nodes.
5. MENSTRUAL STATUS: No restrictions. The apparent imbalance in the treatment allocation was due to a run in the random number table used.

METHOD OF RANDOMIZATION:
By reference to the first digit (odd or even) on a list of 5-digit random numbers with entry order according to that in the ward register. The apparent imbalance in the treatment allocation was due to a run in the random number table used.

RANDOMIZATION STRATIFICATIONS:
None.

YEARS OF ACCRUAL: 1969-1972.

TOTAL NUMBER OF PATIENTS RANDOMIZED: 143. One patient died from sequelae of the mastectomy and was not included in the analysis.

COMPLIANCE AND TOXICITY:
Some form of protocol violation was reported in 23 patients but they have been included in the analysis.

PUBLICATIONS: None.

Table 1. Outcome by allocated treatment for trial 69A (Heidelberg XRT)
Key: T = total free of relevant event at start of relevant year(s)

(i) Death (D) in each separate year,
subdivided by age when randomized (D <50 & D 50+ / T <50 & T 50+)

Year	1. R	2. Nil
1	1+ 9/ 20+ 64	2+ 4/ 17+ 41
2	4+ 9/ 19+ 55	1+ 5/ 15+ 37
3	1+ 8/ 15+ 46	1+ 4/ 14+ 32
4	3+ 5/ 14+ 38	3+ 1/ 13+ 28
5+	3+ 20/ 11+ 33	5+ 14/ 10+ 27

(ii) First recurrence or prior death (F) in each separate year,
subdivided by age when randomized (F <50 & F 50+ / T <50 & T 50+)

Year	1. R	2. Nil

(recurrence data not available)

(iii) Recurrence with survival (R) and Death (D) in all years together,
subdivided by various characteristics when randomized (R & D / T)

Age at Entry	1. R	2. Nil
< 40	0+4/7	0+5/5
40 - 49	0+8/13	0+7/12
50 - 59	0+14/19	0+6/13
60 - 69	0+25/31	0+11/15
70 +	0+12/14	0+11/13
Any age (including "not known")	0+63/84	0+40/58

Nodal Status (N0, N1-3, N4+)

Breakdown by nodal status was not available.

Estrogen Receptors (ER, fmol/mg)

Breakdown by ER status was not available.

TITLE: MANCHESTER REGIONAL BREAST STUDY

ABBREVIATED TRIAL NAME: MANCHESTER RBS
STUDY NUMBER: 70A

STUDY NUMBER: 70A
STUDY TITLE: MANCHESTER RBS

INVESTIGATORS' NAMES AND AFFILIATIONS:
JP Lythgoe (1) and R Swindell (2) in conjunction with the Regional Association of Surgeons and with the staff of the Christie Hospital and Holt Radium Institute, Manchester, England.

(1) Department of Surgery, Royal Preston Hospital, Preston.
(2) Department of Medical Statistics, Christie Hospital, Manchester.

TRIAL DESIGN: TREATMENT GROUPS:
1. R: Radiotherapy by 2 possible techniques each treating the chest wall & the 3 regional nodal areas using 300 Kv to the chest wall and 300 Kv or 4 Mv to nodal areas, 37-45 Gy in 15 fractions over 3 wks.
2. Nil: No immediate postoperative radiotherapy.

ELIGIBILITY CRITERIA FOR ENTRY:
1. AGE: ≤70 years in good general health.
2. CLINICAL STAGES: Operable breast cancer stage I and II (UICC 1968), including Paget's disease of the nipple.
3. PRIMARY TREATMENT: For clinical stage I disease and for clinical stage II disease (nodes palpable) if allocated radiotherapy: simple (total) mastectomy with removal of pectoral fascia, thin skinflaps avoided; for clinical stage II disease if not allocated radiotherapy: total mastectomy with or without the removal of the pectoral muscles.
4. AXILLARY SURGERY: Clearance of axilla if clinical stage II disease and not allocated radiotherapy, otherwise no intentional removal of nodes from the axilla.
5. MENSTRUAL STATUS: No restrictions. However, for those who were within three years of their last menstrual period, ovarian ablation, either by surgery or radiotherapy was offered.

METHOD OF RANDOMIZATION:
Registration and allocation of treatment option by phone (preoperatively in the case of stage II cases) to the Regional Cancer Epidemiology Unit.

RANDOMIZATION STRATIFICATIONS:
1. Participating surgeon.
2. Clinical stage.

YEARS OF ACCRUAL: 1970-1975.

TOTAL NUMBER OF PATIENTS RANDOMIZED: 1022 - (714 Stage I, 308 Stage II). The stage II patients are not relevant to the current overviews.

AIM, RATIONALE, SPECIAL FEATURES:
In the early 1970s, patients with clinical stage I tumors were commonly treated by simple mastectomy and early post-operative radiotherapy. In order to find out whether routine post-operative radiotherapy was beneficial, useless or harmful for stage I patients, simple mastectomy plus early post-operative radiotherapy was compared with simple mastectomy only. This trial was felt to be inappropriate for stage II patients since, at that time, most surgeons believed that some treatment should be directed at palpable axillary nodes (clinical stage II). In a parallel trial, the two most commonly employed treatments (simple mastectomy plus early post-operative radiotherapy and radical mastectomy alone) were compared.

PUBLICATIONS:
Lythgoe JP, Leck I, Swindell R. Manchester Regional Breast Study Preliminary Results. Lancet 1978; i: 744-747.

Lythgoe JP, Palmer MK. Manchester Regional Breast Study - 5 & 10 year results. Br. J. Surg. 1982; 69: 693-696.

Table 1. Outcome by allocated treatment for trial 70A (Manchester RBS1)
Key: T = total free of relevant event at start of relevant year(s)

(i) Death (D) in each separate year,
subdivided by age when randomized (D <50 & D 50+ / T <50 & T 50+)

Year	1. R	2. Nil
1	2+ 9/ 132+ 223	2+ 11/ 137+ 222
2	8+ 14/ 130+ 214	9+ 16/ 135+ 211
3	4+ 18/ 122+ 200	10+ 16/ 126+ 195
4	8+ 15/ 118+ 182	7+ 12/ 116+ 179
5+	22+ 69/ 110+ 167	27+ 73/ 109+ 167

(ii) First recurrence or prior death (F) in each separate year,
subdivided by age when randomized (F <50 & F 50+ / T <50 & T 50+)

Year	1. R	2. Nil
1	12+ 25/ 132+ 223	26+ 40/ 137+ 222
2	9+ 23/ 120+ 198	13+ 34/ 111+ 182
3	6+ 13/ 111+ 175	7+ 22/ 98+ 148
4	7+ 14/ 105+ 162	10+ 19/ 91+ 125
5+	20+ 57/ 98+ 146	18+ 40/ 81+ 106

(iii) Recurrence with survival (R) and Death (D) in all years together,
subdivided by various characteristics when randomized (R & D / T)

Age at Entry	1. R	2. Nil
< 40	1+14/39	1+18/41
40 - 49	9+30/93	18+37/96
50 - 59	1+48/100	15+54/103
60 - 69	5+75/120	12+71/116
70 +	1+2/3	0+3/3
Any age (including "not known")	17+169/355	46+183/359

Nodal Status (N0, N1-3, N4+)

Breakdown by nodal status was not available.

Estrogen Receptors (ER, fmol/mg)

Breakdown by ER status was not available.

Short reports

TITLE: THE CRC (KING'S CAMBRIDGE) TRIAL FOR EARLY BREAST CANCER

STUDY NUMBER: 70B
STUDY TITLE: KING'S/CAMBRIDGE

ABBREVIATED TRIAL NAME: KING'S/CAMBRIDGE
STUDY NUMBER: 70B

INVESTIGATORS' NAMES AND AFFILIATIONS:
Joan Houghton (1), M. Baum (2) J. Haybittle, D Brinkley and CRC Breast Trial Working Party.
 (1) CRC Clinical Trials Centre, London, England.
 (2) King's College Hospital Medical School, London, England.

TRIAL DESIGN: TREATMENT GROUPS:
1. R: Routine post-mastectomy radiotherapy to chest wall and regional nodal areas.
2. Nil: No adjuvant radiotherapy; delayed treatment as appropriate for loco-regional progression.

ELIGIBILITY CRITERIA FOR ENTRY:
1. AGE: <70 yrs.
2. CLINICAL STAGES: Tumor 5 cm or less without fixation to skin or muscle, with or without mobile axillary nodes (Stages I and II by 1964 classification).
3. PRIMARY TREATMENT: Simple (total) mastectomy without adjuvant systemic therapy, regardless of nodal status.
4. AXILLARY SURGERY: Surgical interference with axillary lymph nodes avoided even when palpable.
5. MENSTRUAL STATUS: No restrictions.

METHOD OF RANDOMIZATION:
By opening the next in sequence of a set of sealed and numbered envelopes supplied by the central office to each individual participating centre; randomly selected treatment option and patient's trial number identified from card inside; option drawn locally after result of mastectomy known.

RANDOMIZATION STRATIFICATIONS:
Participating centre.

YEARS OF ACCRUAL: 1970-1975.

TOTAL NUMBER OF PATIENTS RANDOMIZED: 2800.

PUBLICATIONS:
Anonymous. Management of early cancer of the breast. Report on an international multicentre trial supported by the Cancer Research Campaign. Br. Med. J. 1976; i: 1035-1038.

Brinkley D, Haybittle JL, Houghton J. The Cancer Research Campaign (King's/Cambridge) trial for early breast cancer: an analysis of the radiotherapy data. Br. J. Radiol. 1984; 57: 309-316.

Berstock DA, Houghton J, Haybittle J and Baum M. The role of radiotherapy following total mastectomy for patients with early breast cancer. World J. Surg. 1985: 9: 667-670.

Haybittle JL, Brinkley D, Houghton J, A'Hern RP, and Baum M. Postoperative radiotherapy and late mortality evidence from the Cancer Research Campaign trial for early breast cancer. Br. Med. J. 1989; 298:1611-1614.

Table 1. Outcome by allocated treatment for trial 70B (Kings/Cambridge)
Key: T = total free of relevant event at start of relevant year(s)

(i) Death (D) in each separate year,
subdivided by age when randomized (D <50 & D 50+ / T <50 & T 50+)

Year	1. R	2. Nil
1	16+ 38/ 496+ 880	15+ 27/ 482+ 942
2	36+ 55/ 480+ 842	24+ 62/ 467+ 915
3	32+ 70/ 444+ 787	38+ 71/ 443+ 853
4	25+ 49/ 412+ 717	31+ 74/ 405+ 782
5+	100+255/ 387+ 668	93+242/374+ 708

(ii) First recurrence or prior death (F) in each separate year,
subdivided by age when randomized (F <50 & F 50+ / T <50 & T 50+)

Year	1. R	2. Nil
1	35+ 50/ 496+ 880	41+ 64/ 482+ 942
2	46+ 75/ 460+ 827	50+ 96/ 440+ 878
3	38+ 88/ 414+ 752	49+ 88/ 390+ 782
4	26+ 44/ 376+ 663	27+ 63/ 341+ 694
5+	104+263/ 350+ 618	100+280/314+ 631

(iii) Recurrence with survival (R) and Death (D) in all years together,
subdivided by various characteristics when randomized (R & D / T)

Age at Entry	1. R	2. Nil
< 40	16+51/114	15+50/109
40 - 49	24+158/382	51+151/373
50 - 59	21+202/404	64+204/455
60 - 69	31+255/458	50+263/469
70 +	1+10/17	1+6/14
Any age (including "not known")	93+676/1376	181+677/1424

Nodal Status (N0, N1-3, N4+)

		1. R	2. Nil
<50	N0 (by clinical	30+145/369	53+141/369
50+	evidence only)	34+329/664	85+336/702
<50	Remainder (i.e.	10+64/127	13+60/113
50+	other N+ or N?)	19+138/216	30+140/240

Estrogen Receptors (ER, fmol/mg)

Breakdown by ER status was not available.

TITLE: OVARIAN IRRADIATION IN POST-MENOPAUSAL WOMEN WITH BREAST CANCER AND POSITIVE AXILLARY NODES

ABBREVIATED TRIAL NAME: CRFB CAEN A
STUDY NUMBER: 71A

INVESTIGATORS' NAMES AND AFFILIATIONS:
T Delozier, P Juret[+], JE Couette, J Mace-Lesech,
Centre François Baclesse, Caen, France.
[+] Deceased.

TRIAL DESIGN: TREATMENT GROUPS:
1. Ov Irr: Ovarian irradiation: 9 Gy in 2 fractions, 48 hours apart or 14 Gy in 7 fractions, 5 fractions a week over 9 days.
2. Nil: No ovarian irradiation.

ELIGIBILITY CRITERIA FOR ENTRY:
1. AGE: ≤70 yrs.
2. CLINICAL STAGES: All except M1 (distant metastases), N2 (fixed axillary lymph nodes), N3 (supraclavicular nodes), T4, or inflammatory tumors.
3. PRIMARY TREATMENT: Tumorectomy or mastectomy, radiotherapy to the breast or to the breast and nodes at the discretion of the clinician.
4. NODE DISSECTION: Axillary sampling.
5. HISTOLOGY OF AXILLARY NODES: Positive.
6. MENSTRUAL STATUS: Postmenopausal (no menses for 1 year).

METHOD OF RANDOMIZATION:
By sealed numbered envelopes. Allocation based on a table of random numbers.

RANDOMIZATION STRATIFICATIONS: None.

YEARS OF ACCRUAL: 1971-1976.

TOTAL NUMBER OF PATIENTS RANDOMIZED: 52.

AIM, RATIONALE, SPECIAL FEATURES:
Ovarian function does not stop abruptly after the last menses, and Nissen-Meyer had suggested a possible efficacy of castration in postmenopausal women with breast cancer in terms of disease-free and overall survival.

ADDITIONAL ELIGIBILITY OR STRATIFICATION CRITERIA:
Disease was initially assessed by clinical examination with TNM staging, chest X-ray, X-rays of head, spine and pelvis, and vaginal smear to measure the proportion of acidophil cells.

COMPLIANCE AND TOXICITY:
The follow-up examinations, at least every 6 months, included a physical examination and X-ray investigations, including a skeletal survey.

DISCUSSION:
The trial plan was to include at least 350 patients, but disinterest in the study lead to premature termination.

REFERENCES TO OTHER RELEVANT STUDIES:
Nissen-Meyer, R. The role of prophylactic castration in the therapy of human mammary cancer. Eur. J. Cancer 1967; 3:395-403.

PUBLICATIONS: None.

Table 1. Outcome by allocated treatment for trial 71A (CRFB Caen A)
Key: T = total free of relevant event at start of relevant year(s)

(i) Death (D) in each separate year,
subdivided by age when randomized (D <50 & D 50+ / T <50 & T 50+)

Year	1. OvIrr	2. Nil
1	0+ 1/ 1+ 26	0+ 2/ 0+ 25
2	0+ 6/ 1+ 25	0+ 5/ 0+ 23
3	0+ 8/ 1+ 19	0+ 1/ 0+ 18
4	0+ 1/ 1+ 11	0+ 1/ 0+ 17
5+	0+ 3/ 1+ 10	0+ 6/ 0+ 16

(ii) First recurrence or prior death (F) in each separate year,
subdivided by age when randomized (F <50 & F 50+ / T <50 & T 50+)

Year	1. OvIrr	2. Nil
1	0+ 5/ 1+ 26	0+ 4/ 0+ 25
2	0+ 8/ 1+ 21	0+ 4/ 0+ 21
3	0+ 5/ 1+ 13	0+ 4/ 0+ 17
4	0+ 1/ 1+ 8	0+ 3/ 0+ 13
5+	1+ 3/ 1+ 7	0+ 2/ 0+ 10

(iii) Recurrence with survival (R) and Death (D) in all years together,
subdivided by various characteristics when randomized (R & D / T)

Age at Entry	1. OvIrr	2. Nil
40 - 49	1+0/1	0+0/0
50 - 59	1+4/7	2+4/9
60 - 69	2+14/18	0+11/16
70 +	0+1/1	0+0/0
Any age (including "not known")	4+19/27	2+15/25

Nodal Status (N0, N1-3, N4+)

Breakdown by nodal status was not available.

Estrogen Receptors (ER, fmol/mg)

Breakdown by ER status was not available.

TITLE: THE STOCKHOLM TRIAL ON ADJUVANT RADIO-THERAPY IN OPERABLE BREAST CANCER

STUDY NUMBER: 71B
STUDY TITLE: STOCKHOLM A

ABBREVIATED TRIAL NAME: STOCKHOLM A
STUDY NUMBER: 71B

INVESTIGATORS' NAMES AND AFFILIATIONS:
LE Rutquist (1), O Arner (2), J Bergstrom (3), B Blomstedt (4), PO Granberg (5), L Raf (3), C Silfversward (6).

(1) Dept of Oncology, Karolinska Hospital, Stockholm.
(2) Dept of Surgery, St. Erik Hospital, Stockholm.
(3) Dept of Surgery, St. Goran Hospital, Stockholm.
(4) Dept of Surgery, Sabbatsberg Hospital, Stockholm.
(5) Dept of Surgery, Karolinska Hospital, Stockholm.
(6) Dept of Pathology, Karolinska Hospital, Stockholm.

TRIAL DESIGN: TREATMENT GROUPS:
1. R(preop): 45 Gy via megavoltage (cobalt) delivered preoperatively to breast, axilla & ipsilateral internal mammary chain in 25 fractions by parallel opposed fields but a single supraclavicular field.
2. R(postop): As above for nodal therapy, but postoperatively by high energy electrons, to same dose, to chest wall fields.
3. Nil: No pre- or post-operative adjuvant therapy.

ELIGIBILITY CRITERIA FOR ENTRY:
1. AGE: ≤ 70 years.
2. CLINICAL STAGES: T_{1-3}, N_{0-1}, M_0 operable and unilateral breast cancer confirmed by positive FNA cytology.
3. PRIMARY TREATMENT: Modified radical mastectomy with removal of pectoral fascia but sparing both muscles.
4. AXILLARY SURGERY: Clearance of axillary nodes.
5. MENSTRUAL STATUS: No restrictions.

METHOD OF RANDOMIZATION:
Randomization was performed using lists which were kept centrally. Treatment allocations were based on random number tables and were balanced in blocks of nine.

RANDOMIZATION STRATIFICATIONS:
1. AGE: <50 yrs or ≥50 yrs.
2. CLINICAL NODE STATUS: N_0 or N_1.
3. CLINICAL TUMOR SIZE: <20 mm, 21-50 mm, or >50 mm.

YEARS OF ACCRUAL: 1971-1976.

TOTAL NUMBER OF PATIENTS RANDOMIZED: 960.

RESULTS:
This trial contributes to the comparison of radiotherapy as an adjunct to radical surgery versus the same surgical procedure without irradiation. Previous analyses of this trial have indicated a significantly increased recurrence-free survival with radiotherapy, mainly due to a reduced frequency of locoregional relapses but also a trend towards fewer distant metastases. Between the two types of radiotherapy, there was no difference. Up to at least 10 years, there was an increasing gap between the recurrence-free survival of the irradiated patients and the surgical controls. This indicates that radiotherapy not only postponed but prevented recurrence. At a median follow-up of 11 years, there were significantly fewer distant metastases in node positive patients treated with postoperative radiotherapy, compared to the surgical controls (p<0.01) and a 13% reduction of mortality, this difference was, however, not statistically significant.

DISCUSSION:
This study demonstrated that adjuvant radiotherapy with high-voltage technique is an effective method to prevent locoregional recurrences and hence to prolong the recurrence-free interval. Other studies have yielded similar results.

On the other hand, the effect from adjuvant radiotherapy on survival remains controversial. Several randomized studies — including the current trial — have failed to demonstrate significant overall survival benefits from a more radical local therapy of early breast cancer. The addition of various types of radiotherapy to different surgical procedures, for instance, has generally not increased survival. However, many radiotherapy trials have employed treatment techniques that are inadequate with, e.g. insufficient dose to the target volume or large fraction sizes which are conducive to increased late side effects. Consequently, the divergence of the procedures in the trials included in the current overview and the obvious differences in local efficiency cast some doubt on the summary results concerning the effects from adjuvant radiotherapy on long-term survival.

PUBLICATIONS:
Rutqvist LE, Cedermark B, Glas U, Johansson H, Rotstein S, Skoog L, Somell A, Theve T, Askergren J, Friberg S, Bergström J, Blomstedt B, Räf L, Silfverswärd C, Einhorn J. Radiotherapy, chemotherapy, and tamoxifen as adjuncts to surgery in early breast cancer: A summary of three randomised trials. Int. J. Radiation Oncology Biol. Phys. 1989; 16:629-639.

Table 1. Outcome by allocated treatment for trial 71B (Stockholm A)
Key: T = total free of relevant event at start of relevant year(s)

(i) Death (D) in each separate year,
subdivided by age when randomized (D <50 & D 50+ / T <50 & T 50+)

Year	1. R(pre)	2. R(post)	3. Nil
1	1+ 5/ 80+ 236	1+ 7/ 91+ 232	3+ 10/ 83+ 238
2	4+ 10/ 79+ 231	5+ 12/ 90+ 225	4+ 10/ 80+ 228
3	3+ 9/ 75+ 221	4+ 15/ 85+ 213	4+ 9/ 76+ 218
4	1+ 15/ 72+ 212	3+ 15/ 81+ 198	5+ 20/ 72+ 209
5+	22+ 52/ 71+ 197	11+ 46/ 78+ 183	15+ 52/ 67+ 189

(ii) First recurrence or prior death (F) in each separate year,
subdivided by age when randomized (F <50 & F 50+ / T <50 & T 50+)

Year	1. R(pre)	2. R(post)	3. Nil
1	5+ 13/ 80+ 236	7+ 21/ 91+ 232	15+ 32/ 83+ 238
2	13+ 18/ 75+ 223	9+ 16/ 84+ 211	8+ 24/ 68+ 206
3	3+ 21/ 62+ 205	3+ 14/ 75+ 195	5+ 19/ 60+ 182
4	2+ 9/ 59+ 184	2+ 13/ 72+ 181	0+ 7/ 55+ 163
5+	13+ 40/ 57+ 175	6+ 39/ 70+ 168	15+ 49/ 55+ 156

(iii) Recurrence with survival (R) and Death (D) in all years together,
subdivided by various characteristics when randomized (R & D / T)

Age at Entry	1. R(pre)	2. R(post)	3. Nil
< 40	1+4/13	0+2/18	3+6/12
40 - 49	4+27/67	3+22/73	9+25/71
50 - 59	6+46/123	5+34/101	18+45/121
60 - 69	4+38/105	3+56/124	12+48/109
70 +	0+7/8	0+5/7	0+8/8
Any age (including "not known")	15+122/316	11+119/323	42+132/321

	Nodal Status (N0, N1-3, N4+)			
<50	N0 (by axillary	0+0/0	2+10/58	7+11/44
50+	clearance)	0+0/0	4+45/146	18+43/153
<50	N1-N3 (by axillary	0+0/0	0+2/10	2+2/13
50+	clearance)	0+0/0	0+21/33	5+18/28
<50	Remainder (i.e.	5+31/80	1+12/23	3+18/26
50+	other N+ or N?)	10+91/236	4+29/53	7+40/57

	Estrogen Receptors (ER, fmol/mg)			
≤50	ER+ or ER 10-99	0+0/0	0+0/0	1+0/1
50+		0+0/0	0+0/0	0+0/1
50+	ER++ or ER 100+	1+1/2	0+0/0	0+0/0

TITLE: THE NATIONAL SURGICAL ADJUVANT BREAST PROJECT TRIAL B-04 COMPARING RADICAL MASTECTOMY AND TOTAL MASTECTOMY WITH OR WITHOUT RADIATION.

STUDY NUMBER: 71C
STUDY TITLE: NSABP B-04

ABBREVIATED TRIAL NAME: NSABP B-04
STUDY NUMBER: 71C

INVESTIGATORS' NAMES AND AFFILIATIONS:
B Fisher (1), C Redmond (2), E Montague (3) and 33 participating institutions in USA and Canada.

(1) Department of Surgery, University of Pittsburgh, Pennsylvania.
(2) NSABP Headquarters, Pittsburgh, Pennsylvania.
(3) M.D. Anderson Cancer Center, Houston, Texas.

TRIAL DESIGN: TREATMENT GROUPS:
1. Halsted Mast: Halsted radical mastectomy only.
2. Total Mast: Total mastectomy only.
3. Total Mast + R: Total mastectomy plus 50 Gy were given to the chest wall, 45 Gy at 3 cm to the supraclavicular and internal mammary nodes and 50 Gy to the axilla by supervoltage beam in 25 fractions. A boost of 10-20 Gy was given to clinically positive axillae.

ELIGIBILITY CRITERIA FOR ENTRY:
1. AGE: No restrictions.
2. CLINICAL STAGES: Clinically operable untreated breast cancer with no deep fixation, no skin involvement >2 cm and no fixation of axillary nodes. No tumor size restriction.
3. PRIMARY TREATMENT: As randomized following informed consent.
4. NODE DISSECTION: Performed only in Halsted mastectomy arm.
5. MENSTRUAL STATUS: No restrictions.

METHOD OF RANDOMIZATION:
Randomization was done by telephone call to a designated person at each participating institution, who assigned the patient to a treatment group from a previously prepared serial file.

RANDOMIZATION STRATIFICATIONS:
Clinical nodal status.

YEARS OF ACCRUAL: 1971-1974.

TOTAL NUMBER OF PATIENTS RANDOMIZED. 1765. 100 patients were ineligible and were not followed up, leaving 1665 available for analysis. Clinical node negative patients (n=1079) were randomized between all 3 treatment groups. Clinical node positive patients (n=586) were randomized between arms 1 and 3 only and are thus not relevant to the current overview.

PUBLICATIONS:
Fisher B, Montague E, Redmond C et al. Comparison of radical mastectomy with alternative treatments for primary breast cancer; a first report of results from a prospective randomised clinical trial. Cancer 1977; 39:2827-2839.

Fisher B, Montague E, Redmond C et al. Findings from NSABP Protocol No. B-04 - Comparison of radical mastectomy with alternative treatments for primary breast cancer. Cancer 1980; 46:1-13.

Fisher B, Redmond C, Fisher E et al. Ten-year results of a randomised clinical trial comparing radical mastectomy and total mastectomy with or without radiation. N. Engl. J. Med. 1985; 312:674-681.

Table 1. Outcome by allocated treatment for trial 71C (NSABP B-04)
Key: T = total free of relevant event at start of relevant year(s)

N.B: published results from this trial have been used since individual patient data were not available in 1985.

(i) Death (D) in each separate year, subdivided by age when randomized (D <50 & D 50+ / T <50 & T 50+)

Year	1. s2	2. s1	3. Rs1
Unknown	43+ 114/ 111+ 251	50+ 120/ 118+ 247	43+ 102/ 117+ 235

(ii) First recurrence or prior death (F) in each separate year, subdivided by age when randomized (F <50 & F 50+ / T <50 & T 50+)

Year	1. s2	2. s1	3. Rs1

(recurrence data not available)

(iii) Recurrence with survival (R) and Death (D) in all years together, subdivided by various characteristics when randomized (R & D / T)

Age at Entry	1. s2	2. s1	3. Rs1
< 50	?+43/111	?+50/118	?+43/117
50 +	?+114/251	?+120/247	?+102/235
Any age (including "not known")	?+157/362	?+170/365	?+145/352

Nodal Status (N0, N1-3, N4+)

		1. s2	2. s1	3. Rs1
< 50	N0 (by clinical evidence only)	?+43/111	?+50/118	?+43/117
50 +		?+114/251	?+120/247	?+102/235

Estrogen Receptors (ER, fmol/mg)

Breakdown by ER status was not available.

TITLE: WEST OF SCOTLAND STUDY OF TREATMENT OF MAMMARY CARCINOMA.

STUDY NUMBER: 72A
STUDY TITLE: WSSA GLASGOW

ABBREVIATED TRIAL NAME: WSSA GLASGOW
STUDY NUMBER: 72A

INVESTIGATORS' NAMES AND AFFILIATIONS:
A Litton (1) for the West of Scotland Surgical Association and in association with the Institute of Radiotherapy, Glasgow.

(1) Division of Surgery, Southern General Hospital, Glasgow, Scotland.

TRIAL DESIGN: TREATMENT GROUPS:
1. Mast + R_1: Simple mastectomy with radiotherapy to the chest wall but not the nodal areas.
2. Mast + R_1 + R_2: Simple mastectomy with radiotherapy to both the chest wall and nodal areas including the axilla and supra-clavicular fossa.
3. Mast + R_1 + Axillary Clearance: Simple mastectomy with axillary clearance plus radiotherapy to the chest wall but not the nodal areas.

ELIGIBILITY CRITERIA FOR ENTRY:
1. AGE: ≤76 years.
2. CLINICAL STAGES: Clinically early operable breast cancer, with no deep fixation, no skin involvement and no fixation of axillary lymph nodes. No size restrictions.
3. PRIMARY TREATMENT: Simple mastectomy defined as the removal of breast tissue with no disturbance of nodes.
4. AXILLARY SURGERY: For those allocated to axillary clearance, the axillary contents were removed with division but not the removal of pectoralis minor muscle. For other treatment groups the axilla was not entered.
5. MENSTRUAL STATUS: No restrictions.

METHOD OF RANDOMIZATION:
By opening the next in sequence of a set of sealed and numbered envelopes supplied to each individual participating surgeon by the central office. Random treatment option identified on card after diagnosis confirmed.

RANDOMIZATION STRATIFICATIONS:
Participating surgeons only.

YEARS OF ACCRUAL: 1972-1977

TOTAL NUMBER OF PATIENTS RANDOMIZED: 335.

PUBLICATIONS: None.

Table 1. Outcome by allocated treatment for trial 72A (WSSA Glasgow)
Key: T = total free of relevant event at start of relevant year(s)

(i) Death (D) in each separate year,
subdivided by age when randomized (D <50 & D 50+ / T <50 & T 50+)

Year	1. R1s1	2. R2s1	3. R1s2
1	2+ 5/ 48+ 75	0+ 1/ 31+ 63	2+ 7/ 41+ 77
2	9+ 3/ 46+ 70	1+ 3/ 31+ 62	8+ 4/ 39+ 70
3	2+ 9/ 37+ 67	2+ 7/ 30+ 59	6+ 8/ 31+ 66
4	3+ 3/ 35+ 58	1+ 4/ 28+ 52	1+ 5/ 25+ 58
5+	8+ 23/ 32+ 55	5+ 24/ 27+ 48	7+ 21/ 24+ 53

(ii) First recurrence or prior death (F) in each separate year,
subdivided by age when randomized (F <50 & F 50+ / T <50 & T 50+)

Year	1. R1s1	2. R2s1	3. R1s2
1	7+ 14/ 48+ 75	5+ 4/ 31+ 63	9+ 16/ 41+ 77
2	14+ 17/ 41+ 61	3+ 10/ 26+ 59	13+ 15/ 32+ 61
3	4+ 11/ 27+ 44	1+ 9/ 23+ 47	2+ 10/ 19+ 46
4	4+ 6/ 23+ 33	4+ 5/ 22+ 38	4+ 7/ 17+ 36
5+	3+ 13/ 18+ 27	9+ 19/ 17+ 33	2+ 14/ 13+ 28

(iii) Recurrence with survival (R) and Death (D) in all years together,
subdivided by various characteristics when randomized (R & D / T)

Age at Entry	1. R1s1	2. R2s1	3. R1s2
< 40	2+5/12	1+1/2	0+6/6
40 - 49	6+19/36	12+8/29	6+18/35
50 - 59	10+17/34	4+16/28	10+23/41
60 - 69	7+21/34	3+17/28	5+18/24
70 +	1+5/7	1+6/7	2+4/11
Any age (including "not known")	26+67/123	21+48/94	23+69/118

Nodal Status (N0, N1-3, N4+)

		1. R1s1	2. R2s1	3. R1s2
< 50	N0 (by clinical	6+18/39	13+7/29	4+11/25
50 +	evidence only)	17+37/68	6+34/56	7+18/35
< 50	Remainder (i.e.	2+6/9	0+2/2	2+13/16
50 +	other N+ or N?)	1+6/7	2+5/7	10+27/42

Estrogen Receptors (ER, fmol/mg)

Breakdown by ER status was not available.

TITLE: L-PHENYLALANINE MUSTARD (L-PAM) IN THE MANAGEMENT OF PRIMARY BREAST CANCER

ABBREVIATED TRIAL NAME: NSABP B-05
STUDY NUMBER: 72B

INVESTIGATORS' NAMES AND AFFILIATIONS:
B Fisher, C Redmond, E R Fisher, N Wolmark, and 37 participating institutions in USA and Canada.
National Surgical Adjuvant Breast Project, Pittsburgh, PA, USA.

TRIAL DESIGN: TREATMENT GROUPS:
1. Mel: Melphalan 0.15 mg/kg, orally, for 5 consecutive days every 6 weeks for 2 years.
2. Nil: Placebo.

ELIGIBILITY CRITERIA FOR ENTRY:
1. AGE: ≤ 75 years.
2. CLINICAL/PATHOLOGICAL STAGES: Operable breast cancers with one or more histologically involved axillary lymph nodes.
3. PRIMARY TREATMENT: Conventional or radical mastectomy.
4. AXILLARY NODE DISSECTION: Required
5. NUMBER OF NODES REMOVED: ?
6. MENSTRUAL STATUS: No restrictions
7. RECEPTOR STATUS: No restrictions

METHOD OF RANDOMIZATION:
Randomization was done by telephone call to a central office.

RANDOMIZATION STRATIFICATIONS:
Number of positive axillary lymph nodes (1-3, ≥4); age (≤49, ≥50 years); institution.

YEARS OF ACCRUAL: 1972 - 1974.

TOTAL NUMBER OF PATIENTS RANDOMIZED: 380. Data on 31 ineligible or lost patients are not included in the overview.

AIM, RATIONALE, SPECIAL FEATURES:
This study was the first randomized clinical trial utilizing a prolonged course of adjuvant chemotherapy. The single agent, melphalan, was chosen because of its relatively modest toxicity, since toxicity was felt to be an extremely important variable when this study was first initiated and the potential benefits from adjuvant therapy were still largely unknown.

FOLLOW-UP, COMPLIANCE AND TOXICITY:
86% of all patients received 95-100% of the amount of Mel prescribed in the original protocol, and no patient randomized to the protocol treatment received less than 75% of the scheduled doses. One patient on each arm actually received the treatment prescribed in the other arm of the study. Patients were examined every 6 weeks while on protocol and had a complete blood count and platelet count every 3 weeks. A blood urea nitrogen, creatinine, transaminases, and serum calcium were measured every 3 months. A chest X-ray was performed every 3 months and a bone survey every 6 months. The frequency of these studies was decreased during the follow-up period. The major acute toxicity was myelo-suppression, which was seen in all melphalan-treated patients. However, no patient had life-threatening complications, and only 2 patients were treated for stomatitis. Nausea and vomiting was observed in 30% of the melphalan-treated patients and 11% of the placebo patients. However, most of these occurrences were mild. Delayed toxicity has consisted of 2 cases of leukemia and 2 cases of myelodysplastic syndrome.

DISCUSSION:
After 10 years of follow-up there was an 8% difference in disease-free survival (p=0.06) and a 5% difference in overall survival (p=0.4). However, for women ≤ 49 years of age, the improvement in disease-free survival was statistically significant (p=0.03) and there was a 37% reduction in mortality which was also significant (p=0.05).

STUDY NUMBER: 72B
STUDY TITLE: NSABP B-05

PUBLICATIONS:
Fisher B, Carbone P, Economou SG et al. L-phenylalanine mustard (L-PAM) in the management of primary breast cancer. N. Engl. J. Med. 1975; 292:117-122.

Fisher B, Rockette H, Fisher E et al. Leukemia in breast cancer patients following adjuvant chemotherapy or postoperative radiation: the NSABP experience. J. Clin. Oncol. 1985; 3:1640-1658.

Fisher B, Fisher ER, Redmond C, and participating NSABP investigators. Ten-year results from the National Surgical Adjuvant Breast and Bowel Project (NSABP) clinical trial evaluating the use of L-phenylalanine mustard (L-PAM) in the management of primary breast cancer. J. Clin. Oncol. 1986; 4:929-941.

Table 1. Outcome by allocated treatment for trial 72B (NSABP B-05)
Key: T = total free of relevant event at start of relevant year(s)

(i) Death (D) in each separate year,
subdivided by age when randomized (D <50 & D 50+ / T <50 & T 50+)

Year	1. Mel	2. Nil
1	0+ 6/ 68+ 130	1+ 5/ 61+ 121
2	3+ 12/ 59+ 114	5+ 14/ 60+ 104
3	5+ 13/ 56+ 102	4+ 11/ 55+ 90
4	4+ 8/ 51+ 89	5+ 11/ 51+ 79
5+	12+ 34/ 47+ 81	23+ 22/ 46+ 68

(ii) First recurrence or prior death (F) in each separate year,
subdivided by age when randomized (F <50 & F 50+ / T <50 & T 50+)

Year	1. Mel	2. Nil

(recurrence data not available)

(iii) Recurrence with survival (R) and Death (D) in all years together,
subdivided by various characteristics when randomized (R & D / T)

Age at Entry	1. Mel	2. Nil
< 40	0+9/24	0+12/17
40 – 49	0+15/44	0+26/44
50 – 59	0+35/71	0+32/64
60 – 69	0+28/46	0+26/46
70 +	0+10/13	0+5/11
Any age (including "not known")	0+97/198	0+101/182

Nodal Status (N0, N1-3, N4+)

		1. Mel	2. Nil
< 50	N1–N3 (by axillary clearance)	0+6/40	0+16/31
50 +		0+28/73	0+22/66
< 50	N4+ (by axillary sample/clearance)	0+18/28	0+22/30
50 +		0+45/57	0+41/55

Estrogen Receptors (ER, fmol/mg)

Breakdown by ER status was not available.

Short reports

TITLE: SOUTHAMPTON BREAST DISEASE STUDY

ABBREVIATED TRIAL NAME: WESSEX
STUDY NUMBER: 73A

INVESTIGATORS' NAMES AND AFFILIATIONS:
RB Buchanan (1), JD Fraser, JM Shepherd with members of the University Department of Surgery, Southampton, England.

(1) Wessex Radiotherapy Centre, Southampton, England.

TRIAL DESIGN: TREATMENT GROUPS:
1. R: With a megavoltage (cobalt) beam, 46 Gy delivered to chest wall and internal mammary region via glancing fields, 55 Gy to the axilla and supraclavicular regions by a single anterior field in 22 fractions over 4.5 weeks and an additional posterior boost to the axilla to bring the central dose to 45 Gy.
2. Nil: No postoperative radiotherapy or other adjuvant therapy.

ELIGIBILITY CRITERIA FOR ENTRY:
1. AGE: ≤70 years.
2. CLINICAL STAGES: Operable breast cancer, stages I and II.
3. PRIMARY TREATMENT: Simple (total) mastectomy.
4. AXILLARY SURGERY: Biopsy of at least one node from the lower axilla for histological examination.
5. MENSTRUAL STATUS: No restrictions.

ADDITIONAL ELIGIBILITY OR STRATIFICATION CRITERIA
Registration was by attendance at the Royal Southampton Breast Clinic, and in majority, included preoperative immunological tests.

METHOD OF RANDOMIZATION:
By reference to cards inside sealed envelopes held, with the registration book, in the surgical department.

RANDOMIZATION STRATIFICATIONS:
Menstrual status.

YEARS OF ACCRUAL: 1973-1976.

TOTAL NUMBER OF PATIENTS RANDOMIZED: 146.

PUBLICATIONS:
Turnbull AR, Chant ADB, Buchanan RB, Turner DLT, Shepherd JM, and Fraser JD. Treatment of early breast cancer. Lancet 1978; ii: 7-9.

STUDY NUMBER: 73A
STUDY TITLE: WESSEX

Table 1. Outcome by allocated treatment for trial 73A (Wessex)
Key: T = total free of relevant event at start of relevant year(s)

(i) Death (D) in each separate year,
subdivided by age when randomized (D <50 & D 50+ / T <50 & T 50+)

Year	1. R	2. Nil
1	0+ 1/ 26+ 45	1+ 3/ 24+ 51
2	1+ 3/ 26+ 44	3+ 6/ 23+ 48
3	2+ 4/ 25+ 41	1+ 2/ 20+ 42
4	2+ 1/ 23+ 37	2+ 2/ 19+ 40
5+	4+ 9/ 21+ 36	8+ 12/ 17+ 38

(ii) First recurrence or prior death (F) in each separate year,
subdivided by age when randomized (F <50 & F 50+ / T <50 & T 50+)

Year	1. R	2. Nil
1	1+ 6/ 26+ 45	7+ 15/ 24+ 51
2	4+ 3/ 25+ 39	4+ 2/ 17+ 36
3	1+ 3/ 21+ 36	0+ 2/ 13+ 34
4	0+ 2/ 20+ 33	4+ 4/ 13+ 32
5+	7+ 9/ 20+ 31	5+ 5/ 9+ 28

(iii) Recurrence with survival (R) and Death (D) in all years together,
subdivided by various characteristics when randomized (R & D / T)

Age at Entry	1. R	2. Nil
< 40	1+2/5	1+5/7
40 - 49	3+7/21	4+10/17
50 - 59	3+10/28	0+16/28
60 - 69	2+7/16	3+8/22
70 +	0+1/1	0+1/1
Any age (including "not known")	9+27/71	8+40/75

Nodal Status (N0, N1-3, N4+)

		1. R	2. Nil
<50	N0 (by clinical	3+4/11	1+7/10
50+	evidence only)	1+9/22	2+6/21
<50	Remainder (i.e.	1+5/15	4+8/14
50+	other N+ or N?)	4+9/23	1+19/30

Estrogen Receptors (ER, fmol/mg)

Breakdown by ER status was not available.

TITLE: CMF ADJUVANT TREATMENT FOR OPERABLE BREAST CANCER WITH POSITIVE AXILLARY LYMPH NODES

STUDY NUMBER: 73B
STUDY NAME: INT MILAN 7205

ABBREVIATED TRIAL NAME: INT MILAN 7205
STUDY NUMBER: 73B

INVESTIGATORS' NAMES AND AFFILIATIONS:
G Bonadonna, P Valagussa, A Rossi, G Tancini, C Brambilla, U Veronesi.
Istituto Nazionale Tumori, Milan.

TRIAL DESIGN: TREATMENT GROUPS:
1. CMF: Cyclophosphamide 100 mg/M^2, orally, days 1-14 of a 4 week cycle; methotrexate 40 mg/M^2 and 5-fluorouracil 600 mg/M^2, intravenously, on days 1 and 8. Treatment continued for 12 cycles.
2. Nil: No adjuvant chemotherapy.

ELIGIBILITY CRITERIA FOR ENTRY:
1. AGE: ≤ 75 years.
2. CLINICAL/PATHOLOGICAL STAGES: T1-3a,N0-1b with at least 1 histologically positive lymph node.
3. PRIMARY TREATMENT: Halsted radical mastectomy or extended radical mastectomy. No adjuvant radiotherapy.
4. AXILLARY NODE DISSECTION: Full axillary node dissection required.
5. NUMBER OF NODES REMOVED: Not specified.
6. MENSTRUAL STATUS: No restriction.
7. RECEPTOR STATUS: No restriction.

METHOD OF RANDOMIZATION:
By block randomization. An excess number of patients were randomized to the CMF arm to balance patients randomized to CMF who refused to complete treatment.

RANDOMIZATION STRATIFICATIONS:
Age (≤49 and 50-75 years), number of positive lymph nodes (1-3, ≥4), type of mastectomy (Halsted or extended).

YEARS OF ACCRUAL: 6/1/73-9/11/75

TOTAL NUMBER OF PATIENTS RANDOMIZED: 391.

AIM, RATIONALE, SPECIAL FEATURES:
The goals of this study were to determine the benefits, if any, of prolonged adjuvant therapy with a chemotherapy combination. This trial was begun before any previous study had demonstrated a significant improvement in either time to recurrence or overall survival from the use of adjuvant chemotherapy.

FOLLOW-UP, COMPLIANCE AND TOXICITY:
Five patients were considered inevaluable in the initial reports; 2 patients, one in each study arm, died of cardiovascular disease without treatment failure. Three patients had major protocol violations. Although included in the analysis, 23 patients on the CMF arm refused to complete therapy after 2 (2 patients), 3 (5), 4 (3), 5 (2), 6 (2), 7 (1), 8 (2), 9 (2), 10 (1), and 11 (3) cycles of CMF. In six other patients CMF was discontinued at some time for 2-3 months before treatment was restarted.

CMF induced amenorrhea occurred in 73% of the premenopausal patients, 41% of those ≤40 years and 96% >40 years. However this was reversible in 12% of the patients, 39% of those ≤40 years and 4.5% of those >40 years. Most patients experienced some subjective side effects, but in most cases these were mild to moderate. Some alopecia occurred in 69% of the women, but alopecia severe enough to require a wig occurred in only 5%. Although patients had prolonged nausea with daily oral administration of cyclophosphamide, myelosuppression was ultimately dose-limiting.

DISCUSSION:
The results of this trial have been reported on numerous occasions with up to 14 years of follow-up. A significant improvement in disease-free survival was observed when the analysis included all patients, and there was a trend towards improved survival as well. However when the study was analyzed for pre- and postmenopausal women separately, a statistically significant improvement in disease-free survival was observed only among premenopausal patients, and an overall survival benefit was also observed in this group. (p<0.02)

PUBLICATIONS:
Bonadonna G, Brusamolino E, Valagussa P, et al. Combination chemotherapy as an adjuvant treatment in operable breast cancer. N. Engl. J. Med. 1976; 294: 405-410.

Bonadonna G, Rossi A, Valagussa P, et al. The CMF program for operable breast cancer with positive axillary nodes. Cancer 1977; 39:2904-2915.

Bonadonna G, Valagussa P, Rossi A, et al. Ten year experience with CMF-based adjuvant chemotherapy in resectable breast cancer. Breast Cancer Res. and Treat. 1985; 5:95-115.

Table 1. Outcome by allocated treatment for trial 73B (INT Milan 7205)
Key: T = total free of relevant event at start of relevant year(s)

(i) Death (D) in each separate year, subdivided by age when randomized (D <50 & D 50+ / T <50 & T 50+)

Year	1. CMF	2. Nil
1	2+ 2/ 95+ 115	1+ 3/ 75+ 106
2	1+ 6/ 93+ 113	4+ 10/ 74+ 103
3	9+ 10/ 92+ 107	6+ 10/ 70+ 93
4	4+ 10/ 83+ 97	4+ 10/ 64+ 83
5+	25+ 35/ 79+ 87	31+ 26/ 60+ 73

(ii) First recurrence or prior death (F) in each separate year, subdivided by age when randomized (F <50 & F 50+ / T <50 & T 50+)

Year	1. CMF	2. Nil
1	7+ 8/ 95+ 115	18+ 18/ 75+ 106
2	11+ 14/ 88+ 107	12+ 20/ 57+ 88
3	8+ 19/ 77+ 93	9+ 12/ 45+ 68
4	6+ 11/ 69+ 74	3+ 4/ 36+ 56
5+	20+ 25/ 63+ 63	13+ 22/ 33+ 52

(iii) Recurrence with survival (R) and Death (D) in all years together, subdivided by various characteristics when randomized (R & D / T)

Age at Entry	1. CMF	2. Nil
< 40	3+15/29	1+18/22
40 – 49	8+26/66	8+28/53
50 – 59	5+28/54	8+28/53
60 – 69	8+33/57	9+26/43
70 +	1+2/4	0+5/10
Any age (including "not known")	25+104/210	26+105/181

Nodal Status (N0, N1-3, N4+)

		1. CMF	2. Nil
< 50	N1–N3 (by axillary clearance)	5+23/63	4+31/53
50 +		12+39/80	14+36/74
< 50	N4+ (by axillary sample/clearance)	6+18/32	5+15/22
50 +		2+24/35	3+23/32

Estrogen Receptors (ER, fmol/mg)

Breakdown by ER status was not available.

TITLE: ADJUVANT CHEMOTHERAPY WITH MELPHALAN VERSUS CFPr WITH OR WITHOUT IRRADIATION FOLLOWING MASTECTOMY IN PATIENTS WITH PROGNOSTICALLY UNFAVORABLE BREAST CANCER

STUDY NUMBER: 73C1-4
STUDY TITLE: MAYO CLINIC

ABBREVIATED TRIAL NAME: MAYO CLINIC
STUDY NUMBER: 73C1-4

INVESTIGATORS' NAMES AND AFFILIATIONS:
DL Ahmann (1)*, JR O'Fallon (2), PW Scanlon (3), WS Payne (4), HF Bisel (1) JH Edmonson (1), S Frytak (1), RG Hahn (1), JN Ingle (1), J Rubin (1), ET Creagan (1).

(1) Div. of Medical Oncology, Mayo Clinic, Rochester, MN.
(2) Cancer Center Stat. Unit, Mayo Clinic, Rochester, MN.
(3) Div. of Radiation Oncology, Mayo Clinic, Rochester, MN.
(4) Dept. of Surgery, Mayo Clinic, Rochester, MN.

* Principal Investigator

TRIAL DESIGN: TREATMENT GROUPS:
1. CFPr: Cyclophosphamide was given at a dose of 150 mg/M^2 and 5-fluorouracil at 300 mg/M^2 was given intravenously on 5 successive days every 6 weeks. Prednisone 30 mg orally was given daily for 7 days of every 6 week cycle for 10 cycles.
2. CFPr + R: CFPr, as above, with adjuvant radiotherapy to axillary and clavicular nodes administered concurrently.
3. Nil: Surgery alone.
4. R: Radiotherapy as above.
5. Mel: Melphalan was given orally at a dose of 6 mg/M^2 daily for 5 days, repeated every 6 weeks.

ELIGIBILITY CRITERIA FOR ENTRY:
1. AGE: ≤70 years.
2. CLINICAL STAGE: Unilateral breast disease with evidence of lymph node involvement or unfavorable local signs such as skin, muscle, fascia or nipple involvement (excluding isolated Paget's disease) and no distant metastases.
3. PRIMARY TREATMENT: Total mastectomy with preservation of the pectoral muscle when possible.
4. NODE DISSECTION: Required.
5. MENSTRUAL STATUS: No restrictions.

METHOD OF RANDOMIZATION:
Randomization was by telephone to a central office.

RANDOMIZATION STRATIFICATIONS:
By: tumor size- (≤3.0 cm vs ≥3.0 cm); presence or absence of unfavorable local signs; number of positive lymph nodes- ≤3 vs ≥4; menopausal status; and place of surgery.

YEARS OF ACCRUAL: 1973-1980.

TOTAL NUMBER OF PATIENTS RANDOMIZED: 328.

AIM, RATIONALE, SPECIAL FEATURES:
This trial (73C1) was instituted initially to study 4 regimens: surgery or radiotherapy alone, CFPr, and CFPr with radiotherapy. After 34 patients were entered, however, early evidence from other studies indicated that chemotherapy was beneficial in this patient group, and ethical considerations prompted us to amend the study (73C2 postmenopausal, 73C3 pre-menopausal) to: continue with the CFPr and CFPr + R arms; add Mel; and close the surgery only and radiotherapy only arms. In September 1977, the melphalan arm was dropped for pre-menopausal patients (73C4) who were subsequently randomised to CFPr ± R only.

In patients with cutaneous involvement at the time of surgery or with involvement of the muscle or fascia, a radiation dose was delivered to the anterior chest wall regardless of the patient's study arm.

COMPLIANCE AND TOXICITY:
Of the 294 patients entered into the redesigned 3-arm study, 2 were subsequently ruled ineligible, and only 13 (4.4%) withdrew from treatment prior to completing the planned 10 treatment courses: 3 in the Mel arm and 5 in each of the CFPr arms.

DISCUSSION:
In premenopausal patients, there appears to be a significant increase in disease-free interval and survival associated with the use of CFPr rather than Mel. In the postmenopausal group, no distinct survival or recurrence-free advantage was associated with any of the treatment arms.

PUBLICATIONS:
Ahmann DL, O'Fallon JR, Scanlon PW, Bisel HF, Edmonson JH, Frytak S, Payne WS, O'Fallon JR, Hahn RG, Ingle JN, O'Connell MJ, Rubin J. Repeated adjuvant chemotherapy with phenylalanine mustard or 5-fluorouracil, cyclophosphamide and prednisone with or without radiation, after mastectomy for breast cancer. Lancet 1978; i: 893-896.

Ahmann DL, O'Fallon JR, Scanlon PW, Payne WS, Bisel HF, Edmonson JH, Frytak S, Hahn RG, Ingle JN, Rubin J, Creagan ET. A preliminary assessment of factors associated with recurrent disease in a surgical adjuvant clinical trial for patients with breast cancer with special emphasis on the aggressiveness of therapy. Am. J. Clin. Oncol. (CCT) 1982; 5:371-381.

Ahmann DL. Status of adjuvant chemotherapy in patients with breast cancer. Cancer (Supp.) 1984; 53: 724-728.

Table 1. Outcome by allocated treatment for trial 73C1 (Mayo Clinic)
Key: T = total free of relevant event at start of relevant year(s)

(i) Death (D) in each separate year,
subdivided by age when randomized (D <50 & D 50+ / T <50 & T 50+)

Year	1. CFPr	2. CFPrR	3. Nil	4. R
1	0+ 0/2+ 8	0+ 0/3+ 8	0+ 0/3+ 5	0+ 0/0+ 5
2	0+ 1/2+ 8	0+ 0/3+ 8	1+ 0/3+ 5	0+ 0/0+ 5
3	1+ 2/2+ 7	1+ 1/3+ 8	0+ 1/2+ 5	0+ 0/0+ 5
4	1+ 0/1+ 5	0+ 0/2+ 7	0+ 0/2+ 4	0+ 0/0+ 5
5+	0+ 3/0+ 5	0+ 3/2+ 7	0+ 2/2+ 4	0+ 1/0+ 5

(ii) First recurrence or prior death (F) in each separate year,
subdivided by age when randomized (F <50 & F 50+ / T <50 & T 50+)

Year	1. CFPr	2. CFPrR	3. Nil	4. R
1	0+ 2/2+ 8	0+ 1/3+ 8	2+ 1/3+ 5	0+ 1/0+ 5
2	2+ 2/2+ 6	1+ 1/3+ 7	0+ 1/1+ 4	0+ 0/0+ 4
3	0+ 1/0+ 4	0+ 1/2+ 6	0+ 1/1+ 3	0+ 1/0+ 4
4	0+ 0/0+ 3	0+ 0/2+ 5	0+ 1/1+ 2	0+ 1/0+ 3
5+	0+ 2/0+ 3	0+ 1/2+ 5	0+ 1/1+ 1	0+ 0/0+ 2

(iii) Recurrence with survival (R) and Death (D) in all years together,
subdivided by various characteristics when randomized (R & D / T)

Age at Entry	1. CFPr	2. CFPrR	3. Nil	4. R
< 40	0+0/0	0+1/2	1+0/1	0+0/0
40 - 49	0+2/2	0+0/1	0+1/2	0+0/0
50 - 59	1+3/5	0+2/4	0+2/2	0+1/4
60 - 69	0+3/3	0+1/3	1+1/3	0+0/1
70 +	0+0/0	0+1/1	0+0/0	0+0/0
Any age (including "not known")	1+8/10	0+5/11	2+4/8	0+1/5

Nodal Status (N0, N1-3, N4+)

		1. CFPr	2. CFPrR	3. Nil	4. R
< 50	N1-N3 (by axillary	0+0/0	0+1/2	1+0/1	0+0/0
50 +	clearance)	1+2/4	0+1/3	1+1/3	0+0/2
< 50	N4+ (by axillary	0+2/2	0+0/1	0+1/2	0+0/0
50 +	sample/clearance)	0+4/4	0+3/5	0+2/2	0+1/3

Estrogen Receptors (ER, fmol/mg)

Breakdown by ER status was not available.

Trial continued overleaf

Table 1. Outcome by allocated treatment for trial 73C2 (Mayo Clinic)
Key: T = total free of relevant event at start of relevant year(s)

(i) Death (D) in each separate year,
subdivided by age when randomized (D <50 & D 50+ / T <50 & T 50+)

Year	1. CFPr	2. CFPrR	5. Mel
1	0+ 1/ 2+ 56	0+ 1/ 5+ 62	0+ 3/ 2+ 65
2	0+ 4/ 2+ 55	0+ 5/ 5+ 61	0+ 5/ 2+ 62
3	1+ 11/ 2+ 51	1+ 6/ 5+ 56	2+ 7/ 2+ 57
4	0+ 1/ 1+ 40	0+ 5/ 4+ 50	0+ 8/ 0+ 50
5+	1+ 6/ 1+ 39	0+ 13/ 3+ 45	0+ 11/ 0+ 42

(ii) First recurrence or prior death (F) in each separate year,
subdivided by age when randomized (F <50 & F 50+ / T <50 & T 50+)

Year	1. CFPr	2. CFPrR	5. Mel
1	0+ 10/ 2+ 56	1+ 8/ 5+ 62	0+ 13/ 2+ 65
2	2+ 12/ 2+ 46	0+ 10/ 4+ 54	1+ 9/ 2+ 52
3	0+ 7/ 0+ 34	1+ 9/ 4+ 44	1+ 10/ 1+ 43
4	0+ 2/ 0+ 27	0+ 2/ 3+ 35	0+ 6/ 0+ 33
5+	0+ 5/ 0+ 25	1+ 6/ 3+ 33	0+ 9/ 0+ 27

(iii) Recurrence with survival (R) and Death (D) in all years together,
subdivided by various characteristics when randomized (R & D / T)

Age at Entry	1. CFPr	2. CFPrR	5. Mel
< 40	0+0/0	1+1/3	0+0/0
40 - 49	0+2/2	0+1/2	0+2/2
50 - 59	5+11/24	3+8/27	5+10/27
60 - 69	3+12/30	2+21/34	6+23/37
70 +	0+0/2	0+1/1	0+1/1
Any age (including "not known")	8+25/58	6+32/67	11+36/67

Nodal Status (N0, N1-3, N4+)

		1. CFPr	2. CFPrR	5. Mel
<50	N1-N3 (by axillary	0+0/0	0+0/1	0+1/1
50 +	clearance)	5+5/27	3+11/31	7+11/29
<50	N4+ (by axillary	0+2/2	1+2/4	0+1/1
50 +	sample/clearance)	3+18/29	2+19/31	4+23/36

Table 1. Outcome by allocated treatment for trial 73C3 (Mayo Clinic)
Key: T = total free of relevant event at start of relevant year(s)

(i) Death (D) in each separate year,
subdivided by age when randomized (D <50 & D 50+ / T <50 & T 50+)

Year	1. CFPr	2. CFPrR	5. Mel
1	0+ 0/ 15+ 2	1+ 0/ 12+ 4	2+ 0/ 18+ 2
2	0+ 0/ 15+ 2	0+ 1/ 11+ 4	3+ 0/ 16+ 2
3	2+ 0/ 15+ 2	0+ 0/ 11+ 3	3+ 0/ 13+ 2
4	0+ 0/ 13+ 2	1+ 0/ 11+ 3	2+ 0/ 10+ 2
5+	2+ 0/ 13+ 2	3+ 2/ 10+ 3	3+ 0/ 8+ 2

(ii) First recurrence or prior death (F) in each separate year,
subdivided by age when randomized (F <50 & F 50+ / T <50 & T 50+)

Year	1. CFPr	2. CFPrR	5. Mel
1	1+ 0/ 15+ 2	1+ 2/ 12+ 4	7+ 0/ 18+ 2
2	2+ 0/ 14+ 2	2+ 0/ 11+ 2	6+ 0/ 11+ 2
3	0+ 0/ 12+ 2	0+ 0/ 9+ 2	0+ 0/ 5+ 2
4	1+ 0/ 12+ 2	3+ 1/ 9+ 2	2+ 0/ 5+ 2
5+	4+ 0/ 11+ 2	0+ 0/ 6+ 1	1+ 1/ 3+ 2

(iii) Recurrence with survival (R) and Death (D) in all years together,
subdivided by various characteristics when randomized (R & D / T)

Age at Entry	1. CFPr	2. CFPrR	5. Mel
< 40	1+0/3	0+3/5	0+7/8
40 - 49	0+4/12	1+2/7	2+6/10
50 - 59	0+0/2	0+3/4	1+0/2
Any age (including "not known")	1+4/17	1+8/16	3+13/20

Nodal Status (N0, N1-3, N4+)

		1. CFPr	2. CFPrR	5. Mel
<50	N1-N3 (by axillary	0+1/6	0+0/5	1+3/5
50+	clearance)	0+0/2	0+1/1	0+0/0
<50	N4+ (by axillary	1+3/9	1+5/7	1+10/13
50+	sample/clearance)	0+0/0	0+2/3	1+0/2

Table 1. Outcome by allocated treatment for trial 73C4 (Mayo Clinic)
Key: T = total free of relevant event at start of relevant year(s)

(i) Death (D) in each separate year,
subdivided by age when randomized (D <50 & D 50+ / T <50 & T 50+)

Year	1. CFPr	2. CFPrR
1	1+ 0/ 23+ 4	0+ 0/ 17+ 5
2	2+ 0/ 22+ 4	3+ 2/ 17+ 5
3	3+ 0/ 20+ 4	0+ 0/ 14+ 3
4	4+ 0/ 17+ 4	0+ 0/ 14+ 3
5+	0+ 2/ 13+ 4	4+ 1/ 14+ 3

(ii) First recurrence or prior death (F) in each separate year,
subdivided by age when randomized (F <50 & F 50+ / T <50 & T 50+)

Year	1. CFPr	2. CFPrR
1	7+ 0/ 23+ 4	1+ 0/ 17+ 5
2	5+ 2/ 16+ 4	2+ 2/ 16+ 5
3	0+ 0/ 11+ 2	1+ 1/ 14+ 3
4	4+ 0/ 11+ 2	3+ 0/ 13+ 2
5+	2+ 0/ 7+ 2	3+ 0/ 10+ 2

(iii) Recurrence with survival (R) and Death (D) in all years together,
subdivided by various characteristics when randomized (R & D / T)

Age at Entry	1. CFPr	2. CFPrR
< 40	3+3/9	2+4/8
40 - 49	1+7/14	1+3/9
50 - 59	0+2/4	0+3/5
Any age (including "not known")	4+12/27	3+10/22

Nodal Status (N0, N1-3, N4+)

		1. CFPr	2. CFPrR
<50	N1-N3 (by axillary	4+2/12	2+3/9
50+	clearance)	0+1/2	0+0/1
<50	N4+ (by axillary	0+8/11	1+4/8
50+	sample/clearance)	0+1/2	0+3/4

Estrogen Receptors (ER, fmol/mg)

Breakdown by ER status was not available.

TITLE: BERLIN-BUCH INDIVIDUALIZED ADJUVANT CHEMOHORMONAL TREATMENT (IACHT) TRIAL IN HIGH RISK BREAST CANCER

STUDY NUMBER: 74A1-2
STUDY TITLE: BERLIN-BUCH

ABBREVIATED TRIAL NAME: BERLIN-BUCH
STUDY NUMBER: 74A1-2

INVESTIGATORS' NAMES AND AFFILIATIONS:
U Peek, St Tanneberger, E Heise, E Nissen, M Görlich, G Marx, D Kunde, R Winkler
  Acad. Sci. GDR, Research Center for Molecular Biology and Medicine, Berlin-Buch.

TRIAL DESIGN: TREATMENT GROUPS:
1. Various: Adriamycin, 5-fluorouracil, methotrexate, melphalan, trenimon, vinblastine. The choice of drug based on *in vitro* sensitivity. Drugs used either as single agents or in a sequential manner. When there was no growth *in vitro*, cyclophosphamide was given. Patients with receptor positive tumors were also offered hormonal therapy: nandralone for 2 years with oophorectomy for premenopausal women.
2. Nil: Surgery alone with no adjuvant therapy.

ELIGIBILITY CRITERIA FOR ENTRY:
1. AGE: <70.
2. CLINICAL/PATHOLOGICAL CATEGORIES: All except M1, $T4_a$, $T4_c$, $T4_i$, pN_0, N3 (but IMN+ included).
3. PRIMARY TREATMENT: Standard radical mastectomy without irradiation.
4. AXILLARY NODE LOCALISATION: Required.
5. INTERNAL MAMMORY NODE BIOPSY/DISSECTION: Required.
6. MENSTRUAL STATUS: No restriction.
7. RECEPTOR STATUS: No restriction.

METHOD OF RANDOMIZATION:
The envelope method was used from 1/74-5/78 (first period), on a 1:1 basis (74A1). Thereafter (second period) for ethical reasons, randomization was on a 2:1 basis in favor of the adjuvant treatment group (74A2). Randomisations were obtained post-operatively (in the first period before results of drug sensitivity tests and receptor assays were available) and so all patients are included in the analyses even though patients whose assays showed drug-resistant tumors did not receive cytotoxic therapy and most receptor positive patients received hormonal therapy.

RANDOMIZATION STRATIFICATIONS:
By menopausal and receptor status.

YEARS OF ACCRUAL: 1/74 - 3/81

TOTAL NUMBER OF PATIENTS RANDOMIZED: 200

ADDITIONAL ELIGIBILITY OR STRATIFICATION CRITERIA:
Patients accepted to the study predominantly had more than 4 positive axillary with or without internal mammary nodes involved. In some cases, patients had less than 4 positive nodes with involvement of the apex and/or internal mammary nodes. Patients were ineligible if they had: palliative operations, severe concomitant diseases, prior cancer within past 5 years.

FOLLOW-UP, COMPLIANCE AND TOXICITY:
One patient refused chemotherapy in the treatment period. Two did not receive the planned dose and treatment duration because of difficulties in the venous access; both relapsed. Thirty premenopausal patients in the adjuvant treatment arm had oophorectomies - usually followed by nandralone - and 14 postmenopausal patients had nandralone (first period). Three patients committed suicide; three patients were lost to follow-up. One patient died of drug-related complications. One case of neurotoxicity with vinblastine, one steroid ulcer, 3 tumors stimulated, and 2 cases of hypercalcemia in postmenopausal patients with nandralone were reported. There were also 4 second tumors in the contralateral breast and 1 ovarian carcinoma.

DISCUSSION:
The relapse rate in the control group was 67/85 (79%): 45/55 (82%) in the first period of the study (74A1 in the overview) and 22/30 (79%) in the second period. The relapse rate for the treated patients was 59/115 (51%), 32/55 (64%) and 27/60 (45%) in the first and second study periods respectively (p=0.001 by log-rank). In the first period, more tumors were found to be resistant or not grown *in vitro*. Survival benefits were apparent after the fifth year (p=0.01). Among tumors sensitive to one or more drugs, differences in recurrence rate between the control (29/35, 83%) and treated (34/77, 44%) groups were greater than differences between all randomized patients in the two groups. Adjuvant therapy benefited both ER positive and ER negative premenopausal patients, but only ER positive postmenopausal patients experienced a reduction in recurrence rate from adjuvant therapy. We conclude that *in vitro* tests for drug sensitivity are helpful in the selection of drugs for adjuvant therapy and that ER positivity is prediction of response to both endocrine therapy and cytotoxic therapy, at least among postmenopausal patients.

PUBLICATIONS:
Peek, U et al. High Risk Breast Cancer - Long-Term Surgical Adjuvant Therapy Based on Production Tests - Preliminary Report. Arch. Geschwulstforsch 1981; 91:139-151.

Table 1. Outcome by allocated treatment for trial 74A1 (Berlin-Buch)
Key: T = total free of relevant event at start of relevant year(s)

(i) Death (D) in each separate year, subdivided by age when randomized (D <50 & D 50+ / T <50 & T 50+)

Year	1. Chem	2. Nil
1	3+ 4/ 23+ 32	1+ 0/ 19+ 36
2	3+ 2/ 20+ 28	1+ 4/ 18+ 36
3	3+ 1/ 17+ 26	4+ 0/ 17+ 32
4	2+ 7/ 14+ 25	0+ 5/ 13+ 32
5+	3+ 6/ 12+ 18	6+ 17/ 13+ 27

(ii) First recurrence or prior death (F) in each separate year, subdivided by age when randomized (F <50 & F 50+ / T <50 & T 50+)

Year	1. Chem	2. Nil
1	4+ 7/ 23+ 32	5+ 7/ 19+ 36
2	7+ 1/ 19+ 25	3+ 7/ 14+ 29
3	0+ 6/ 12+ 24	4+ 8/ 11+ 22
4	2+ 3/ 12+ 18	2+ 1/ 7+ 14
5+	4+ 5/ 10+ 15	2+ 7/ 5+ 13

(iii) Recurrence with survival (R) and Death (D) in all years together, subdivided by various characteristics when randomized (R & D / T)

Age at Entry		1. Chem	2. Nil
< 40		0+3/5	0+1/2
40 - 49		3+11/18	4+11/17
50 - 59		0+6/10	2+11/17
60 - 69		2+14/22	2+14/18
70 +		0+0/0	0+1/1
Any age (including "not known")		5+34/55	8+38/55

	Nodal Status (N0, N1-3, N4+)		
<50	N1-N3 (by axillary	1+3/5	0+3/4
50 +	clearance)	0+3/6	3+4/8
<50	N4+ (by axillary	2+11/18	4+9/15
50 +	sample/clearance)	2+16/24	1+21/27
50 +	Remainder (i.e. other N+ or N?)	0+1/2	0+1/1

	Estrogen Receptors (ER, fmol/mg)		
<50	ER poor or ER < 10	1+7/12	1+6/8
50 +		0+9/11	0+8/8
<50	ER+ or ER 10-99	2+7/11	2+5/8
50 +		2+11/21	3+14/23

Table 1. Outcome by allocated treatment for trial 74A2 (Berlin-Buch)
Key: T = total free of relevant event at start of relevant year(s)

(i) Death (D) in each separate year, subdivided by age when randomized (D <50 & D 50+ / T <50 & T 50+)

Year	1. Chem	2. Nil
1	0+ 0/ 36+ 24	0+ 1/ 9+ 21
2	3+ 7/ 36+ 24	2+ 0/ 9+ 20
3	4+ 2/ 33+ 17	1+ 3/ 7+ 20
4	0+ 3/ 29+ 15	1+ 4/ 6+ 17
5+	1+ 1/ 29+ 12	0+ 3/ 5+ 13

(ii) First recurrence or prior death (F) in each separate year, subdivided by age when randomized (F <50 & F 50+ / T <50 & T 50+)

Year	1. Chem	2. Nil
1	1+ 6/ 36+ 24	2+ 4/ 9+ 21
2	5+ 4/ 35+ 18	0+ 5/ 7+ 17
3	3+ 4/ 30+ 14	2+ 5/ 7+ 12
4	0+ 1/ 27+ 10	0+ 1/ 5+ 7
5+	0+ 0/ 24+ 6	0+ 0/ 4+ 6

(iii) Recurrence with survival (R) and Death (D) in all years together, subdivided by various characteristics when randomized (R & D / T)

Age at Entry		1. Chem	2. Nil
< 40		0+3/12	0+0/1
40 - 49		1+5/24	0+4/8
50 - 59		2+9/14	2+6/11
60 - 69		0+4/10	2+5/10
Any age (including "not known")		3+21/60	4+15/30

	Nodal Status (N0, N1-3, N4+)		
<50	N1-N3 (by axillary	0+2/11	0+2/4
50 +	clearance)	0+2/4	1+1/3
<50	N4+ (by axillary	1+5/24	0+2/5
50 +	sample/clearance)	2+11/18	1+10/16
<50	Remainder (i.e.	0+1/1	0+0/0
50 +	other N+ or N?)	0+0/2	2+0/2

	Estrogen Receptors (ER, fmol/mg)		
<50	ER poor or ER < 10	0+5/18	0+4/7
50 +		1+8/13	1+4/8
<50	ER+ or ER 10-99	1+3/17	0+0/2
50 +		1+5/11	2+7/12

TITLE: SOUTH EAST SCOTLAND TRIAL OF POST-OPERATIVE RADIOTHERAPY

STUDY NUMBER: 74B
STUDY TITLE: EDINBURGH I

ABBREVIATED TRIAL NAME: EDINBURGH I
STUDY NUMBER: 74B

INVESTIGATORS' NAMES AND AFFILIATIONS:
HJ Stewart (1), APM Forrest (2), A Rodger (3), RJ Prescott (4) and 39 participating clinicians from the S.E. of Scotland.

(1) The Scottish Cancer Trials Office (MRC), Edinburgh.
(2) The Department of Surgery (RIE), Edinburgh.
(3) The Department of Radiation Oncology, Edinburgh.
(4) The Medical Statistics Unit, Edinburgh.

TRIAL DESIGN: TREATMENT GROUPS:
1. R: Routine post-mastectomy radiotherapy by the McWhirter technique 42.5-45 Gy maximum tissue dose in 10 fractions over 4 weeks.
2. Nil: No immediate radiotherapy; delayed radiotherapy when indicated by isolated local or regional recurrence.

ELIGIBILITY CRITERIA FOR ENTRY:
1. AGE: < 70 yrs.
2. CLINICAL STAGES: Stage I and II disease without nodal involvement (10 level sectioning) or for which no node was identified for examination (UICC 1968).
3. PRIMARY TREATMENT: Simple (total) mastectomy.
4. AXILLARY SURGERY: Sample of at least one sub-pectoral node.
5. MENSTRUAL STATUS: Unrestricted, but only natural menopause.

ADDITIONAL ELIGIBILITY OR STRATIFICATION CRITERIA:
Registration included a preoperative examination by radiotherapist.

METHOD OF RANDOMIZATION:
Removal from envelope of a previously prepared and numbered registration card, blank except for treatment option. Envelopes held in radiotherapy department and only used when eligibility confirmed by trial staff. Options balanced in blocks of six.

RANDOMIZATION STRATIFICATIONS:
1. MENSTRUAL STATUS: Regular periods; irregular periods or last menstrual period within the preceding 5 years; last menstrual period more than 5 years past.
2. TUMOR SIZE: Up to and including 2 cm (T_1); >2-5 cm (T_2).
3. TUMOR SITE: Involvement or no involvement of the medial half of breast.

YEARS OF ACCRUAL: 1974-1979.

TOTAL NUMBER OF PATIENTS RANDOMIZED: 348.

PUBLICATIONS:
Duncan W, Forrest APM, Gray N, Hamilton T, Langlands AO, Prescott RJ, Shivas AA, Stewart HJ. New Edinburgh primary breast cancer trials: report by co-ordinating committee. Br. J. Cancer 1975; 32: 628-630.

Stewart HJ, Forrest APM, Rodger A, Duncan W, Langlands AO, Gunn JM, Songhorabadi S. and 39 participating clinicians. Mastectomy and node sampling: the need for postoperative radiotherapy. Br. J. Surg. 1983; 70: 692, (Abstract).

Table 1. Outcome by allocated treatment for trial 74B (Edinburgh I)
Key: T = total free of relevant event at start of relevant year(s)

(i) Death (D) in each separate year, subdivided by age when randomized (D <50 & D 50+ / T <50 & T 50+)

Year	1. R	2. Nil
1	2+ 1/ 73+ 100	0+ 0/ 65+ 110
2	0+ 5/ 71+ 99	1+ 3/ 65+ 110
3	5+ 8/ 71+ 94	4+ 3/ 64+ 107
4	4+ 4/ 66+ 86	2+ 8/ 60+ 104
5+	7+ 14/ 62+ 82	8+ 21/ 58+ 96

(ii) First recurrence or prior death (F) in each separate year, subdivided by age when randomized (F <50 & F 50+ / T <50 & T 50+)

Year	1. R	2. Nil

(recurrence data not available)

(iii) Recurrence with survival (R) and Death (D) in all years together, subdivided by various characteristics when randomized (R & D / T)

Age at Entry	1. R	2. Nil
< 40	0+7/19	0+3/14
40 - 49	0+11/54	0+12/51
50 - 59	0+15/49	0+16/56
60 - 69	0+17/51	0+19/54
Any age (including "not known")	0+50/173	0+50/175

Nodal Status (N0, N1-3, N4+)

Breakdown by nodal status was not available.

Estrogen Receptors (ER, fmol/mg)

Breakdown by ER status was not available.

Short reports

TITLE: SOUTH EAST SCOTLAND TRIAL OF ADJUVANT CHEMOTHERAPY

STUDY NUMBER: 74C
STUDY TITLE: EDINBURGH II

ABBREVIATED TRIAL NAME: EDINBURGH II
STUDY NUMBER: 74C

INVESTIGATORS' NAMES AND AFFILIATIONS:
HJ Stewart (1), APM Forrest (2), A Rodger (3), RJ Prescott (4) and 39 participating clinicians from the S.E. of Scotland.

(1) The Scottish Cancer Trials Office (MRC), Edinburgh.
(2) The Department of Surgery, Royal Infirmary, Edinburgh.
(3) The Department of Radiation Oncology, Edinburgh.
(4) The Medical Statistics Unit, Edinburgh.

TRIAL DESIGN: TREATMENT GROUPS:
1. F: 5-fluorouracil 700 mg/M^2 intravenously every four weeks for 12 injections.
2. Nil: No adjuvant chemotherapy.

ELIGIBILITY CRITERIA FOR ENTRY:
1. AGE: <70 years of age.
2. CLINICAL STAGES: Stage I and II disease with positive node histology or stage III disease (UICC 1968).
3. PRIMARY TREATMENT: Simple mastectomy and postoperative radiotherapy; radical radiotherapy if no mastectomy (inoperable disease); or previous mastectomy with recent axillary recurrence treated by delayed radiotherapy.
4. AXILLARY SURGERY: Sample at least one sub-pectoral node.
5. MENSTRUAL STATUS: Unrestricted, but only natural menopause.

ADDITIONAL ELIGIBILITY OR STRATIFICATION CRITERIA:
Registration included preoperative examination by radiotherapist.

METHOD OF RANDOMIZATION:
Removal from envelope of a previously prepared and numbered registration card, blank but for treatment option. Envelopes held in radiotherapy department and only used when eligibility confirmed by trial staff. Options balanced in blocks of six.

RANDOMIZATION STRATIFICATIONS:
1. MENSTRUAL STATUS: <2 yrs or 2-5 yrs or >5 yrs from last menstrual period.
2. OPERABILITY: Mastectomy or no mastectomy.
3. STAGE OF OPERABLE: Histological stage II or operable stage III.
4. TIMING OF RADIOTHERAPY: Postoperative or delayed.

YEARS OF ACCRUAL: 1974-1978.

TOTAL NUMBER OF PATIENTS RANDOMIZED: 345.

PUBLICATIONS:
Duncan W, Forrest APM, Gray N, Hamilton T, Langlands AO, Prescott RJ, Shivas AA, Stewart HJ. New Edinburgh primary breast cancer trials: report by co-ordinating committee. Br. J. Cancer 1975; 32:628-630.

Stewart HJ, Forrest APM, Rodger A, Duncan W, Langlands AO, Gunn JM, Songhorabadi S. and 39 participating clinicians. Adjuvant chemotherapy: the results of a randomized trial in breast cancer. Br. J. of Surgery 1983; 70:687, (Abstract).

Table 1. Outcome by allocated treatment for trial 74C (Edinburgh II)
Key: T = total free of relevant event at start of relevant year(s)

(i) Death (D) in each separate year,
subdivided by age when randomized (D <50 & D 50+ / T <50 & T 50+)

Year	1. F	2. Nil
1	3+ 9/ 59+ 116	10+ 10/ 56+ 114
2	9+ 12/ 56+ 107	6+ 12/ 46+ 104
3	10+ 11/ 47+ 95	5+ 20/ 40+ 92
4	3+ 11/ 37+ 84	6+ 4/ 35+ 72
5+	14+ 26/ 34+ 73	15+ 28/ 29+ 68

(ii) First recurrence or prior death (F) in each separate year,
subdivided by age when randomized (F <50 & F 50+ / T <50 & T 50+)

Year	1. F	2. Nil
1	14+ 23/ 59+ 116	19+ 27/ 56+ 114
2	11+ 21/ 45+ 93	15+ 23/ 37+ 87
3	8+ 14/ 34+ 72	5+ 13/ 22+ 64
4	6+ 9/ 26+ 58	2+ 9/ 17+ 51
5+	7+ 16/ 20+ 49	5+ 15/ 15+ 42

(iii) Recurrence with survival (R) and Death (D) in all years together,
subdivided by various characteristics when randomized (R & D / T)

Age at Entry		1. F	2. Nil
<40		2+10/14	0+10/11
40-49		5+29/45	4+32/45
50-59		4+30/52	4+40/60
60-69		10+39/64	9+34/54
Any age (including "not known")		21+108/175	17+116/170

	Nodal Status (N0, N1-3, N4+)		
<50	N0 (by clinical	1+8/13	2+2/5
50+	evidence only)	3+14/29	4+12/25
<50	Remainder (i.e.	6+31/46	2+40/51
50+	other N+ or N?)	11+55/87	9+62/89

	Estrogen Receptors (ER, fmol/mg)		
<50	ER poor or ER < 10	0+4/5	1+3/4
50+		0+6/7	1+8/11
<50	ER+ or ER 10-99	3+7/14	0+9/10
50+		1+2/5	2+6/12
<50	ER++ or ER 100+	0+1/1	0+3/3
50+		2+7/14	4+7/12

TITLE: MULTIMODALITY THERAPY OF EARLY BREAST CANCER

ABBREVIATED TRIAL NAME: DFCI 74063

STUDY NUMBER: 74D1-2

INVESTIGATORS' NAMES AND AFFILIATIONS:
IC Henderson (1), CL Shapiro (1), RS Gelman (2), JR Harris (3), GP Canellos (1), E. Frei III (1).

(1) Division of Clinical Oncology, Dana-Farber Cancer Institute and Harvard Medical School, Boston.
(2) Division of Biostatistics and Epidemiology, Dana-Farber Cancer Institute and Harvard Medical School, Boston.
(3) Joint Center for Radiation Therapy, Dana-Farber Cancer Institute and Harvard Medical School, Boston.

TRIAL DESIGN: TREATMENT GROUPS:
1. Mel: Melphalan 5.25 mg/M^2 (15 mg per kg daily) given orally for 5 days every 6 weeks for 2 years.
2. AC x 15 w: Doxorubicin 45 mg/M^2 and cyclophosphamide 500 mg/M^2 given intravenously every 21 days for 15 weeks.
3. AC x 30 w: AC, as above, for 30 weeks.
4. Nil: No adjuvant therapy.

ELIGIBILITY CRITERIA FOR ENTRY:
1. AGE: No restrictions.
2. CLINICAL/PATHOLOGICAL STAGES: Stages II and III, including T1-3 lesions with ≥4 histologically positive nodes or ≥1 node in axillary zone III; central or medial tumors with any number of histologically positive axillary nodes; and patients < age 30 with 1-3 positive nodes and no distant metastases.
3. PRIMARY TREATMENT: Radical or modified radical mastectomy or tylectomy/simple mastectomy with radiotherapy.
4. AXILLARY NODE DISSECTION: Required.
5. NUMBER OF NODES REMOVED: Not specified.
6. MENSTRUAL STATUS: No restriction.
7. RECEPTOR STATUS: No restriction.

METHOD OF RANDOMIZATION:
By telephone to a central office using computer-produced stratification sheets.

RANDOMIZATION STRATIFICATIONS:
By primary therapy: radical or modified mastectomy; or tylectomy/simple mastectomy and radiotherapy.

By interval from primary therapy: 6 months, 6-12 months, >12 but <24 months.

YEARS OF ACCRUAL: 1/75-9/75.

TOTAL NUMBER OF PATIENTS RANDOMIZED: 40.

AIM, RATIONALE, SPECIAL FEATURES:
The study was designed to determine the effect of adjuvant chemotherapy on the time to recurrence and survival in patients with primary breast cancer and high risk of relapse. The study was begun before the demonstration by the NSABP that melphalan prolonged time to recurrence, and patients were initially randomized to AC 15 weeks v. AC 30 weeks v. no adjuvant therapy. With publication of the results from NSABP B05 (study number 72B), the no-therapy arm was dropped and melphalan was used in the control arm. The entire study was eventually replaced by DFCI 75-122 (study number 74D3+4). Patients in DFCI 75-122 were randomized to AC 15 weeks v. AC 30 weeks, and patients in these two arms of DFCI 74-003 have been combined with those in 75-122 for analysis of the effect of therapy duration.

DISCUSSION:
In 1989 the median follow-up of this trial is 9.2 years and the maximum 15 years. Sixty percent of all patients have died and 61% have relapsed. The median survival and time to first failure (in months) are 72.9 vs. 70.5 (p = 0.42) and 40.4 vs. 39.9 (p = 0.64) for AC 15 vs. AC 30 respectively.

PUBLICATIONS:
Henderson IC, Gelman RS, Parker LM, Skarin AT, Mayer RJ, Garnick MB, Canellos GP, Frei E. 15 vs. 30 weeks of adjuvant chemotherapy for breast cancer patients with a high risk of recurrence: a randomized trial. Proc. Am. Soc. Clin. Oncol. 1982, 1:C-283:5.

Henderson IC, Gelman RS, Harris JR, Canellos GP. Duration of therapy in adjuvant chemotherapy trials. NCI Monogr. 1986, 1:95-98.

Table 1. Outcome by allocated treatment for trial 74D1 (DFCI 74-063)
Key: T = total free of relevant event at start of relevant year(s)

(i) Death (D) in each separate year, subdivided by age when randomized (D <50 & D 50+ / T <50 & T 50+)

Year	2. AC	3. AC	4. Nil
1	0+ 0/ 1+ 0	0+ 0/ 1+ 2	0+ 0/ 1+ 3
2	0+ 0/ 1+ 0	0+ 0/ 1+ 2	0+ 0/ 1+ 3
3	0+ 0/ 1+ 0	0+ 0/ 1+ 2	0+ 0/ 1+ 3
4	0+ 0/ 1+ 0	0+ 0/ 1+ 2	0+ 1/ 1+ 3
5+	0+ 0/ 1+ 0	1+ 1/ 1+ 2	1+ 1/ 1+ 2

(ii) First recurrence or prior death (F) in each separate year, subdivided by age when randomized (F <50 & F 50+ / T <50 & T 50+)

Year	2. AC	3. AC	4. Nil
1	0+ 0/ 1+ 0	1+ 0/ 1+ 2	0+ 1/ 1+ 3
2	0+ 0/ 1+ 0	0+ 0/ 0+ 2	0+ 1/ 1+ 1
3	0+ 0/ 1+ 0	0+ 0/ 0+ 2	0+ 0/ 1+ 0
4	0+ 0/ 1+ 0	0+ 0/ 0+ 2	0+ 0/ 1+ 0
5+	0+ 0/ 1+ 0	0+ 1/ 0+ 2	1+ 0/ 1+ 0

(iii) Recurrence with survival (R) and Death (D) in all years together, subdivided by various characteristics when randomized (R & D / T)

Age at Entry		2. AC	3. AC	4. Nil
< 40		0+0/0	0+1/1	0+0/0
40 - 49		0+0/1	0+0/0	0+1/1
50 - 59		0+0/0	0+1/2	0+0/0
70 +		0+0/0	0+0/0	0+2/2
Any age (including "not known")		0+0/1	0+2/3	0+3/4

Nodal Status (N0, N1-3, N4+)

		2. AC	3. AC	4. Nil
< 50	N0 (by axillary clearance)	0+0/1	0+0/0	0+0/0
< 50	N1-N3 (by axillary clearance)	0+0/0	0+1/1	0+0/0
50 +		0+0/0	0+0/1	0+0/0
< 50	N4+ (by axillary sample/clearance)	0+0/0	0+0/0	0+1/1
50 +		0+0/0	0+1/1	0+1/1
50 +	Remainder (i.e. other N+ or N?)	0+0/0	0+0/0	0+1/2

Estrogen Receptors (ER, fmol/mg)

		2. AC	3. AC	4. Nil
50 +	ER+ or ER 10-99	0+0/0	0+0/0	0+1/1

Table 1. Outcome by allocated treatment for trial 74D2 (DFCI 74-063)
Key: T = total free of relevant event at start of relevant year(s)

(i) Death (D) in each separate year, subdivided by age when randomized (D <50 & D 50+ / T <50 & T 50+)

Year	1. Mel	2. AC	3. AC
1	0+ 0/ 2+ 3	0+ 1/ 6+ 7	0+ 0/ 7+ 7
2	0+ 0/ 2+ 3	0+ 2/ 6+ 6	2+ 1/ 7+ 7
3	0+ 0/ 2+ 3	0+ 0/ 6+ 4	1+ 0/ 5+ 6
4	0+ 1/ 2+ 3	0+ 0/ 6+ 4	0+ 1/ 4+ 6
5+	1+ 0/ 2+ 2	4+ 1/ 6+ 4	1+ 3/ 4+ 5

(ii) First recurrence or prior death (F) in each separate year, subdivided by age when randomized (F <50 & F 50+ / T <50 & T 50+)

Year	1. Mel	2. AC	3. AC
1	0+ 1/ 2+ 3	0+ 3/ 6+ 7	2+ 1/ 7+ 7
2	0+ 0/ 2+ 2	1+ 0/ 6+ 4	1+ 1/ 5+ 6
3	1+ 0/ 2+ 2	0+ 0/ 5+ 4	1+ 1/ 4+ 5
4	0+ 0/ 1+ 2	0+ 0/ 5+ 4	0+ 0/ 3+ 4
5+	0+ 0/ 1+ 2	4+ 2/ 5+ 4	1+ 4/ 2+ 4

(iii) Recurrence with survival (R) and Death (D) in all years together, subdivided by various characteristics when randomized (R & D / T)

Age at Entry		1. Mel	2. AC	3. AC
< 40		0+0/1	0+1/1	0+3/4
40 - 49		0+1/1	1+3/5	1+1/3
50 - 59		0+1/2	0+1/2	0+2/2
60 - 69		0+0/1	1+3/5	2+2/4
70 +		0+0/0	0+0/0	0+1/1
Any age (including "not known")		0+2/5	2+8/13	3+9/14

Nodal Status (N0, N1-3, N4+)

		1. Mel	2. AC	3. AC
50 +	N0 (by axillary clearance)	0+0/0	0+1/1	0+0/0
< 50	N1-N3 (by axillary clearance)	0+0/0	0+0/0	0+0/2
50 +		0+0/2	1+0/1	0+1/1
< 50	N4+ (by axillary sample/clearance)	0+1/2	1+3/5	1+3/4
50 +		0+1/1	0+3/5	2+3/5
< 50	Remainder (i.e. other N+ or N?)	0+1/1	0+1/1	0+1/1
50 +		0+0/0	0+0/0	0+1/1

Estrogen Receptors (ER, fmol/mg)

		1. Mel	2. AC	3. AC
< 50	ER poor or ER < 10	0+1/1	0+1/1	0+0/0
< 50	ER+ or ER 10-99	0+0/0	0+0/0	0+1/1
50 +		0+1/1	0+1/1	0+1/1

TITLE: CYCLOPHOSPHAMIDE AND DOXORUBICIN 15 VS 30 WEEKS AS ADJUVANT CHEMOTHERAPY FOR HIGH RISK BREAST CANCER PATIENTS

STUDY NUMBER: 74D4
STUDY TITLE: DFCI 75-122

ABBREVIATED TRIAL NAME: DFCI 75-122
STUDY NUMBER: 74D4

INVESTIGATORS' NAMES AND AFFILIATIONS:
IC Henderson (1), RS Gelman (2), CL Shapiro (1), JR Harris (3), GP Canellos (1), E. Frei III (1).
  (1) Division of Clinical Oncology, Dana-Farber Cancer Institute and Harvard Medical School, Boston, MA.
  (2) Division of Biostatistics and Epidemiology, Dana-Farber Cancer Institute and Harvard Medical School, Boston, MA.
  (3) Joint Center for Radiation Therapy, Dana-Farber Cancer Institute and Harvard Medical School, Boston, MA.

TRIAL DESIGN: TREATMENT GROUPS:
1. AC: Doxorubicin, 45 mg/M^2, and cyclophosphamide, 500 mg/M^2, intravenously, every 21 days for 15 weeks.
2. AC: AC, as above, for 30 weeks.

ELIGIBILITY CRITERIA FOR ENTRY:
1. AGE: < 75 years.
2. CLINICAL/PATHOLOGICAL STAGES: T1-3 with ≥4 histologically positive lymph nodes or ≥1 positive lymph node in axillary zone III. No distant metastasis.
3. PRIMARY TREATMENT: Radical or modified radical mastectomy or tylectomy with radiotherapy.
4. AXILLARY NODE DISSECTION: Required.
5. NUMBER OF NODES REMOVED: Not specified.
6. MENSTRUAL STATUS: No restriction.
7. RECEPTOR STATUS: Not specified.

METHOD OF RANDOMIZATION:
By telephone to a Central Office using computer-produced stratification sheets.

RANDOMIZATION STRATIFICATIONS:
By primary therapy:
  1. Radical or modified mastectomy
  2. Tylectomy and radiotherapy

By interval from primary therapy:
  1. <3 months;
  2. 3-6 months;
  3. 6-12 months

YEARS OF ACCRUAL: 1/8/76-2/28/85

TOTAL NUMBER OF PATIENTS RANDOMIZED: 268

AIM, RATIONALE, SPECIAL FEATURES:
The study was initiated to determine if short term therapy (15 weeks) is as effective as a longer course of therapy (30 weeks). It is a successor study to the 4-arm study DFCI 74-063 (study number 74D1-2). Two arms of these studies are identical, and the 31 patients randomized to 15 or 30 weeks of AC in study 74-063 are combined with the 268 patients in study 75-122 for analysis. 256 N1-3 patients were randomized between CMF and MF in a parallel study (74D3) but are not relevant to the present overview.

ADDITIONAL ELIGIBILITY OR STRATIFICATION CRITERIA:
Although a few patients were treated with a modified radical mastectomy and adjuvant radiotherapy before entry onto the study, most were not. These patients were asked to undergo a second randomization after the completion of adjuvant chemotherapy to receive either no adjuvant radiotherapy or radiotherapy to the chest wall, axilla, supraclavicular nodes, and internal mammary nodes.

FOLLOW-UP, COMPLIANCE AND TOXICITY:
As of 1989 follow-up information is available for 97% of patients. Information on cardiotoxicity is available from an analysis performed in 1985-86. At that time the incidence of cardiotoxicity was 1.4% for AC 15 and 6.6% for AC 30. More recent information is presently being analyzed.

DISCUSSION:
In 1989 the median follow-up of this trial is 9.2 years and the maximum 15 years. 60% of all patients have died and 61% have relapsed. The median survival and time to first failure (in months) are 72.9 vs 70.5 (p=0.42) and 40.4 vs 39.9 (p=0.64) for AC 15 vs AC 30, respectively.

PUBLICATIONS:
Henderson IC, Gelman RS, Parker LM, Skarin AT, Mayer RJ, Garnick MB, Canellos GP, Frei E. 15 vs 30 weeks (wks) of adjuvant chemotherapy for breast cancer patients (pts) with a high risk of recurrence: a randomized trial. Proc. ASCO 1982; 1:C-290:75.

Henderson IC, Gelman RS, Harris JR, Canellos GP. Duration of therapy in adjuvant chemotherapy trials. NCI Mongr.1986; 1:95-98.

Griem KL, Henderson IC, Gelman RS, Ascoli D, Silver B, Recht A, Goodman S, Hellman S, Harris JR. The 5-year results of a randomized trial of adjuvant radiation therapy after chemotherapy in breast cancer patients treated with mastectomy. J. Clin. Oncol. 1987; 5:1546-1555.

Table 1. Outcome by allocated treatment for trial 74D4 (DFCI 75-122B)
Key: T = total free of relevant event at start of relevant year(s)

(i) Death (D) in each separate year,
subdivided by age when randomized (D <50 & D 50+ / T <50 & T 50+)

Year	1. AC	2. AC
1	0+ 3/ 54+ 83	1+ 1/ 50+ 81
2	8+ 6/ 51+ 76	4+ 6/ 45+ 78
3	3+ 5/ 39+ 64	11+ 4/ 35+ 64
4	2+ 6/ 32+ 53	3+ 7/ 21+ 53
5+	6+ 8/ 26+ 44	8+ 9/ 17+ 42

(ii) First recurrence or prior death (F) in each separate year,
subdivided by age when randomized (F <50 & F 50+ / T <50 & T 50+)

Year	1. AC	2. AC
1	6+ 14/ 54+ 83	9+ 5/ 50+ 81
2	13+ 12/ 47+ 68	8+ 16/ 37+ 73
3	3+ 9/ 28+ 49	7+ 9/ 22+ 49
4	2+ 7/ 20+ 35	3+ 3/ 14+ 36
5+	3+ 6/ 14+ 23	4+ 5/ 10+ 26

(iii) Recurrence with survival (R) and Death (D) in all years together,
subdivided by various characteristics when randomized (R & D / T)

Age at Entry	1. AC	2. AC
< 40	4+9/24	2+9/15
40 - 49	4+10/30	2+18/35
50 - 59	8+15/46	7+15/48
60 - 69	12+11/33	4+12/31
70 +	0+2/4	0+0/2
Any age (including "not known")	28+47/137	15+54/131

	Nodal Status (N0, N1-3, N4+)		
50 +	N0 (by axillary clearance)	0+0/1	0+0/0
<50	N1-N3 (by axillary	0+2/5	0+0/2
50+	clearance)	0+0/1	0+0/1
<50	N4+ (by axillary	8+15/45	2+24/41
50 +	sample/clearance)	18+25/72	10+25/74
<50	Remainder (i.e.	0+2/4	2+3/7
50 +	other N+ or N?)	2+3/9	1+2/6

	Estrogen Receptors (ER, fmol/mg)		
<50	ER poor or ER < 10	0+11/25	2+9/17
50 +		3+5/20	2+9/19
<50	ER+ or ER 10-99	5+4/15	1+9/17
50 +		5+9/22	4+8/34
<50	ER++ or ER 100+	1+0/2	0+1/2
50 +		4+0/14	3+2/11

TITLE: MCCG STUDY OF ADJUVANT CHEMOTHERAPY FOR EARLY BREAST CANCER

STUDY NUMBER: 74E
STUDY TITLE: UK MCCG 003

ABBREVIATED TRIAL NAME: UK MCCG 003
STUDY NUMBER: 74E

INVESTIGATORS NAMES AND AFFILIATIONS:
MF Spittle (1), MJ Ostrowski (2), TD Bates (3), BT Hill (4), KD MacRae (5).

(1) Meyerstein Institute of Radiotherapy & Oncology, Middlesex Hospital, London.
(2) Dept. Radiotherapy & Oncology, Norfolk & Norwich Hospital, Norwich.
(3) South East London Radiotherapy Centre, St. Thomas' Hospital, London.
(4) Imperial Cancer Research Fund Laboratories, London.
(5) Dept. Medical Statistics, Charing Cross & Westminster Medical School, London.

TRIAL DESIGN: TREATMENT GROUPS:
1. CMFV: Cyclophosphamide 300 mg, vincristine 65 mg, 5-fluorouracil 500 mg on day 1 and cyclophosphamide 300 mg, vincristine 0.65 mg, methotrexate 37.55 mg on day 8 of 4 week cycle. Cyclophosphamide dose reduced to 200 mg and methotrexate to 25 mg if patient weight was less than 54 kg. All chemotherapy was intravenous and courses were repeated for a 6 month period.
2. Nil: No adjuvant chemotherapy.

ELIGIBILITY CRITERIA FOR ENTRY:
1. AGE: ≤70 years of age.
2. CLINICAL STAGES: Primary tumor ≤5 cm.
3. PRIMARY TREATMENT: Any radical surgical procedure with no demonstrable residual local or distant disease. Radiotherapy optional.
4. NODE DISSECTION: Optional.
5. NUMBER OF NODES REMOVED: Not specified.
6. MENSTRUAL STATUS: No restrictions.

METHOD OF RANDOMIZATION:
Each centre was issued its own randomization envelopes. It was theoretically possible to determine patient allocation before randomization. The trial was subject to external audit. Patients withdrawn after randomization due to ineligibility have been flagged with the Office of Population Censuses and Surveys for mortality data.

RANDOMIZATION STRATIFICATIONS: None.

YEARS OF ACCRUAL: 1975 - 1977.

TOTAL NUMBER OF PATIENTS RANDOMIZED: 290.

AIM, RATIONALE, SPECIAL FEATURES:
Any surgical procedure was accepted as were patients of any nodal status provided a local "cure" had been obtained.

ADDITIONAL ELIGIBILITY OR STRATIFICATION CRITERIA:
Patients were required to have a normal complete blood count, normal liver and renal functions. Chest X-ray also had to show no evidence of disease. Isotope scans and liver ultrasound were only mandatory if clinically indicated and again had to show no evidence of disease. Entry to the study had to be within 10 weeks of primary surgery.

FOLLOW-UP, COMPLIANCE AND TOXICITY:
Follow-up was usually arranged at 3 month intervals for the first year, every 4 months for the second year, every 6 months for the next 2 years and annually thereafter.

Approximately 10% of patients allocated to the chemotherapy arm failed for a variety of reasons to receive the complete course of treatment. The chemotherapy regimen was well tolerated when compared to the more traditional CMF regimens.

Of the 290 patients randomized, 38 patients failed to fulfil the eligibility criteria. Of the 252 remaining entrants a further 5 patients in arm 2 were considered ineligible at the time of the external audit, as it was felt that their inclusion might bias the published results in favor of the chemotherapy limb. However, they have been included for the purposes of the overview.

DISCUSSION:
The main finding of this study was that an extended disease free interval could be obtained at the expense of minimal toxicity, using a relatively non toxic chemotherapy regimen, suitable for administration on an outpatient basis both in specialized centres and regional hospitals.

PUBLICATIONS:
Spittle, MF, Ostrowski, MJ, Hill, BT, MacRae, KD, Bates, TD, et al. Adjuvant chemotherapy in the treatment of breast cancer: results of a multicentre study. Eur. J. Surg. Oncol. 1986; 12:109-116.

Table 1. Outcome by allocated treatment for trial 74E (UK MCCG 003)
Key: T = total free of relevant event at start of relevant year(s)

(i) Death (D) in each separate year, subdivided by age when randomized (D <50 & D 50+ /T <50 & T 50+)

Year	1. CMFV	2. Nil
1	1+ 4/ 61+ 88	3+ 3/ 47+ 94
2	8+ 6/ 60+ 83	2+ 12/ 44+ 91
3	7+ 7/ 52+ 77	7+ 9/ 42+ 79
4	2+ 9/ 45+ 70	6+ 4/ 35+ 70
5+	14+ 16/ 43+ 61	7+ 21/ 29+ 66

(ii) First recurrence or prior death (F) in each separate year, subdivided by age when randomized (F <50 & F 50+ /T <50 & T 50+)

Year	1. CMFV	2. Nil
1	8+ 13/ 61+ 88	11+ 16/ 47+ 94
2	12+ 12/ 48+ 70	12+ 18/ 35+ 70
3	8+ 9/ 34+ 58	4+ 9/ 23+ 49
4	4+ 8/ 26+ 49	0+ 5/ 19+ 40
5+	6+ 11/ 22+ 40	3+ 16/ 19+ 34

(iii) Recurrence with survival (R) and Death (D) in all years together, subdivided by various characteristics when randomized (R & D / T)

Age at Entry	1. CMFV	2. Nil
< 40	1+10/16	0+7/8
40 - 49	5+22/45	5+18/39
50 - 59	5+25/48	9+31/55
60 - 69	6+13/31	6+14/32
70 +	0+3/7	0+3/5
Any age (including "not known")	17+74/149	20+74/141

Nodal Status (N0, N1-3, N4+)

		1. CMFV	2. Nil
< 50	N0 (by clinical evidence only)	1+5/10	2+1/9
50 +		2+5/10	1+5/12
< 50	Remainder (i.e. other N+, or N?)	5+27/51	3+24/38
50 +		9+37/78	14+44/82

Estrogen Receptors (ER. fmol/mg)

Breakdown by ER status was not available.

TITLE: WESTERN CANCER STUDY GROUP TRIAL OF CMF VS. 5-FU AS ADJUVANT BREAST CANCER THERAPY

STUDY NUMBER: 74F
STUDY TITLE: WESTERN CSG

ABBREVIATED TRIAL NAME: WESTERN CSG
STUDY NUMBER: 74F

INVESTIGATORS' NAMES AND AFFILIATIONS:
RT Chlebowski* (1), JM Weiner (2), R Reynolds (3), J Luce (4), L Bulcavage (1), JR Bateman (2)

(1) UCLA School of Medicine, Department of Medicine, Harbor/UCLA Medical Center, Torrance.
(2) University of Southern California School of Medicine, Department of Medicine, Los Angeles.
(3) Travis Air Force Base, Fairfield.
(4) Mountain States Tumor Institute, Boise.

* Principal Investigator

TRIAL DESIGN: TREATMENT GROUPS:
1. F: 5-fluorouracil at a dose of 500 mg/M^2 given intravenously every week for 12 months.
2. CMF: Cyclophosphamide 400 mg/M^2, methotrexate 30 mg/M^2 and 5-fluorouracil 500 mg/M^2 given intravenously every 2 weeks for 12 months.

ELIGIBILITY CRITERIA FOR ENTRY:
1. AGE: No age limitation.
2. CLINICAL/PATHOLOGICAL STAGES: Tumor confined to the breast and/or axilla with at least 4 lymph nodes histologically involved with tumor.
3. PRIMARY TREATMENT: Radical mastectomy (conventional or modified) for potentially curable breast carcinoma.
4. AXILLARY NODE DISSECTION: Mandatory.
5. NUMBER OF NODES REMOVED: Minimum of ten.
6. MENSTRUAL STATUS: No restriction.
7. RECEPTOR STATUS: Not specified.

METHOD OF RANDOMIZATION:
Randomization was by telephone to the central office and was based on random number tables. It was not possible for clinicians to know what treatment would be allocated until after the patient identifiers had been recorded centrally and no patient for whom treatment was allocated was subsequently withdrawn prior to receiving treatment.

RANDOMIZATION STRATIFICATIONS:
Menopausal status and post-surgical radiotherapy.

YEARS OF ACCRUAL: 1974-1976

TOTAL NUMBER OF PATIENTS RANDOMIZED: 62.

AIM, RATIONALE, SPECIAL FEATURES:
The study was begun to compare the efficiency of a single agent with a combination of drugs as well as the early projection of patient survival based on actuarial estimates with observed long term survival. Survival after relapse among treated and control patients has been evaluated, too.

ADDITIONAL ELIGIBILITY OR STRATIFICATION CRITERIA:
Patients were ineligible for: (a) histology other than infiltrating ductal or lobular carcinoma; (b) prior carcinoma in either breast; or (c) prior chemotherapy or hormonal therapy. Patients more than six weeks postmastectomy were also ineligible. Patients who received post-surgical chest wall and regional lymph node irradiation were eligible provided all other entry criteria requirements were met.

FOLLOW-UP, COMPLIANCE AND TOXICITY:
All patients were seen weekly during the initial 12 months. Liver function studies were done every 3 months; chest X-ray and bone scan were repeated after 6 months and at the end of treatment. After the first year, follow-up was every 3 months.

Overall, 42% of all patients received ≤85% of the scheduled dose level; 38% received ≤85% but >65% of the scheduled dose level, and 20% of patients received ≤65% of the scheduled dose level.

DISCUSSION:
Although patients randomized to CMF had a better recurrence free survival than patients randomized to 5-FU alone, the patients who initially received 5-FU had a significantly better survival after relapse than patients initially treated with CMF ($p<0.05$). As a result, no overall survival differences were observed for patients on the two arms of the trial. In addition, a subset of long-surviving patients were seen among the 5-FU treated patients, and these constituted 34% of all relapsed patients on that study arm.

The fact that in our experience long-term survival can be achieved after failure of at least 5-FU adjuvant therapy indicates that breast cancer relapse does not necessarily preclude long-term survival, and that therapy given after relapse may influence adjuvant results.

PUBLICATIONS:
Chlebowski RT, Weiner JM, Luce J, et al. Significance of relapse after adjuvant treatment with combination chemotherapy or 5-fluorouracil alone in high risk breast cancer. Cancer Res.1981; 41:4399-4403.

Chlebowski RT, Weiner JM, Reynolds R, Luce J, Bulcavage L, Bateman JR. Long-term survival following relapse after 5-FU but not CMF adjuvant breast cancer therapy. Breast Cancer Res. Treat. 1986; 7:23-29

Table 1. Outcome by allocated treatment for trial 74F (Western CSG)
Key: T = total free of relevant event at start of relevant year(s)

(i) Death (D) in each separate year, subdivided by age when randomized (D <50 & D 50+ / T <50 & T 50+)

Year	1. F	2. CMF
1	0+ 2/ 12+ 20	0+ 0/ 12+ 18
2	3+ 0/ 12+ 18	1+ 4/ 12+ 18
3	2+ 2/ 9+ 18	2+ 1/ 11+ 14
4	0+ 1/ 7+ 16	2+ 2/ 9+ 13
5+	3+ 2/ 7+ 15	2+ 3/ 7+ 11

(ii) First recurrence or prior death (F) in each separate year, subdivided by age when randomized (F <50 & F 50+ / T <50 & T 50+)

Year	1. F	2. CMF
1	3+ 7/ 12+ 20	1+ 2/ 12+ 18
2	0+ 1/ 9+ 13	5+ 5/ 11+ 15
3	4+ 1/ 9+ 12	0+ 3/ 6+ 10
4	2+ 0/ 4+ 11	1+ 0/ 6+ 7
5+	1+ 3/ 2+ 11	1+ 1/ 5+ 7

(iii) Recurrence with survival (R) and Death (D) in all years together, subdivided by various characteristics when randomized (R & D / T)

Age at Entry	1. F	2. CMF
< 40	0+4/5	0+1/1
40 - 49	2+4/7	1+6/11
50 - 59	2+2/8	0+7/11
60 - 69	2+3/9	1+2/5
70 +	1+2/3	0+1/2
Any age (including "not known")	7+15/32	2+17/30

Nodal Status (N0, N1-3, N4+)

		1. F	2. CMF
< 50	N4+ (by axillary	2+8/12	1+7/12
50 +	sample/clearance)	5+7/20	1+10/18

Estrogen Receptors (ER, fmol/mg)

Breakdown by ER status was not available.

TITLE: CASE WESTERN RESERVE UNIVERSITY MULTI-INSTITUTIONAL STUDY ON THERAPY OF PATIENTS WITH STAGE II & III CARCINOMA OF THE BREAST.

STUDY NUMBER: 74G
STUDY TITLE: CASE WESTERN A

ABBREVIATED TRIAL NAME: CASE WESTERN A
STUDY NUMBER: 74G

INVESTIGATORS' NAMES AND AFFILIATIONS:
CA Hubay (1), OH Pearson (3), JS Marshall (3), JP Crowe (1), NH Gordon (1,2).

(1) Department of Surgery, Case Western Reserve University, School of Medicine, Cleveland.
(2) Department of Epidemiology and Biostatistics, Case Western Reserve University School of Medicine, Cleveland.
(3) Department of Medicine, Case Western Reserve University, School of Medicine, Cleveland.

TRIAL DESIGN: TREATMENT GROUPS:
1. CMF: Cyclophosphamide 60 mg/M^2 orally in two divided doses, days 1-14 of 12 monthly cycles. Methotrexate 25 mg/M^2 and 5-fluorouracil 400 mg/M^2 given intravenously, days 1 and 8 of 12 monthly cycles. Dosage was escalated until myelosuppression occurred.
2. CMFTam: CMF, as above, plus tamoxifen 20 mg twice daily, orally for one year.
3. CMFTam + BCG: CMFTam, as above, plus BCG vaccination (TICE) given during the second year.

ELIGIBILITY CRITERIA FOR ENTRY:
1. AGE: < 76.
2. CLINICAL STAGES: All except M1 (distant metastases), and N0 (node negative).
3. PRIMARY TREATMENT: Radical or modified radical mastectomy.
4. AXILLARY NODE DISSECTION: Yes.
5. NUMBER OF NODES REMOVED: Not specified.
6. MENOPAUSAL STATUS: No restrictions.

METHOD OF RANDOMIZATION:
Randomization was done after patient had signed consent and strata characteristics had been recorded. Envelopes prepared and sealed by the department of biostatistics were used.

RANDOMIZATION STRATIFICATIONS:
Nodal status: 1-3 or 4+ nodes positive.
Estrogen receptor: negative (<3 fmol/mg), or positive.

YEARS OF ACCRUAL: 1974-1979.

TOTAL NUMBER OF PATIENTS RANDOMIZED: 318.

FOLLOW-UP, COMPLIANCE AND TOXICITY:
Physical examinations and blood chemistry tests were performed at 3 month intervals. Chest X-rays were done at 6 and 12 month intervals and bone scans and mammograms at yearly intervals. Determination of death was by patient records and/or death certificate.

TABLE A: ELIGIBLE PATIENTS

Treatment groups:		1	2	3
Eligible patients:		99	102	110
Compliance:	Poor:	6	3	10
	Fair:	5	4	3
	Good:	88	95	97

TABLE B: INELIGIBLE PATIENTS

Treatment groups:		1	2	3
Ineligible patients:		2	1	4
Compliance:	Poor:	2	1	1
	Fair:	0	0	1
	Good:	0	0	2

PUBLICATIONS:
Hubay CA, Pearson OH, Marshall JS, Rhodes RS, Debanne SM, Rosenblatt J, Mansour EG, Hermann RE, Jones JC, Flynn WJ, Eckert C, McGuire WL. Adjuvant chemotherapy, antiestrogen therapy and immunotherapy for stage II breast cancer: 45 months follow-up of a prospective randomized clinical trial. Cancer 1980; 46:2805-2808.

Hubay CA, Pearson OH, Marshall JS, Stellato TA, Rhodes RS, Debanne SM, Rosenblatt J, Mansour EG, Hermann RE, Jones JC, Flynn WJ, Eckert C, McGuire WL. Adjuvant therapy of stage II breast cancer. Breast Cancer Res. Treat. 1981; 1:77-82.

Hubay CA, Gordon NH, Crowe JP, Guyton SP, Pearson OH, Marshall JS, Mansour EG, Hermann RE, Jones JC, Flynn WJ, Eckert C, Sponzo RW, McGuire WL, Evans D, et al. Antiestrogen cytotoxic chemotherapy and bacillus Calmette-Guerin vaccination in stage II breast cancer: 72 month follow-up. Surgery 1984; 96:61-71.

Hubay CA, Gordon NH, Pearson OH, Marshall JS, McGuire WL, et al. Eight-year follow-up of adjuvant therapy for stage II breast cancer. World J. of Surgery 1985; 9: 738-749.

Marshall JS, Gordon NH, Hubay CA, Pearson OH. Assessment of tamoxifen as adjuvant therapy in stage II breast cancer: A long-term follow-up. J. Lab. Clin. Med. 1987; 109:300-307.

Table 1. Outcome by allocated treatment for trial 74G (Case Western A)
Key: T = total free of relevant event at start of relevant year(s)

(i) Death (D) in each separate year,
subdivided by age when randomized (D <50 & D 50+ / T <50 & T 50+)

Year	1. CMF	2. TamCMF	3. TamCMFBCG
1	0+ 1/ 42+ 59	3+ 2/ 31+ 72	0+ 3/ 37+ 77
2	5+ 6/ 42+ 58	2+ 4/ 28+ 70	1+ 3/ 37+ 74
3	4+ 6/ 37+ 52	0+ 5/ 26+ 66	5+ 6/ 36+ 71
4	2+ 5/ 33+ 46	4+ 5/ 26+ 61	1+ 7/ 31+ 65
5+	10+ 13/ 31+ 41	3+ 14/ 22+ 56	7+ 17/ 30+ 58

(ii) First recurrence or prior death (F) in each separate year,
subdivided by age when randomized (F <50 & F 50+ / T <50 & T 50+)

Year	1. CMF	2. TamCMF	3. TamCMFBCG
1	3+ 7/ 42+ 59	3+ 5/ 31+ 72	2+ 5/ 37+ 77
2	8+ 12/ 39+ 52	5+ 8/ 28+ 67	5+ 9/ 35+ 71
3	4+ 8/ 31+ 40	2+ 10/ 23+ 59	5+ 11/ 30+ 62
4	4+ 6/ 27+ 32	2+ 6/ 21+ 49	3+ 4/ 25+ 51
5+	8+ 6/ 23+ 26	2+ 10/ 19+ 43	3+ 20/ 22+ 47

(iii) Recurrence with survival (R) and Death (D) in all years together,
subdivided by various characteristics when randomized (R & D / T)

Age at Entry		1. CMF	2. TamCMF	3. TamCMFBCG
< 40		4+10/16	1+9/14	1+4/8
40 - 49		2+11/26	1+3/17	3+10/29
50 - 59		5+20/37	6+13/37	12+16/43
60 - 69		1+7/14	3+13/29	0+16/28
70 +		2+4/8	0+4/6	1+4/6
Any age (including "not known")		14+52/101	11+42/103	17+50/114

Nodal Status (N0, N1-3, N4+)

		1. CMF	2. TamCMF	3. TamCMFBCG
< 50	N1-N3 (by axillary	1+6/20	2+2/14	2+4/19
50 +	clearance)	3+11/26	5+10/35	7+9/34
< 50	N4+ (by axillary	5+15/22	0+10/17	2+10/18
50 +	sample/clearance)	5+20/33	4+20/37	6+27/43

Estrogen Receptors (ER, fmol/mg)

		1. CMF	2. TamCMF	3. TamCMFBCG
< 50	ER poor or ER < 10	1+11/16	0+7/15	0+7/16
50 +		0+12/19	1+12/24	4+13/28
< 50	ER+ or ER 10-99	5+9/23	2+4/14	3+6/19
50 +		2+12/24	4+9/26	3+12/22
< 50	ER++ or ER 100+	0+1/3	0+1/2	1+1/2
50 +		6+7/16	4+9/22	6+11/27

TITLE: COMBINATION CHEMOTHERAPY (CMFVP) VERSUS L-PHENYLALANINE MUSTARD (MELPHALAN) FOR OPERABLE BREAST CANCER WITH POSITIVE NODES: A SOUTHWEST ONCOLOGY GROUP STUDY

STUDY NUMBER: 75A
STUDY TITLE: SWOG7436

ABBREVIATED TRIAL NAME: SWOG 7436
STUDY NUMBER: 75A

INVESTIGATORS' NAMES AND AFFILIATIONS:
CK Osborne (1), SE Rivkin (2), SJ Green (3), B Tranum (4), J Costanzi (5), J Athens (6), and C Vaughn (7).
  (1) Department of Medicine, University of Texas Health Science Center, San Antonio, TX.
  (2) Tumor Institute of Swedish Hospital Medical Center, Seattle, WA.
  (3) SWOG Statistical Center, Fred Hutchinson Cancer Center, Seattle, WA.
  (4) University of Arkansas for Medical Sciences, Little Rock, AR.
  (5) Department of Medicine, University of Medicine, University of Texas Medical Branch, Galveston, TX.
  (6) Department of Medicine, University of Utah Medical Center, Salt Lake City, UT.
  (7) Providence Hospital, Southfield, MI.

TRIAL DESIGN: TREATMENT GROUPS:
1. Mel: Melphalan 5 mg/M^2 orally days 1-5 of 6 week cycle for 24 months.
2. CMFVPr: Cyclophosphamide 60 mg/M^2 orally, methotrexate 15 mg/M^2, 5-fluorouracil 300 mg/M^2 given intravenously weekly for 12 months and vincristine 0.625 mg/M^2 given intravenously weekly for 10 weeks. Prednisone given orally 30 mg/M^2 days 1-14, 20 mg/M^2 days 1-14, 20 mg/M^2 days 15-28, and 10 mg/M^2 days 29-42.

ELIGIBILITY CRITERIA FOR ENTRY:
1. AGE: No restrictions.
2. PATHOLOGIC STAGES: T_{1-3a}, N_1, M_0.
3. PRIMARY TREATMENT: Radical or modified radical mastectomy. Postoperative radiotherapy at discretion of physician.
4. NODE DISSECTION: Required for all patients.
5. MENSTRUAL STATUS: No restrictions.

ADDITIONAL ELIGIBILITY OR STRATIFICATION CRITERIA:.
Chemotherapy was initiated within 42 days of mastectomy. Eligibility also included: WBC≥4000; platelets ≥100,000; and BUN ≤25 mg%.

METHOD OF RANDOMIZATION:
To register a patient, the institution telephoned the Statistical Center and verified eligibility. A permuted block design was used to randomize patients within stratification levels, with randomization assignment overridden when necessary to maintain institution balance.

RANDOMIZATION STRATIFICATIONS:
Menopausal status, number of positive nodes, (1-3 and >4), and planned postoperative radiotherapy (yes or no).

YEARS OF ACCRUAL: 1975-1978.

TOTAL NUMBER OF PATIENTS RANDOMIZED: 440.

PUBLICATIONS:
Glucksberg H, Rivkin SE, Rasmussen S, Tranum B, Gad-el-Mawla N, Costanzi J, Hoogstraten B, Athens J, Maloney T, McCracken J, and Vaughn C. Combination chemotherapy (CMFVP) versus L-phenylalanine mustard (L-PAM) for operable breast cancer with positive axillary nodes. Cancer 1982; 50: 423-434.

Rivkin SE, Glucksberg H, and Foulkes M. Adjuvant therapy of breast cancer: a Southwest Oncology Group experience. Recent Results in Cancer Research. 1984; 96:166-174

Rivkin SE, Green S, Metch B, Glucksberg H, Gad-el-Mawla N, Costanzi J, Hoogstraten B, Athens J, Maloney T, and Osborne CK. Adjuvant combination chemotherapy (CMFVP) versus L-phenylalanine mustard (L-PAM) for operable breast cancer with positive axillary nodes: 10 year results of a Southwest Oncology Group study. J. Clin. Oncol. 1989; 7:1229-1238.

Table 1. Outcome by allocated treatment for trial 75A (SWOG 7436)

The results from this trial have been blinded.

TITLE: RANDOMIZED TRIAL OF FIVE AND THREE DRUG CHEMOTHERAPY AND CHEMO-IMMUNOTHERAPY IN WOMEN WITH OPERABLE NODE POSITIVE BREAST CANCER

STUDY NUMBER: 75B1-2
STUDY TITLE: CALGB 7581

ABBREVIATED TRIAL NAME: CALGB 7581
STUDY NUMBER: 75B1-2

INVESTIGATORS' NAMES AND AFFILIATIONS:
R Weiss (1), D Tormey (2), A Hughes Korzun (3), W. Wood (4)

(1) Walter Reed Army Medical Center, Department of Medical Oncology, Washington, DC.
(2) Wisconsin Clinical Cancer Center, Department of Human Oncology, Madison, WI.
(3) Harvard School of Public Health, Department of Biostatistics, Boston, MA.
(4) Massachusetts General Hospital, Department of Surgery, Boston, MA.

TRIAL DESIGN: TREATMENT GROUPS:
1. CMFVPr — Cyclophosphamide (100 mg/M^2/day p.o.) on days 1 to 14; methotrexate (40 mg/M^2/week i.v.) on days 1 and 8; 5-fluorouracil (500 mg/M^2/week i.v.) on days 1 and 8; vincristine (1 mg/M^2/week i.v.) on days 1 and 8; prednisone (40 mg/M^2/day p.o.) on days 1 to 14. Given for ten 28 day cycles. Then repeated for another ten 28 day cycles but without vincristine and prednisone.
2. CMF — 2 years of chemotherapy as above but without vincristine and prednisone.
3. CMF + MER — Chemotherapy as in group 2 but with methanol extraction residue of BCG (1 mg intradermally on day 8 of each cycle) added during the first year.

ELIGIBILITY CRITERIA FOR ENTRY:
1. AGE: <76 yrs.
2. CLINICAL STAGES: T_{1-3a}, N_0 or N_1.
3. PRIMARY TREATMENT: Conventional or modified radical mastectomy with removal of the breast and at least the lower axillary contents not more than 4 weeks prior to the start of chemotherapy.
4. NODE DISSECTION: ?
5. NUMBER OF NODES REMOVED: ?
6. MENSTRUAL STATUS: No restriction.

METHOD OF RANDOMIZATION
By a sealed envelope technique using a "Latin square" design balancing within and across institutions and stratification categories. The sealed envelopes were available at the institution, and the investigator was to complete the attached on-study form before opening to reveal the randomized therapy. The top of the form was then immediately mailed to the CALGB central office.

RANDOMIZATION STRATIFICATIONS:
Age (<50 vs. ≥50 years). Clinical size of tumor (<3 vs. ≥3 cm).

YEARS OF ACCRUAL: May 1975 to October 1980. Randomisation to group 3 was stopped on 28 October 1978.

TOTAL NUMBER OF PATIENTS RANDOMIZED: 906. Prior to October 1978, 560 patients were randomized; after October 1978, 346 patients were randomized.

FOLLOW-UP:
Post-treatment follow-up was performed at least every 4 months for 2 years, then every 6 months for 4 years, then annually. Required assessments at follow-up were history and physical examination, complete blood count, liver function tests, chest X-ray, bone scan (PRN), and liver scan (if liver function tests were abnormal).

PUBLICATIONS:
Tormey DC, Weinberg VE, Holland JF, Weiss RB, Glidewell OJ, Perloff M, Falkson G, Falkson HC, Henry PH, Leone LA, Rafla S, Ginsberg SJ, Silver RT, Blom J, Carey RW, Schein PS, and Lesnick GJ. A randomized trial of five and three drug chemotherapy and chemoimmunotherapy in women with operable node positive breast cancer: A CALGB study. J. Clin. Oncol. 1983; 1:138-145.

Weiss RB, Korzun AH, Tormey DC, Holland JF, Weinberg VE, and Wood WC. Adjuvant chemotherapy for breast cancer using CMFVP vs. CMF vs. CMF-MER: A CALGB study. Breast Cancer Res. and Treat., 1984; 4:339.

Table 1. Outcome by allocated treatment for trial 75B$_1$ (CALGB 7581)
Key: T = total free of relevant event at start of relevant year(s)

(i) Death (D) in each separate year, subdivided by age when randomized (D <50 & D 50+ / T <50 & T 50+)

Year	1. CMFVPr	2. CMF	3. CMFBCG
1	0+ 7/ 74+ 107	3+ 1/ 80+ 107	1+ 2/ 91+ 101
2	5+ 4/ 74+ 100	11+ 14/ 77+ 106	6+ 14/ 90+ 99
3	3+ 5/ 69+ 96	7+ 9/ 66+ 92	10+ 9/ 84+ 85
4	4+ 7/ 66+ 91	2+ 6/ 59+ 83	7+ 8/ 74+ 76
5+	7+ 28/ 62+ 84	9+ 16/ 57+ 77	11+ 20/ 67+ 68

(ii) First recurrence or prior death (F) in each separate year, subdivided by age when randomized (F <50 & F 50+ / T <50 & T 50+)

Year	1. CMFVPr	2. CMF	3. CMFBCG

(recurrence data not available)

(iii) Recurrence with survival (R) and Death (D) in all years together, subdivided by various characteristics when randomized (R & D / T)

Age at Entry	1. CMFVPr	2. CMF	3. CMFBCG
< 40	0+6/23	0+17/32	0+10/27
40 – 49	0+13/48	0+13/44	0+19/53
50 – 59	0+31/58	0+26/63	0+20/46
60 – 69	0+15/37	0+13/31	0+27/42
70 +	0+1/4	0+3/4	0+2/5
Any age (including "not known")	0+70/181	0+78/187	0+88/192

Nodal Status (N0, N1-3, N4+)

		1. CMFVPr	2. CMF	3. CMFBCG
50 +	N0 (by clinical evidence only)	0+0/0	0+0/0	0+0/1
< 50	N4+ (by axillary sample/clearance)	0+15/36	0+20/42	0+18/36
50 +		0+34/68	0+34/64	0+45/71
< 50	Remainder (i.e. other N+, or N?)	0+4/35	0+10/34	0+11/44
50 +		0+17/42	0+14/47	0+14/40

Estrogen Receptors (ER, fmol/mg)

		1. CMFVPr	2. CMF	3. CMFBCG
< 50	ER poor or ER < 10	0+6/10	0+5/9	0+5/11
50 +		0+5/8	0+7/11	0+8/18
< 50	ER+ or ER 10-99	0+1/2	0+3/10	0+2/8
50 +		0+3/9	0+7/14	0+6/12

Table 1. Outcome by allocated treatment for trial 75B$_2$ (CALGB 7581)
Key: T = total free of relevant event at start of relevant year(s)

(i) Death (D) in each separate year, subdivided by age when randomized (D <50 & D 50+ / T <50 & T 50+)

Year	1. CMFVPr	2. CMF
1	3+ 4/ 84+ 89	0+ 1/ 71+ 100
2	7+ 9/ 81+ 85	12+ 7/ 71+ 99
3	5+ 12/ 74+ 76	4+ 10/ 59+ 92
4	3+ 7/ 69+ 64	7+ 3/ 55+ 82
5+	2+ 1/ 66+ 57	4+ 10/ 48+ 79

(ii) First recurrence or prior death (F) in each separate year, subdivided by age when randomized (F <50 & F 50+ / T <50 & T 50+)

Year	1. CMFVPr	2. CMF

(recurrence data not available)

(iii) Recurrence with survival (R) and Death (D) in all years together, subdivided by various characteristics when randomized (R & D / T)

Age at Entry	1. CMFVPr	2. CMF
< 40	0+10/30	0+9/21
40 – 49	0+7/35	0+13/38
50 – 59	0+19/45	0+12/41
60 – 69	0+10/23	0+7/25
70 +	0+0/2	0+2/7
Any age (including "not known")	0+53/173	0+58/171

Nodal Status (N0, N1-3, N4+)

		1. CMFVPr	2. CMF
< 50	N4+ (by axillary sample/clearance)	0+10/38	0+12/23
50 +		0+18/54	0+29/61
< 50	Remainder (i.e. other N+, or N?)	0+7/27	0+10/36
50 +		0+18/54	0+7/51

Estrogen Receptors (ER, fmol/mg)

		1. CMFVPr	2. CMF
< 50	ER poor or ER < 10	0+9/28	0+12/25
50 +		0+13/34	0+17/34
< 50	ER+ or ER 10-99	0+4/21	0+6/19
50 +		0+16/48	0+10/48

TITLE: ALABAMA BREAST CANCER PROJECT STUDY OF ADJUVANT CHEMOTHERAPY AFTER MASTECTOMY

STUDY NUMBER: 75C
STUDY TITLE: ALABAMA BCP

ABBREVIATED TRIAL NAME: ALABAMA BCP
STUDY NUMBER: 75C

INVESTIGATORS' NAMES AND AFFILIATIONS:
JT Carpenter (1), WA Maddox (2), HL Laws (3), DD Wirtschafter (4), GA Cloud (5), SJ Soong (5).
  (1) Division of Hematology/Oncology, University of Alabama at Birmingham, Birmingham, AL.
  (2) Department of Surgery, University of Alabama at Birmingham, Birmingham, AL.
  (3) Department of Surgery, Carraway Methodist Medical Center, Birmingham, AL.
  (4) Regional Coordinator of Perinatal Services, Southern California Kaiser-Permanente Medical Care Program, Pasadena, CA.
  (5) Biostatistics Unit, Comprehensive Cancer Center, University of Alabama at Birmingham, Birmingham, AL.

TRIAL DESIGN: TREATMENT GROUPS:
1. Mel: Melphalan 7 mg/M^2 by mouth for 5 consecutive days every 6 weeks for 8 cycles.
2. CMF: Cyclophosphamide 300 mg/M^2, Methotrexate 30 mg/M^2 and 5-fluorouracil 300 mg/M^2 all given intravenously every 2 weeks for 24 cycles.

ELIGIBILITY CRITERIA FOR ENTRY:
1. AGE: ≤70 years.
2. CLINICAL/PATHOLOGICAL STAGES: T1-3, N0 or N1 and M0 disease with positive axillary lymph nodes. No clinical evidence of distant metastasis.
3. PRIMARY TREATMENT: Mastectomy and axillary dissection.
4. NODE DISSECTION: Mandatory.
5. NUMBER OF NODES REMOVED: Not specified.
6. MENSTRUAL STATUS: No restriction.

METHODS OF RANDOMISATION:
Randomization was accomplished at the time of registration using the month of birthdate for each patient; those born in odd-numbered months were allocated to melphalan and those in even-numbered months to combination chemotherapy. Prior knowledge of each patient's allocation was therefore present or possible in every instance. No patient was withdrawn after randomization; every randomized patient started chemotherapy. All are included in this analysis.

RANDOMIZATION STRATIFICATIONS: None.

YEARS OF ACCRUAL: 1975-1978.

TOTAL NUMBER OF PATIENTS RANDOMIZED: 171.

ADDITIONAL ELIGIBILITY OR STRATIFICATION CRITERIA:
Serum creatinine ≤1.5 mg/dl, white blood cell count ≥4.0 x 10^9/l, platelet count ≥150 x 10^9/l, no previous or concurrent second malignancy, and signed informed consent. Disease was assessed by clinical examination with TNM staging, chest X-ray, and liver function tests including bilirubin, SGOT, and alkaline phosphatase. Bone scintigraphy was not done routinely. Estrogen and progesterone receptor analyses were not done on most patients because the assays were not performed in most hospital or commercial laboratories in Alabama until the last year or two of the trial.

FOLLOW-UP:
Chest X-ray, liver function studies and blood counts were done every 6 months for 3 years. Clinical examination was done every 3 months for 3 years, every 6 months for 2 additional years, and annually thereafter. National death records were not used. The central office requested follow-up information from each treating physician every 6 months for 5 years after entry into the trial and annually thereafter. When no information was obtained by that method, we contacted patients or patients' families directly for information.

COMPLIANCE BEFORE FIRST RECURRENCE:
Physicians followed the advice given by the visit-specified form provided at about 96% of visits. Eleven patients experienced severe nausea and vomiting on the CMF arm; they were switched to melphalan to complete approximately 1 year of therapy. Of the remaining 65, 44 completed the year of CMF without recurrence; the mean number of cycles completed for the group was 20.1. Of 94 Mel patients 74 completed the year of therapy without recurrence; the mean number of cycles in this group was 7.2. Of the total projected dose of chemotherapy, the median percent of dose received was 83% for the CMF group and 84% for the Mel group.

DISCUSSION:
Follow-up information is available through 1985 on all but 2 patients, 1 in each arm of the study. There is no significant difference in relapse-free survival (p=0.25) or overall survival (p=0.52) in the 2 arms. In the Mel group there continues to be a significant difference in survival between those patients who experienced one or more episodes of modest leukopenia (white blood cell count <3.0 x 10^9/l) as compared to those who did not (23/38, 60% vs. 25/56, 44%, p=0.03).

PUBLICATIONS:
Laws HL, Carpenter JT, Maddox WA. Breast Cancer Management, Part II. J. Med. Assoc. State of Ala. 1975; 44:293-302.

Carpenter JT, Maddox WA, Laws HL, Wirtschafter D, Durant JR, Soong SJ. Adjuvant chemotherapy of node-positive breast cancer in the community. In: Adjuvant Therapy of Cancer. Salmon SE, Jones SE (Eds.), Elsevier/North-Holland Biomedical Press, Amsterdam, 1977, pp.115-122.

Wirtschafter D, Carpenter J, Mesel S. A consultant-extender system for breast cancer adjuvant chemotherapy. Ann. Int. Med. 1979; 90:396-401.

Carpenter JT, Maddox WA, Laws HL, Wirtschafter DL, Soong SK. Favorable factors in the adjuvant therapy of breast cancer. Cancer 1982; 50:18-23.

Table 1. Outcome by allocated treatment for trial 75C (Alabama BCP)
Key: T = total free of relevant event at start of relevant year(s)

(i) Death (D) in each separate year,
subdivided by age when randomized (D <50 & D 50+ / T <50 & T 50+)

Year	1. Mel	2. CMF
1	3+ 0/ 36+ 58	4+ 4/ 40+ 37
2	4+ 9/ 33+ 58	5+ 4/ 36+ 33
3	3+ 6/ 29+ 49	3+ 3/ 31+ 29
4	2+ 3/ 26+ 43	2+ 2/ 28+ 26
5+	4+ 9/ 24+ 40	6+ 3/ 26+ 24

(ii) First recurrence or prior death (F) in each separate year,
subdivided by age when randomized (F <50 & F 50+ / T <50 & T 50+)

Year	1. Mel	2. CMF
1	5+ 4/ 36+ 58	6+ 5/ 40+ 37
2	8+ 9/ 31+ 54	7+ 9/ 34+ 32
3	2+ 9/ 23+ 45	7+ 1/ 27+ 23
4	1+ 2/ 20+ 36	0+ 1/ 20+ 22
5+	4+ 7/ 19+ 34	4+ 7/ 20+ 21

(iii) Recurrence with survival (R) and Death (D) in all years together,
subdivided by various characteristics when randomized (R & D / T)

Age at Entry		1. Mel	2. CMF
< 40		0+5/9	3+2/9
40 - 49		4+11/27	1+18/31
50 - 59		3+15/37	7+10/20
60 - 69		1+10/18	0+6/17
70 +		0+2/3	0+0/0
Any age (including "not known")		8+43/94	11+36/77

Nodal Status (N0, N1-3, N4+)

		1. Mel	2. CMF
< 50	N1-N3 (by axillary	1+6/10	2+12/18
50 +	clearance)	3+13/23	3+9/16
< 50	N4+ (by axillary	3+10/26	2+8/22
50 +	sample/clearance)	1+14/35	4+7/21

Estrogen Receptors (ER, fmol/mg)

Breakdown by ER status was not available.

TITLE: ADJUVANT ENDOCRINE THERAPY IN PRE- AND POSTMENOPAUSAL WOMEN WITH OPERABLE BREAST CANCER

STUDY NUMBER: 75D1-2
STUDY TITLE: COPENHAGEN

ABBREVIATED TRIAL NAME: Copenhagen
STUDY NUMBER: 75D1-2

INVESTIGATORS' NAMES AND AFFILIATIONS:
T Palshof (1), HT Mouridsen (2), JL Daehnfeldt (3), P Dombernowsky (4)
1. Copenhagen County Hospitals, Gentoft and Glostrup, Copenhagen Municipal Hospital, Bispebjerg.
2. Finsen Institute, Copenhagen.
3. Fibiger Laboratory, Copenhagen.
4. Copenhagen County Hospital, Herlev; Denmark.

TRIAL DESIGN: TREATMENT GROUPS:
1. Tam: Tamoxifen, 10 mg, 3 x daily, orally, x 2 years.
2. DES: Diethylstilbestrol, 1 mg, 3 x daily, orally, x 2 years. Only postmenopausal patients were randomized to this arm.
3. Nil: Placebo x 2 years.

ELIGIBILITY CRITERIA FOR ENTRY:
1. AGE: < 70 years.
2. CLINICAL/PATHOLOGICAL STAGES: Operable patients with T1-T4, N0-N3, M0.
3. PRIMARY TREATMENT: Simple mastectomy and postoperative irradiation.
4. AXILLARY NODE DISSECTION: Not required.
5. NUMBER OF NODES REMOVED: ?
6. MENSTRUAL STATUS: No restrictions.
7. RECEPTOR STATUS: Estrogen receptor measurement not required but determined in 78% of the cases. Both those with receptor-positive and receptor-negative tumors were enrolled.

METHOD OF RANDOMIZATION:
?

RANDOMIZATION STRATIFICATIONS:
By menopausal status.

YEARS OF ACCRUAL: 1975-1978.

TOTAL NUMBER OF PATIENTS RANDOMIZED: 382. Data were not available on 5 patients who refused to take any tablets, 8 who were lost to follow-up and 1 major protocol deviant. Thus 368 patients were included in the overview analyses.

AIM, RATIONALE, SPECIAL FEATURES:
This trial was designed to evaluate the efficacy of adjuvant endocrine therapy based on the well-known demonstration that endocrine manipulation is effective in patients with metastatic disease. This was one of the earliest studies utilizing adjuvant tamoxifen as a single agent and the only study in which tamoxifen was compared with DES. Treatment was double-blind.

FOLLOW-UP, COMPLIANCE AND TOXICITY:
Thirty-six patients discontinued treatment within 3 months due to side effects, including 8 premenopausal and 28 postmenopausal women. During the first 2 years of follow-up, patients were examined every 3 months and subsequently every 6 months for the first 5 years. Follow-up was continued yearly thereafter for the majority of patients, but some discontinued follow-up after 5 years. Follow-up evaluation included physical examination, complete blood count, biochemical profile, annual chest X-rays, bone scans, and mammograms. More toxicity was observed among patients randomized to the DES arm. Among postmenopausal women, the percent *without* side effects was 48%, 15%, and 53% respectively for patients treated with tamoxifen, DES, and placebo. Treatment was discontinued in 17% of the postmenopausal women, including 42% of the patients treated with DES but only 13% and 10%, respectively, of the patients treated with tamoxifen and placebo. Among premenopausal women, 36% of those randomized to tamoxifen and 45% of those in the placebo group had no toxicity. Seventeen percent of the patients on DES had thromboembolic events, but none of these were fatal. Other side effects were generally mild. A total of 26% of the patients on DES but only 8% and 7% of the patients on tamoxifen and placebo, respectively, had nausea and vomiting. The incidence of vaginal bleeding in postmenopausal women was also much higher among those randomized to DES.

DISCUSSION:
Although there was a slight trend towards improved disease-free survival and overall survival among premenopausal women randomized to tamoxifen, this did not reach statistical significance. However, among postmenopausal women, a statistically significant improvement in disease-free survival was seen for both those randomized to tamoxifen (p=0.04) and diethylstilbestrol (p=0.02). The five-year relapse-free survival rates with 95% confidence interval for patients on placebo, tamoxifen, and diethylstilbestrol were 49% (35-61%), 68% (53-78%), and 71% (56-81%) respectively. Trends towards improved overall survival were not statistically significant. Although tamoxifen and diethylstilbestrol appeared to be equally effective, tamoxifen was considered superior to DES because of the high frequency of toxicity associated with DES.

PUBLICATIONS:
Palshof T, Mouridsen HT, Daehnfeldt JL. Adjuvant endocrine therapy of primary operable breast cancer. Report on the Copenhagen breast cancer trials. In: Breast Cancer, Experimental and Clinical Aspects. Mouridsen HT, Palshof T, eds. Pergamon Press, Oxford, 1980, pp. 183-187.

Palshof T, Carstensen B, Mouridsen HT, Dombernowsky P. Adjuvant endocrine therapy in pre- and postmenopausal women with operable breast cancer. Reviews on Endocrine-Related Cancer 1985; 17s:43-49.

Table 1. Outcome by allocated treatment for trial 75D1 (Copenhagen)
Key: T = total free of relevant event at start of relevant year(s)

(i) Death (D) in each separate year, subdivided by age when randomized (D <50 & D 50+ / T <50 & T 50+)

Year	1. Tam	3. Nil
1	0+ 0/ 68+ 44	1+ 1/ 60+ 41
2	6+ 6/ 68+ 44	2+ 4/ 59+ 40
3	2+ 1/ 62+ 38	3+ 2/ 57+ 36
4	1+ 1/ 60+ 37	5+ 6/ 54+ 34
5+	9+ 8/ 59+ 36	10+ 7/ 49+ 28

(ii) First recurrence or prior death (F) in each separate year, subdivided by age when randomized (F <50 & F 50+ / T <50 & T 50+)

Year	1. Tam	3. Nil
1	6+ 1/ 68+ 44	6+ 3/ 60+ 41
2	5+ 11/ 62+ 43	6+ 9/ 54+ 38
3	4+ 1/ 57+ 32	7+ 4/ 48+ 29
4	5+ 0/ 53+ 31	2+ 2/ 41+ 25
5+	5+ 7/ 48+ 31	4+ 5/ 39+ 23

(iii) Recurrence with survival (R) and Death (D) in all years together, subdivided by various characteristics when randomized (R & D / T)

Age at Entry	1. Tam	3. Nil
< 40	4+10/21	0+3/11
40 - 49	3+8/47	4+18/49
50 - 59	4+16/43	3+20/41
60 - 69	0+0/1	0+0/0
Any age (including "not known")	11+34/112	7+41/101

Nodal Status (N0, N1-3, N4+)
Breakdown by nodal status was not available.

Estrogen Receptors (ER, fmol/mg)

		1. Tam	3. Nil
<50	ER poor or ER < 10	2+15/44	2+17/43
50+		2+12/31	1+10/19
<50	ER+ or ER 10-99	5+3/23	2+3/16
50+		1+2/9	1+8/18
<50	ER++ or ER 100+	0+0/1	0+1/1
50+		1+2/4	1+2/4

Table 1. Outcome by allocated treatment for trial 75D2 (Copenhagen)
Key: T = total free of relevant event at start of relevant year(s)

(i) Death (D) in each separate year, subdivided by age when randomized (D <50 & D 50+ / T <50 & T 50+)

Year	1. Tam	2. DES	3. Nil
1	0+ 1/ 0+ 52	0+ 1/ 0+ 51	0+ 2/ 0+ 52
2	0+ 2/ 0+ 51	0+ 4/ 0+ 50	0+ 5/ 0+ 50
3	0+ 5/ 0+ 49	0+ 2/ 0+ 46	0+ 1/ 0+ 45
4	0+ 0/ 0+ 44	0+ 2/ 0+ 44	0+ 1/ 0+ 38
5+	0+ 11/ 0+ 44	0+ 8/ 0+ 42	0+ 10/ 0+ 37

(ii) First recurrence or prior death (F) in each separate year, subdivided by age when randomized (F <50 & F 50+ / T <50 & T 50+)

Year	1. Tam	2. DES	3. Nil
1	0+ 5/ 0+ 52	0+ 3/ 0+ 51	0+ 4/ 0+ 52
2	0+ 5/ 0+ 47	0+ 4/ 0+ 48	0+ 14/ 0+ 48
3	0+ 1/ 0+ 42	0+ 2/ 0+ 44	0+ 4/ 0+ 34
4	0+ 5/ 0+ 41	0+ 3/ 0+ 42	0+ 2/ 0+ 30
5+	0+ 6/ 0+ 36	0+ 9/ 0+ 39	0+ 8/ 0+ 28

(iii) Recurrence with survival (R) and Death (D) in all years together, subdivided by various characteristics when randomized (R & D / T)

Age at Entry	1. Tam	2. DES	3. Nil
50 - 59	0+4/15	1+4/12	1+6/11
60 - 69	3+15/37	3+12/37	6+17/37
70 +	0+0/0	0+1/1	0+2/4
Any age (including "not known")	3+19/52	4+17/51	7+25/52

Nodal Status (N0, N1-3, N4+)
Breakdown by nodal status was not available.

Estrogen Receptors (ER, fmol/mg)

		1. Tam	2. DES	3. Nil
50+	ER poor or ER < 10	1+12/27	2+10/29	2+16/26
50+	ER+ or ER 10-99	1+3/11	0+4/5	2+5/15
50+	ER++ or ER 100+	1+4/14	2+3/16	3+4/11

Short reports

TITLE: CONTROLLED TRIAL OF ADJUVANT CHEMO-THERAPY WITH MELPHALAN FOR BREAST CANCER

STUDY NUMBER: 75E1-2
STUDY TITLE: GUYS MELPHALAN, MANCHESTER I

ABBREVIATED TRIAL NAME: GUY'S MELPHALAN, MANCHESTER I
STUDY NUMBER: 75E1-2

INVESTIGATORS' NAMES AND AFFILIATIONS:
RD Rubens (1), A Howell (2), JL Hayward (1), RA Sellwood (3).

(1) Imperial Cancer Research Fund Clinical Oncology Unit, Guy's Hospital, London, UK.
(2) Department of Medical Oncology, University Hospital of South Manchester, Manchester, UK.
(3) Department of Surgery, University Hospital of South Manchester, Manchester, UK.

TRIALS DESIGN: TREATMENT GROUPS:
1. Mel: Melphalan 6 mg/M^2 (maximum 10 mg) orally days 1-5 of 6 week cycle for 24 months.
2. Nil: No adjuvant chemotherapy.
3. CMF: Cyclophosphamide 80 mg/M^2 orally, days 1-14; methotrexate 32 mg/M^2 and 5-fluorouracil 480 mg/M^2 intravenously, day 1 and 8 of 4 week cycle.

ELIGIBILITY CRITERIA FOR ENTRY:
1. AGE: <75 Guy's, <70 Manchester.
2. PATHOLOGICAL STAGES: Involved axillary nodes (Stage II).
3. PRIMARY TREATMENT: Total mastectomy with axillary clearance and no radiotherapy.
4. NODE DISSECTION: Mandatory.
5. NUMBER OF NODES REMOVED: 25.4 ± 9.4 per patient.
6. MENSTRUAL STATUS: No restrictions.
7. RECEPTOR STATUS: No restrictions.

METHOD OF RANDOMIZATION:
Separate for each centre. Telephone to respective administrative offices where allocations were based on random number tables and were balanced in blocks of 9 (Manchester) and blocks of 25 (Guys). Allocation not released until patient identifiers recorded. There were no post randomization exclusions.

RANDOMIZATION STRATIFICATIONS:
Premenopausal, postmenopausal.

YEARS OF ACCRUAL: 1975-1979 Guy's
1976-1979 Manchester.

TOTAL NUMBER OF PATIENTS RANDOMIZED: 425.

AIM, RATIONALE, SPECIAL FEATURES:
The preliminary results of a trial of adjuvant melphalan as chemotherapy by the NSABP, suggested that this therapy could significantly prolong relapse-free survival. Because of the importance of these findings, it was decided in March 1975, to repeat the trial at Guy's Hospital (75E1). The publication of results from the Instituto Nazionale Tumori in Milan on the use of adjuvant CMF in patients with involved axillary nodes led to the initiation of a 3-armed trial at the University Hospital of South Manchester (75E2), in March 1976, which compared either melphalan or CMF with no adjuvant treatment. (Results of the CMF vs. control comparison are analysed in conjunction with those from 79E1-2.)

Because of the similarity in the protocols at Guy's Hospital and in Manchester, we decided in 1979 to amalgamate the trials. The results of the combined randomized trial comparing adjuvant melphalan with no adjuvant therapy are reported.

FOLLOW-UP, COMPLIANCE AND TOXICITY:
Follow-up included physical exam on day 1 of each treatment cycle and at precisely the same time in controls. White blood cell and platelet counts were made on day 1 of each treatment cycle in patients randomized to receive adjuvant chemotherapy. For two years after mastectomy biochemical screening and chest radiography were carried out every 3 months and isotopic bone scans every 6 months. Thereafter follow-up was every 3 months by physical exam only and other investigations were repeated when indicated. From 5 years onwards follow-up was annual.

The trial was assessed by post-operative relapse-free survival, overall survival, pattern of recurrent disease, and toxicity due to treatment. Relapse-free survival was taken as the time from the date of mastectomy to the date of first relapse. Disease status was assessed annually by external review (Dr. JW Meakin).

The patients randomized to receive melphalan received no other treatment before first relapse. The mean percent projected dose of melphalan received is 77.35% (range 50-100%).

PUBLICATIONS:
Rubens RD, Hayward JL, Knight RK, Bulbrook RD, Fentiman IS, Chaudary MA, Howell A, Bush H, Crowther D, Sellwood RA, George WD, Howatt JMT. Controlled trial of adjuvant chemotherapy with melphalan for breast cancer. Lancet 1983; i: 839-843.

Table 1. Outcome by allocated treatment for trial 75E1 (Guy's L-Pam)
Key: T = total free of relevant event at start of relevant year(s)

(i) Death (D) in each separate year,
subdivided by age when randomized (D <50 & D 50+ / T <50 & T 50+)

Year	1. Mel	2. Nil
1	2+ 4/ 47+ 83	0+ 3/ 50+ 80
2	4+ 6/ 45+ 79	4+ 3/ 50+ 77
3	4+ 5/ 41+ 73	4+ 8/ 46+ 74
4	5+ 7/ 37+ 68	4+ 4/ 42+ 66
5+	2+ 24/ 32+ 61	13+ 19/ 38+ 62

(ii) First recurrence or prior death (F) in each separate year,
subdivided by age when randomized (F <50 & F 50+ / T <50 & T 50+)

Year	1. Mel	2. Nil
1	7+ 12/ 47+ 83	11+ 16/ 50+ 80
2	5+ 13/ 40+ 71	9+ 11/ 39+ 64
3	4+ 9/ 35+ 58	3+ 5/ 29+ 53
4	0+ 5/ 31+ 49	2+ 7/ 26+ 48
5+	4+ 19/ 31+ 44	3+ 8/ 24+ 41

(iii) Recurrence with survival (R) and Death (D) in all years together,
subdivided by various characteristics when randomized (R & D / T)

Age at Entry		1. Mel	2. Nil
<40		0+3/8	0+5/10
40-49		3+14/39	3+20/40
50-59		3+24/39	6+15/40
60-69		7+18/33	4+15/29
70+		2+4/11	0+7/11
Any age (including "not known")		15+63/130	13+62/130

	Nodal Status (N0, N1-3, N4+)		
<50	N1-N3 (by axillary	2+8/28	3+15/36
50+	clearance)	9+22/53	3+14/45
<50	N4+ (by axillary	1+9/19	0+10/14
50+	sample/clearance)	3+24/30	7+23/35

	Estrogen Receptors (ER, fmol/mg)		
<50	ER poor or ER < 10	0+4/8	0+4/7
50+		2+15/20	0+10/14
<50	ER+ or ER 10-99	3+8/23	1+16/27
50+		2+18/26	5+16/32
<50	ER++ or ER 100+	0+2/2	0+3/4
50+		4+11/25	4+8/22

Table 1. Outcome by allocated treatment for trial 75E2 (Manchester I)
Key: T = total free of relevant event at start of relevant year(s)

(i) Death (D) in each separate year,
subdivided by age when randomized (D <50 & D 50+ / T <50 & T 50+)

Year	1. Mel	2. Nil	3. CMF
1	0+ 2/ 22+ 35	0+ 0/ 24+ 30	0+ 1/ 21+ 33
2	1+ 1/ 22+ 33	2+ 1/ 24+ 30	2+ 2/ 21+ 31
3	3+ 5/ 21+ 32	2+ 1/ 22+ 29	2+ 2/ 21+ 31
4	1+ 2/ 18+ 27	3+ 1/ 20+ 28	4+ 1/ 19+ 29
5+	4+ 7/ 17+ 25	2+ 6/ 17+ 27	2+ 5/ 15+ 28

(ii) First recurrence or prior death (F) in each separate year,
subdivided by age when randomized (F <50 & F 50+ / T <50 & T 50+)

Year	1. Mel	2. Nil	3. CMF
1	1+ 7/ 22+ 35	2+ 5/ 24+ 30	0+ 3/ 21+ 33
2	4+ 6/ 21+ 28	8+ 5/ 22+ 25	2+ 2/ 21+ 30
3	3+ 4/ 17+ 22	3+ 1/ 14+ 20	8+ 4/ 19+ 28
4	1+ 2/ 14+ 18	1+ 2/ 11+ 18	2+ 3/ 11+ 24
5+	1+ 2/ 13+ 15	1+ 0/ 10+ 16	0+ 2/ 9+ 21

(iii) Recurrence with survival (R) and Death (D) in all years together,
subdivided by various characteristics when randomized (R & D / T)

Age at Entry		1. Mel	2. Nil	3. CMF
<40		0+5/8	2+1/5	2+4/7
40-49		1+4/14	3+8/18	2+4/14
50-59		2+9/15	3+6/20	2+5/18
60-69		2+6/16	1+3/10	2+5/15
70+		0+1/1	0+0/0	0+0/0
Any age (including "not known")		5+26/57	10+18/54	8+18/54

	Nodal Status (N0, N1-3, N4+)			
50+	N0 (by axillary clearance)	0+0/1	0+0/0	2+1/5
<50	N1-N3 (by axillary	1+1/10	4+6/16	3+4/14
50+	clearance)	1+6/18	2+5/21	1+5/19
<50	N4+ (by axillary	0+8/12	1+3/7	1+4/7
50+	sample/clearance)	3+11/16	3+4/10	1+4/9

	Estrogen Receptors (ER, fmol/mg)			
<50	ER poor or ER < 10	0+6/10	2+4/9	1+3/6
50+		0+6/10	2+5/12	0+3/8
<50	ER+ or ER 10-99	1+0/4	3+2/8	2+3/7
50+		2+5/11	1+1/8	0+1/6
<50	ER++ or ER 100+	0+0/0	0+0/0	0+0/1
50+		0+1/4	1+2/6	1+3/6

TITLE: ADJUVANT CHEMOTHERAPY FOR STAGE II AND III BREAST CANCER

ABBREVIATED TRIAL NAME: EVANSTON USA
STUDY NUMBER: 75F

INVESTIGATORS NAMES AND AFFILIATIONS:
JA Caprini (1,2), HA Oviedo (1,3), MP Cunningham (4), E Cohen (5), RS Trueheart (4), JD Khandekar (1,2), EF Scanlon (1,2).

(1) Northwestern University Medical School, Chicago, IL.
(2) Evanston Hospital, Evanston, IL.
(3) Northwestern Memorial Hospital, Chicago, IL.
(4) University of Illinois, Abraham Lincoln School of Medicine and Saint Francis Hospital, Evanston, IL.
(5) Northwestern University, Vogelback Computing Center, Evanston, IL.

TRIAL DESIGN: TREATMENT GROUPS
1. Mel: Melphalan 0.15 mg/kg orally, daily, for 5 days, every 6 weeks.
2. CFPr: Cytoxan 4.0 mg/kg, 5-fluorouracil 8.0 mg/kg, intravenously, daily for 5 days every 6 weeks, prednisone 30 mg, orally, daily for 7 days, every 6 weeks for 9 courses.
3. CFPr + BCG: CFPr, as above, + BCG, Tice Strain, Tine technique, days 2-12 & 2-8 first 2 cycles, and day 28 next 7 cycles.

ELIGIBILITY CRITERIA FOR ENTRY:
1. AGE: < 70 years of age.
2. CLINICAL/PATHOLOGICAL STAGE: Positive lymph nodes and/or unfavorable signs: skin, muscle, fascia or nipple invasion. No demonstrable metastatic disease elsewhere.
3. PRIMARY TREATMENT: Total mastectomy with axillary dissection.
4. NODE DISSECTION: Mandatory.
5. NUMBER OF NODES DISSECTED: Not specified.
6. MENSTRUAL STATUS: Pre- and postmenopausal.
7. RECEPTOR STATUS: No restriction.

METHOD OF RANDOMIZATION:
Telephone to central office.

RANDOMIZATION STRATIFICATIONS:
Tumor size (<3.0 cm, ≥3.0 cm); degree of localized disease (≤3, ≥4 positive nodes); menopausal status (premenopausal, post-menopausal); and participating hospitals (1,2,3).

YEARS OF ACCRUAL: 1975-1979. Randomization to group 1 was stopped in October 1977.

TOTAL NUMBER OF PATIENTS RANDOMIZED: 194 (165 3-way and 29 2-way).

ADDITIONAL ELIGIBILITY OR STRATIFICATION CRITERIA:
The following were also necessary to meet eligibility requirements: unilateral breast cancer; WBC >4,100; platelets >130,000; not pregnant or lactating and no previous chemotherapy.

AIM, RATIONALE, SPECIAL FEATURES:
This study was designed to compare the efficacy of a single drug treatment versus polychemotherapy with and without a specific immunostimulant. The comparison between the polychemotherapy and a single agent was discontinued after 28 months because of a high recurrence rate among those receiving melphalan alone, but the study comparing a combination chemotherapy regimen with and without an immunostimulant continued for 2 years beyond. Adjuvant chemotherapy dosage was adjusted at each cycle to achieve a total white blood count nadir of 1500–2000/mg/mm^3. Immunologic surveillance was performed by the use of phytohemagglutinin assays and skin tests.

FOLLOW-UP, COMPLIANCE AND TOXICITY:
Patients were followed at 6-week intervals during the first year when they were receiving active therapy. A staging evaluation including a chest X-ray, bone scan, and complete physical examination was performed at the end of the first year, and physicians were encouraged to follow their patients at 3-monthly intervals thereafter. All but 14 patients completed the entire protocol: 2 died of unrelated causes, 6 discontinued for personal reasons, and 6 were removed from the protocol by their physicians because of other medical problems. Reduction of dosage occurred at least once in 35%, 28%, and 31%, respectively, of patients in the 3 arms of the study. Patients receiving polychemotherapy had a significantly higher incidence of nausea, vomiting, diarrhea, alopecia, and stomatitis. Patients on the melphalan arm had a significantly higher frequency of thrombocytopenia. There was no difference in the incidence of cystitis and leukopenia.

DISCUSSION:
At the end of 240 weeks, the time to recurrence was significantly longer among women on the 2 polychemotherapy arms compared to those randomized to melphalan alone. However, the total *number* of patients who had recurred in the 3 arms of the study was not significantly different.

PUBLICATIONS:
Caprini JA, Oviedo MA, Cunningham MP, Cohen E.Trueheart RS, Khandekar JD, Scanlon, EP: Advanced chemotherapy for Stage II and III breast carcinoma. JAMA 1980; 244:243-246.

Cohen E, Scanlon EF, Caprini JA, Cunningham MP, Oviedo MA, Robinson B, Knox KL. Follow-up adjuvant chemotherapy and chemoimmunotherapy for stage II and III carcinoma of the breast. Cancer 1982; 49:1754-1761.

Rabadi SJ, Haid M, Scanlon EF, Khandekar JD, Caprini JA, Oviedo MA, Cunningham MP, Grizenko KK, Cohen E. Treatment of breast carcinoma recurrent after adjuvant chemoimmunotherapy. J. Surg. Oncol. 1984; 26:233-237.

Table 1. Outcome by allocated treatment for trial 75F (Evanston USA)
Key: T = total free of relevant event at start of relevant year(s)

(i) Death (D) in each separate year,
subdivided by age when randomized (D <50 & D 50+ / T <50 & T 50+)

Year	1. Mel	2. CFPr	3. CFPrBCG
1	0+ 1/ 9+ 29	0+ 0/ 9+ 28	0+ 1/ 8+ 30
2	1+ 3/ 9+ 28	0+ 3/ 9+ 28	1+ 0/ 8+ 29
3	0+ 3/ 8+ 25	1+ 0/ 9+ 25	0+ 0/ 7+ 29
4	1+ 4/ 8+ 22	1+ 1/ 8+ 25	1+ 5/ 7+ 29
5+	2+ 6/ 7+ 18	3+ 8/ 7+ 24	1+ 6/ 6+ 24

(ii) First recurrence or prior death (F) in each separate year,
subdivided by age when randomized (F <50 & F 50+ / T <50 & T 50+)

Year	1. Mel	2. CFPr	3. CFPrBCG
1	1+ 6/ 9+ 29	0+ 1/ 9+ 28	1+ 1/ 8+ 30
2	2+ 5/ 8+ 23	2+ 3/ 9+ 27	1+ 4/ 7+ 29
3	0+ 4/ 6+ 18	1+ 4/ 7+ 24	1+ 1/ 6+ 25
4	0+ 1/ 6+ 14	3+ 3/ 6+ 20	1+ 6/ 5+ 24
5+	1+ 3/ 6+ 13	1+ 6/ 3+ 17	0+ 3/ 4+ 18

(iii) Recurrence with survival (R) and Death (D) in all years together,
subdivided by various characteristics when randomized (R & D / T)

Age at Entry	1. Mel	2. CFPr	3. CFPrBCG
< 40	0+1/2	0+2/3	0+1/3
40 - 49	0+3/7	2+3/6	1+2/5
50 - 59	1+7/13	5+5/15	1+4/10
60 - 69	0+7/8	0+6/9	2+5/15
70 +	1+3/8	0+1/4	0+3/5
Any age (including "not known")	2+21/38	7+17/37	4+15/38

Nodal Status (N0, N1-3, N4+)

		1. Mel	2. CFPr	3. CFPrBCG
< 50	N1-N3 (by axillary	0+2/6	1+1/3	0+1/3
50 +	clearance)	1+9/17	3+7/17	2+6/20
< 50	N4+ (by axillary	0+2/3	1+4/6	1+2/5
50 +	sample/clearance)	1+8/12	2+5/11	1+6/10

Estrogen Receptors (ER, fmol/mg)

		1. Mel	2. CFPr	3. CFPrBCG
< 50	ER poor or ER < 10	0+2/4	0+1/2	0+2/4
50 +		1+12/19	2+5/13	2+4/13
< 50	ER+ or ER 10-99	0+2/4	1+3/5	1+1/3
50 +		1+5/9	1+5/10	0+5/11

TITLE: A CONTROLLED PROSPECTIVE RANDOMIZED TRIAL OF ADJUVANT VINCRISTINE AND MELPHALAN IN STAGE I AND II BREAST CANCER

STUDY NUMBER: 75G
STUDY TITLE: NORTHWICK PARK

ABBREVIATED TRIAL NAME: NORTHWICK PARK
STUDY NUMBER: 75G

INVESTIGATORS' NAMES AND AFFILIATIONS:
MW Kissin, D Pinto, G Wardle, C Dore*, D Beggs, B MacIntyre, C Kibbin, A E Kark.
Department of Surgery and Division of Biostatistics*, Northwick Park Hospital, Harrow, UK.

TRIAL DESIGN: TREATMENT GROUPS:
1. Mel + V: Melphalan 0.15 mg/kg orally (escalated to 0.2 mg after 3 cycles) on days 1-5 of 6 week cycle for 12 months. Vincristine 0.02 mg/kg iv on day 1 of cycle.
2. Nil: No adjuvant chemotherapy.

ELIGIBILITY CRITERIA FOR ENTRY:
1. AGE: < 75; later altered to <70.
2. CLINICAL STAGES: Operable breast cancer (T1a-2b, N0-1b, M0).
3. PRIMARY TREATMENT: Total mastectomy with axillary clearance. Adjuvant radiotherapy not used.
4. AXILLARY NODE DISSECTION: Mandatory.
5. NUMBER OF NODES REMOVED: Not specified.
6. MENSTRUAL STATUS: No restriction.
7. RECEPTOR STATUS: Not specified.

METHOD OF RANDOMIZATION:
Randomization was by sealed envelopes prepared from random number tables. Prior knowledge was not possible.

RANDOMIZATION STRATIFICATIONS:
Pre-, peri- and postmenopausal.

YEARS OF ACCRUAL: 1975-1980.

TOTAL NUMBER OF PATIENTS RANDOMIZED: 140.

ADDITIONAL ELIGIBILITY OR STRATIFICATION CRITERIA:
Exclusions for eligibility included: a) previous or co-existing malignancy; (b) pregnancy or lactation; (c) high risk factors for surgical or chemotherapeutic morbidity; (d) distant domicile. The disease was assessed by clinical examination with TNM staging supplemented by the results of chest radiograph, xeromammography, bone scintigraphy, liver function tests and liver ultrasound.

FOLLOW-UP, COMPLIANCE AND TOXICITY:
Clinical assessment plus blood counts and liver function tests were carried out every 6 weeks for the first year, every 3 months for the second year, every 6 months for 3 more years and annually thereafter. For the first 2 years a chest radiograph, bone scan and liver ultrasound were carried out at 3 month intervals. Two patients emigrated five years after randomization and both are now classified as lost to follow-up (both were free of recurrence at that time, one was a control and the other received chemotherapy). Median follow-up was 7.2 years (range 5.2 to 10.0 years).

Of the 65 patients allocated to treatment, 43 (66%) received at least 80% of the expected total dose of chemotherapy, 13 (20%) received 50% to 70%, and 9 (14%) <40%. The commonest causes of default were excessive nausea and vomiting, and marrow toxicity.

Three patients were randomized after major protocol violations (one was pregnant and two had positive bone scans); 1 patient treated by mastectomy plus axillary sampling constituted a minor protocol violation. All 4 of these patients were allocated to receive chemo-therapy. 65 patients (46%) were randomized to the treatment arm and the remaining 75 patients were randomized as controls. 87 patients (62%) were node negative, 40 (29%) had 1-3 positive nodes and 13 (9%) had >4 positive nodes. 90 women (64%) were postmenopausal and 95 women (68%) were >50 years old.

DISCUSSION:
Entry was prematurely closed in January 1980 because analysis at that time failed to detect any benefit for node negative patients allocated to receive chemotherapy. Despite meticulous 10 year follow up, no significant benefit of adjuvant melphalan and vincristine was detected in the group as a whole or after stratification for nodal status and age. This trial suffered from two major deficiencies. Firstly, too few patients were entered, and secondly, too many low risk patients (node negative) were included. It is interesting to note that analysis of the data after seven years showed a statistically significant relapse free survival benefit for treated patients with 1-3 positive nodes (p=0.04), which was no longer apparent for node positive patients (p=0.43) at ten years. This highlights the inherent dangers of over enthusiastic early interpretation of breast cancer trials and the inappropriate use of sub-set analysis. The actual relapse free survival in node positive patients was 33% for controls versus 50% for treated patients.

In patients allocated to the treatment arm, the incidence of contralateral breast cancers was less than in controls.

PUBLICATIONS:
Kissin M, Kibbin C, MacIntyre B, et al. Prospective trial of adjuvant chemotherapy for stage I and II breast cancer: 7 year results. In: Proceedings of International Conference on Advances in Adjuvant Therapy of Cancer, London 16th - 18th June 1982.

Table 1. Outcome by allocated treatment for trial 75G (Northwick Park)
Key: T = total free of relevant event at start of relevant year(s)

(i) Death (D) in each separate year,
subdivided by age when randomized (D <50 & D 50+ / T <50 & T 50+)

Year	1. MelV	2. Nil
1	1+ 1/ 21+ 44	1+ 0/ 24+ 51
2	2+ 2/ 20+ 43	1+ 2/ 23+ 51
3	1+ 2/ 18+ 41	1+ 2/ 22+ 49
4	0+ 1/ 17+ 39	2+ 3/ 21+ 47
5+	3+ 6/ 17+ 38	3+ 9/ 19+ 44

(ii) First recurrence or prior death (F) in each separate year,
subdivided by age when randomized (F <50 & F 50+ / T <50 & T 50+)

Year	1. MelV	2. Nil
1	1+ 3/ 21+ 44	2+ 3/ 24+ 51
2	4+ 4/ 20+ 41	2+ 5/ 22+ 48
3	1+ 1/ 16+ 37	2+ 3/ 20+ 43
4	3+ 1/ 15+ 36	1+ 5/ 18+ 40
5+	1+ 6/ 12+ 35	2+ 5/ 17+ 35

(iii) Recurrence with survival (R) and Death (D) in all years together,
subdivided by various characteristics when randomized (R & D / T)

Age at Entry	1. MelV	2. Nil
< 40	1+2/4	0+0/3
40 - 49	2+5/17	1+8/21
50 - 59	1+6/17	2+9/21
60 - 69	2+2/20	3+6/28
70 +	0+4/7	0+1/2
Any age (including "not known")	6+19/65	6+24/75

Nodal Status (N0, N1-3, N4+)

		1. MelV	2. Nil
< 50	N0 (by axillary	2+2/12	0+1/14
50 +	clearance)	1+6/26	3+7/34
< 50	N0 (by clinical evidence only)	0+0/1	0+0/0
< 50	N1-N3 (by axillary	0+3/5	1+5/8
50 +	clearance)	2+4/16	2+4/11
< 50	N4+ (by axillary	1+2/3	0+2/2
50 +	sample/clearance)	0+2/2	0+5/6

Estrogen Receptors (ER, fmol/mg)

Breakdown by ER status was not available.

TITLE: 12 VERSUS 6 CYCLES OF ADJUVANT CMF IN PATIENTS WITH RESECTABLE BREAST CANCER AND POSITIVE AXILLARY LYMPH NODES

STUDY NUMBER: 75H1-2
STUDY NAME: INT MILAN 7502

ABBREVIATED TRIAL NAME: INT MILAN 7502
STUDY NUMBER: 75H1-2

INVESTIGATORS' NAMES AND AFFILIATIONS:
G Bonadonna, G Tancini, P Valagussa, U Veronesi,
Istituto Nazionale Tumori, Milan.

TRIAL DESIGN: TREATMENT GROUPS:
1. CMF x 6 m: Cyclophosphamide 100 mg/M^2, orally, days 1-14 of a 28 day cycle; methotrexate 40 mg/M^2 and 5-fluorouracil 600 mg/M^2, intravenously, on days 1 and 8. Treatment continued for 6 cycles.
2. CMF x 12 m: CMF, as above, for 12 cycles.

ELIGIBILITY CRITERIA FOR ENTRY:
1. AGE: ≤ 70 years.
2. CLINICAL/PATHOLOGICAL STAGES: T1-3a with histologically positive lymph nodes.
3. PRIMARY TREATMENT: Halsted radical, extended radical (16%) or modified radical mastectomy. No postoperative radiotherapy.
4. AXILLARY NODE DISSECTION: Full axillary node dissection was required.
5. NUMBER OF NODES REMOVED: Not specified.
6. MENSTRUAL STATUS: No restriction.
7. RECEPTOR STATUS: No restriction.

METHOD OF RANDOMIZATION:
Block randomization with a permuted block of length 4 for the 2 regimens.

RANDOMIZATION STRATIFICATIONS:
Number of positive axillary nodes (1-3 or ≥4).

YEARS OF ACCRUAL:
9/75 - 5/78 for premenopausal (75H1);
9/75 - 11/76 for postmenopausal (75H2).

TOTAL NUMBER OF PATIENTS RANDOMIZED: 434.

AIM, RATIONALE, SPECIAL FEATURES:
After the demonstration that adjuvant CMF was effective, especially in premenopausal women, this study was designed to evaluate the feasibility of reducing the duration of adjuvant CMF without compromising therapeutic efficacy. The study initially included premenopausal and postmenopausal women, but when the first CMF study failed to show an advantage for the use of CMF in postmenopausal women, this study was closed to further accrual of postmenopausal women in December 1976. For this reason the study includes 331 premenopausal and only 103 postmenopausal women.

FOLLOW-UP, COMPLIANCE AND TOXICITY:
19 patients refused to complete the adjuvant therapy because of negative psychological responses. 16 of these women were randomized to 12 months of CMF and only 3 to the 6 month course. In addition, treatment was temporarily discontinued for periods of 2-3 months in 4 patients. All of these patients were included in the analysis of the study.

DISCUSSION:
No statistically significant differences in either disease-free survival or overall survival have been observed among patients randomized to these two different durations of treatment. However, trends favor the shorter, or 6 month treatment rather than the 12 month course of therapy. These trends were seen in all subsets of patients whether defined by menopausal or nodal status.

PUBLICATIONS:
Tancini G, Bonadonna G, Valagussa P, et al. Adjuvant CMF in breast cancer: Comparative 5 year results of 12 versus 6 cycles. J. Clin. Oncol. 1983; 1:2-10.

Bonadonna G, Valagussa P, Rossi A, et al. Ten year experience with CMF-based adjuvant chemotherapy in resectable breast cancer. Breast Cancer Res. and Treat. 1985; 5:95-115.

Table 1. Outcome by allocated treatment for trial 75H1 (INT Milan 7502)
Key: T = total free of relevant event at start of relevant year(s)

(i) Death (D) in each separate year, subdivided by age when randomized (D <50 & D 50+ / T <50 & T 50+)

Year	1. CMF	2. CMF
1	3+ 0/ 146+ 21	0+ 0/ 142+ 22
2	6+ 0/ 143+ 21	7+ 2/ 142+ 22
3	10+ 4/ 137+ 21	14+ 1/ 135+ 20
4	5+ 0/ 127+ 17	12+ 4/ 121+ 19
5+	25+ 1/ 122+ 17	26+ 6/ 109+ 15

(ii) First recurrence or prior death (F) in each separate year, subdivided by age when randomized (F <50 & F 50+ / T <50 & T 50+)

Year	1. CMF	2. CMF
1	11+ 0/ 146+ 21	6+ 1/ 142+ 22
2	18+ 2/ 135+ 21	18+ 7/ 136+ 21
3	9+ 2/ 117+ 19	16+ 2/ 118+ 14
4	9+ 1/ 108+ 17	11+ 1/ 102+ 12
5+	17+ 5/ 99+ 16	20+ 4/ 91+ 11

(iii) Recurrence with survival (R) and Death (D) in all years together, subdivided by various characteristics when randomized (R & D / T)

Age at Entry	1. CMF	2. CMF
< 40	5+23/49	1+24/48
40 - 49	10+26/97	11+35/94
50 - 59	5+5/21	2+13/22
Any age (including "not known")	20+54/167	14+72/164

Nodal Status (N0, N1-3, N4+)

		1. CMF	2. CMF
< 50	N1-N3 (by axillary clearance)	11+20/88	5+28/89
50 +		3+4/13	1+3/11
< 50	N4+ (by axillary sample/clearance)	4+29/58	7+31/53
50 +		2+1/8	1+10/11

Estrogen Receptors (ER, fmol/mg)

		1. CMF	2. CMF
< 50	ER poor or ER < 10	1+10/22	1+15/27
50 +		2+2/5	0+3/3
< 50	ER+ or ER 10-99	10+22/66	9+21/51
50 +		3+1/10	0+7/11

Table 1. Outcome by allocated treatment for trial 75H2 (INT Milan 7502)
Key: T = total free of relevant event at start of relevant year(s)

(i) Death (D) in each separate year, subdivided by age when randomized (D <50 & D 50+ / T <50 & T 50+)

Year	1. CMF	2. CMF
1	0+ 0/ 2+ 50	0+ 0/ 3+ 59
2	0+ 2/ 2+ 50	0+ 5/ 3+ 59
3	0+ 7/ 2+ 48	1+ 3/ 3+ 54
4	0+ 2/ 2+ 41	0+ 5/ 2+ 51
5+	1+ 8/ 2+ 39	2+ 17/ 2+ 46

(ii) First recurrence or prior death (F) in each separate year, subdivided by age when randomized (F <50 & F 50+ / T <50 & T 50+)

Year	1. CMF	2. CMF
1	0+ 1/ 2+ 50	0+ 3/ 3+ 59
2	0+ 10/ 2+ 49	0+ 7/ 3+ 56
3	0+ 4/ 2+ 39	1+ 9/ 3+ 49
4	0+ 1/ 2+ 35	0+ 5/ 2+ 40
5+	1+ 14/ 2+ 34	2+ 12/ 2+ 35

(iii) Recurrence with survival (R) and Death (D) in all years together, subdivided by various characteristics when randomized (R & D / T)

Age at Entry	1. CMF	2. CMF
40 - 49	0+1/2	0+3/3
50 - 59	8+10/29	3+8/18
60 - 69	3+9/21	2+22/38
70 +	0+0/0	1+0/3
Any age (including "not known")	11+20/52	6+33/62

Nodal Status (N0, N1-3, N4+)

		1. CMF	2. CMF
< 50	N1-N3 (by axillary clearance)	0+0/1	0+2/2
50 +		8+12/32	3+14/36
< 50	N4+ (by axillary sample/clearance)	0+1/1	0+1/1
50 +		3+7/18	3+16/23

Estrogen Receptors (ER, fmol/mg)

		1. CMF	2. CMF
< 50	ER poor or ER < 10	0+0/0	0+1/1
50 +		0+1/8	0+4/7
< 50	ER+ or ER 10-99	0+0/1	0+0/0
50 +		4+6/12	3+8/19

TITLE: KING'S ADJUVANT TRIAL FOR EARLY BREAST CANCER.

ABBREVIATED TRIAL NAME: KING'S CRC M/M
STUDY NUMBER: 75J

STUDY NUMBER: 75J
STUDY TITLE: KING'S CRC M/M

INVESTIGATORS' NAMES AND AFFILIATIONS:
M Baum (1), DA Berstock (2), DM Brinkley (1), J Cuzick (3), AR Lyons (4), TJ McElwain (5), J MacIntyre (6), K MacRae (7), NW M Orr (8), R Peto (9), TJ Powles (5), AD Roy (10), J Houghton (11).

(1) King's College Hospital, UK.
(2) Clatterbridge General Hospital.
(3) Imperial Cancer Research Fund.
(4) N. Ireland Radiotherapy Centre.
(5) Royal Marsden Hospital.
(6) Perth Royal Infirmary.
(7) Charing Cross Hospital.
(8) Essex County Hospital.
(9) Radcliffe Infirmary, Oxford.
(10) Royal Victoria Hospital, Belfast.
(11) CRC Clinical Trials Centre, London, UK.

TRIAL DESIGN: TREATMENT GROUPS:
1. Mel + M: Melphalan 10 mg on days 1-5 of 6 week cycle and methotrexate 15 mg on day 1 of cycle given orally for 24 months.
2. Nil: No adjuvant chemotherapy.

ELIGIBILITY CRITERIA FOR ENTRY:
1. AGE: < 70.
2. PATHOLOGICAL STAGES: T1 or T2 tumors with at least one histologically involved pectoral, axillary or axillary tail nodes.
3. PRIMARY TREATMENT: Simple mastectomy with axillary sampling ± radiotherapy or radical mastectomy.
4. AXILLARY NODE DISSECTION: Optional.
5. NUMBER OF NODES REMOVED: Not specified.
6. MENSTRUAL AND RECEPTOR STATUS: No restrictions.

METHOD OF RANDOMIZATION:
By sealed envelopes held locally at participating center.

YEARS OF ACCRUAL: 1976-1979.

TOTAL NUMBER OF PATIENTS RANDOMIZED: 434.

AIM, RATIONALE, SPECIAL FEATURES:
Preliminary results from the NSABP trial B05 of oral melphalan indicated an increase in disease-free survival following surgery for patients with early breast carcinoma and that this was achieved without great morbidity. Since studies in advanced disease suggested that combination chemotherapy achieved a greater anti-tumor effect, it was felt that trials of systemic adjuvant chemotherapy should be repeated in the UK, and a regimen was sought that would be more effective than melphalan alone.

ADDITIONAL ELIGIBILITY OR STRATIFICATION CRITERIA:
Patients were excluded for the following reasons: previous treatment for other malignant disease; pregnancy; abnormal liver and renal function; or blood counts (including platelets). The protocol required all patients to have negative X-ray of chest, spine or pelvis, (bone scan evidence alone did not exclude patients).

FOLLOW-UP, COMPLIANCE AND TOXICITY:
The protocol required follow-up every 3 months the first two years, 6 months the third year, and annually thereafter. Upon detection of recurrent disease, follow-up every 6 months was requested.

Patients receiving chemotherapy were seen every six weeks.

COMPLIANCE BEFORE FIRST RECURRENCE:
None of the control-allocated patients received chemotherapy prior to relapse. Compliance with this oral therapy was poor; the number of patients taking a full course steadily declined, and by course 6 almost 30% of the patients who were eligible for a further full-dose course did not receive it, although only 13% had recorded hematological toxicity to account for this. Marked marrow depression with a count below 2,500 was not a major problem. Thrombocytopenia was seen less frequently. Approximately 40% of patients reported nausea and vomiting during each course.

DISCUSSION:
This adjuvant cytotoxic regimen was chosen because it was judged to have minimal toxicity and could be given on an outpatient basis. In fact, subjective toxicity was significant and as a result patient compliance was poor. The overall results of the trial did not reproduce those from the NSABP for melphalan alone, and, in retrospect, it would have been more sensible to have exactly repeated the earlier trial, since no conclusions can be drawn from the current results on the addition of methotrexate. The fact that the results mirror those obtained in this Overview, particularly in respect of the relapse-free survival of premenopausal patients, implies that a small non-significant benefit was obtained by use of this regimen but that such a difference, even if of clinical importance, could never be detected by a trial of this size.

PUBLICATIONS:
Liebermann DP, Berstock DA, Houghton J, Kearney G - Oral adjuvant therapy in breast carcinoma - a multicentre trial. Cancer Treat. Rev. 1979; 6 suppl: 91-96.

MacIntyre J, Baum M, Houghton J. Cancer Research Campaign Clinical Trials Centre. A multicentre trial of oral adjuvant chemotherapy in node positive breast cancer. Br. J. Surg. 1986; 70: 692.

Table 1. Outcome by allocated treatment for trial 75J (King's CRC M/M)
Key: T = total free of relevant event at start of relevant year(s)

(i) Death (D) in each separate year,
subdivided by age when randomized (D <50 & D 50+ / T <50 & T 50+)

Year	1. MelM	2. Nil
1	3+ 8/ 85+ 137	2+ 5/ 62+ 150
2	8+ 16/ 82+ 129	7+ 19/ 60+ 145
3	10+ 11/ 74+ 113	6+ 11/ 53+ 126
4	4+ 12/ 64+ 102	6+ 9/ 47+ 115
5+	10+ 15/ 60+ 90	8+ 24/ 41+ 106

(ii) First recurrence or prior death (F) in each separate year,
subdivided by age when randomized (F <50 & F 50+ / T <50 & T 50+)

Year	1. MelM	2. Nil
1	7+ 21/ 85+ 137	17+ 25/ 62+ 150
2	11+ 22/ 76+ 108	6+ 26/ 45+ 118
3	12+ 16/ 64+ 84	9+ 12/ 38+ 87
4	2+ 11/ 48+ 63	3+ 12/ 29+ 69
5+	10+ 15/ 42+ 48	6+ 20/ 24+ 50

(iii) Recurrence with survival (R) and Death (D) in all years together,
subdivided by various characteristics when randomized (R & D / T)

Age at Entry	1. MelM	2. Nil
< 40	2+11/20	2+11/15
40 - 49	5+24/65	10+18/47
50 - 59	7+38/69	17+45/87
60 - 69	16+22/63	10+23/59
70 +	0+2/4	0+0/4
Any age (including "not known")	30+97/222	39+97/212

Nodal Status (N0, N1-3, N4+)

		1. MelM	2. Nil
<50	N0 (by axillary	0+0/1	0+0/0
50+	clearance)	0+0/0	0+1/1
<50	N0 (by clinical	3+1/8	1+3/5
50+	evidence only)	3+3/14	2+5/16
<50	N1-N3 (by axillary	2+14/30	3+9/21
50+	clearance)	6+6/28	6+13/36
<50	N4+ (by axillary	1+4/12	3+5/11
50+	sample/clearance)	3+11/16	4+13/21
<50	Remainder (i.e.	1+16/34	5+12/25
50+	other N+ or N?)	11+42/79	15+36/76

Estrogen Receptors (ER, fmol/mg)
Breakdown by ER status was not available.

TITLE: A RANDOMIZED COMPARATIVE TRIAL OF CHEMOTHERAPY AND IRRADIATION THERAPY FOR STAGE II BREAST CANCER

STUDY NUMBER: 75K
STUDY TITLE: PIEDMONT OA

ABBREVIATED TRIAL NAME: PIEDMONT OA
STUDY NUMBER: 75K

INVESTIGATORS' NAMES AND AFFILIATIONS:
R Cooper, H Muss, C Ferree, F Richards II, J Stuart, D White, B Wells, C Spurr, D Jackson, R Capizzi.
Piedmont Oncology Association and Bowman-Gray School of Medicine, Wake Forest University, Winston-Salem, NC, USA.

TRIAL DESIGN: TREATMENT GROUPS:
1. Mel: Melphalan 0.15 mg/kg, orally, for 5 consecutive days every 6 weeks x 2 years.
2. Mel + R: Melphalan, as above, plus adjuvant radiotherapy to the chest wall + supraclavicular, low cervical, and internal mammary lymph nodes.
3. CMF: Cyclophosphamide 100 mg/M^2, orally, days 1-14 of a 28-day cycle, methotrexate 40 mg/M^2 and 5-fluorouracil 600 mg/M^2, intravenously, on days 1 and 8. Treatment continued for 2 years.
4. CMF + R: CMF as in arm 3 + adjuvant radiotherapy as in arm 2.

ELIGIBILITY CRITERIA FOR ENTRY:
1. AGE: ≤ 75 years.
2. CLINICAL/PATHOLOGICAL STAGES: T 2-5 cm with at least one histologically positive axillary lymph node.
3. PRIMARY TREATMENT: Radical or modified radical mastectomy.
4. AXILLARY NODE DISSECTION: Required.
5. NUMBER OF NODES REMOVED: ?
6. MENSTRUAL STATUS: No restrictions.
7. RECEPTOR STATUS: No restrictions.

METHOD OF RANDOMIZATION:
Patients were randomized initially for radiotherapy and subsequently between 2 different types of chemotherapy.

RANDOMIZATION STRATIFICATIONS:
Number of positive nodes (1-3, ≥4); menopausal status; age (<50, ≥50 years).

YEARS OF ACCRUAL: 1975-1979.

TOTAL NUMBER OF PATIENTS RANDOMIZED: 281.

AIM, RATIONALE, SPECIAL FEATURES:
This study was designed to test the effectiveness of melphalan with and without postoperative radiation therapy and to compare these two arms to CMF with and without postoperative radiation therapy. The study was undertaken in part because of several studies which suggested that the use of adjuvant radiotherapy plus chemotherapy might be less effective than chemotherapy alone, at least in terms of overall survival. Patients randomized to receive postoperative radiation therapy began this therapy 2 weeks or more after surgery, and chemotherapy was begun within 2 weeks after the completion of the adjuvant irradiation.

FOLLOW-UP, COMPLIANCE AND TOXICITY:
During therapy, patients had routine blood tests every 3 months, a chest X-ray every 6 months, and a bone scan every 12 months. At the end of 2 years, each patient was fully staged with physical examination, chest X-ray, liver and bone scans. Subsequently patients were followed at 6-month intervals with physical examination and blood tests and at yearly intervals with chest X-ray and bone scan. The average amount of drug delivered as a percent of that called for in the protocol was greater in the patients treated with chemotherapy alone compared to those treated with chemotherapy plus adjuvant radiotherapy. This was true whether the patients received melphalan or CMF, and these differences were statistically significant during the first 6 cycles. The trends were not significant during the last year to 1^{1}/$_2$ years of treatment. However, the degree of leukopenia and anemia were similar in patients in the four arms. Nausea, vomiting, oral mucositis, and cystitis were seen significantly more frequently in patients on the CMF arms of the study.

DISCUSSION:
The relapse-free survival at 5 years was similar for all 4 treatments. There was no evidence of an adverse effect on the survival of patients randomized to radiotherapy arms compared to those receiving chemotherapy alone.

PUBLICATIONS:
Cooper MR, Rhyne AL, Muss HB, et al. A randomized comparative trial of chemotherapy and radiation therapy for Stage II breast cancer. Cancer 1989; 47:2833-2839.

Cooper R, Muss H, Ferree C. A six and one-half year follow-up of a randomized adjuvant study of chemotherapy (CT) with and without radiation therapy (RT) for Stage II breast cancer. Breast Cancer Res. and Treat. 1985; 6:169.

Table 1. Outcome by allocated treatment for trial 75K (Piedmont OA)
Key: T = total free of relevant event at start of relevant year(s)

(i) Death (D) in each separate year,
subdivided by age when randomized (D <50 & D 50+ / T <50 & T 50+)

Year	1. Mel	2. MelR	3. CMF	4. CMFR
1	0+ 9/ 29+ 41	2+ 5/ 25+ 51	0+ 3/ 24+ 42	0+ 4/ 26+ 43
2	3+ 0/ 25+ 30	2+ 5/ 23+ 44	0+ 2/ 24+ 39	5+ 4/ 26+ 39
3	2+ 2/ 22+ 30	3+ 3/ 21+ 39	2+ 5/ 23+ 37	3+ 1/ 21+ 35
4	0+ 2/ 20+ 28	1+ 7/ 18+ 36	2+ 3/ 21+ 32	1+ 4/ 18+ 34
5+	3+ 9/ 20+ 26	3+ 3/ 17+ 29	2+ 7/ 19+ 29	6+ 9/ 17+ 30

(ii) First recurrence or prior death (F) in each separate year,
subdivided by age when randomized (F <50 & F 50+ / T <50 & T 50+)

Year	1. Mel	2. MelR	3. CMF	4. CMFR
1	1+ 10/ 29+ 41	2+ 5/ 25+ 51	0+ 3/ 24+ 42	0+ 4/ 26+ 43
2	3+ 0/ 25+ 30	2+ 5/ 23+ 44	0+ 2/ 21+ 38	5+ 4/ 26+ 39
3	2+ 2/ 22+ 30	3+ 3/ 21+ 39	2+ 5/ 21+ 36	3+ 1/ 21+ 35
4	0+ 2/ 19+ 28	1+ 7/ 18+ 36	2+ 3/ 19+ 31	1+ 4/ 18+ 34
5+	4+ 12/ 19+ 26	5+ 9/ 17+ 29	4+ 10/ 17+ 28	6+ 11/ 16+ 30

(iii) Recurrence with survival (R) and Death (D) in all years together,
subdivided by various characteristics when randomized (R & D / T)

Age at Entry	1. Mel	2. MelR	3. CMF	4. CMFR
< 40	0+5/9	1+7/13	0+2/7	0+6/12
40 - 49	2+3/20	1+4/12	2+4/17	0+9/14
50 - 59	1+10/19	5+15/32	0+9/20	2+12/25
60 - 69	2+10/18	0+4/13	2+10/17	0+7/14
70 +	1+2/4	1+4/5	1+1/5	0+2/3
Any age (including "not known")	6+30/70	8+34/76	5+26/66	2+37/69

Nodal Status (N0, N1-3, N4+)

		1. Mel	2. MelR	3. CMF	4. CMFR
<50	N0 (by clinical	0+1/3	0+0/1	0+0/1	0+0/1
50+	evidence only)	0+1/3	0+1/2	0+0/0	0+1/1
<50	N1-N3 (by axillary	0+2/10	0+2/6	0+1/7	0+5/11
50+	clearance)	2+3/12	0+2/10	0+5/17	1+5/9
<50	N4+ (by axillary	1+1/5	1+1/6	2+3/6	0+6/6
50+	sample/clearance)	1+9/13	2+5/10	3+8/13	1+4/12
<50	Remainder (i.e.	1+4/11	1+8/12	0+2/10	0+4/8
50+	other N+ or N?)	1+9/13	4+15/29	0+7/12	0+12/21

Estrogen Receptors (ER, fmol/mg)

Breakdown by ER status was not available.

Short reports

TITLE: PREVENTION OF METASTASES IN OPERABLE BREAST CANCER: SHORT OR LONG-TERM ADJUVANT CHEMOTHERAPY?

STUDY NUMBER: 75L
STUDY TITLE: SAKK 27/76

ABBREVIATED TRIAL NAME: SAKK 27/76
STUDY NUMBER: 75L

INVESTIGATORS' NAMES AND AFFILIATIONS:
F Jungi (1), F Cavalli (2), P Alberto (3), G Martz (4).

(1) Med. Klinik C, Kantonsspital, St. Gallen, CH.
(2) Servizio Oncologico, Ospedale San Giovanni, Bellinzona, CH.
(3) Div. Onco-Hématologie, Hôpital Cantonal Universitaire, Genève, CH.
(4) Abt. Onkologie, Dept. Innere Medizin, Universitätsspital, Zürich, CH.

TRIAL DESIGN: TREATMENT GROUPS:
1. LeuMF: 6 months of leukeran (chlorambucil) 5 mg/M^2/day (rounded to the next lower mg), orally, on days 1-14 of a 4 week cycle; methotrexate 10 mg/M^2/week (rounded to the next lower unit of 1.25 mg), orally, any day of week 1 and 2; and 5-fluorouracil 500 mg/M^2/week (rounded to the next upper unit of 125 mg), orally, any day of week 1 and 2.
2. LeuMF: 24 months of LeuMF at the same dose schedule, monthly for the first 12 months and every other month for the second 12 months.

ELIGIBILITY CRITERIA FOR ENTRY:
1. AGE: <70 years.
2. CLINICAL STAGES: T1-3a and N0, N1a, N1b (UICC 1974).
3. PRIMARY TREATMENT: Treatment varied from the minimum of ablatio simplex and excision of axillary tail to the maximum of radical mastectomy and axillary clearance.
4. AXILLARY NODE DISSECTION: Either axillary tail or axillary clearance acceptable.
5. NUMBER OF NODES REMOVED: Not specified.
6. MENSTRUAL STATUS: No restriction.
7. RECEPTOR STATUS: No restriction.

METHOD OF RANDOMIZATION:
Randomization was performed by phone to the central office; balanced blocks of 6 patients were used.

RANDOMIZATION STRATIFICATIONS:
Six institutions were involved in the study.

YEARS OF ACCRUAL: 1976-1978.

TOTAL NUMBER OF PATIENTS RANDOMIZED: 421. Twenty-two ineligible patients are not included in these analyses (see reference 3 for reasons for ineligibility).

PUBLICATIONS:
Cavalli F, Jungi F, Albert P, Martz G, Brunner KW. Praeliminaere Resultate eines laufenden Studieprotokolles mit einer adjuvanten Chemotherapie beim primaer operablen Mammakarzinom. Adjuvante Zytostatische Chemotherapie 1978; 108:61-64.

Senn HJ. Behandlungsfortschritte durch multimodale Therapie beim Mammakarzinom. Schweiz. Med. Wschr. 1978; 108:1938-1947.

Jungi WF, Alberto P et al. Short- or long-term adjuvant chemotherapy for breast cancer. In: Salmon SE, Jones SE (Eds) Adjuvant treatment of cancer III. Grune & Stratton, New York, London, Toronto, Sydney, San Francisco, 1981, p.395.

Jungi WF, Brunner KW, Cavalli F, Martz G, Rosset G, Barrelet L. Short or long-term adjuvant chemotherapy for breast cancer? Rec. Res. in Cancer Res. 1984; 96:175-177.

Table 1. Outcome by allocated treatment for trial 75L (SAKK 27/76)
Key: T = total free of relevant event at start of relevant year(s)

(i) Death (D) in each separate year,
subdivided by age when randomized (D <50 & D 50+ / T <50 & T 50+)

Year	1. LeuMF	2. LeuMF
1	1+ 4/ 76+ 121	2+ 3/ 102+ 100
2	6+ 12/ 75+ 117	11+ 7/ 100+ 97
3	9+ 15/ 69+ 105	9+ 9/ 89+ 90
4	3+ 4/ 60+ 90	6+ 10/ 80+ 81
5+	19+ 27/ 57+ 86	19+ 25/ 74+ 71

(ii) First recurrence or prior death (F) in each separate year,
subdivided by age when randomized (F <50 & F 50+ / T <50 & T 50+)

Year	1. LeuMF	2. LeuMF
1	11+ 16/ 76+ 121	16+ 14/ 102+ 100
2	14+ 28/ 65+ 105	18+ 17/ 85+ 86
3	9+ 8/ 51+ 77	5+ 12/ 67+ 69
4	4+ 11/ 42+ 69	7+ 10/ 61+ 57
5+	13+ 22/ 38+ 58	17+ 17/ 52+ 47

(iii) Recurrence with survival (R) and Death (D) in all years together,
subdivided by various characteristics when randomized (R & D / T)

Age at Entry	1. LeuMF	2. LeuMF
< 40	3+9/20	2+13/25
40 - 49	10+29/56	14+34/77
50 - 59	10+35/66	9+26/52
60 - 69	12+26/51	6+25/42
70 +	1+1/4	1+3/6
Any age (including "not known")	36+100/197	32+101/202

Nodal Status (N0, N1-3, N4+)

		1. LeuMF	2. LeuMF
< 50	N1-N3 (by axillary	8+15/41	8+14/46
50 +	clearance)	11+15/47	6+18/41
< 50	N4+ (by axillary	3+14/20	5+19/34
50 +	sample/clearance)	8+31/48	6+24/34
< 50	Remainder (i.e.	2+9/15	3+14/22
50 +	other N+ or N?)	4+16/26	4+12/25

Estrogen Receptors (ER, fmol/mg)

Breakdown by ER status was not available.

TITLE: A COMPARISON OF SINGLE-AGENT L-PHENYLALANINE MUSTARD L-PAM) WITH THE COMBINATION L-PAM + 5-FLUOROURACIL

STUDY NUMBER: 75M
STUDY TITLE: NSABP B-07

ABBREVIATED TRIAL NAME: NSABP B-07
STUDY NUMBER: 75M

INVESTIGATORS' NAMES AND AFFILIATIONS:
B Fisher, C Redmond, E R Fisher, N Wolmark, and 37 participating institutions in USA and Canada.
National Surgical Adjuvant Breast Project, Pittsburgh, PA, USA.

TRIAL DESIGN: TREATMENT GROUPS:
1. Mel: Melphalan, 6 mg/M^2/day, orally, for 5 consecutive days every 6 weeks for 2 years.
2. MelF: Melphalan, 4 mg/M^2/day, orally, and 5-fluorouracil, 300 mg/M^2/day, intravenously, each for 5 consecutive days every 6 weeks for 2 years.

ELIGIBILITY CRITERIA FOR ENTRY:
1. AGE: ≤70 years.
2. CLINICAL/PATHOLOGICAL STAGES: Operable breast cancers with one or more histologically involved axillary lymph nodes.
3. PRIMARY TREATMENT: Conventional or modified radical mastectomy.
4. AXILLARY NODE DISSECTION: Required.
5. NUMBER OF NODES REMOVED: ?
6. MENSTRUAL STATUS: No restrictions.
7. RECEPTOR STATUS: No restrictions.

RANDOMIZATION STRATIFICATIONS:
Age (≤49, ≥50 years); number of histologically involved nodes (1-3, ≥4); institution.

METHOD OF RANDOMIZATION:
Randomization was done by telephone call to a central office.

YEARS OF ACCRUAL: 1975-1976.

TOTAL NUMBER OF PATIENTS RANDOMIZED: 741.

AIM, RATIONALE, SPECIAL FEATURES:
This study was designed to follow NSABP B-05 (72B), in which it had been previously demonstrated that the use of a prolonged course of melphalan alone significantly improved disease-free survival. This trial was designed to determine if a combination of drugs would be superior to a single agent, both administered for the same duration of time. All but 57 patients (7.7%) were judged fully eligible for this protocol.

FOLLOW-UP, COMPLIANCE AND TOXICITY:
Of the 2077 courses of treatment for which patients were eligible, only 78 (3.8%) or 0.3 courses per patient were missed in the melphalan arm, and 102 of 1986 (5.1%), representing 0.4 courses per patient, for the melphalan + 5-FU arm. Patients treated with both melphalan and 5-fluorouracil had a slightly lower nadir white blood count but also a more complete recovery of white count between courses. There was no difference in nadir platelet counts between the two arms of the protocol. Patients on melphalan + 5-fluorouracil also had more nausea and vomiting and a slightly greater incidence of alopecia. Twenty patients in this study have had leukemia or myeloproliferative disorders, 9 on the melphalan arm and 11 on the melphalan + 5-fluorouracil arm.

DISCUSSION:
Although the survival of patients randomized to melphalan + 5-fluorouracil was slightly better than that of patients randomized to melphalan alone, this difference was significant at 10 years in the patients who were 50 years or older (p=0.08), and especially in the subset with more than 4 positive nodes.

PUBLICATIONS:
Fisher B, Glass A, Redmond C, et al. L-phenylalanine mustard in the management of primary breast cancer: an update of earlier findings in a comparison with those utilizing melphalan + 5-fluorouracil (5-FU). Cancer 1977; 39:2883-2903.

Fisher B, Redmond CK, Wolmark N, and NSABP investigators. Long-term results from NSABP trials of adjuvant therapy for breast cancer. In Salmon SE, Adjuvant Therapy of Breast Cancer V, Grune & Stratton, Orlando, 1987, pp 283-295.

Table 1. Outcome by allocated treatment for trial 75M (NSABP B-07)
Key: T = total free of relevant event at start of relevant year(s)

(i) Death (D) in each separate year,
subdivided by age when randomized (D <50 & D 50+ / T <50 & T 50+)

Year	1. Mel	2. FMel
1	7+ 4/ 151+ 222	5+ 4/ 160+ 208
2	13+ 13/ 136+ 198	8+ 10/ 140+ 177
3	13+ 24/ 123+ 185	18+ 17/ 131+ 167
4	8+ 23/ 109+ 161	13+ 8/ 112+ 150
5+	29+ 45/ 100+ 138	24+ 50/ 99+ 142

(ii) First recurrence or prior death (F) in each separate year,
subdivided by age when randomized (F <50 & F 50+ / T <50 & T 50+)

Year	1. Mel	2. FMel

(recurrence data not available)

(iii) Recurrence with survival (R) and Death (D) in all years together,
subdivided by various characteristics when randomized (R & D / T)

Age at Entry	1. Mel	2. FMel
< 40	0+27/56	0+28/50
40 - 49	0+43/95	0+40/110
50 - 59	0+62/138	0+59/139
60 - 69	0+42/76	0+26/60
70 +	0+5/8	0+4/8
Any age (including "not known")	0+179/373	0+157/368

Nodal Status (N0, N1-3, N4+)

		1. Mel	2. FMel
<50	N1-N3 (by axillary	0+26/80	0+24/81
50+	clearance)	0+37/114	0+30/107
<50	N4+ (by axillary	0+44/71	0+44/79
50+	sample/clearance)	0+72/108	0+59/101

Estrogen Receptors (ER, fmol/mg)

Breakdown by ER status was not available.

Short reports

TITLE: POSTSURGICAL ADJUVANT CHEMOTHERAPY WITH OR WITHOUT RADIOTHERAPY IN WOMEN WITH BREAST CANCER AND POSITIVE AXILLARY NODES

STUDY NUMBER: 76A1-3
STUDY TITLE: SECSG 1

ABBREVIATED TRIAL NAME: SECSG 1
STUDY NUMBER: 76A1-3

INVESTIGATORS' NAMES AND AFFILIATIONS:
E Velez-Garcia (1), JT Carpenter (2), M Moore (3),
CL Vogel (4), V Marcial (1), A Ketcham (5), M Raney (2), R Smalley (6).

(1) University of Puerto Rico, School of Medicine, San Juan, Puerto Rico.
(2) University of Alabama, School of Medicine, Birmingham, AL, USA.
(3) Emory University, School of Medicine, Atlanta, GA.
(4) AMI-Kendall Hospital Comprehensive Cancer Center, Miami, FL.
(5) University of Miami, School of Medicine, Miami, FL.
(6) University of Wisconsin, Madison, WI.

TRIAL DESIGN: TREATMENT GROUPS:
1. CMF x 12: Cyclophosphamide 100 mg/M^2, days 1-14 of a 4-week cycle; methotrexate, 40 mg/M^2, and 5-fluorouracil, 600 mg/M^2, intravenously, on days 1 and 8. Treatment continued for 12 cycles.
2. CMF x 6: As above, but repeated for 6 cycles.
3. CMF x 6 + R Radiotherapy given to the ipsilateral chest wall and regional lymph nodes over 5 weeks following mastectomy. Chemotherapy begun after radiotherapy with the first 2 cycles at 25% of the doses given in the other groups. Cycles 3-6 as above. This group included only patients with ≥4 positive lymph nodes.

ELIGIBILITY CRITERIA FOR ENTRY:
1. AGE: No restrictions.
2. CLINICAL/PATHOLOGICAL STAGES: Operable breast cancer with T1-T3 lesions and one or more histologically positive axillary lymph node.
3. PRIMARY TREATMENT: Radical or modified radical mastectomy.
4. AXILLARY NODE DISSECTION: Required.
5. NUMBER OF NODES REMOVED: ≥10.
6. MENSTRUFAL STATUS: No restrictions.
7. RECEPTOR STATUS: No restrictions.

METHOD OF RANDOMIZATION:
By telephone to the central statistical office. Treatment assigned from computer-generated list.

RANDOMIZATION STRATIFICATIONS:
Treating institution; number of histologically positive lymph nodes (1-3, ≥4); menopausal status (pre- or postmenopausal); mastectomy type (radical vs. modified radical); elapsed number of days from operation to onset of therapy (<28, ≥28 days).

YEARS OF ACCRUAL:
1976-83.

TOTAL NUMBER OF PATIENTS RANDOMIZED:
645. Data on 622 were available for the overview (see below).

AIM, RATIONALE, SPECIAL FEATURES:
This study was designed to address the question of the optimal duration of adjuvant chemotherapy and the role of post-mastectomy radiotherapy when given along with chemotherapy. In the initial design of the trial, patients with 1-3 positive lymph nodes are randomized to 2 durations of chemotherapy (treatment groups 1 + 2) and patients with 4 or more positive lymph nodes were randomized to all 3 groups. However, accrual among patients with 4 or more positive nodes was slow, and for this reason one of the three treatment groups (CMF x 12, group 1) was closed to these patients in February 1980.

FOLLOW-UP, COMPLIANCE AND TOXICITY:
Patients were examined and had complete blood counts at 6-month intervals following completion of therapy. An annual chest X-ray was performed. A liver scan was obtained 1 year after entry into the study, and an annual bone scan obtained for the first 3 years following enrolment. Twenty-three patients were randomized but not treated in the study and were not followed by the statistical office, including 3 patients from group 1, 8 patients from group 2, and 12 patients from group 3. Compliance was generally high; 78% of the patients in group 1 received 12 courses, and 83% and 81%, respectively, of those in groups 2 and 3 completed 6 courses of treatment. The percentage of patients who received 85% or more of the dose prescribed by the protocol was 27% in group 1, 51% in group 2, and 43% in group 3. The median time from mastectomy to initiation of chemotherapy was 37 days for patients in trial groups 1 and 2 and 97 days for patients in trial group 3.

DISCUSSION:
The results have been published with almost 10 years of follow-up. No significant difference in disease-free survival or overall survival were seen among patients randomized to the 2 different durations of therapy or to receive adjuvant radiotherapy. The only group in which there appeared to be a significant survival benefit from more prolonged chemotherapy was the premenopausal women with 1-3 positive lymph nodes. The addition of adjuvant radiotherapy reduced the proportion with locoregional recurrences from 11% among those given 6 months of CMF without adjuvant radiotherapy to 5% among those given both CMF and adjuvant radiotherapy (p=0.17).

PUBLICATIONS:
Velez-Garcia E, Moore M, Vogel CL, et al. Post-mastectomy adjuvant chemotherapy with or without radiation therapy in women with operable breast cancer and positive axillary nodes: the Southeastern Cancer Study Group experience. Breast Cancer Research and Treatment 1983; 3:49-60.

Velez-Garcia E, Carpenter JT, Moore M, et al. Postsurgical adjuvant chemotherapy with or without radiotherapy in women with breast cancer and positive axillary nodes: progress report of a Southeastern Cancer Study Group (SEG) trial. In Adjuvant Therapy of Cancer V, Salmon SE, ed., Grune & Stratton, Orlando, 1987, pp 347-355.

Table 1. Outcome by allocated treatment for trial 76A1 (SECSG 1)
Key: T = total free of relevant event at start of relevant year(s)

(i) Death (D) in each separate year,
subdivided by age when randomized (D <50 & D 50+ / T <50 & T 50+)

Year	1. CMF	2. CMF
1	0+ 0/ 68+ 80	1+ 1/ 73+ 86
2	1+ 6/ 68+ 80	3+ 3/ 72+ 85
3	1+ 10/ 67+ 74	5+ 3/ 69+ 82
4	2+ 3/ 66+ 64	2+ 8/ 64+ 79
5+	2+ 5/ 56+ 50	5+ 11/ 56+ 61

(ii) First recurrence or prior death (F) in each separate year,
subdivided by age when randomized (F <50 & F 50+ / T <50 & T 50+)

Year	1. CMF	2. CMF
1	2+ 3/ 68+ 80	6+ 3/ 73+ 86
2	4+ 11/ 64+ 75	6+ 11/ 65+ 80
3	4+ 11/ 59+ 63	5+ 9/ 56+ 68
4	1+ 5/ 51+ 48	3+ 5/ 50+ 57
5+	3+ 10/ 44+ 34	12+ 9/ 43+ 44

(iii) Recurrence with survival (R) and Death (D) in all years together,
subdivided by various characteristics when randomized (R & D / T)

Age at Entry	1. CMF	2. CMF
< 40	0+3/23	7+8/29
40 - 49	3+3/44	3+8/44
50 - 59	6+14/49	1+14/41
60 - 69	4+10/24	3+7/29
70 +	3+0/7	1+4/15
Any age (including "not known")	17+30/148	15+42/159

Nodal Status (N0, N1-3, N4+)

		1. CMF	2. CMF
< 50	N1-N3 (by axillary	3+6/67	10+16/73
50 +	clearance)	14+24/81	5+26/86

Estrogen Receptors (ER, fmol/mg)

		1. CMF	2. CMF
< 50	ER poor or ER < 10	0+2/9	1+7/14
50 +		1+9/15	2+1/8
< 50	ER+ or ER 10-99	2+2/23	4+2/21
50 +		6+6/32	0+10/37

Trial continued overleaf

Table 1. Outcome by allocated treatment for trial 76A2 (SECSG 1)
Key: T = total free of relevant event at start of relevant year(s)

(i) Death (D) in each separate year,
subdivided by age when randomized (D <50 & D 50+ / T <50 & T 50+)

Year	1. CMF	2. CMF	3. CMFR
1	1+ 1/ 30+ 29	2+ 1/ 27+ 34	2+ 3/ 21+ 28
2	3+ 4/ 29+ 28	2+ 2/ 25+ 33	1+ 4/ 19+ 25
3	1+ 1/ 26+ 24	6+ 2/ 23+ 31	1+ 1/ 18+ 21
4	4+ 6/ 25+ 23	1+ 3/ 17+ 29	1+ 0/ 17+ 20
5+	7+ 2/ 21+ 17	2+ 9/ 16+ 26	1+ 4/ 16+ 20

(ii) First recurrence or prior death (F) in each separate year,
subdivided by age when randomized (F <50 & F 50+ / T <50 & T 50+)

Year	1. CMF	2. CMF	3. CMFR
1	4+ 3/ 30+ 29	6+ 3/ 27+ 34	3+ 3/ 21+ 28
2	4+ 8/ 26+ 25	6+ 6/ 20+ 30	2+ 5/ 17+ 24
3	4+ 2/ 22+ 17	3+ 4/ 14+ 22	1+ 4/ 13+ 19
4	4+ 1/ 18+ 15	2+ 7/ 11+ 18	0+ 0/ 12+ 14
5+	4+ 4/ 14+ 12	2+ 1/ 9+ 11	3+ 4/ 12+ 14

(iii) Recurrence with survival (R) and Death (D) in all years together,
subdivided by various characteristics when randomized (R & D / T)

Age at Entry		1. CMF	2. CMF	3. CMFR
< 40		0+8/10	2+10/15	0+2/5
40 - 49		3+8/20	3+3/12	3+3/15
50 - 59		3+4/12	3+6/18	1+7/18
60 - 69		1+9/13	0+8/12	1+3/8
70 +		0+1/4	0+3/4	0+2/2
Any age (including "not known")		7+30/59	8+30/61	5+18/49

Nodal Status (N0, N1-3, N4+)

		1. CMF	2. CMF	3. CMFR
< 50	N4+ (by axillary	3+16/30	5+13/27	3+5/20
50 +	sample/clearance)	4+14/29	3+17/34	2+13/29

Estrogen Receptors (ER, fmol/mg)

		1. CMF	2. CMF	3. CMFR
< 50	ER poor or ER < 10	1+1/3	0+2/3	1+1/4
50 +		0+0/1	0+1/2	0+1/5
< 50	ER+ or ER 10-99	0+2/4	3+4/9	1+1/4
50 +		1+5/12	1+7/12	1+2/5

Table 1. Outcome by allocated treatment for trial 76A3 (SECSG 1)
Key: T = total free of relevant event at start of relevant year(s)

(i) Death (D) in each separate year,
subdivided by age when randomized (D <50 & D 50+ / T <50 & T 50+)

Year	2. CMF	3. CMFR
1	0+ 1/ 28+ 40	4+ 1/ 33+ 45
2	3+ 8/ 28+ 39	1+ 5/ 29+ 44
3	2+ 4/ 24+ 29	7+ 0/ 26+ 36
4	3+ 0/ 20+ 17	1+ 1/ 13+ 28
5+	0+ 1/ 13+ 12	0+ 1/ 7+ 16

(ii) First recurrence or prior death (F) in each separate year,
subdivided by age when randomized (F <50 & F 50+ / T <50 & T 50+)

Year	2. CMF	3. CMFR
1	3+ 7/ 28+ 40	7+ 3/ 33+ 45
2	7+ 15/ 23+ 33	5+ 8/ 23+ 40
3	1+ 1/ 14+ 16	2+ 3/ 16+ 30
4	4+ 0/ 13+ 10	1+ 0/ 8+ 22
5+	2+ 1/ 8+ 7	0+ 1/ 3+ 15

(iii) Recurrence with survival (R) and Death (D) in all years together,
subdivided by various characteristics when randomized (R & D / T)

Age at Entry		2. CMF	3. CMFR
< 40		4+2/9	1+6/13
40 - 49		3+6/19	1+7/20
50 - 59		5+7/20	4+5/23
60 - 69		5+5/18	3+2/15
70 +		0+2/2	0+1/6
Any age (including "not known")		17+22/68	9+21/78

Nodal Status (N0, N1-3, N4+)

		2. CMF	3. CMFR
50 +	N1-N3 (by axillary clearance)	0+0/0	0+0/1
< 50	N4+ (by axillary	7+8/28	2+13/33
50 +	sample/clearance)	10+14/40	7+8/44

Estrogen Receptors (ER, fmol/mg)

		2. CMF	3. CMFR
< 50	ER poor or ER < 10	0+2/4	0+2/6
50 +		0+4/8	2+3/8
< 50	ER+ or ER 10-99	4+2/10	1+5/11
50 +		5+4/16	3+1/20

Short reports

TITLE: THE GLASGOW ADJUVANT CHEMOTHERAPY BREAST CANCER TRIAL

STUDY NUMBER: 76C
STUDY TITLE: GLASGOW

ABBREVIATED TRIAL NAME: GLASGOW
STUDY NUMBER: 76C

INVESTIGATORS' NAMES AND AFFILIATIONS:
CS McArdle (1) and DC Smith (2) with surgeons at The Victoria Infirmary, The Royal Infirmary and Gartnavel General Hospital and in association with the West of Scotland Institute of Radiotherapy, the University Department of Clinical Oncology and the Cancer Surveillance Unit, Ruchill Hospital Glasgow.

(1) Department of Surgery, Royal Infirmary, Glasgow, Scotland.
(2) Division of Surgery, Victoria Infirmary, Glasgow, Scotland.

TRIAL DESIGN: TREATMENT GROUPS:

1. CMF: Intravenous cyclophosphamide (300 mg/M^2), methotrexate (40 mg/M^2) & 5-fluorouracil (600 mg/M^2) on day 1 and 8 every 28 days for 1 year.
2. R + CMF: CMF as above and orthovoltage radiotherapy to chest wall and axillary, supraclavicular and internal mammary nodal areas in 15 fractions over 3 weeks (mean tumor dose 37.8 Gy).
3. R: Radiotherapy as above.

ELIGIBILITY CRITERIA FOR ENTRY:
1. AGE: ≤70 years.
2. CLINICAL/PATHOLOGICAL STAGES: Operable invasive breast cancer without fixation to chest wall and without fixed or matted axillary nodes. Histological confirmation of axillary node involvement.
3. PRIMARY TREATMENT: Simple (total) mastectomy.
4. AXILLARY SURGERY: Axillary clearance to level of vein.
5. MENSTRUAL STATUS: No restrictions.

METHOD OF RANDOMIZATION:
Treatment allocated by reference to lists of random codes prepared for and held in each of the three referral centres.

RANDOMIZATION STRATIFICATIONS:
By centre of referral.

YEARS OF ACCRUAL: 1976-1982.

TOTAL NUMBER OF PATIENTS RANDOMIZED: 322.

PUBLICATIONS:
McArdle CS, Calman KC, Cooper AF, Hughson AVM, Russel AR, and Smith DC. The social, emotional and financial implications of adjuvant chemotherapy in breast cancer. Br. J. Surg. 1981; 68: 261-264.

McArdle CS, Crawford D, Dykes EH, Calman KC, Hole D, Russel AR, and Smith DC. Adjuvant radiotherapy and chemotherapy in breast cancer. Br. J. Surg. 1986; 73: 264-266.

Table 1. Outcome by allocated treatment for trial 76C (Glasgow)
Key: T = total free of relevant event at start of relevant year(s)

(i) Death (D) in each separate year, subdivided by age when randomized (D <50 & D 50+ / T <50 & T 50+)

Year	1. CMF	2. CMFR	3. R
1	0+ 3/ 33+ 69	0+ 5/ 47+ 65	0+ 8/ 34+ 66
2	2+ 7/ 33+ 66	6+ 9/ 47+ 60	7+ 8/ 34+ 58
3	4+ 7/ 31+ 59	4+ 2/ 41+ 51	7+ 4/ 27+ 50
4	1+ 6/ 24+ 50	1+ 7/ 35+ 44	2+ 2/ 18+ 43
5+	2+ 10/ 16+ 37	5+ 5/ 31+ 30	2+ 9/ 14+ 32

(ii) First recurrence or prior death (F) in each separate year, subdivided by age when randomized (F <50 & F 50+ / T <50 & T 50+)

Year	1. CMF	2. CMFR	3. R
1	4+ 7/ 33+ 69	3+ 10/ 47+ 65	9+ 15/ 34+ 66
2	0+ 19/ 29+ 60	10+ 10/ 43+ 54	8+ 15/ 25+ 51
3	4+ 8/ 29+ 40	1+ 7/ 33+ 43	1+ 5/ 17+ 34
4	2+ 6/ 19+ 30	1+ 5/ 28+ 31	0+ 3/ 14+ 24
5+	2+ 5/ 11+ 18	5+ 7/ 26+ 19	1+ 3/ 11+ 19

(iii) Recurrence with survival (R) and Death (D) in all years together, subdivided by various characteristics when randomized (R & D / T)

Age at Entry		1. CMF	2. CMFR	3. R
< 40		1+1/5	1+5/11	1+9/13
40 - 49		2+8/28	3+10/35	0+9/22
50 - 59		6+15/33	6+11/31	6+15/35
60 - 69		6+16/35	5+15/32	3+16/31
70 +		0+2/2	0+2/2	1+0/2
Any age (including "not known")		15+42/107	15+44/112	11+49/103

	Nodal Status (N0, N1-3, N4+)			
< 50	N0 (by axillary	0+0/0	0+0/0	0+2/2
50 +	clearance)	1+0/1	0+1/1	0+1/3
< 50	N1-N3 (by axillary	2+3/21	2+7/30	1+8/25
50 +	clearance)	9+18/48	3+14/40	9+16/41
< 50	N4+ (by axillary	1+6/11	2+8/16	0+8/8
50 +	sample/clearance)	2+15/20	7+14/24	1+13/22
< 50	Remainder (i.e.	0+0/1	0+0/0	0+0/0
50 +	other N+ or N?)	0+0/5	1+0/1	0+1/2

	Estrogen Receptors (ER, fmol/mg)			
< 50	ER poor or ER < 10	2+7/15	0+8/16	1+7/13
50 +		5+10/20	3+13/24	2+8/15
< 50	ER+ or ER 10-99	0+2/13	2+5/15	0+4/11
50 +		5+17/34	3+8/24	5+13/35

TITLE: EDINBURGH PRIMARY BREAST TRIAL

ABBREVIATED TRIAL NAME: DUBLIN
STUDY NUMBER: 76D

STUDY NUMBER: 76D
STUDY TITLE: DUBLIN

INVESTIGATORS NAMES AND AFFILIATIONS:
C Smith, N Corcoran
St. Luke's Hospital, Dublin, Eire.

TRIAL DESIGN: TREATMENT GROUPS:
1. F: 5-fluorouracil 700 mg/M^2, intravenously day 1 of 4 week cycle for 12 cycles.
2. Nil: No adjuvant chemotherapy.

ELIGIBILITY CRITERIA FOR ENTRY:
1. AGE: < 75.
2. CLINICAL STAGES: All patients presenting with primary breast cancer, except stage IV or bilateral disease.
3. PRIMARY TREATMENT: Simple mastectomy + node sampling; radiotherapy for all Stage II and III.
4. NODE DISSECTION: Optional.
5. NUMBER OF NODES REMOVED: Not specified.
6. MENSTRUAL STATUS: All except those who had previous bilateral oophorectomy or hysterectomy.
7. RECEPTOR STATUS: No restriction.

METHOD OF RANDOMIZATION:
Randomization by sealed envelopes which were held at a central office by the coordinator. Allocations were based on random number tables. Prior knowledge of treatment allocation was not possible. Patients withdrawn from the trial have continued follow-up.

RANDOMIZATION STRATIFICATIONS:
Stage (N0, N1-3, N4+), menopausal status.

YEARS OF ACCRUAL: 1976 - 1978.

TOTAL NUMBER OF PATIENTS RANDOMIZED: 41.

AIM, RATIONALE, SPECIAL FEATURES:
All female patients who presented for treatment of primary breast cancer were eligible for entry into the study.

ADDITIONAL ELIGIBILITY CRITERIA OR STRATIFICATION METHODS:
Patients were ineligible for entry for the following reasons: pregnancy; age >75 yrs; bilateral breast cancer; stage IV disease; previous history of malignant disease at any site; and previous bilateral oophorectomy or hysterectomy. Staging was assessed by clinical examination with TNM staging, X-rays of the chest, lumbar spine and pelvis, and liver function tests, including alkaline phosphatase, liver enzymes and bilirubin.

Primary treatment was to be simple mastectomy and axillary node sampling. However, axillary node sampling was not done in all cases and clinical assessment was sufficient.

FOLLOW-UP, COMPLIANCE AND TOXICITY:
Clinical assessment, with full blood count and liver function tests, was carried out every two months for the first year, then every three months for the second. Thereafter, follow-up was every four months for the third year and every six month until five years. Annual follow-up after five years. In a few cases, however, follow-up after two years was carried out by local hospitals when domiciliary distance was a problem. Follow-up was by means of a questionnaire to the local hospital at the intervals stated. Total number of patients randomized was 41. No patient randomized was subsequently withdrawn.

PUBLICATIONS: None.

Table 1. Outcome by allocated treatment for trial 76D (Dublin)
Key: T = total free of relevant event at start of relevant year(s)

(i) Death (D) in each separate year,
subdivided by age when randomized (D <50 & D 50+ / T <50 & T 50+)

Year	1. F	2. Nil
1	0+ 0/ 8+ 12	0+ 1/ 10+ 11
2	1+ 3/ 8+ 12	1+ 1/ 10+ 10
3	1+ 1/ 7+ 9	1+ 1/ 9+ 9
4	3+ 1/ 6+ 8	1+ 0/ 8+ 8
5+	0+ 2/ 3+ 7	2+ 3/ 7+ 8

(ii) First recurrence or prior death (F) in each separate year,
subdivided by age when randomized (F <50 & F 50+ / T <50 & T 50+)

Year	1. F	2. Nil
1	2+ 2/ 8+ 12	3+ 2/ 10+ 11
2	2+ 4/ 6+ 10	0+ 1/ 7+ 9
3	0+ 0/ 4+ 6	1+ 1/ 7+ 8
4	1+ 0/ 4+ 6	2+ 1/ 6+ 7
5+	0+ 3/ 3+ 6	0+ 1/ 4+ 6

(iii) Recurrence with survival (R) and Death (D) in all years together,
subdivided by various characteristics when randomized (R & D / T)

Age at Entry	1. F	2. Nil
< 40	0+1/1	0+2/2
40 - 49	0+4/7	1+3/8
50 - 59	0+3/4	0+5/7
60 - 69	2+2/5	0+1/4
70 +	0+2/3	0+0/0
Any age (including "not known")	2+12/20	1+11/21

Nodal Status (N0, N1-3, N4+)

		1. F	2. Nil
<50	N0 (by clinical	0+0/2	0+0/4
50 +	evidence only)	0+2/4	0+3/6
<50	Remainder (i.e.	0+5/6	1+5/6
50 +	other N+ or N?)	2+5/8	0+3/5

Estrogen Receptors (ER, fmol/mg)

Breakdown by ER status was not available.

TITLE: DUTCH EORTC STUDY ON LOW-DOSE CMF ADJUVANT
CHEMOTHERAPY IN OPERABLE BREAST CANCER

STUDY NUMBER: 76E
STUDY TITLE: EORTC 09771

ABBREVIATED TRIAL NAME: EORTC 09771
STUDY NUMBER: 76E

INVESTIGATORS' NAMES AND AFFILIATIONS:
K Welvaart (Study Coordinator), OJ Repelaer van Driel, CJH van de Velde.
Dept. of Surgery at University Hospital, P.O. Box 9600, 2300; RC Leiden, The Netherlands; and Surgeons and Medical Oncologists of 20 Dutch hospitals.

TRIAL DESIGN: TREATMENT GROUPS:
1. CMF: Cyclophosphamide 50 mg/M^2/day orally on days 1-14 inclusive; methotrexate 5 mg/M^2/day intravenously on days 1 and 8; and 5-fluorouracil intravenously 350 mg/M^2 on days 1 and 8. Treatment given for 2 years on a monthly schedule.
2. Nil: No adjuvant chemotherapy.

ELIGIBILITY CRITERIA FOR ENTRY:
1. AGE: < 70.
2. CLINICAL/PATHOLOGICAL STAGES: All operable stages; M1 excluded. One or more positive nodes; top node-free.
3. PRIMARY TREATMENT: Modified radical mastectomy and loco-regional radiotherapy.
4. AXILLARY NODE DISSECTION: Required.
5. NUMBER OF NODES REMOVED: Optional.
6. MENSTRUAL STATUS: No restriction.
7. RECEPTOR STATUS: No restriction.

METHOD OF RANDOMIZATION:
Telephone to a central office where allocations were based on randomization sheets, using permuted blocks, provided by EORTC Data Center. Patient allocation not released until patient identifiers recorded in central office.

YEARS OF ACCRUAL: 1976-1980.

TOTAL NUMBER OF PATIENTS RANDOMIZED: 452.

ADDITIONAL ELIGIBILITY OR STRATIFICATION CRITERIA:
Additional eligibility criteria included: WBC >4000/mm^3; platelet count >30,000/mm^3; and serum creatinine <115 µmol/l.

Patients were excluded for the following reasons: previous treatment for other malignancy; bilaterality; pregnancy and lactation.

Only 17 patients (3.8%) were ineligible.

FOLLOW-UP, COMPLIANCE AND TOXICITY:
All patients were followed up every 3 months for the first 2 years, every 6 months for three more years, and annually thereafter. National death records were not used.

PUBLICATIONS:
Repelaer van Driel OJ, Welvaart K, Zwaveling A, van der Velde CJM, and Co-op investigators. Influence of low-dose adjuvant CMF chemotherapy on survival and locoregional recurrences in operable breast cancer (EORTC 09771). Eur. Surg. Res. 1987; 19s:93.

Repelaer van Driel OJ. Acjuvante chemotherapie bij mammacarcinoom. Thesis, Leiden; 1987.

Repelaer van Driel OJ, Welvaart K, and Co-op. investigators. Prolonged low-dose CMF as adjuvant chemotherapy in operable breast cancer: Treatment results (EORTC 09771). Cancer Chem. Pharm. 1989; 18:17-59.

Table 1. Outcome by allocated treatment for trial 76E (EORTC 09771)
Key: T = total free of relevant event at start of relevant year(s)

(i) *Death (D) in each separate year,*
subdivided by age when randomized (D <50 & D 50+ / T <50 & T 50+)

Year	1. CMF	2. Nil
1	0+ 0/ 96+ 133	1+ 3/ 90+ 133
2	3+ 2/ 96+ 132	3+ 6/ 89+ 130
3	4+ 7/ 93+ 130	10+ 7/ 86+ 124
4	3+ 7/ 89+ 123	1+ 6/ 76+ 117
5+	5+ 9/ 86+ 116	8+ 9/ 75+ 111

(ii) *First recurrence or prior death (F) in each separate year,*
subdivided by age when randomized (F <50 & F 50+ / T <50 & T 50+)

Year	1. CMF	2. Nil
1	6+ 6/ 96+ 133	5+ 17/ 90+ 133
2	7+ 14/ 69+ 99	18+ 16/ 65+ 81
3	8+ 7/ 60+ 83	7+ 9/ 47+ 63
4	2+ 4/ 49+ 72	8+ 5/ 40+ 53
5+	7+ 15/ 41+ 56	4+ 12/ 26+ 42

(iii) *Recurrence with survival (R) and Death (D) in all years together,*
subdivided by various characteristics when randomized (R & D / T)

Age at Entry	1. CMF	2. Nil
< 40	4+6/26	6+8/29
40 – 49	11+9/71	13+15/61
50 – 59	12+15/73	17+13/64
60 – 69	9+10/56	10+18/62
70 +	0+0/1	0+0/2
Any age (including "not known")	36+40/229	47+54/223

Nodal Status (N0, N1-3, N4+)

		1. CMF	2. Nil
< 50	N1-N3 (by axillary clearance)	11+9/70	15+14/66
50 +		16+15/100	21+19/99
< 50	N4+ (by axillary sample/clearance)	4+6/27	4+9/24
50 +		5+10/32	7+12/33
50 +	Remainder (i.e. other N+, or N?)	0+0/0	0+0/1

Estrogen Receptors (ER, fmol/mg)

Breakdown by ER status was not available.

TITLE: ADJUVANT TAMOXIFEN FOR OPERABLE CARCINOMA OF THE BREAST

STUDY NUMBER: 76F
STUDY TITLE: CHRISTIE B

ABBREVIATED TRIAL NAME: CHRISTIE B
STUDY NUMBER: 76F

INVESTIGATORS NAMES AND AFFILIATIONS:
G Ribeiro, MK Palmer, R Swindell
Christie Hospital and the Holt Radium Institute, Withington, Manchester, UK.

TRIAL DESIGN: TREATMENT GROUPS:
1. Tam: Tamoxifen 10 mg orally, twice daily for one year.
2. Nil: No adjuvant endocrine therapy.

ELIGIBILITY CRITERIA FOR ENTRY:
1. AGE: 38-70 yrs.
2. CLINICAL STAGES: T1-3, N0 - lb, M0.
3. PRIMARY TREATMENT: Simple mastectomy and axillary node sampling, or radical mastectomy with total axillary clearance or adjuvant radiotherapy if axillary lymph nodes histologically positive or tumor T3.
4. NODE DISSECTION: Optional.
5. NUMBER OF NODES REMOVED: Not specified.
6. MENSTRUAL STATUS: Postmenopausal (more than 2 years after natural menopause or over age 55 if previous hysterectomy).
7. RECEPTOR STATUS: Not specified.

METHOD OF RANDOMIZATION:
Balanced randomizations were generated. Each centre had sealed envelopes containing allocations made from this list. These were opened as each patient was entered into the trial. A form was then sent into the Central Office where a master list was held. Prior knowledge of allocations was theoretically possible.

RANDOMIZATION STRATIFICATIONS: None.

YEARS OF ACCRUAL: 1976-1982.

TOTAL NUMBER OF PATIENTS RANDOMIZED: 588. The slight shortfall in the number of patients allocated Tam (282 vs. 306) was counterbalanced by an excess (199 vs. 174) in a parallel study for premenopausal patients of Tam vs. ovarian irradiation that used the same envelopes.

PUBLICATIONS:
Ribeiro G, Swindell R. The Christie Hospital Tamoxifen Adjuvant Trial for Operable Breast Carcinoma 7 year results. Eur. J. Cancer Clin. Oncol. 1985; 21: 897-900.

Ribeiro G, Swindell R. The Christie Hospital adjuvant tamoxifen trial: status at 10 years. Br. J. Cancer 1988; 57:601-603.

Table 1. Outcome by allocated treatment for trial 76F (Christie B)
Key: T = total free of relevant event at start of relevant year(s)

(i) Death (D) in each separate year, subdivided by age when randomized (D <50 & D 50+ / T <50 & T 50+)

Year	1. Tam	2. Nil
1	2+ 8/ 12+ 270	1+ 15/ 9+ 297
2	0+ 20/ 10+ 262	1+ 22/ 8+ 282
3	2+ 20/ 10+ 242	0+ 33/ 7+ 260
4	0+ 17/ 8+ 222	0+ 18/ 7+ 227
5+	1+ 31/ 6+ 176	1+ 29/ 7+ 186

(ii) First recurrence or prior death (F) in each separate year, subdivided by age when randomized (F <50 & F 50+ / T <50 & T 50+)

Year	1. Tam	2. Nil
1	3+ 19/ 12+ 270	1+ 37/ 9+ 297
2	2+ 33/ 9+ 251	2+ 37/ 8+ 260
3	1+ 23/ 7+ 218	0+ 30/ 6+ 223
4	0+ 13/ 6+ 195	0+ 18/ 6+ 193
5+	0+ 36/ 6+ 179	2+ 27/ 6+ 173

(iii) Recurrence with survival (R) and Death (D) in all years together, subdivided by various characteristics when randomized (R & D / T)

Age at Entry	1. Tam	2. Nil
< 40	1+0/1	0+0/0
40 - 49	0+5/11	2+3/9
50 - 59	15+49/130	12+56/136
60 - 69	13+47/140	20+60/160
70 +	0+0/0	0+1/1
Any age (including "not known")	29+101/282	34+120/306

Nodal Status (N0, N1-3, N4+)

		1. Tam	2. Nil
<50	N0 (by axillary clearance)	0+0/2	0+0/1
50 +		4+9/31	2+10/33
<50	N0 (by clinical evidence only)	1+0/3	1+2/4
50 +		6+10/55	5+13/56
<50	N1-N3 (by axillary clearance)	0+0/1	0+0/0
50 +		1+14/24	1+10/22
<50	N4+ (by axillary sample/clearance)	0+3/3	0+0/1
50 +		5+23/38	2+24/32
<50	Remainder (i.e. other N+ or N?)	0+2/3	1+1/3
50 +		12+40/122	22+60/154

Estrogen Receptors (ER, fmol/mg)

Breakdown by ER status was not available.

TITLE: THE STOCKHOLM TRIAL ON ADJUVANT TAMOXIFEN IN EARLY BREAST CANCER

STUDY NUMBER: 76G1-2
STUDY TITLE: STOCKHOLM B

ABBREVIATED TRIAL NAME: STOCKHOLM B
STUDY NUMBER: 76G1-2

INVESTIGATORS' NAMES AND AFFILIATIONS:
LE Rutqvist (1)*, B Cedermark (2), U Glas (3), A Somell (4), N-O Theve (5), ML Hjalmer (6), J Askergren (7), L Skoog (8), S. Rotstein (6).

(1) Radiumhemmet (Dept. of Oncology), Karolinska Hosp., Stockholm.
(2) Dept. of Surgery, Karolinska Hosp., Stockholm.
(3) Dept. of Oncology, Södersjukhuset, Stockholm.
(4) Dept. of Surgery, Södersjukhuset, Stockholm.
(5) Dept. of Surgery, Sabbatsberg Hosp., Stockholm.
(6) Dept. of Oncology, Danderyd Hosp., Stockholm.
(7) Dept. of Surgery, Danderyd Hosp., Stockholm.
(8) Dept. of Cytology, Karolinska Hosp., Stockholm.

* Principal Investigator.

TRIAL DESIGN: TREATMENT GROUPS:
1. TamR: Tamoxifen 40 mg daily begun 4-6 weeks postoperatively for either 2 or 5 (determined by a second randomization) years plus radiotherapy 46 Gy in 4.5 to 5 weeks, given with high voltage machines.
2. TamCMF: Tamoxifen as above plus 12 courses of cyclophosphamide 100 mg/M^2, orally, days 1 to 14; methotrexate 40 mg/M^2 i.v. and 5-fluorouracil 600 mg/M^2 i.v. on day 1 and 8. Until June 1978, chlorambucil 10-15 mg orally on day 1-8 was used instead of cyclophosphamide; about 15% of chemotherapy patients were treated with the LMF regimen.
3. R: Radiotherapy as above.
4. CMF: CMF as above.

ELIGIBILITY CRITERIA FOR ENTRY:
1. AGE: ≤70 years initially, ≤65 after June 1978.
2. CLINICAL STAGES: Operable unilateral breast cancer with positive nodes and/or tumor exceeding 30 mm (i.e. "high-risk").
3. PRIMARY TREATMENT: Modified radical mastectomy.
4. MENSTRUAL STATUS: Postmenopausal (no menstrual period for 6 months or after a hysterectomy if the woman was ≥50 years).
5. RECEPTOR STATUS: Not required. However, estrogen receptor values were available on 75% of trial patients. They were determined by means of isoelectric focusing on polyacrylamide gel and were expressed as femtomoles per microgram of DNA. For the purposes of the EBCTCG overview, values of <0.04 were considered to be ER-, 0.04 to <1.2 were ER+ and ≥1.2 were ER++.

METHOD OF RANDOMIZATION:
By telephone to a central office, where allocation was performed using balanced lists. Allocations were balanced in blocks of 24. Between March 1982 and September 1984 (76G2), patients were randomised to arms 2 and 4 in a 3:1 ratio because of a shortage of radiation treatment capacity. It was not possible for clinicians to know what treatment would be allocated until after the patient identifiers had been recorded centrally. No patient for whom a treatment was allocated was subsequently withdrawn from the analysis.

RANDOMIZATION STRATIFICATIONS:
Age, menstrual status, tumor size, nodal status.

YEARS OF ACCRUAL: 1976-1984.

TOTAL NUMBER OF PATIENTS RANDOMIZED: 442.

AIM, RATIONALE, SPECIAL FEATURES:
Patients were secondarily randomised to receive either adjuvant chemotherapy or postoperative radiotherapy, but this comparison is not relevant to the current overview.

ADDITIONAL ELIGIBILITY OR STRATIFICATION CRITERIA:
Patients in 76G were stratified into four groups according to extent of disease and primary therapy. This group (76G1-2) consisted of "high risk" patients who could be considered for chemotherapy. Patients with previous malignancies or debilitating conditions were not eligible.

FOLLOW-UP, COMPLIANCE AND TOXICITY:
Follow-up visits took place every 3 months during the first 2 years, every 6 months during the next 2-5 years and annually thereafter. The median follow-up period in the current analysis is about 5 years. Less than 1% of the patients were lost to follow-up. Patients who did not come to a scheduled follow-up visit were investigated using official death registers, a computerised hospital register, or were contacted by letter.

DISCUSSION:
Previous interim analyses have demonstrated an increased recurrence-free interval in the tamoxifen group, mainly due to a significantly reduced frequency of loco-regional relapses during the first 3 years. No significant differences in the efficacy of the tamoxifen treatment have been observed in the different stages or treatment groups. So far, tamoxifen treatment has not significantly improved survival.

PUBLICATIONS:
Rutqvist LE, Cedermark B, Glas U, Johansson H, Nordenskjöld B, Skoog L, Somell A, Theve T, Friberg S, Askergren J. The Stockholm trial on adjuvant tamoxifen in early breast cancer. Correlation between estrogen receptor level and treatment effect. Breast Cancer Res. Treat. 1987, 10:255-266.

TITLE: THE STOCKHOLM TRIAL ON ADJUVANT TAMOXIFEN IN EARLY BREAST CANCER

STUDY NUMBER: 76G3
STUDY TITLE: STOCKHOLM B

ABBREVIATED TRIAL NAME: STOCKHOLM B
STUDY NUMBER: 76G3

INVESTIGATORS' NAMES AND AFFILIATIONS:
LE Rutqvist (1)*, B Cedermark (2), U Glas (3), A Somell (4), N-O Theve (5), ML Hjalmer (6), J Askergren (7), L Skoog (8), S. Rotstein (6).

(1) Radiumhemmet (Dept. of Oncology), Karolinska Hosp., Stockholm.
(2) Dept. of Surgery, Karolinska Hosp., Stockholm.
(3) Dept. of Oncology, Södersjukhuset, Stockholm.
(4) Dept. of Surgery, Södersjukhuset, Stockholm.
(5) Dept. of Surgery, Sabbatsberg Hosp., Stockholm.
(6) Dept. of Oncology, Danderyd Hosp., Stockholm.
(7) Dept. of Surgery, Danderyd Hosp., Stockholm.
(8) Dept. of Cytology, Karolinska Hosp., Stockholm.

* Principal Investigator.

TRIAL DESIGN: TREATMENT GROUPS:
1. Tam: Tamoxifen 40 mg daily for either 2 or 5 (determined by a second randomization) years. Therapy begun 4-6 weeks postoperatively.
2. Nil: No adjuvant endocrine therapy.

ELIGIBILITY CRITERIA FOR ENTRY:
1. AGE: ≤70 years.
2. CLINICAL STAGES: Operable unilateral breast cancer with positive nodes.and/or tumor exceeding 30 mm (i.e. "high-risk").
3. PRIMARY TREATMENT: Modified radical mastectomy and postoperative radiotherapy (46 Gy in 4.5 to 5 weeks).
4. MENSTRUAL STATUS: Postmenopausal (no menstrual period for 6 months or after a hysterectomy if the woman was ≥50 years).
5. RECEPTOR STATUS: Not required However, estrogen receptor values were available on 75% of trial patients. They were determined by means of isoelectric focusing on polyacrylamide gel and were expressed as femtomoles per microgram of DNA. For the purposes of the EBCTCG overview, values of <0.04 were considered to be ER-, 0.04 to <1.2 were ER+ and ≥1.2 were ER++.

METHOD OF RANDOMIZATION:
By telephone to a central office, where allocation was performed using balanced lists. Allocations were balanced in blocks of 6. It was not possible for clinicians to know what treatment would be allocated until after the patient identifiers had been recorded centrally. No patient for whom a treatment was allocated was subsequently withdrawn from the analysis.

RANDOMIZATION STRATIFICATIONS:
Age, menstrual status, tumor size, nodal status.

YEARS OF ACCRUAL: 1976-1984.

TOTAL NUMBER OF PATIENTS RANDOMIZED: 92.

ADDITIONAL ELIGIBILITY OR STRATIFICATION CRITERIA:
Patients with previous malignancies or debilitating conditions were not eligible. Patients in 76G were stratified into four groups according to extent of disease and primary therapy. This group (76G3) consisted of "high-risk" patients who could not be considered for CMF chemotherapy (e.g. because of poor general health or patient refusal to be secondarily randomized to chemotherapy or radiotherapy).

FOLLOW-UP, COMPLIANCE AND TOXICITY:
Follow-up visits took place every 3 months during the first 2 years, every 6 months during the next 2-5 years and annually thereafter. The median follow-up period in the current analysis is about 5 years. Less than 1% of the patients were lost to follow-up.

Patients who did not come to a scheduled follow-up visit were investigated using official death registers, a computerised hospital register, or were contacted by letter.

DISCUSSION:
Previous interim analyses have demonstrated an increased recurrence-free interval in the tamoxifen group, mainly due to a significantly reduced frequency of loco-regional relapses during the first 3 years. No significant differences in the efficacy of the tamoxifen treatment have been observed in the different stages or treatment groups. So far, tamoxifen treatment has not significantly improved survival.

PUBLICATIONS:
Rutqvist LE, Cedermark B, Glas U, Johansson H, Nordenskjöld B, Skoog L, Somell A, Theve T, Friberg S, Askergren J. The Stockholm trial on adjuvant tamoxifen in early breast cancer. Correlation between estrogen receptor level and treatment effect. Breast Cancer Res. Treat. 1987, 10:255-266.

Table 1. Outcome by allocated treatment for trial 76G3 (Stockholm B)
Key: T = total free of relevant event at start of relevant year(s)

(i) Death (D) in each separate year,
subdivided by age when randomized (D <50 & D 50+ / T <50 & T 50+)

Year	1. TamR	2. R
1	0+ 0/ 0+ 44	0+ 0/ 0+ 48
2	0+ 4/ 0+ 42	0+ 1/ 0+ 48
3	0+ 0/ 0+ 25	0+ 1/ 0+ 28
4	0+ 0/ 0+ 16	0+ 0/ 0+ 15
5+	0+ 0/ 0+ 6	0+ 0/ 0+ 5

(ii) First recurrence or prior death (F) in each separate year,
subdivided by age when randomized (F <50 & F 50+ / T <50 & T 50+)

Year	1. TamR	2. R
1	0+ 3/ 0+ 44	0+ 1/ 0+ 48
2	0+ 3/ 0+ 37	0+ 0/ 0+ 39
3	0+ 1/ 0+ 21	0+ 2/ 0+ 24
4	0+ 0/ 0+ 11	0+ 0/ 0+ 9
5+	0+ 0/ 0+ 2	0+ 0/ 0+ 1

(iii) Recurrence with survival (R) and Death (D) in all years together,
subdivided by various characteristics when randomized (R & D / T)

Age at Entry		1. TamR	2. R
50 - 59		1+0/8	0+1/7
60 - 69		2+2/33	1+1/39
70 +		0+2/3	0+0/2
Any age (including "not known")		3+4/44	1+2/48
	Nodal Status (N0, N1-3, N4+)		
50 +	N0 (by axillary clearance)	0+1/6	0+0/7
50 +	N1-N3 (by axillary clearance)	2+0/25	0+0/28
50 +	N4+ (by axillary sample/clearance)	1+3/13	1+1/11
50 +	Remainder (i.e. other N+ or N?)	0+0/0	0+1/2
	Estrogen Receptors (ER, fmol/mg)		
50 +	ER poor or ER < 10	1+1/5	0+0/7
50 +	ER+ or ER 10-99	1+2/21	1+1/22
50 +	ER++ or ER 100+	1+1/14	0+0/14

Short reports

TITLE: THE STOCKHOLM TRIAL ON ADJUVANT TAMOXIFEN IN EARLY BREAST CANCER

STUDY NUMBER: 76G4-5
STUDY TITLE: STOCKHOLM B

ABBREVIATED TRIAL NAME: STOCKHOLM B
STUDY NUMBER: 76G4-5

INVESTIGATORS' NAMES AND AFFILIATIONS:
LE Rutqvist (1)*, B Cedermark (2), U Glas (3), A Somell (4), N-O Theve (5), ML Hjalmer (6), J Askergren (7), L Skoog (8), S. Rotstein (6).

(1) Radiumhemmet (Dept. of Oncology), Karolinska Hosp., Stockholm.
(2) Dept. of Surgery, Karolinska Hosp., Stockholm.
(3) Dept. of Oncology, Södersjukhuset, Stockholm.
(4) Dept. of Surgery, Södersjukhuset, Stockholm.
(5) Dept. of Surgery, Sabbatsberg Hosp., Stockholm.
(6) Dept. of Oncology, Danderyd Hosp., Stockholm.
(7) Dept. of Surgery, Danderyd Hosp., Stockholm.
(8) Dept. of Cytology, Karolinska Hosp., Stockholm.

* Principal Investigator.

TRIAL DESIGN: TREATMENT GROUPS:
1. Tam: Tamoxifen 40 mg daily for either 2 or 5 (determined by a second randomization) years. Therapy begun 4-6 weeks postoperatively.
2. Nil: No adjuvant endocrine therapy.

ELIGIBILITY CRITERIA FOR ENTRY:
1. AGE: ≤70 years.
2. CLINICAL STAGES: Operable unilateral breast cancer without positive nodes. Tumor size less than 30 mm (i.e. "low-risk").
3. PRIMARY TREATMENT: Modified radical mastectomy (76G4) or breast-conserving surgery and post-operative radiotherapy (50 Gy/5 weeks) to the breast (76G5).
4. MENSTRUAL STATUS: Postmenopausal (no menstrual period for 6 months or after a hysterectomy if the woman was ≥50 years).
5. RECEPTOR STATUS: Not required. However, estrogen receptor values were available on 75% of trial patients. They were determined by means of isoelectric focusing on polyacrylamide gel and were expressed as femtomoles per microgram of DNA. For the purposes of the EBCTCG overview, values of <0.04 were considered to be ER-, 0.04 to <1.2 were ER+ and ≥1.2 were ER++.

METHOD OF RANDOMIZATION:
By telephone to a central office, where allocation was performed using balanced lists. Allocations were balanced in blocks of 6. It was not possible for clinicians to know what treatment would be allocated until after the patient identifiers had been recorded centrally. No patient for whom a treatment was allocated was subsequently withdrawn from the analysis.

RANDOMIZATION STRATIFICATIONS:
Age, menstrual status, tumor size, nodal status.

YEARS OF ACCRUAL: 1976-1984.

TOTAL NUMBER OF PATIENTS RANDOMIZED: 790 (76G4) and 149 (76G5).

ADDITIONAL ELIGIBILITY OR STRATIFICATION CRITERIA:
Patients with previous malignancies or debilitating conditions were not eligible. Patients in 76G were stratified into four groups according to extent of disease and primary therapy. These two groups consisted of "low-risk" patients who were treated with mastectomy (76G4) or breast-conserving surgery and radiotherapy (76G5).

FOLLOW-UP, COMPLIANCE AND TOXICITY:
Follow-up visits took place every 3 months during the first 2 years, every 6 months during the next 2-5 years and annually thereafter. The median follow-up period in the current analysis is about 5 years. Less than 1% of the patients were lost to follow-up.

Patients who did not come to a scheduled follow-up visit were investigated using official death registers, a computerised hospital register, or were contacted by letter.

DISCUSSION:
Previous interim analyses have demonstrated an increased recurrence-free interval in the tamoxifen group, mainly due to a significantly reduced frequency of loco-regional relapses during the first 3 years. No significant differences in the efficacy of the tamoxifen treatment have been observed in the different stages or treatment groups. So far, tamoxifen treatment has not significantly improved survival.

PUBLICATIONS:
Rutqvist LE, Cedermark B, Glas U, Johansson H, Nordenskjöld B, Skoog L, Somell A, Theve T, Friberg S, Askergren J. The Stockholm trial on adjuvant tamoxifen in early breast cancer. Correlation between estrogen receptor level and treatment effect. Breast Cancer Res. Treat. 1987, 10:255-266.

Table 1. Outcome by allocated treatment for trial 76G4 (Stockholm B)
Key: T = total free of relevant event at start of relevant year(s)

(i) Death (D) in each separate year, subdivided by age when randomized (D <50 & D 50+ / T <50 & T 50+)

Year	1. Tam	2. Nil
1	0+ 2/3+ 393	0+ 3/5+ 389
2	0+ 0/3+ 373	0+ 6/5+ 368
3	0+ 4/2+ 318	0+ 5/3+ 315
4	0+ 14/2+ 272	0+ 4/3+ 272
5+	0+ 9/0+ 198	0+ 8/2+ 200

(ii) First recurrence or prior death (F) in each separate year, subdivided by age when randomized (F <50 & F 50+ / T <50 & T 50+)

Year	1. Tam	2. Nil
1	0+ 7/3+ 393	0+ 14/5+ 389
2	0+ 10/3+ 347	0+ 18/5+ 345
3	0+ 11/2+ 284	0+ 17/3+ 274
4	0+ 8/1+ 228	0+ 5/3+ 213
5+	0+ 10/0+ 165	0+ 9/1+ 150

(iii) Recurrence with survival (R) and Death (D) in all years together, subdivided by various characteristics when randomized (R & D / T)

Age at Entry	1. Tam	2. Nil
40 - 49	0+0/3	0+0/5
50 - 59	5+10/143	14+12/144
60 - 69	10+15/229	21+12/225
70 +	2+4/21	2+2/20
Any age (including "not known")	17+29/396	37+26/394

	Nodal Status (N0, N1-3, N4+)	1. Tam	2. Nil
<50 / 50+	N0 (by clinical evidence only)	15+28/382	34+25/381
50+	Remainder (i.e. other N+ or N?)	2+1/11	3+1/8

	Estrogen Receptors (ER, fmol/mg)	1. Tam	2. Nil
<50	ER poor or ER < 10	0+0/1	0+0/1
50+		6+7/60	5+4/58
<50	ER+ or ER 10-99	0+0/1	0+0/2
50+		6+8/141	18+8/162
<50	ER++ or ER 100+	0+0/0	0+0/1
50+		4+8/121	12+7/107

Table 1. Outcome by allocated treatment for trial 76G5 (Stockholm B)
Key: T = total free of relevant event at start of relevant year(s)

(i) Death (D) in each separate year, subdivided by age when randomized (D <50 & D 50+ / T <50 & T 50+)

Year	1. TamR	2. R
1	0+ 0/3+ 70	0+ 0/0+ 76
2	0+ 0/3+ 69	0+ 1/0+ 73
3	0+ 1/1+ 51	0+ 1/0+ 56
4	0+ 0/1+ 34	0+ 2/0+ 44
5+	0+ 1/0+ 22	0+ 0/0+ 22

(ii) First recurrence or prior death (F) in each separate year, subdivided by age when randomized (F <50 & F 50+ / T <50 & T 50+)

Year	1. TamR	2. R
1	0+ 0/3+ 70	0+ 0/0+ 76
2	0+ 1/2+ 60	0+ 4/0+ 68
3	0+ 1/1+ 42	0+ 2/0+ 50
4	0+ 1/1+ 28	0+ 1/0+ 31
5+	0+ 0/0+ 18	0+ 0/0+ 17

(iii) Recurrence with survival (R) and Death (D) in all years together, subdivided by various characteristics when randomized (R & D / T)

Age at Entry	1. TamR	2. R
40 - 49	0+0/3	0+0/0
50 - 59	0+1/36	1+1/36
60 - 69	0+0/33	2+3/40
70 +	0+1/1	0+0/0
Any age (including "not known")	0+2/73	3+4/76

	Nodal Status (N0, N1-3, N4+)	1. TamR	2. R
<50 / 50+	N0 (by clinical evidence only)	0+0/3 / 0+2/68	0+0/0 / 2+4/75
50+	Remainder (i.e. other N+ or N?)	0+0/2	1+0/1

	Estrogen Receptors (ER, fmol/mg)	1. TamR	2. R
50+	ER poor or ER < 10	0+1/11	0+1/11
<50 / 50+	ER+ or ER 10-99	0+0/3 / 0+0/30	0+0/0 / 1+1/29
50+	ER++ or ER 100+	0+0/15	2+1/18

TITLE: WMOA TRIAL OF ADJUVANT CHEMOTHERAPY FOR OPERABLE BREAST CANCER WITH POSITIVE AXILLARY NODES.

STUDY NUMBER: 76H1
STUDY TITLE: WEST MIDLANDS A

ABBREVIATED TRIAL NAME: WEST MIDLANDS A
STUDY NUMBER: 76H1

INVESTIGATORS' NAMES AND AFFILIATIONS:
JM Morrison (1), A Howell (2), KA Kelly (3), RJ Grieve (4), IJ Monypenny (5), RA Walker (6), JAH Waterhouse (7).

(1) Department of Surgery, Selly Oak Hospital, Birmingham, U.K.
(2) Department of Medical Oncology, Christie Hospital, Manchester.
(3) West Midlands Cancer Research Campaign Clinical Trials Unit, Birmingham.
(4) Walsgrave Hospital, Coventry.
(5) Llandough Hospital, Cardiff.
(6) Department of Pathology, University of Leicester.
(7) West Midlands Regional Cancer Registry, Birmingham.

TRIAL DESIGN: TREATMENT GROUPS:
1. CMFVALvor: Doxorubicin 50 mg, vincristine 1 mg, and cyclophosphamide 250 mg, intravenously, methotrexate 150 mg via 12 hour infusion, followed by 5-fluorouracil 250 mg and leucovorin rescue. Treatment given every 3 weeks for 6 months.
2. Nil: No adjuvant therapy.

ELIGIBILITY CRITERIA FOR ENTRY:
1. AGE: < 65.
2. CLINICAL STAGES: T1a-T3a, N1, M0.
3. PRIMARY TREATMENT: Simple mastectomy with axillary node sampling; no radiotherapy.
4. AXILLARY NODE DISSECTION: Not required.
5. NUMBER OF NODES REMOVED: Not specified.
6. MENSTRUAL STATUS: No restrictions.
7. RECEPTOR STATUS: No restrictions.

METHOD OF RANDOMIZATION:
Telephone call to a central office where allocations were based on random number tables and balanced in pairs.

RANDOMIZATION STRATIFICATIONS:
Menopausal status: pre, peri, post, hysterectomy and tumor size: <5 cm, or 5 cm or larger.

YEARS OF ACCRUAL: 1976-1984.

TOTAL NUMBER OF PATIENTS RANDOMIZED: 569.

ADDITIONAL ELIGIBILITY CRITERIA OR STRATIFICATION METHODS:
Patients were eligible if they had a white blood cell count greater than 4×10^9/l, platelets greater than 100×10^9/l and normal liver function tests. Patients were ineligible if they were pregnant or lactating, had previous malignancy, serious intercurrent disease or psychiatric disorder.

FOLLOW UP, COMPLIANCE AND TOXICITY:
Patients were examined every three months for 18 months and then every six months until recurrence or death. Investigations included complete blood count and biochemistry profile at each visit, chest radiograph, skeletal radiographs and radio-isotope bone scan six monthly for two years followed by annual examinations to five years. Toxicity was recorded for each treatment cycle.

569 patients were randomized of whom 540 (277 treated and 263 control) were eligible. All randomized patients are available for the analysis of survival, since all patients were flagged with the West Midlands Regional Cancer Registry who notified the Trial Secretariat when any patient died. Other analyses, including the analysis of relapse-free survival, can be based on only the eligible patients since the 29 (17 treated and 12 control) ineligible patients were not followed up. One treated patient was completely lost to follow-up.

The regime was well tolerated at a high level of compliance. Of the eligible patients, 80% received all eight courses, 88% received seven courses or more and 91% four courses or more. Only 3% received no chemotherapy. The reasons for failure to receive a full course of chemotherapy were patient refusal (37%), toxicity (22%), illness not due to treatment (15%), and administrative error, mainly involving the 8th course (26%).

PUBLICATIONS:
Morrison JM, Howell A, Grieve RJ, et al. West Midlands Oncology Association trials of adjuvant chemotherapy for operable breast cancer. In Adjuvant Therapy of Cancer V. Ed. Salmon SE and Jones SE 1987, Grune and Stratton.

Morrison JM, Howell A, Kelly KA, et al. West Midlands Oncology Association trials of adjuvant chemotherapy in operable breast cancer: results after a median follow-up of 7 years. I. Patients with involved axillary lymph nodes. Br. J. Cancer 1989; 60:911-918.

Table 1. Outcome by allocated treatment for trial 76H1 (West Midlands)
Key: T = total free of relevant event at start of relevant year(s)

(i) Death (D) in each separate year,
subdivided by age when randomized (D <50 & D 50+ / T <50 & T 50+)

Year	1. CMFVALvor	2. Nil
1	4+ 8/ 119+ 175	5+ 9/ 119+ 156
2	12+ 20/ 115+ 167	16+ 19/ 114+ 147
3	17+ 17/ 99+ 143	17+ 19/ 95+ 123
4	8+ 17/ 80+ 115	10+ 13/ 72+ 95
5+	7+ 18/ 60+ 87	11+ 15/ 54+ 73

(ii) First recurrence or prior death (F) in each separate year,
subdivided by age when randomized (F <50 & F 50+ / T <50 & T 50+)

Year	1. CMFVALvor	2. Nil
1	19+ 36/ 119+ 175	35+ 46/ 119+ 156
2	26+ 35/ 99+ 134	27+ 32/ 84+ 104
3	14+ 18/ 71+ 98	12+ 12/ 54+ 71
4	4+ 8/ 54+ 74	6+ 12/ 39+ 54
5+	7+ 15/ 44+ 53	8+ 10/ 28+ 37

(iii) Recurrence with survival (R) and Death (D) in all years together,
subdivided by various characteristics when randomized (R & D / T)

Age at Entry	1. CMFVALvor	2. Nil
< 40	6+17/30	8+11/26
40 - 49	16+31/89	21+48/93
50 - 59	19+60/125	27+55/108
60 - 69	13+20/49	10+20/47
70 +	0+0/1	0+0/1
Any age (including "not known")	54+128/294	66+134/275

Nodal Status (N0, N1-3, N4+)

Breakdown by nodal status was not available.

Estrogen Receptors (ER, fmol/mg)

		1. CMFVALvor	2. Nil
<50	ER poor or ER < 10	2+23/32	4+24/38
50 +		2+18/29	4+23/35
<50	ER+ or ER 10-99	7+10/34	13+16/39
50 +		11+21/42	12+23/46
<50	ER++ or ER 100+	5+1/11	3+1/5
50 +		7+7/31	13+11/32

TITLE: WMOA TRIAL OF ADJUVANT CHEMOTHERAPY FOR OPERABLE
BREAST CANCER WITH NEGATIVE AXILLARY NODES.

STUDY NUMBER: 76H2
STUDY TITLE: WEST MIDLANDS B

ABBREVIATED TRIAL NAME: WEST MIDLANDS B
STUDY NUMBER: 76H2

INVESTIGATORS' NAMES AND AFFILIATIONS:
JM Morrison (1), A Howell (2), KA Kelly (3), RJ Grieve (4), IJ Monypenny (5), RA Walker (6), JAH Waterhouse (7).

(1) Department. of Surgery, Selly Oak Hospital, Birmingham, UK.
(2) Department. of Medical Oncology, Christie Hospital, Manchester.
(3) West Midlands Cancer Research Campaign Clinical Trials Unit, Birmingham.
(4) Walsgrave Hospital, Coventry.
(5) Llandough Hospital, Cardiff.
(6) Department of Pathology, University of Leicester.
(7) West Midlands Regional Cancer Registry, Birmingham.

TRIAL DESIGN: TREATMENT GROUPS:
1. LeuMF: Chlorambucil 10 mg orally on days 1 and 2; methotrexate 25 mg orally on day 1; and 5-fluorouracil 500 mg orally on days 1 and 2 of a 3-week cycle for 6 months.
2. Nil: No adjuvant therapy.

ELIGIBILITY CRITERIA FOR ENTRY:
1. AGE: < 65.
2. CLINICAL STAGES: T1a-T3a, N0, M0.
3. PRIMARY TREATMENT: Simple mastectomy with axillary node sampling; no radiotherapy.
4. AXILLARY NODE DISSECTION: Not required.
5. NUMBER OF NODES REMOVED: Not specified.
6. MENSTRUAL STATUS: No restrictions.
7. RECEPTOR STATUS: No restrictions.

METHOD OF RANDOMIZATION:
Telephone call to a central office where allocations were based on random number tables and balanced in pairs.

RANDOMIZATION STRATIFICATIONS:
Menopausal status: pre, peri, post, hysterectomy and tumor size: <5 cm or 5 cm or larger.

YEARS OF ACCRUAL: 1976-1984.

TOTAL NUMBER OF PATIENTS RANDOMIZED: 574.

ADDITIONAL ELIGIBILITY CRITERIA OR STRATIFICATION METHODS:
Patients were eligible if they had a white blood cell count greater than 4 x 10^9/l, platelets greater than 100 x 10^9/l and normal liver function tests. Patients were ineligible if they were pregnant or lactating, had previous malignancy, serious intercurrent disease or psychiatric disorder.

FOLLOW-UP, COMPLIANCE AND TOXICITY:
Patients were examined every three months for 18 months and then every six months until recurrence or death. Investigations included complete blood count and biochemistry profile at each visit, chest radiograph, skeletal radiographs and radio-isotope bone scan six monthly for two years followed by annual examinations to five years. Toxicity was recorded for each treatment cycle.

574 patients were randomized of whom 543 (273 treated and 270 control) were eligible. All randomized patients are available for the analysis of survival, since all patients were flagged with the West Midlands Regional Cancer Registry who notified the Trial Secretariat when any patient died. Other analyses, including the analysis of relapse-free survival, can be based on only the eligible patients since the 31 (13 treated and 18 control) ineligible patients were not followed up. One treated patient was completely lost to follow-up.

The regime was well tolerated with a high level of compliance. Of the eligible patients, 90% received 8 courses, 93% 7 courses or more and 96% 4 courses or more; all patients received at least 2 courses. The reasons for failure to receive a full course of chemotherapy were patient refusal (30%), toxicity (15%), illness not due to treatment (3%) and administrative error, mainly involving the 8th course (52%).

DISCUSSION:
The patients included in this trial had less advanced disease than those in the NSABP B05 (72B) and Milan NCI 7205 (73B) studies. Toxicity was minor and compliance was excellent in this study.

PUBLICATIONS:
Morrison JM, Howell A, Grieve RJ, et al. West Midlands Oncology Association trials of adjuvant chemotherapy for operable breast cancer. In Adjuvant Therapy of Cancer V. Ed. Salmon SE and Jones SE 1987, Grune and Stratton.

Morrison JM, Howell A, Kelly KA, et al. West Midlands Oncology Association trials of adjuvant chemotherapy in operable breast cancer: results after a median follow-up of 7 years. II. Patients without involved axillary lymph nodes. Br. J. Cancer 1989; 60:919-924.

Table 1. Outcome by allocated treatment for trial 76H2 (West Midlands)
Key: T = total free of relevant event at start of relevant year(s)

(i) Death (D) in each separate year,
subdivided by age when randomized (D <50 & D 50+ / T <50 & T 50+)

Year	1. LeuMF	2. Nil
1	2+ 0/ 122+ 163	0+ 2/ 131+ 158
2	3+ 5/ 120+ 163	4+ 3/ 131+ 156
3	1+ 1/ 115+ 150	6+ 4/ 124+ 148
4	4+ 7/ 104+ 139	3+ 3/ 108+ 136
5+	9+ 9/ 89+ 113	7+ 12/ 90+ 119

(ii) First recurrence or prior death (F) in each separate year,
subdivided by age when randomized (F <50 & F 50+ / T <50 & T 50+)

Year	1. LeuMF	2. Nil
1	7+ 11/ 122+ 163	14+ 12/ 131+ 158
2	9+ 6/ 112+ 148	5+ 9/ 115+ 138
3	8+ 14/ 102+ 140	6+ 6/ 108+ 125
4	5+ 4/ 86+ 121	4+ 9/ 100+ 115
5+	6+ 10/ 71+ 101	7+ 11/ 74+ 92

(iii) Recurrence with survival (R) and Death (D) in all years together,
subdivided by various characteristics when randomized (R & D / T)

Age at Entry	1. LeuMF	2. Nil
< 40	3+8/34	9+9/41
40 - 49	13+11/88	7+11/90
50 - 59	16+15/116	15+17/107
60 - 69	7+7/47	8+7/50
70 +	0+0/0	0+0/1
Any age (including "not known")	39+41/285	39+44/289

Nodal Status (N0, N1-3, N4+)

		1. LeuMF	2. Nil
<50	N0 (by clinical	16+19/122	16+20/131
50+	evidence only)	23+22/163	23+24/158

Estrogen Receptors (ER, fmol/mg)

		1. LeuMF	2. Nil
<50	ER poor or ER < 10	6+6/36	4+7/28
50+		4+8/42	3+9/32
<50	ER+ or ER 10-99	7+5/37	5+6/37
50+		5+4/34	1+4/30
<50	ER++ or ER 100+	0+1/12	1+1/14
50+		4+3/21	10+3/34

TITLE: THE SCANDINAVIAN STUDY OF ONE SINGLE, SHORT PERIOPERATIVE ADJUVANT CHEMOTHERAPY COURSE: EFFECT OF CONTINUING THE ADJUVANT CHEMOTHERAPY FOR ONE YEAR

STUDY NUMBER: 77A
STUDY TITLE: SCANDINAVIA 2A

ABBREVIATED TRIAL NAME: SCANDINAVIA 2A
STUDY NUMBER: 77A

INVESTIGATORS' NAMES AND AFFILIATIONS:
R Nissen-Meyer on behalf of The Scandinavian Adjuvant Chemotherapy Study Group, Tyribakken 10, 0280 Oslo 2, Norway.

TRIAL DESIGN: TREATMENT GROUPS:
1. periCMFV: One short course of cyclophosphamide 500 mg, vincristine 1 mg and 5-fluorouracil 750 mg given intravenously immediately after mastectomy, followed by cyclophosphamide 500 mg, vincristine 1 mg and methotrexate 50 mg, all intravenously, 7 days later.
2. periCMFV + CMF: PeriCMFV, as above, followed by cyclophosphamide 500 mg, methotrexate 50 mg and 5-fluorouracil (750 mg) administered intravenously on days 1 and 8 every month for one year.

ELIGIBILITY CRITERIA FOR ENTRY:
1. AGE: Not specified.
2. PATHOLOGICAL STAGES: At least one positive node.
3. PRIMARY TREATMENT: Not specified.
4. AXILLARY NODE DISSECTION: Required.
5. NUMBER OF NODES REMOVED: Not specified.
6. MENSTRUAL STATUS: No restriction.
7. RECEPTOR STATUS: Not required.

METHOD OF RANDOMIZATION:
Randomization was done by sealed envelopes which had been prepared to give an excess of treatment cases during the first period of the trial (without the knowledge of the local clinicians), in order to obtain an evaluation of the tolerance of the drug combination as soon as possible. Half the patients were also randomized to receive immunotherapy with Corynebacterium parvum.

YEARS OF ACCRUAL: 1977-1985.

TOTAL NUMBER OF PATIENTS RANDOMIZED: 383.

ADDITIONAL ELIGIBILITY OR STRATIFICATION CRITERIA:
For unknown reasons two envelopes were taken but not assigned to a patient in the trial. Accordingly, no follow-up for these anonymous or non-existing patients is available. Furthermore, 11 patients have been excluded for the following reasons: 6 were found to be node-negative, 4 patients were found to have advanced disease, and one patient had a cervical cancer diagnosed before mastectomy.

After exclusion of ineligible patients, there were 162 cases in the control group and 208 in the treatment arm.

DISCUSSION:
Analysis of this study was based on intention to treat. The patients were encouraged to continue the treatment for one year but had the option to discontinue the treatment if they felt they could not tolerate it any longer. Eventually 45% insisted the chemotherapy be terminated after a median treatment duration of 6 1/2 months. The reason was nausea and vomiting. The first 3 to 4 courses (months) were usually fairly well tolerated, but later most of the patients felt the continued chemotherapy impaired their quality of life.

However, even if compliance was poor, long-term chemotherapy resulted in significantly greater relapse-free intervals during the first 6 years after mastectomy. With longer follow-up, differences between the two groups diminished.

PUBLICATIONS:
Nissen-Meyer R, Høst H, Kjellgren K, Mansson B, and Norin T. Neoadjuvant chemotherapy in breast cancer: As single perioperative treatment and with supplementary long-term chemotherapy. In: Adjuvant Therapy of Cancer V, Salmon SE, Ed., Grune & Stratton Inc. (1987) New York.

Table 1. Outcome by allocated treatment for trial 77A (Scandinavia 2A)
Key: T = total free of relevant event at start of relevant year(s)

(i) Death (D) in each separate year, subdivided by age when randomized (D <50 & D 50+ / T <50 & T 50+)

Year	1. periCMFV	2. periCMFV,CMF
1	4+ 2/ 55+ 114	2+ 1/ 71+ 141
2	3+ 8/ 50+ 110	0+ 5/ 69+ 135
3	2+ 8/ 39+ 97	2+ 6/ 66+ 120
4	2+ 6/ 35+ 79	3+ 8/ 63+ 104
5+	2+ 6/ 29+ 65	5+ 8/ 46+ 87

(ii) First recurrence or prior death (F) in each separate year, subdivided by age when randomized (F <50 & F 50+ / T <50 & T 50+)

Year	1. periCMFV	2. periCMFV,CMF
1	11+ 19/ 55+ 114	5+ 14/ 71+ 141
2	6+ 13/ 37+ 88	8+ 11/ 65+ 113
3	1+ 10/ 28+ 66	6+ 11/ 55+ 94
4	3+ 8/ 25+ 47	1+ 9/ 38+ 79
5+	3+ 8/ 20+ 34	4+ 7/ 31+ 58

(iii) Recurrence with survival (R) and Death (D) in all years together, subdivided by various characteristics when randomized (R & D / T)

Age at Entry	1. periCMFV	2. periCMFV,CMF
< 40	5+4/16	4+5/28
40 - 49	6+9/39	8+7/43
50 - 59	14+13/57	10+17/64
60 - 69	14+17/56	14+9/74
70 +	0+0/1	0+2/3
Any age (including "not known")	39+43/171	36+40/212

Nodal Status (N0, N1-3, N4+)

		1. periCMFV	2. periCMFV,CMF
<50	N0 (by axillary	0+0/1	0+0/1
50+	clearance)	0+0/3	0+0/1
<50	N1-N3 (by axillary	6+1/30	8+5/47
50+	clearance)	14+14/64	11+10/85
<50	N4+ (by axillary	5+11/23	4+7/23
50+	sample/clearance)	14+15/46	13+18/54
<50	Remainder (i.e.	0+1/1	0+0/0
50+	other N+ or N?)	0+1/3	0+0/1

Estrogen Receptors (ER, fmol/mg)

Breakdown by ER status was not available.

Short reports

TITLE: ADJUVANT SYSTEMIC THERAPY WITH LEVAMISOLE OR CYCLOPHOSPHAMIDE OR CYCLOPHOSPHAMIDE, METHOTREXATE AND 5-FLUOROURACIL IN PREMENOPAUSAL HIGH-RISK BREAST CANCER PATIENTS

STUDY NUMBER 77B1-3
STUDY NAME: DANISH BCG 77b

ABBREVIATED TRIAL NAME: DANISH BCG 77b
STUDY NUMBER: 77B1-3

INVESTIGATORS' NAMES AND AFFILIATIONS:
Danish Breast Cancer Cooperative Group, Finsen Institute, Copenhagen.

TRIAL DESIGN: TREATMENT GROUPS:
1. Nil: No adjuvant systemic therapy was given.
2. Levam: Levamisole 2.5 mg/kg, was given orally days 1 and 2. The first cycle was started 2-4 weeks postoperatively and subsequent cycles were repeated weekly for one year.
3. C: 12 cycles of cyclophosphamide, 130 mg/M^2, were given orally days 1-14 of each cycle. The first cycle was given 2-4 weeks postoperatively and subsequent cycles were repeated every 4 weeks.
4. CMF: 12 cycles were given and included: cyclophosphamide 80 mg/M^2 orally days 1-14; methotrexate 30 mg/M^2; and 5-fluorouracil 500 mg/M^2, intravenously, days 1 and 8. The first cycle was given 2-4 weeks postoperatively and subsequent cycles were repeated every 4 weeks.

ELIGIBILITY CRITERIA FOR ENTRY:
1. AGE: Not specified.
2. CLINICAL/PATHOLOGICAL STAGES: Invasive breast carcinoma without distant metastases; positive axillary lymph nodes and/or tumor >5 cm. and/or skin invasion and/or deep fascia invasion of the primary tumor.
3. PRIMARY TREATMENT: Total mastectomy and axillary sampling. Two to four weeks postoperatively the patients received radiotherapy to the chest wall, the axillary nodes, and the supraclavicular nodes at a dose equivalent to 1335 Rets.
4. AXILLARY NODE DISSECTION: Not required.
5. NUMBER OF NODES REMOVED: No restriction.
6. MENSTRUAL STATUS: Pre- and perimenopausal (≤5 years of spontaneous menostasia).
7. RECEPTOR STATUS: No restriction.

METHOD OF RANDOMIZATION:
Randomization was decentralized and performed by 20 different departments which received randomization cards from the secretariat in numbered closed envelopes.

RANDOMIZATION STRATIFICATIONS:
Institution.

YEARS OF ACCRUAL: 1977-1982.

TOTAL NUMBER OF PATIENTS RANDOMIZED:
1212 patients were randomized: 461 entered 11/77 to 11/79 were randomized to one of 4 arms (77B1). 251 patients entered 12/79 to 12/80 were randomized to C, CMF, or nil (77B2). 500 patients entered 1/81 to 12/82 were randomized to C or CMF (77B3).

AIM, RATIONALE, SPECIAL FEATURES:
This study was carried out to analyze the efficacy of adjuvant therapy in pre- and perimenopausal high risk breast cancer patients. In November 1979 accrual to the levamisole arm was stopped when an increased rate of recurrence became evident. In December 1980, for similar reasons, entry to the control arm was closed.

FOLLOW-UP, COMPLIANCE AND TOXICITY:
Clinical assessment was done every 3 months for the first 2 years, every 6 months for the next 3 years, then annually until 10 years after mastectomy. Liver function and bone studies were repeated 6 and 12 months postoperatively. Chest X-rays were repeated after 6 and 12 months and then annually for another 4 years.

COMPLIANCE BEFORE FIRST RECURRENCE:
Side effects led to discontinuation of treatment in 50% of the patients in group 2. 10% of the patients relapsed during treatment and another 40% received full treatment for one year in group 2. In groups 3 and 4 the average given doses were 59% and 61% respectively. 47% of the group 4 patients received ≥75% of the intended dose in all 12 cycles compared to only 38% of the group 3 patients.

PUBLICATIONS:
Brincker H, Mouridsen HT, Andersen KW, Andersen J, Castberg Th, Fischerman K, Henriksen E, Hou-Jensen K, Johansen H, Rossing N, Rorth M, and DBCG. Increased breast cancer recurrence rate after adjuvant therapy with levamisole. Lancet 1980; 2:824-827.

Brincker H, Mouridsen HT, Andersen KW. Adjuvant chemotherapy with cyclophosphamide or CMF in premenopausal women with stage II breast cancer. Breast Cancer Res. & Treat. 1983; 3:91-95.

Andersen KW, Mouridsen HT. Danish Breast Cancer Cooperative Group (DBCG): a description of the register of the nationwide programme for primary breast cancer. Acta Oncol. 1988; 27:627-647.

Table 1. Outcome by allocated treatment for trial 77B1 (Danish BCG 77b)
Key: T = total free of relevant event at start of relevant year(s)

*(i) Death (D) in each separate year,
subdivided by age when randomized (D <50 & D 50+ / T <50 & T 50+)*

Year	1. Nil	2. Levam	3. C	4. CMF
1	1+ 1/ 70+ 40	6+ 3/ 80+ 41	4+ 2/ 75+ 38	4+ 0/ 80+ 37
2	5+ 4/ 69+ 39	7+ 7/ 74+ 38	8+ 2/ 71+ 36	7+ 3/ 76+ 37
3	7+ 5/ 64+ 35	5+ 6/ 67+ 31	3+ 3/ 63+ 34	4+ 3/ 69+ 34
4	5+ 3/ 57+ 30	6+ 2/ 62+ 25	3+ 5/ 60+ 31	3+ 3/ 65+ 31
5+	12+ 5/ 52+ 27	7+ 4/ 56+ 23	5+ 1/ 57+ 26	9+ 1/ 62+ 28

*(ii) First recurrence or prior death (F) in each separate year,
subdivided by age when randomized (F <50 & F 50+ / T <50 & T 50+)*

Year	1. Nil	2. Levam	3. C	4. CMF
1	13+ 5/ 70+ 40	18+ 7/ 80+ 41	10+ 6/ 75+ 38	8+ 0/ 80+ 37
2	12+ 5/ 57+ 35	10+ 10/ 62+ 34	6+ 4/ 65+ 32	7+ 4/ 72+ 37
3	6+ 6/ 45+ 30	2+ 5/ 52+ 24	5+ 2/ 59+ 28	10+ 5/ 65+ 33
4	6+ 3/ 39+ 24	7+ 3/ 50+ 19	2+ 3/ 54+ 26	7+ 2/ 55+ 28
5+	7+ 5/ 33+ 21	6+ 3/ 43+ 16	4+ 0/ 52+ 23	1+ 1/ 48+ 26

*(iii) Recurrence with survival (R) and Death (D) in all years together,
subdivided by various characteristics when randomized (R & D / T)*

Age at Entry		1. Nil	2. Levam	3. C	4. CMF
<40		7+12/22	4+8/23	1+9/19	2+7/25
40 - 49		7+18/48	8+23/57	3+14/56	4+20/55
50 - 59		6+18/40	6+22/41	2+13/38	2+10/37
Any age (including "not known")		20+48/110	18+53/121	6+36/113	8+37/117
	Nodal Status (N0, N1-3, N4+)				
<50	N0 (by clinical	1+5/13	2+3/20	1+2/13	2+1/20
50+	evidence only)	1+2/5	1+2/6	1+2/15	0+1/10
<50	Remainder (i.e.	13+25/57	10+28/60	3+21/62	4+26/60
50+	other N+ or N?)	5+16/35	5+20/35	1+11/23	2+9/27
	Estrogen Receptors (ER, fmol/mg)				
<50	ER poor or ER < 10	0+0/0	0+1/2	0+0/0	0+0/0
<50	ER+ or ER 10-99	0+0/0	1+0/2	0+0/1	0+0/0
50+		0+0/0	0+0/0	0+1/1	0+0/0

Trial continued overleaf

Table 1. Outcome by allocated treatment for trial 77B2 (Danish BCG 77b)
Key: T = total free of relevant event at start of relevant year(s)

(i) Death (D) in each separate year,
subdivided by age when randomized (D <50 & D 50+ / T <50 & T 50+)

Year	1. Nil	3. C	4. CMF
1	5+ 1/ 60+ 26	2+ 2/ 53+ 28	2+ 0/ 69+ 15
2	3+ 2/ 55+ 25	9+ 0/ 51+ 26	3+ 0/ 67+ 15
3	3+ 2/ 52+ 23	2+ 0/ 42+ 26	6+ 1/ 64+ 15
4	2+ 1/ 49+ 21	2+ 3/ 40+ 26	4+ 3/ 58+ 14
5+	1+ 0/ 46+ 19	1+ 1/ 38+ 23	4+ 2/ 54+ 11

(ii) First recurrence or prior death (F) in each separate year,
subdivided by age when randomized (F <50 & F 50+ / T <50 & T 50+)

Year	1. Nil	3. C	4. CMF
1	13+ 2/ 60+ 26	6+ 3/ 53+ 28	5+ 0/ 69+ 15
2	6+ 5/ 47+ 24	7+ 1/ 47+ 25	9+ 0/ 64+ 15
3	2+ 2/ 41+ 19	1+ 2/ 40+ 24	4+ 4/ 55+ 15
4	1+ 1/ 39+ 16	3+ 4/ 39+ 22	4+ 3/ 51+ 11
5+	1+ 0/ 37+ 15	1+ 1/ 36+ 18	2+ 0/ 47+ 8

(iii) Recurrence with survival (R) and Death (D) in all years together,
subdivided by various characteristics when randomized (R & D / T)

Age at Entry	1. Nil	3. C	4. CMF
< 40	4+5/14	0+6/13	3+3/17
40 - 49	5+9/46	2+10/40	2+16/52
50 - 59	4+6/26	5+6/28	1+6/15
Any age (including "not known")	13+20/86	7+22/81	6+25/84

Nodal Status (N0, N1-3, N4+)

		1. Nil	3. C	4. CMF
<50	N0 (by clinical	0+1/13	0+4/12	0+1/6
50+	evidence only)	0+1/6	0+1/3	0+0/1
<50	Remainder (i.e.	9+13/47	2+12/41	5+18/63
50+	other N+ or N?)	4+5/20	5+5/25	1+6/14

Estrogen Receptors (ER, fmol/mg)

		1. Nil	3. C	4. CMF
<50	ER poor or ER < 10	1+2/4	1+0/5	1+2/4
50+		0+0/1	1+0/2	0+0/0
<50	ER+ or ER 10-99	1+1/6	0+2/4	1+1/9
50+		0+0/0	0+1/2	0+2/3
<50	ER++ or ER 100+	1+0/3	0+0/1	0+1/3
50+		0+0/1	0+0/2	0+0/1

Table 1. Outcome by allocated treatment for trial 77B3 (Danish BCG 77b)
Key: T = total free of relevant event at start of relevant year(s)

(i) Death (D) in each separate year,
subdivided by age when randomized (D <50 & D 50+ / T <50 & T 50+)

Year	3. C	4. CMF
1	4+ 2/ 177+ 77	2+ 2/ 172+ 74
2	18+ 6/ 173+ 75	8+ 5/ 170+ 72
3	14+ 3/ 155+ 69	13+ 4/ 162+ 67
4	2+ 0/ 129+ 59	4+ 0/ 135+ 61
5+	0+ 0/ 50+ 21	0+ 0/ 49+ 22

(ii) First recurrence or prior death (F) in each separate year,
subdivided by age when randomized (F <50 & F 50+ / T <50 & T 50+)

Year	3. C	4. CMF
1	11+ 3/ 177+ 77	10+ 7/ 172+ 74
2	25+ 11/ 166+ 74	17+ 7/ 162+ 67
3	11+ 5/ 141+ 63	13+ 3/ 145+ 60
4	1+ 3/ 77+ 36	6+ 0/ 76+ 36
5+	0+ 0/ 8+ 0	0+ 0/ 5+ 2

(iii) Recurrence with survival (R) and Death (D) in all years together,
subdivided by various characteristics when randomized (R & D / T)

Age at Entry	3. C	4. CMF
< 40	4+11/49	9+9/50
40 - 49	6+27/128	10+18/122
50 - 59	11+11/77	6+11/74
Any age (including "not known")	21+49/254	25+38/246

Nodal Status (N0, N1-3, N4+)

		3. C	4. CMF
<50	N0 (by clinical	2+1/23	3+5/30
50+	evidence only)	2+1/14	0+1/11
<50	Remainder (i.e.	8+37/154	16+22/142
50+	other N+ or N?)	9+10/63	6+10/63

Estrogen Receptors (ER, fmol/mg)

		3. C	4. CMF
<50	ER poor or ER < 10	1+5/13	1+0/7
50+		0+1/2	0+2/8
<50	ER+ or ER 10-99	2+3/22	3+4/28
50+		0+1/9	0+3/9
<50	ER++ or ER 100+	0+1/5	1+0/8
50+		0+1/9	1+0/3

TITLE: ADJUVANT SYSTEMIC THERAPY WITH LEVAMISOLE OR TAMOXIFEN IN POSTMENOPAUSAL HIGH-RISK BREAST CANCER PATIENTS

STUDY NUMBER 77C1-2
STUDY NAME: DANISH BCG 77c

ABBREVIATED TRIAL NAME: DANISH BCG 77c
STUDY NUMBER: 77C1-2

INVESTIGATORS' NAMES AND AFFILIATIONS:
Danish Breast Cancer Cooperative Group, Finsen Institute, Copenhagen.

TRIAL DESIGN: TREATMENT GROUPS:
1. Nil: No adjuvant systemic therapy given.
2. Levam: Levamisole 2.5 mg/kg given orally days 1 and 2. The first cycle was started 2-4 weeks postoperatively and subsequent cycles were repeated weekly for one year.
3. Tam: Tamoxifen 10 mg three times daily, was started 2-4 weeks postoperatively and continued for 1 year.

ELIGIBILITY CRITERIA FOR ENTRY:
1. AGE: Not specified.
2. CLINICAL STAGES: Invasive breast carcinoma without distant metastases: positive axillary lymph nodes and/or tumor > 5 cm. and/or skin invasion and/or deep fascia invasion of the primary tumor.
3. PRIMARY TREATMENT: Total mastectomy and axillary sampling. 2 to 4 weeks postoperatively the patients received radiotherapy to the chest wall, the axillary nodes and supraclavicular nodes.
4. AXILLARY NODE DISSECTION: Not required.
5. NUMBER OF NODES REMOVED: No restriction.
6. MENSTRUAL STATUS: Postmenopausal (>5 years of spontaneous menostasia).
7. RECEPTOR STATUS: No restriction.

METHOD OF RANDOMIZATION:
Randomization was decentralized and performed by 20 different departments which received randomization cards from the secretariat in numbered closed envelopes.

RANDOMIZATION STRATIFICATIONS:
Institute.

YEARS OF ACCRUAL: 1977-1982.

TOTAL NUMBER OF PATIENTS RANDOMIZED:
2016 patients were randomized, 746 entered and randomized to one of 3 arms between 11/77 and 11/79 and 1270 entered and randomized to Tamoxifen or Nil between 12/79 and 12/82.

AIM, RATIONALE, SPECIAL FEATURES:
This study was carried out to analyze the efficacy of adjuvant therapy in postmenopausal high risk breast cancer patients. In November 1979 accrual to the levamisole arm was stopped when an increased rate of recurrence became evident.

FOLLOW-UP, COMPLIANCE AND TOXICITY:
Clinical assessment was done every 3 months for the first 2 years, every 6 months for the next 3 years, then annually until 10 years after mastectomy. Liver function and bone studies were repeated 6 and 12 months postoperatively. Chest X-rays were repeated after 6 and 12 months and then annually for another 4 years.

COMPLIANCE BEFORE FIRST RECURRENCE:
Side effects led to discontinuation of levamisole treatment in 61% of the patients. 10% of the patients experienced relapse during treatment and another 29% received full treatment for one year in group 2.

Nearly 100% compliance was observed among patients randomized to tamoxifen.

PUBLICATIONS:
Brincker H, Mouridsen HT, Andersen KW, Andersen J, Castberg Th, Fischerman K, Henriksen E, Hou-Jensen K, Johansen H, Rossing N, Rorth M, and DBCG. Increased breast cancer recurrence rate after adjuvant therapy with levamisole. Lancet 1980; 2:824-827.

Rose C, Thorpe SM, Anderson KW, Pedersen BV, Mouridsen HT, Blichert-Toft M, Rasmussen BB, on behalf of the Danish Breast Cancer Cooperative Group: Beneficial effect of adjuvant therapy in primary breast cancer patients with high estrogen receptor values. Lancet 1985; 1:16-19.

Rose C, Mouridsen HT, Thorpe SM, Andersen J, Blichert-Toft M, Andersen KW. Antiestrogen treatment of postmenopausal breast cancer patients with analysis and steroid hormone receptor status. World J. Surg. 1985; 9:765-774.

Andersen KW, Mouridsen HT. Danish Breast Cancer Cooperative Group (DBCG): a description of the register of the nationwide programme for primary breast cancer. Acta Oncol. 1988; 27:627-647.

Table 1. Outcome by allocated treatment for trial 77C1 (Danish BCG 77c)
Key: T = total free of relevant event at start of relevant year(s)

(i) Death (D) in each separate year,
subdivided by age when randomized (D <50 & D 50+ / T <50 & T 50+)

Year	1. Nil	2. Levam	3. Tam
1	0+ 13/ 0+ 251	0+ 26/ 0+ 243	0+ 20/ 3+ 249
2	0+ 22/ 0+ 238	0+ 21/ 0+ 217	0+ 20/ 3+ 229
3	0+ 31/ 0+ 216	0+ 26/ 0+ 196	0+ 23/ 3+ 209
4	0+ 20/ 0+ 185	0+ 20/ 0+ 170	1+ 17/ 3+ 186
5+	0+ 43/ 0+ 165	0+ 25/ 0+ 150	0+ 34/ 2+ 169

(ii) First recurrence or prior death (F) in each separate year,
subdivided by age when randomized (F <50 & F 50+ / T <50 & T 50+)

Year	1. Nil	2. Levam	3. Tam
1	0+ 38/ 0+ 251	0+ 54/ 0+ 243	0+ 36/ 3+ 249
2	0+ 33/ 0+ 213	0+ 33/ 0+ 189	0+ 23/ 3+ 213
3	0+ 30/ 0+ 180	0+ 23/ 0+ 156	0+ 20/ 3+ 190
4	0+ 21/ 0+ 150	0+ 11/ 0+ 133	1+ 23/ 3+ 170
5+	0+ 36/ 0+ 129	0+ 22/ 0+ 122	0+ 36/ 2+ 147

(iii) Recurrence with survival (R) and Death (D) in all years together,
subdivided by various characteristics when randomized (R & D / T)

Age at Entry	1. Nil	2. Levam	3. Tam
40 - 49	0+0/0	0+0/0	0+1/3
50 - 59	8+26/54	4+25/56	2+19/50
60 - 69	16+52/117	13+50/109	13+41/103
70 +	5+51/80	8+43/78	9+54/96
Any age (including "not known")	29+129/251	25+118/243	24+115/252

	Nodal Status (N0, N1-3, N4+)	1. Nil	2. Levam	3. Tam
<50	N0 (by clinical	0+0/0	0+0/0	0+1/1
50+	evidence only)	6+26/67	1+18/48	5+25/61
<50	Remainder (i.e.	0+0/0	0+0/0	0+0/2
50+	other N+ or N?)	23+103/184	24+100/195	19+89/188

	Estrogen Receptors (ER, fmol/mg)	1. Nil	2. Levam	3. Tam
50+	ER poor or ER < 10	0+0/2	0+1/1	0+0/0
50+	ER+ or ER 10-99	0+1/5	1+1/3	0+1/4
50+	ER++ or ER 100+	0+1/1	0+2/2	0+1/2

Table 1. Outcome by allocated treatment for trial 77C2 (Danish BCG 77c)
Key: T = total free of relevant event at start of relevant year(s)

(i) Death (D) in each separate year,
subdivided by age when randomized (D <50 & D 50+ / T <50 & T 50+)

Year	1. Nil	3. Tam
1	0+ 51/ 1+ 635	0+ 55/ 7+ 627
2	0+ 70/ 1+ 584	1+ 62/ 7+ 572
3	0+ 58/ 1+ 514	0+ 51/ 6+ 510
4	0+ 33/ 1+ 421	0+ 24/ 6+ 427
5+	0+ 12/ 0+ 235	0+ 10/ 3+ 241

(ii) First recurrence or prior death (F) in each separate year,
subdivided by age when randomized (F <50 & F 50+ / T <50 & T 50+)

Year	1. Nil	3. Tam
1	0+101/ 1+ 635	0+95/ 7+ 627
2	0+105/ 1+ 534	2+76/ 7+ 532
3	0+ 66/ 1+ 429	0+ 48/ 5+ 456
4	0+ 31/ 0+ 258	0+ 25/ 4+ 284
5+	0+ 11/ 0+ 123	0+ 7/ 3+ 132

(iii) Recurrence with survival (R) and Death (D) in all years together,
subdivided by various characteristics when randomized (R & D / T)

Age at Entry	1. Nil	3. Tam
40 - 49	0+0/1	1+1/7
50 - 59	21+46/134	12+44/140
60 - 69	48+91/276	22+87/285
70 +	21+87/225	15+71/202
Any age (including "not known")	90+224/636	50+203/634

	Nodal Status (N0, N1-3, N4+)	1. Nil	3. Tam
50+	N0 (by clinical evidence only)	11+25/102	7+24/90
<50	Remainder (i.e.	0+0/1	1+1/7
50+	other N+ or N?)	79+199/533	42+178/537

	Estrogen Receptors (ER, fmol/mg)	1. Nil	3. Tam
50+	ER poor or ER < 10	2+19/30	3+12/26
<50	ER+ or ER 10-99	0+0/1	0+0/0
50+		7+11/38	2+19/36
50+	ER++ or ER 100+	14+10/69	6+14/69

TITLE: OXFORD STUDY OF PROPHYLACTIC CHEMOTHERAPY IN POTENTIALLY CURABLE BREAST CANCER

STUDY NUMBER: 77E
STUDY TITLE: OXFORD

ABBREVIATED TRIAL NAME: OXFORD
STUDY NUMBER: 77E

INVESTIGATORS' NAMES AND AFFILIATIONS:
KR Durrant[+] (1), MH Gough (2), AM Giraud-Saunders (1), CH Paine (1), K McPherson (3), MP Vessey (3).
(1) Dept. of Radiotherapy and Oncology, Churchill Hospital, Oxford.
(2) Department of Surgery, John Radcliffe Hospital, Oxford.
(3) Department of Community Medicine and General Practice, Radcliffe Infirmary, Oxford.

[+] Deceased

TRIAL DESIGN: TREATMENT GROUPS:
1. Mel: Melphalan 0.2 mg/kg, orally, days 1-5 of 6 week cycle for 24 months.
2. Mel+MF: Melphalan 10 mg, orally, days 1-5 of 6 week cycle. Methotrexate 15 mg and 5-fluorouracil 250 mg, orally, on day 1 of each cycle for 24 months. 5-fluorouracil escalated to 500 mg after first cycle if no gastrointestinal or hematologic toxicity.
3. Nil: No adjuvant chemotherapy.

ELIGIBILITY CRITERIA FOR ENTRY:
1. AGE: < 65 years.
2. CLINICAL STAGES: All except M1 (distant metastases), N3 (supraclavicular nodes), and T1N0M0 (UICC).
3. PRIMARY TREATMENT: Segmental or simple mastectomy + radiotherapy. Axillary radiotherapy for histologically involved nodes.
4. NODE DISSECTION: Optional.
5. NUMBER OF NODES REMOVED: Not specified.
6. MENSTRUAL STATUS: No restrictions.

METHOD OF RANDOMIZATION:
Telephone to a central office where allocations were based on random number tables and balanced in blocks of 9. Patient allocation not released until patient identifiers recorded in central office.

RANDOMIZATION STRATIFICATIONS:
Premenopausal, postmenopausal.

YEARS OF ACCRUAL: 1977-1983.

TOTAL NUMBER OF PATIENTS RANDOMIZED: 306.

DISCUSSION:
The results of the present breast trial considered independently of the overview have already been presented and discussed elsewhere. Those randomised patients who were excluded from the earlier analysis are included. The reasons for the previous exclusions were: TN0M0: 14, N3/M1: 6, age over 65 years: 3, distant domicile: 5, late entry: 2, low platelet count: 2.

Patients included in this trial had on average less advanced disease than those admitted to the trials reported by Fisher and Bonadonna et al where histological evidence of axillary node involvement was mandatory for entry. In addition, compliance was poorer in this study. The results however are not incompatible statistically with those reported by others, although none of the differences in outcome reported in this study approaches statistical significance. Delay in recurrence of breast cancer and minor benefits in terms of prolongation of survival in pre-menopausal women receiving cytotoxic chemotherapy were not excluded. The toxicity of the simple oral cytotoxic chemotherapy schedule was however considerable and any benefits from therapy were obtained at a high price.

PUBLICATIONS:
Gough MH, Durrant KR, Giraud-Saunders AM, Paine CH, McPherson K, Vessey MP. A randomized controlled trial of prophylactic cytotoxic chemotherapy in potentially curable breast cancer. Br. J. Surg. 1985; 72:182-185, .

Table 1. Outcome by allocated treatment for trial 77E (Oxford)
Key: T = total free of relevant event at start of relevant year(s)

(i) Death (D) in each separate year, subdivided by age when randomized (D <50 & D 50+ / T <50 & T 50+)

Year	1. Mel	2. MelMF	3. Nil
1	0+ 2/ 44+ 56	1+ 3/ 45+ 61	1+ 6/ 48+ 52
2	6+ 4/ 44+ 54	4+ 5/ 44+ 58	7+ 5/ 47+ 46
3	3+ 7/ 38+ 50	4+ 3/ 40+ 53	1+ 2/ 40+ 41
4	1+ 2/ 32+ 41	2+ 8/ 33+ 46	5+ 4/ 38+ 37
5+	7+ 10/ 29+ 37	2+ 8/ 27+ 34	5+ 7/ 29+ 30

(ii) First recurrence or prior death (F) in each separate year, subdivided by age when randomized (F <50 & F 50+ / T <50 & T 50+)

Year	1. Mel	2. MelMF	3. Nil
1	4+ 7/ 44+ 56	3+ 9/ 45+ 61	10+ 12/ 48+ 52
2	8+ 11/ 40+ 49	9+ 6/ 42+ 52	9+ 5/ 38+ 40
3	3+ 4/ 32+ 38	6+ 5/ 33+ 46	0+ 6/ 29+ 35
4	3+ 3/ 28+ 34	0+ 6/ 27+ 41	5+ 4/ 29+ 29
5+	2+ 3/ 21+ 30	1+ 7/ 23+ 30	3+ 6/ 24+ 25

(iii) Recurrence with survival (R) and Death (D) in all years together, subdivided by various characteristics when randomized (R & D / T)

Age at Entry	1. Mel	2. MelMF	3. Nil
< 40	2+5/12	1+4/7	3+3/14
40 - 49	1+12/32	5+9/38	5+16/34
50 - 59	3+15/36	3+16/37	5+19/37
60 - 69	0+10/20	2+11/22	4+5/15
70 +	0+0/0	0+0/1	0+0/0
Any age (including "not known")	6+42/100	12+40/106	17+43/100

Nodal Status (N0, N1-3, N4+)

		1. Mel	2. MelMF	3. Nil
< 50	N0 (by clinical	1+4/21	3+6/28	4+5/23
50 +	evidence only)	0+9/29	3+14/35	4+6/22
< 50	Remainder (i.e.	2+13/23	3+7/17	4+14/25
50 +	other N+ or N?)	3+16/27	3+13/26	5+18/30

Estrogen Receptors (ER, fmol/mg)

Breakdown by ER status was not available.

TITLE: PHASE III EARLY BREAST CANCER STUDY (TRIAL 009)

ABBREVIATED TRIAL NAME: UK MCCG 009
STUDY NUMBER: 77F

INVESTIGATORS' NAMES AND AFFILIATIONS:
MF Spittle (1), G Edelstyn (2), T Bates (3), N Nicol (4), MJ Ostrowski (5), R Buchanan (6), B Hill (7), K MacRae (8).

(1) Meyerstein Institute of Radiotherapy and Oncology.
(2) Middlesex Hospital, London.
(3) S.C. Radiotherapy Centre, St. Thomas' Hospital, London.
(4) Leicester Royal Infirmary, Leicester.
(5) Department of Radiotherapy and Oncology, Norfolk and Norwich Hospital, Norwich.
(6) Dept. of Radiotherapy and Oncology, Royal South Hants Hospital, Southampton.
(7) I.C.R.F., Lincoln's Inn Fields, London.
(8) Department of Medical Statistics, Charing Cross Hospital, London.

TRIAL DESIGN: TREATMENT GROUPS:
1. LeuMFV: Chlorambucil 10 mg, orally, every 6 hours x 3, methotrexate 50 mg, intramuscularly, 5-fluorouracil 500 mg and vincristine 1 mg, intravenously, on day 1 of 3 week cycle for 6 months.
2. LeuMFV + Tam: LeuMFV, as above, plus tamoxifen 10 mg orally, twice daily for 6 months.
3. CMFV: Cyclophosphamide 300 mg, 5-fluorouracil 500 mg and vincristine 0.65 mg, intravenously, on day 1, and cyclophosphamide 300 mg, vincristine 0.65 mg and methotrexate 37.5 mg, intravenously, on day 8 of 3 week cycle for 6 months.
4. CMFV + Tam: CMFV, as above, plus tamoxifen 10 mg, orally, twice daily for 6 months.
5. Tam: Tamoxifen 10 mg, orally, twice daily for 12 months. No adjuvant chemotherapy.

STUDY NUMBER: 77F
STUDY TITLE: UK MCCG 009

ELIGIBILITY CRITERIA FOR ENTRY:
1. AGE: <70.
2. CLINICAL STAGES: Stage II: primary tumor 5 cm or less, histologically positive lymph node metastases, and no distant metastases (T1-2, N1b-2, M0).
3. PRIMARY TREATMENT: Any surgery considered by participating clinician to be locally curative. Radiotherapy: optional.
4. NODE DISSECTION: Required.
5. NUMBER OF NODES REMOVED: Not specified.
6. MENSTRUAL STATUS: No restriction.
7. RECEPTOR STATUS: No restriction.

METHOD OF RANDOMIZATION:
Randomization was by sealed envelopes held at each centre. It was theoretically possible to determine treatment allocation before randomization. Allocations were based on random number tables balanced in blocks of 300. Follow-up data is available on all randomized patients. Only those patients entered into the tamoxifen randomization are used in the overview analysis.

RANDOMIZATION STRATIFICATIONS:
Premenopausal, perimenopausal (no menstrual periods for 6 months - 5 years), postmenopausal (>5 years after last menstrual period).

YEARS OF ACCRUAL: 1977-1979.

TOTAL NUMBER OF PATIENTS RANDOMIZED: 348, of whom only 264 in late tamoxifen randomization were included in the overview analyses. After the late tamoxifen randomization was abandoned, the study continued with 3 arms and a further 645 patients were recruited before it closed in May 1985.

PUBLICATIONS: None.

Table 1. Outcome by allocated treatment for trial 77F (UK MCCG 009)
Key: T = total free of relevant event at start of relevant year(s)

(i) Death (D) in each separate year,
subdivided by age when randomized (D <50 & D 50+ /T <50 & T 50+)

Year	1. LeuMFV	2. TamLeuMFV	3. CMFV	4. TamCMFV	5. Tam
1	4+ 1/ 31+ 42	1+ 0/ 33+ 31	0+ 2/ 27+ 41	0+ 2/ 28+ 31	2+ 1/ 36+ 48
2	2+ 4/ 27+ 41	4+ 2/ 32+ 31	0+ 2/ 27+ 39	5+ 5/ 28+ 29	3+ 3/ 34+ 47
3	1+ 4/ 25+ 37	1+ 2/ 28+ 29	0+ 2/ 23+ 24	3+ 1/ 31+ 44	3+ 1/ 31+ 44
4	3+ 3/ 24+ 33	2+ 0/ 27+ 27	3+ 2/ 27+ 35	0+ 1/ 22+ 22	0+ 2/ 28+ 43
5+	1+ 2/ 21+ 30	3+ 1/ 25+ 27	1+ 1/ 23+ 33	4+ 1/ 22+ 21	3+ 3/ 28+ 41

(ii) First recurrence or prior death (F) in each separate year,
subdivided by age when randomized (F <50 & F 50+ /T <50 & T 50+)

Year	1. LeuMFV	2. TamLeuMFV	3. CMFV	4. TamCMFV	5. Tam
1	7+ 3/ 31+ 42	5+ 2/ 33+ 31	3+ 5/ 27+ 41	2+ 4/ 28+ 31	5+ 3/ 36+ 48
2	4+ 11/ 24+ 35	5+ 3/ 27+ 24	2+ 7/ 18+ 33	7+ 8/ 25+ 25	3+ 4/ 29+ 39
3	2+ 7/ 15+ 23	3+ 2/ 19+ 17	2+ 3/ 14+ 24	5+ 0/ 16+ 15	6+ 5/ 22+ 29
4	2+ 1/ 11+ 13	3+ 1/ 14+ 14	2+ 2/ 10+ 19	1+ 4/ 11+ 15	0+ 2/ 15+ 21
5+	0+ 3/ 5+ 12	2+ 3/ 9+ 13	1+ 1/ 7+ 17	2+ 0/ 9+ 9	1+ 4/ 15+ 18

(iii) Recurrence with survival (R) and Death (D) in all years together,
subdivided by various characteristics when randomized (R & D /T)

Age at Entry	1. LeuMFV	2. TamLeuMFV	3. CMFV	4. TamCMFV	5. Tam
< 40	2+5/13	3+5/18	1+0/7	2+3/9	2+6/13
40 - 49	2+6/18	4+6/15	5+4/20	5+7/19	2+5/23
50 - 59	5+9/23	4+3/17	4+4/24	5+7/22	6+5/27
60 - 69	6+5/19	2+2/14	5+4/16	0+4/9	2+5/18
70 +	0+0/0	0+0/0	0+1/1	0+0/0	0+0/3
Any age (including "not known")	15+25/73	13+16/64	15+13/68	12+21/59	12+21/84

Nodal Status (N0, N1-3, N4+)

		1. LeuMFV	2. TamLeuMFV	3. CMFV	4. TamCMFV	5. Tam
< 50	N0 (by clinical evidence only)	1+1/5	1+0/3	2+0/7	1+2/7	3+0/3
50 +		2+1/7	2+0/6	0+1/9	2+2/6	1+1/7
< 50	Remainder (i.e. other N+, or N?)	3+10/26	6+11/30	4+4/20	6+8/21	1+11/33
50 +		9+13/35	4+5/25	9+8/32	3+9/25	7+9/41

Estrogen Receptors (ER, fmol/mg)
Breakdown by ER status was not available.

TITLE: THE RELATIONSHIP BETWEEN ADJUVANT CHEMOTHERAPY AND ESTROGEN RECEPTOR STATUS

STUDY NUMBER: 77G
STUDY TITLE: VIENNA SURG

ABBREVIATED TRIAL NAME: VIENNA SURG
STUDY NUMBER: 77G

INVESTIGATORS' NAMES AND AFFILIATIONS:
R Jakesz (1), R Kolb (1), G Reiner (1), H Rainer (2), C Dittrich (2), K Moser (2), A Reiner (3), M Schemper (1).

(1) First Department of Surgery, University of Vienna, Austria.
(2) Department of Chemotherapy, University of Vienna, Austria.
(3) Institute of Pathology, University of Vienna, Austria.

TRIAL DESIGN: TREATMENT GROUPS:
1. Nil: No adjuvant chemotherapy.
2. CMFV: Cyclophosphamide 100 mg, orally, twice daily on day 1; methotrexate 25 mg, 5-fluorouracil 750 mg, and vinblastine 5 mg, intravenously, on day 7. 4 cycles were administered in the first year, 2 cycles in the second and third years.
3. CMFV + Azimexon: Chemotherapy, as above, plus administration of azimexon, a non-specific immunostimulant.

ELIGIBILITY CRITERIA FOR ENTRY:
1. AGE: ≤70 years old.
2. CLINICAL/PATHOLOGICAL STAGES: Patient with operable breast cancer. Pathological T1, T2, N0. No distant metastasis.
3. PRIMARY TREATMENT: Quadrantectomy plus axillary dissection or modified radical mastectomy. No postoperative radiotherapy.
4. AXILLARY NODE DISSECTION: Mandatory.
5. NUMBER OF NODES REMOVED: Not specified.
6. MENSTRUAL STATUS: No restriction.
7. RECEPTOR STATUS: Not required.

METHOD OF RANDOMIZATION:
Intra- and postoperative randomization was by telephone to a central office using the method of Pocock and Simon. No patient for whom treatment was allocated was subsequently withdrawn.

RANDOMIZATION STRATIFICATIONS:
Stratification included tumor and lymph node stage and menopausal status.

YEARS OF ACCRUAL: 1977-1981.

TOTAL NUMBER OF PATIENTS RANDOMIZED: 241.

ADDITIONAL ELIGIBILITY OR STRATIFICATION CRITERIA:
Patients were ineligible for the following reasons: pregnancy, previous malignancy, bilateral or inflammatory breast cancer, supraclavicular nodes, serious medical or emotional problems.

Patients with T1-3, N1-2 lesions were randomized to modified radical and radical mastectomy as well as to one of 3 adjuvant therapies.

FOLLOW-UP, COMPLIANCE AND TOXICITY:
Complete clinical examination was done every 3 months for the first 3 years, every 6 months for 2 more years and then annually. Liver function studies and chest X-ray were performed every 6 months.

COMPLIANCE BEFORE FIRST RECURRENCE:
10 patients randomized to treatment received no treatment; 46% of patients received all 8 cycles, and 34% received at least 3 cycles. No drug-related death was recorded. Toxicity was quite tolerable. Doses were adjusted in patients with leukopenia <2000 and thrombocytopenia <50,000.

DISCUSSION:
Overall the data revealed no statistically significant benefit of adjuvant chemotherapy. No relationship between menopausal status or lymph node stage and effect of adjuvant chemotherapy could be observed.
However, estrogen receptor status seems to have strongly influenced the therapeutic benefit of polychemotherapy, since treated patients with estrogen receptor-negative tumors had a significant improvement in overall survival.

PUBLICATIONS:
Jakesz R, Smith CA, Aitken S, Huff K, Schuette W, Sheckney S, Lippman M. Influence of cell proliferation and cell cycle phase on expression of estrogen receptor in MCF-7 breast cancer cells. Cancer Res. 1984; 44:619-625.

Jakesz R, Dittrich C, Haunsch J, Kolb R, Lenzhofer R, Moser K, Rainer H, Reiner G, Schemper M, Spona J, Teleky B. Simultaneous and sequential determination of steroid hormone receptors in human breast cancer: Influence of intervening therapy. Ann. Surg. 1985; 305-310.

Reiner A, Kolb R, Reiner G, Schemper M, Spona J. Prognostic significance of steroid hormone receptors and histopathological characterization of human breast cancer. J. Cancer Res. and Clin. Oncol. 1987; 113:285-290.

Table 1. Outcome by allocated treatment for trial 77G (Vienna Surg.)
Key: T = total free of relevant event at start of relevant year(s)

(i) Death (D) in each separate year,
subdivided by age when randomized (D <50 & D 50+ / T <50 & T 50+)

Year	1. Nil	2. CMFV	3. AzimCMFV
1	0+ 2/ 27+ 55	1+ 1/ 32+ 49	1+ 1/ 24+ 54
2	1+ 4/ 27+ 53	1+ 2/ 31+ 48	1+ 3/ 23+ 53
3	4+ 6/ 26+ 49	4+ 4/ 30+ 46	1+ 2/ 22+ 50
4	2+ 1/ 22+ 43	1+ 3/ 26+ 42	0+ 2/ 21+ 48
5+	2+ 5/ 16+ 35	2+ 3/ 20+ 34	2+ 6/ 18+ 41

(ii) First recurrence or prior death (F) in each separate year,
subdivided by age when randomized (F <50 & F 50+ / T <50 & T 50+)

Year	1. Nil	2. CMFV	3. AzimCMFV
1	3+ 6/ 27+ 55	3+ 2/ 32+ 49	2+ 1/ 24+ 54
2	5+ 7/ 24+ 49	4+ 6/ 29+ 47	1+ 7/ 22+ 53
3	3+ 5/ 19+ 42	4+ 6/ 25+ 41	3+ 3/ 21+ 46
4	2+ 2/ 16+ 37	0+ 2/ 21+ 35	1+ 2/ 18+ 43
5+	1+ 3/ 14+ 32	2+ 2/ 19+ 32	0+ 6/ 17+ 41

(iii) Recurrence with survival (R) and Death (D) in all years together,
subdivided by various characteristics when randomized (R & D / T)

Age at Entry		1. Nil	2. CMFV	3. AzimCMFV
< 40		1+5/9	2+5/13	0+3/6
40 - 49		4+4/18	2+4/19	2+2/18
50 - 59		2+5/23	2+6/25	3+5/25
60 - 69		3+12/30	3+7/24	2+9/29
70 +		0+1/2	0+0/0	0+0/0
Any age (including "not known")		10+27/82	9+22/81	7+19/78

Nodal Status (N0, N1-3, N4+)

		1. Nil	2. CMFV	3. AzimCMFV
< 50	N0 (by axillary	2+5/17	1+3/16	2+0/12
50 +	clearance)	1+7/29	1+2/27	1+5/27
< 50	N1-N3 (by axillary	2+0/4	1+2/7	0+2/6
50 +	clearance)	4+4/15	2+5/12	3+4/17
< 50	N4+ (by axillary	1+4/6	2+2/7	0+3/5
50 +	sample/clearance)	0+7/9	2+5/9	1+5/9
< 50	Remainder (i.e.	0+0/0	0+2/2	0+0/1
50 +	other N+ or N?)	0+0/2	0+1/1	0+0/1

Estrogen Receptors (ER, fmol/mg)

		1. Nil	2. CMFV	3. AzimCMFV
< 50	ER poor or ER < 10	0+4/7	1+3/9	1+1/6
50 +		0+7/12	1+3/10	1+5/14
< 50	ER+ or ER 10-99	4+2/13	1+4/10	0+3/11
50 +		2+3/15	1+1/10	0+2/13
< 50	ER++ or ER 100+	0+0/0	1+0/5	0+1/3
50 +		2+5/17	1+3/12	3+4/21

Short reports

TITLE: NOLVADEX ADJUVANT BREAST CANCER TRIAL

ABBREVIATED TRIAL NAME: NATO
STUDY NUMBER: 77H

STUDY NUMBER: 77H
STUDY TITLE: NATO

INVESTIGATORS' NAMES AND AFFILIATIONS:
M Baum, DM Brinkley, JA Dossett, K McPherson, JS Patterson, RD Rubens, FG Smiddy, BA Stoll, A Wilson, I Jackson, SH Ellis of the Nolvadex Adjuvant Trial Organization.

TRIAL DESIGN: TREATMENT GROUPS:
1. Tam: Tamoxifen 20 mg, orally, daily for 24 months.
2. Nil: No adjuvant endocrine therapy.

ELIGIBILITY CRITERIA FOR ENTRY:
1. AGE: ≤75 years old.
2. CLINICAL/PATHOLOGICAL STAGES: All except M1 (distant metastases), and N3 (supraclavicular nodes); node positive if premenopausal.
3. PRIMARY TREATMENT: Total mastectomy with full axillary clearance, or total mastectomy with axillary node sampling with post-operative radiotherapy if histologically involved nodes.
4. AXILLARY NODE DISSECTION: Optional.
5. NUMBER OF NODES REMOVED: Not specified.
6. MENSTRUAL/NODAL STATUS: Premenopausal if node positive; and postmenopausal.
7. RECEPTOR STATUS: Encouraged but optional.

METHOD OF RANDOMIZATION:
By telephone to a central office where allocations were based on random number tables and balanced in blocks.

YEARS OF ACCRUAL: 1977-1981.

TOTAL NUMBER OF PATIENTS RANDOMIZED: 1,285 (1131 available for analysis).

ADDITIONAL ELIGIBILITY OR STRATIFICATION CRITERIA:
The disease was assessed by clinical examination with TNM staging, chest X-ray and at minimum X-rays of the lumbar spine and pelvis. Bone scintigraphy was not mandatory and liver function tests, including alkaline phosphatase, were routinely performed. Patients were ineligible if they were pregnant. Clinicians had to determine in advance what their standard primary surgical treatment policy would be and adhere to that policy throughout the duration of the trial. The patients receiving radiotherapy were treated according to conventional local policies, which usually included radiation to the chest wall and internal mammary nodes as well as the axilla.

Estrogen receptor measurements were available for 524 patients using the dextran coated charcoal method in 5 laboratories. Quality control was not part of the study but good agreement between laboratories was subsequently shown in the UK quality control study.

FOLLOW-UP, COMPLIANCE AND TOXICITY:
Clinical assessment was carried out every three months for the first two years, every six months for the next three years, and annually thereafter.

Compliance within the trial was excellent based on patients' reports. In addition a subset of patients was randomly chosen for measurement of serum tamoxifen and desmethyl tamoxifen levels. None of the control patients had detectable levels in their blood and all but one of the treated patients had levels well within the therapeutic range.

Toxicity was minimal and only 4% failed to complete the prescribed course of tamoxifen due to putative side effects. Of these the most common included nausea, vomiting, depression, and "hot flashes". Twelve percent (154) of randomized patients were excluded. The majority of these cases were ineligible (76 Tam, 75 Nil) and included pre-menopausal women without node involvement or patients with T4 lesions. Two patients in the treated and one patient in the control group had no data available at follow-up which left 564 patients for analysis in the tamoxifen group and 567 in the control group.

DISCUSSION:
The results from this trial demonstrate a highly significant prolongation of the disease free interval associated with a highly significant improvement in absolute survival. Subgroup analyses by age, menopausal status, nodal status or estrogen receptor status failed to define a group in which tamoxifen had no effect. The result concerning estrogen receptor status of the primary tumor is counterintuitive, yet can not be dismissed out of hand as an artifact of inaccurate measurement technique, particularly as the estrogen receptor measurements were significantly correlated with survival and thus were measuring something of biological significance.

PUBLICATIONS:
Controlled Trial of Tamoxifen as Adjuvant Agent in Management of Early Breast Cancer: Interim Analysis at Four Years by Nolvadex Adjuvant Trial Organization. Lancet 1983; i:257-61.

Controlled Trial of Tamoxifen as Adjuvant Agent in Management of Early Breast Cancer: Analysis at Six Years by Nolvadex Adjuvant Trial Organization. Lancet 1985; i:836-840.

Controlled Trial of Tamoxifen as a Single Adjuvant Agent in the Management of Early Breast Cancer: Analysis at Eight Years by Nolvadex Adjuvant Trial Organization. Br J Cancer 1988; 57:608-611.

Table 1. Outcome by allocated treatment for trial 77H (NATO)
Key: T = total free of relevant event at start of relevant year(s)

(i) Death (D) in each separate year,
    subdivided by age when randomized (D <50 & D 50+ / T <50 & T 50+)

Year	1. Tam	2. Nil
1	5+ 15/ 69+ 495	4+ 23/ 57+ 510
2	5+ 27/ 64+ 480	5+ 32/ 53+ 487
3	6+ 31/ 59+ 453	5+ 42/ 48+ 455
4	5+ 17/ 53+ 422	6+ 38/ 43+ 413
5+	7+ 45/ 48+ 405	9+ 44/ 37+ 375

(ii) First recurrence or prior death (F) in each separate year,
     subdivided by age when randomized (F <50 & F 50+ / T <50 & T 50+)

Year	1. Tam	2. Nil
1	9+ 36/ 69+ 495	18+ 49/ 57+ 510
2	8+ 36/ 60+ 459	10+ 75/ 38+ 461
3	7+ 29/ 51+ 419	4+ 42/ 28+ 386
4	8+ 29/ 44+ 389	2+ 30/ 24+ 343
5+	3+ 44/ 36+ 351	4+ 41/ 21+ 311

(iii) Recurrence with survival (R) and Death (D) in all years together,
      subdivided by various characteristics when randomized (R & D / T)

Age at Entry	1. Tam	2. Nil
< 40	0+9/16	3+7/14
40 - 49	8+19/53	6+22/43
50 - 59	19+46/200	26+71/215
60 - 69	20+64/213	25+65/213
70 +	3+25/82	7+43/82
Any age (including "not known")	50+163/564	67+208/567

Nodal Status (N0, N1-3, N4+)

Age	Status	1. Tam	2. Nil
< 50	N0 (by axillary clearance)	0+0/2	0+0/0
50 +		14+33/179	20+41/173
< 50	N0 (by clinical evidence only)	0+1/2	1+1/5
50 +		13+19/118	11+34/127
< 50	N1-N3 (by axillary clearance)	0+3/10	0+5/10
50 +		4+15/54	8+23/54
< 50	N4+ (by axillary sample/clearance)	4+4/9	0+5/5
50 +		4+14/25	5+11/21
< 50	Remainder (i.e. other N+ or N?)	4+20/46	8+18/37
50 +		7+54/119	14+70/135

Estrogen Receptors (ER, fmol/mg)

Age	Status	1. Tam	2. Nil
< 50	ER poor or ER < 10	3+2/13	2+5/7
50 +		3+31/103	8+41/88
< 50	ER+ or ER 10-99	2+3/9	0+4/5
50 +		1+15/63	14+31/94
< 50	ER++ or ER 100+	1+3/5	0+2/3
50 +		6+17/68	5+22/67

TITLE: AN ADJUVANT CLINICAL TRIAL TO COMPARE CMF TO CMF+PREDNISONE WITH OR WITHOUT TAMOXIFEN IN PRE-MENOPAUSAL WOMEN WITH STAGE II BREAST CANCER

STUDY NUMBER: 77J1
STUDY TITLE: ECOG EST 5177

ABBREVIATED TRIAL NAME: ECOG EST 5177
STUDY NUMBER: 77J1

INVESTIGATORS' NAMES AND AFFILIATIONS:
DC Tormey (1), FJ Cummings (2), T Grage (3), and the Eastern Cooperative Oncology Group, Madison, WI.

(1) University of Wisconsin, Madison, WI.
(2) Brown University, Providence, RI.
(3) University of Minnesota, Minneapolis, MN.

TRIAL DESIGN: TREATMENT GROUPS
1. CMF: Cyclophosphamide 100 mg/M^2, orally, days 1-14; methotrexate 40 mg/M^2 and 5-fluorouracil 600 mg/M^2, intravenously, days 1 and 8 of a monthly cycle for 12 months.
2. CMFPr: CMF, as above, plus prednisone 40 mg/M^2 orally days 1-14 of a monthly cycle for 12 months.
3. CMFPrTam: CMFPr, as above, plus tamoxifen 10 mg orally, twice daily.

ELIGIBILITY CRITERIA FOR ENTRY:
1. AGE: No restrictions if menstruating.
2. PRIMARY TREATMENT: Modified radical or radical mastectomy for potentially curable breast cancer within prior 10 weeks.
3. CLINICAL/PATHOLOGICAL STAGE: ≥1 histologically positive ipsilateral axillary lymph node; axillary nodes clinically movable; no distant metastases.
4. AXILLARY NODE DISSECTION: Required.
5. NUMBER OF NODES REMOVED: Not specified.
6. MENSTRUAL STATUS: Premenopausal (if prior hysterectomy, <52 years old and no menopausal symptoms).
7. RECEPTOR STATUS: Required.

RANDOMIZATION METHOD:
For North American patients, via phone to ECOG Central Office; for South African patients, via sealed envelope. Treatment allocations were assigned at random by a method of permuted blocks within strata with block size 2.

RANDOMIZATION STRATIFICATION:
Axillary node involvement: 1-3 or ≥4; estrogen receptor: positive (≥10 fmol/mg) or negative/borderline (<10 fmol/mg).

ACCRUAL PERIOD: March 1978 to February 1982.

TOTAL NUMBER OF PATIENTS RANDOMIZED: 662.

ADDITIONAL ELIGIBILITY OR STRATIFICATION CRITERIA:
Patients were ineligible for the following reasons: preoperative ulceration or infiltration of skin >2 cm; peau d'orange involving >1/3 of the breast; satellite or parasternal nodules; edema of the arm; palpable supraclavicular or infraclavicular nodes; fixation of primary tumor to skin, underlying muscle, or chest wall; inflammatory carcinoma; bilateral malignancy; prior radiation, chemotherapy, or hormone therapy for breast cancer; psychiatric disorder; diabetes mellitus; peptic ulcer disease; congestive heart failure; hematologic, hepatic or renal abnormality or dysfunction. Patients whose initial biopsy was more than 4 weeks prior to mastectomy and patients with prior or concomitant malignancy were also ineligible.

AIM, RATIONALE, SPECIAL FEATURES:
The specific aims of this study were: 1) to compare CMF, CMFPr, and CMFPrTam with respect to disease-free interval, survival, and toxicity and 2) to examine the impact of pretreatment characteristics on these outcome measures.

COMPLIANCE, FOLLOW-UP, AND TOXICITY:
Follow-up included monthly examinations and blood tests for the first year and every 3 months thereafter. Chest X-rays were obtained every three months for the first year and every 6 months thereafter. Bone scans were done every 6 months for the first year and annually thereafter.

During therapy, overall worst degree of toxicity was significantly higher on CMFPr or CMFPrTam than on CMF. (39% of patients on CMF had severe or worse toxicity vs. 50% on CMFPr and 54% on CMFPrTam.) Individual toxicities were significantly less common on CMF than on CMFPr and CMFPrTam. CMF was associated with significantly worse nausea and vomiting than the other two regimens and significantly worse thrombocytopenia than CMFPr.

DISCUSSION:
The overall survival difference is not significant, with 5-year survival percents of 72% on CMF, 70% on CMFPr, and 70% on CMFPrTam.

PUBLICATIONS:
Tormey DC, Gray R, Taylor SG, Knuiman M, Olson JE, Cummings FJ. Postoperative chemotherapy and chemohormonal therapy in women with node-positive breast cancer. NCI Monographs 1986; 1:75-80.

Tormey DC, Gray R, Gilchrist K, Grange T, Carbone PP, Wolter J, Woll JE, Cummings FJ. Adjuvant chemohormonal therapy with cyclophosphamide, methotrexate, fluorouracil, prednisone (CMFP) or CMFP plus tamoxifen compared to CMF for premenopausal breast cancer patients. An Eastern Cooperative Oncology Group trial. Cancer 1990; 65:200-206.

Table 1. Outcome by allocated treatment for trial 77J1 (ECOG EST5177)
Key: T = total free of relevant event at start of relevant year(s)

(i) Death (D) in each separate year, subdivided by age when randomized (D <50 & D 50+ / T <50 & T 50+)

Year	1. CMF	2. CMFPr	3. TamCMFPr
1	3+ 0/ 194+ 28	1+ 1/ 189+ 31	3+ 0/ 186+ 34
2	23+ 3/ 189+ 28	17+ 1/ 187+ 30	12+ 0/ 182+ 33
3	16+ 2/ 164+ 25	18+ 6/ 167+ 29	20+ 3/ 167+ 33
4	6+ 1/ 146+ 23	10+ 1/ 148+ 21	15+ 2/ 143+ 30
5+	9+ 0/ 116+ 18	11+ 2/ 110+ 18	10+ 0/ 103+ 23

(ii) First recurrence or prior death (F) in each separate year, subdivided by age when randomized (F <50 & F 50+ / T <50 & T 50+)

Year	1. CMF	2. CMFPr	3. TamCMFPr

(recurrence data not available)

(iii) Recurrence with survival (R) and Death (D) in all years together, subdivided by various characteristics when randomized (R & D / T)

Age at Entry		1. CMF	2. CMFPr	3. TamCMFPr
40 - 49		0+57/194	0+57/189	0+60/186
50 - 59		0+6/28	0+11/31	0+5/34
Any age (including "not known")		0+63/222	0+68/220	0+65/220

Nodal Status (N0, N1-3, N4+)

		1. CMF	2. CMFPr	3. TamCMFPr
<50	N0 (by clinical	0+0/1	0+0/0	0+0/0
50+	evidence only)	0+0/0	0+0/1	0+0/0
<50	N4+ (by axillary	0+40/92	0+39/85	0+39/95
50+	sample/clearance)	0+4/15	0+10/18	0+2/14
<50	Remainder (i.e.	0+17/101	0+18/104	0+21/91
50+	other N+ or N?)	0+2/13	0+1/12	0+3/20

Estrogen Receptors (ER, fmol/mg)

		1. CMF	2. CMFPr	3. TamCMFPr
<50	ER poor or ER < 10	0+40/92	0+34/86	0+37/91
50+		0+4/12	0+6/16	0+2/13
<50	ER+ or ER 10-99	0+16/96	0+20/91	0+23/84
50+		0+2/15	0+4/12	0+2/18
<50	ER++ or ER 100+	0+1/5	0+2/10	0+0/9
50+		0+0/1	0+1/2	0+1/2

TITLE: A CLINICAL TRIAL TO COMPARE ADJUVANT THERAPY WITH CMF AND PREDNISONE (CMFPr) VERSUS CMFP PLUS TAMOXIFEN VERSUS NO ADJUVANT THERAPY IN POST-MENOPAUSAL PATIENTS WITH STAGE II BREAST CANCER

STUDY NUMBER: 77J2
STUDY TITLE: ECOG EST 6177

ABBREVIATED TRIAL NAME: ECOG EST 6177
STUDY NUMBER: 77J2

INVESTIGATORS' NAMES AND AFFILIATIONS:
SG Taylor IV (1), R Gray (2), FJ Cummings (3), JE Olson (4), and the Eastern Cooperative Oncology Group, Madison, WI.

(1) Rush-Presbyterian-St. Luke's Hospital, Chicago, IL.
(2) Harvard School of Public Health, Boston, MA.
(3) Brown University, Providence, RI.
(4) Bassett Hospital, Cooperstown, NY.

TRIAL DESIGN: TREATMENT GROUPS
1. Nil: No adjuvant therapy.
2. CMFPr: Cyclophosphamide 100 mg/M^2, orally, days 1-14; methotrexate 40 mg/M^2 and 5-fluorouracil 600 mg/M^2, intravenously, day 1 and 8; prednisone 40 mg/M^2, orally, days 1-14 of a monthly cycle for one year.
3. CMFPrTam: CMFPr, as above, plus tamoxifen 10 mg, orally, twice daily.

ELIGIBILITY:
1. AGE: ≤65 years old (if prior hysterectomy must be ≥52 years).
2. PRIMARY TREATMENT: Modified radical or radical mastectomy for potentially curable breast cancer within prior 10 weeks.
3. CLINICAL/PATHOLOGICAL STAGE: ≥1 histologically positive ipsilateral axillary lymph node; axillary nodes clinically movable; no distant metastases.
4. AXILLARY NODE DISSECTION: Required.
5. NUMBER OF NODES REMOVED: Not specified.
6. MENSTRUAL STATUS: Postmenopausal (if prior hysterectomy, ≥ 52 years old).
7. RECEPTOR STATUS: Required.

RANDOMIZATION METHOD:
For North American patients, via phone to Central Office of ECOG; for South African patients, via sealed envelope; treatment allocations were assigned at random by a method of permuted blocks within strata with block size 2.

RANDOMIZATION STRATIFICATION:
Axillary node involvement: 1-3 or ≥4; estrogen receptor: positive (≥10 fmol/mg) or negative/borderline (<10 fmol/mg).

YEARS OF ACCRUAL: March 1978 to July 1981.

TOTAL NUMBER OF PATIENTS RANDOMIZED: 265.

ADDITIONAL ELIGIBILITY OR STRATIFICATION CRITERIA:
Patients were ineligible for the following reasons: preoperative ulceration or infiltration of skin >2 cm; peau d'orange involving >1/3 of the breast; satellite or parasternal nodules; edema of the arm; palpable supraclavicular or infraclavicular nodes; fixation of primary tumor to skin, underlying muscle, or chest wall; inflammatory carcinoma; bilateral malignancy; prior radiation, chemotherapy, or hormone therapy for breast cancer; psychiatric disorder; diabetes mellitus; peptic ulcer disease; congestive heart failure; hematologic, hepatic or renal abnormality or dysfunction. Patients whose initial biopsy was more than 4 weeks prior to mastectomy and patients with prior or concomitant malignancy were also ineligible.

AIM, RATIONALE, SPECIAL FEATURES:
The specific aims of this study were: 1) to compare one year of adjuvant therapy with CMFPr to no treatment control with regard to disease-free interval, survival, and toxicity and 2) to evaluate any beneficial effect of adding tamoxifen to CMFPr in both ER-positive and ER-negative patients.

COMPLIANCE, FOLLOW-UP, AND TOXICITY:
Follow-up included monthly examinations and blood tests for the first year and every 3 months thereafter. Chest X-rays were obtained every three months for the first year and every 6 months thereafter. Bone scans were done every 6 months for the first year and annually thereafter.

There were no significant differences in toxicity between CMFPr and CMFPrTam except for thrombophlebitis (5% without tamoxifen, 19% with tamoxifen).

DISCUSSION:
The treatment groups did not differ significantly in survival, neither overall nor in any subgroups defined by the prognostic factors. The percent alive at 5 years is 68% on observation, 60% on CMFPr, and 66% on CMFPrTam.

PUBLICATIONS:
Taylor SG, Kalish LA, Olson JE, Cummings F, Bennett JM, Falkson G, Tormey DC, Carbone PP. Adjuvant CMFP versus CMFP plus tamoxifen versus observation alone in postmenopausal, node-positive breast cancer patients: Three year results of an Eastern Cooperative Oncology Group Study, J. Clin. Oncol. 1985; 3:144-154.

Taylor SG IV, Knuiman M, Olson JE, Tormey DC, Gilchrist KW, Falkson G, Rosenthal SN, Sleeper LA, Carbone PP, Cummings FJ. Six year results of the Eastern Cooperative Oncology Group trial of observation versus CMFP versus CMFPT in postmenopausal patients with node positive breast cancer. J. Clin. Oncol. 1989; 7:879-889.

Table 1. Outcome by allocated treatment for trial 77J2 (ECOG EST6177)
Key: T = total free of relevant event at start of relevant year(s)

(i) Death (D) in each separate year,
subdivided by age when randomized (D <50 & D 50+ / T <50 & T 50+)

Year	1. Nil	2. CMFPr	3. TamCMFPr
1	0+ 2/ 8+ 87	0+ 3/ 1+ 86	0+ 2/ 5+ 78
2	0+ 10/ 8+ 84	0+ 8/ 1+ 82	1+ 4/ 5+ 76
3	0+ 8/ 8+ 74	0+ 7/ 1+ 73	0+ 6/ 4+ 71
4	0+ 4/ 8+ 64	0+ 5/ 1+ 66	0+ 10/ 4+ 65
5+	1+ 7/ 7+ 60	0+ 14/ 1+ 60	0+ 8/ 4+ 53

(ii) First recurrence or prior death (F) in each separate year,
subdivided by age when randomized (F <50 & F 50+ / T <50 & T 50+)

Year	1. Nil	2. CMFPr	3. TamCMFPr

(recurrence data not available)

(iii) Recurrence with survival (R) and Death (D) in all years together,
subdivided by various characteristics when randomized (R & D / T)

Age at Entry		1. Nil	2. CMFPr	3. TamCMFPr
40 - 49		0+1/8	0+/1	0+1/5
50 - 59		0+31/86	0+37/86	0+30/78
Any age (including "not known")		0+32/95	0+37/87	0+31/83

	Nodal Status (N0, N1-3, N4+)			
50 +	N0 (by clinical evidence only)	0+0/1	0+0/0	0+0/0
<50	N4+ (by axillary sample/clearance)	0+1/6	0+0/0	0+1/2
50 +		0+20/40	0+25/41	0+18/39
<50	Remainder (i.e. other N+ or N?)	0+0/2	0+0/1	0+0/3
50 +		0+11/46	0+12/45	0+12/39

	Estrogen Receptors (ER, fmol/mg)			
<50	ER poor or ER < 10	0+1/1	0+0/1	0+0/1
50 +		0+16/32	0+16/30	0+12/26
<50	ER+ or ER 10-99	0+0/6	0+0/0	0+1/3
50 +		0+10/32	0+12/37	0+13/37
<50	ER++ or ER 100+	0+0/1	0+0/0	0+0/1
50 +		0+5/20	0+9/19	0+5/14

TITLE: ADJUVANT CHEMOTHERAPY WITH AND WITHOUT TAMOXIFEN IN THE TREATMENT OF PRIMARY BREAST CANCER

STUDY NUMBER: 77K
STUDY TITLE: NSABP B-09

ABBREVIATED TRIAL NAME: NSABP B-09
STUDY NUMBER: 77K

INVESTIGATORS' NAMES AND AFFILIATIONS:
B Fisher, C Redmond, ER Fisher, N Wolmark, and 40 participating institutions in USA and Canada.
National Surgical Adjuvant Breast Project, Pittsburgh, PA, USA.

TRIAL DESIGN: TREATMENT GROUPS:
1. MelF: Melphalan, 4 mg/M^2, orally, and 5-fluorouracil, 300 mg/M^2, intravenously, daily for 5 days of a 6-week cycle for 17 cycles.
2. MelFT: Melphalan and 5-fluorouracil, as above, plus tamoxifen, 10 mg, orally, twice a day for 2 years.

ELIGIBILITY CRITERIA FOR ENTRY:
1. AGE: ≤70.
2. CLINICAL/PATHOLOGICAL STAGES: Operable breast cancer with at least one histologically positive axillary lymph node
3. PRIMARY TREATMENT: Conventional or modified radical mastectomy.
4. AXILLARY NODE DISSECTION: Required.
5. NUMBER OF NODES REMOVED: ?
6. MENSTRUAL STATUS: No restrictions.
7. RECEPTOR STATUS: Estrogen receptor was not initially required for entry on study but became a requirement after the first 449 patients were entered. Later both estrogen receptor and progesterone receptor values were required. However, patients were admitted regardless of the receptor value.

METHOD OF RANDOMIZATION:
Randomization was done by telephone calls to a central office.

RANDOMIZATION STRATIFICATIONS:
Age (≤49, ≥50 years); number of positive nodes (1-3, ≥4); institution.

YEARS OF ACCRUAL: 1977-1980.

TOTAL NUMBER OF PATIENTS RANDOMIZED: 1891.

AIM, RATIONALE, SPECIAL FEATURES:
Previous studies had demonstrated that disease-free survival could be prolonged by the use of adjuvant chemotherapy, and of 3 separate programs evaluated by the NSABP, the combination of melphalan + 5-fluorouracil seemed to be the most effective. This study was begun to determine whether the addition of tamoxifen to melphalan + 5-fluorouracil would result in a better disease-free survival or overall survival than melphalan + 5-fluorouracil alone.

FOLLOW-UP, COMPLIANCE AND TOXICITY:
All but 2% of the patients were deemed fully eligible for the study. Protocol compliance was measured by calculating the number of treatment forms that would be expected under the optimal conditions of no delay in therapy and complete submission of required forms and comparing this with the number of treatment forms actually received. These computations yielded estimated compliance rates of 86% and 87% for the 2 treatment arms. Toxicity was generally similar, with 23% and 27% of the patients experiencing a white cell nadir below 2500 on the MelF and MelFT arms, respectively. Nausea was experienced by 80% and 83% of the patients in the two arms, and 15% of the patients on each arm had alopecia. The only major toxicity experienced by patients receiving adjuvant tamoxifen that was not experienced to the same degree among patients given MelF was hot flashes, which occurred in 36% of the patients on MelFT, compared to only 23% of those on MelF.

DISCUSSION:
After 8 years of follow-up, a statistically significant disease-free survival advantage was seen for all patients treated with MelFT compared to those treated with MelF (p=0.001) This resulted in a small, nonsignificant trend toward improved survival. However, among women aged 50 and over, the improvement in disease-free survival and overall survival was statistically significant regardless of receptor status. Relative to receptor status, the greatest benefits were seen among the patients with an estrogen receptor or progesterone receptor ≥10 fmol. No significant improvement in disease-free survival or overall survival was seen in women aged ≤49 years, even among the receptor-positive group.

PUBLICATIONS:
Fisher B, Redmond C, Brown A, et al. Treatment of primary breast cancer with chemotherapy and tamoxifen. N. Engl. J. Med. 1981; 305:1-6.

Fisher B, Redmond C, Brown A, et al. Adjuvant chemotherapy with and without tamoxifen in the treatment of primary breast cancer: five-year results from the National Surgical Adjuvant Breast and Bowel Project trial. J. Clin. Oncol. 1986; 4:459-471.

Fisher B, Redmond CK, Wolmark N, and NSABP investigators. Long-term results from NSABP trials of adjuvant therapy for breast cancer. In Salmon SE, Adjuvant Therapy of Breast Cancer V, Grune & Stratton, Orlando, 1987, pp 283-295.

Table 1. Outcome by allocated treatment for trial 77K (NSABP B-09)
Key: T = total free of relevant event at start of relevant year(s)

(i) Death (D) in each separate year,
subdivided by age when randomized (D <50 & D 50+ / T <50 & T 50+)

Year	1. FMel	2. TamFMel
1	5+ 15/ 398+ 543	16+ 7/ 391+ 559
2	32+ 35/ 385+ 519	36+ 43/ 373+ 538
3	34+ 48/ 352+ 483	32+ 31/ 337+ 494
4	39+ 40/ 318+ 433	28+ 35/ 303+ 461
5+	38+ 77/ 276+ 391	44+ 87/ 274+ 423

(ii) First recurrence or prior death (F) in each separate year,
subdivided by age when randomized (F <50 & F 50+ / T <50 & T 50+)

Year	1. FMel	2. TamFMel
1	38+ 63/ 398+ 543	52+ 44/ 391+ 559
2	53+ 88/ 352+ 471	51+ 50/ 337+ 501
3	59+ 70/ 299+ 383	35+ 54/ 286+ 450
4	23+ 47/ 240+ 311	35+ 55/ 249+ 394
5+	40+ 56/ 215+ 262	31+ 74/ 213+ 337

(iii) Recurrence with survival (R) and Death (D) in all years together,
subdivided by various characteristics when randomized (R & D / T)

Age at Entry	1. FMel	2. TamFMel
< 40	17+56/129	19+61/140
40 – 49	48+92/269	29+95/251
50 – 59	59+125/312	40+112/328
60 – 69	50+86/221	33+87/220
70 +	0+4/10	1+4/11
Any age (including "not known")	174+363/941	122+359/950

Nodal Status (N0, N1-3, N4+)

		1. FMel	2. TamFMel
< 50	N1–N3 (by axillary clearance)	32+55/214	25+57/209
50 +		49+64/257	30+79/271
< 50	N4+ (by axillary sample/clearance)	33+93/184	23+99/182
50 +		60+151/286	44+124/288

Estrogen Receptors (ER, fmol/mg)

(breakdown of recurrence data by ER status was not available)

		1. FMel	2. TamFMel
< 50	ER poor or ER < 10	0+67/156	0+91/170
50 +		0+80/152	0+71/148
< 50	ER+ or ER 10–99	0+50/145	0+45/140
50 +		0+59/181	0+64/187
< 50	ER++ or ER 10?+	0+5/21	0+3/17
50 +		0+36/120	0+40/116

TITLE: SOUTHERN SWEDISH STUDY OF ADJUVANT THERAPY FOR PREMENOPAUSAL PATIENTS WITH STAGE II BREAST CANCER

ABBREVIATED TRIAL NAME: S. SWEDEN BCG A
STUDY NUMBER: 78A1

INVESTIGATORS' NAMES AND AFFILIATIONS:
S Ryden, K Aspegren, S Borgström, LO Hafström, D Killander, T Landberg, T Möller, L Risholm, O Wiklander.
The Southern Swedish Breast Cancer Group.

TRIAL DESIGN: TREATMENT GROUPS:
1. R: High voltage postoperative radiotherapy by external beam to the axilla, supraclavicular fossa and parasternal nodes with a target dose of 48 Gy in 20 fractions over 48 days, and to the thoracic wall with a dose of 38 Gy in 20 fractions over 48 days. In cases of periglandular growth in the axilla, the dose to the axilla and supraclavicular fossa was elevated to 60 Gy in 25 fractions over 55 days.
2. R + C: Radiotherapy, as above, plus cyclophosphamide, 130 mg/M^2, orally, day 1-14 of a 28 day cycle for 12 cycles.
3. C: Cyclophosphamide, as above, without radiotherapy.

ELIGIBILITY CRITERIA FOR ENTRY:
1. AGE: No restrictions if pre- or perimenopausal.
2. CLINICAL STAGES: T1b,N0,M0; T2a-b,N0-1,M0.
3. PRIMARY TREATMENT: Modified radical mastectomy with no residual tumor.
4. AXILLARY NODE DISSECTION: Required.
5. NUMBER OF NODES REMOVED: Not specified.
6. MENSTRUAL STATUS: Premenopausal or perimenopausal (up to 5 years after menopause).
7. RECEPTOR STATUS: Not required.

METHOD OF RANDOMIZATION:
By telephone to the trial secretariat where randomization was performed with closed envelopes and permuted blocks of 6 for each stratum.

RANDOMIZATION STRATIFICATIONS:
By hospital; tumor size (≤2 cm, >2 cm); number of positive nodes (0, 1-3, ≥4).

YEARS OF ACCRUAL: 1978-1983.

TOTAL NUMBER OF PATIENTS RANDOMIZED: 429.

ADDITIONAL ELIGIBILITY OR STRATIFICATION CRITERIA:
Criteria for ineligibility included pregnancy or lactation; history of other malignant disease; and previous radiotherapy or cytotoxic drugs.

FOLLOW-UP, COMPLIANCE AND TOXICITY:
Clinical assessment with full blood count and liver function tests were done monthly during chemotherapy. Patients not receiving chemotherapy were checked by clinical examination 5 times during the first year. Clinical assessment was done every 3 months during the second and third year, every 4 months during the fourth year, every 6 months during the fifth year, and annually thereafter. Bone scans were performed every 6 months during the first 2 years, and then annually until the fifth year. Mammography of the remaining breast was performed annually. Blood counts and liver function tests were performed every 6 months during the first 2 years and then annually until the fifth year.

COMPLIANCE BEFORE FIRST RECURRENCE:
All patients received radiotherapy in accordance with the allocation. Chemotherapy was discontinued due to side effects in 12/147 (8.2%) of patients in the combined treatment group and in 9/139 (8.5%) of patients receiving chemotherapy alone. The proportion of patients who received 75% or more of their intended doses of cyclophosphamide were 73% in the radiotherapy and chemotherapy group and 77% in the cyclophosphamide group.

DISCUSSION:
The trial did not reveal any benefit from adjuvant chemotherapy with single-agent cyclophosphamide, given in combination with or instead of postoperative radiotherapy, when compared to postoperative radiotherapy as the only treatment after surgery for early breast cancer.

STUDY NUMBER: 78A1
STUDY TITLE: S. SWEDEN BCG A

PUBLICATIONS:
Ryden S, Möller T, Hafström L, Ranstam J, Westrup C, Wiklander O for the Southern Swedish Breast Cancer Group. Adjuvant therapy of breast cancer: Compliance and data validity in a multicenter trial. Controlled Clin. Trials 1986; 7:290-305.

Ryden S, Aspegren K, Borgström S, Hafström L, Killander D, Landberg T, Möller T, Risholm L, Wiklander O. Adjuvant therapy for premenopausal patients with stage II breast cancer. ECCO 3 Congress, Stockholm, 1985, 174. (Abstract)

Table 1. Outcome by allocated treatment for trial 78A1 (S Swedish BCG)
Key: T = total free of relevant event at start of relevant year(s)

(i) Death (D) in each separate year, subdivided by age when randomized (D <50 & D 50+ / T <50 & T 50+)

Year	1. R	2. CR	3. C
1	4+ 1/ 90+ 53	6+ 3/ 98+ 49	5+ 2/ 91+ 48
2	0+ 3/ 86+ 52	5+ 4/ 92+ 46	5+ 4/ 86+ 46
3	3+ 2/ 82+ 48	3+ 2/ 76+ 41	4+ 2/ 81+ 42
4	1+ 1/ 69+ 43	2+ 0/ 58+ 28	1+ 1/ 70+ 33
5+	2+ 4/ 49+ 35	1+ 1/ 48+ 23	1+ 2/ 50+ 22

(ii) First recurrence or prior death (F) in each separate year, subdivided by age when randomized (F <50 & F 50+ / T <50 & T 50+)

Year	1. R	2. CR	3. C
1	8+ 2/ 90+ 53	8+ 5/ 98+ 49	10+ 5/ 91+ 48
2	5+ 5/ 82+ 51	8+ 5/ 87+ 43	6+ 4/ 79+ 43
3	7+ 3/ 69+ 44	5+ 3/ 66+ 30	4+ 3/ 69+ 35
4	1+ 3/ 54+ 34	1+ 0/ 45+ 20	1+ 1/ 54+ 27
5+	3+ 2/ 41+ 23	1+ 2/ 36+ 14	0+ 1/ 34+ 12

(iii) Recurrence with survival (R) and Death (D) in all years together, subdivided by various characteristics when randomized (R & D / T)

Age at Entry		1. R	2. CR	3. C
< 40		2+2/18	3+5/28	1+5/17
40 - 49		12+8/72	3+12/70	4+11/74
50 - 59		4+11/51	6+8/47	3+11/46
60 - 69		0+0/2	0+1/2	0+0/2
Any age (including "not known")		18+21/143	12+26/147	8+27/139

	Nodal Status (N0, N1-3, N4+)			
< 50	N0 (by axillary	3+2/32	1+3/36	2+1/36
50 +	clearance)	0+2/15	1+2/13	1+1/13
< 50	N1-N3 (by axillary	5+4/41	4+3/36	1+5/35
50 +	clearance)	2+6/25	3+3/25	2+2/25
< 50	N4+ (by axillary	6+4/17	1+11/26	2+10/20
50 +	sample/clearance)	2+3/13	2+4/11	0+8/10

	Estrogen Receptors (ER, fmol/mg)			
< 50	ER poor or ER < 10	3+6/28	1+6/24	1+6/30
50 +		1+7/19	0+4/10	1+4/13
< 50	ER+ or ER 10-99	5+2/27	4+2/31	2+3/19
50 +		1+0/8	4+1/13	1+2/10
< 50	ER++ or ER 100+	0+0/1	0+0/6	0+1/3
50 +		1+1/9	0+0/5	1+0/5

TITLE: SOUTHERN SWEDISH STUDY OF ADJUVANT THERAPY FOR POSTMENOPAUSAL PATIENTS WITH STAGE II BREAST CANCER

STUDY NUMBER: 78A2
STUDY TITLE: S. SWEDEN BCG B

ABBREVIATED TRIAL NAME: S. SWEDEN BCG B
STUDY NUMBER: 78A2

INVESTIGATORS' NAMES AND AFFILIATIONS:
S Ryden, K Aspegren, S Borgström, LO Hafström, D Killander, T Landberg, T Möller, L Risholm, O Wiklander.
The Southern Swedish Breast Cancer Group.

TRIAL DESIGN: TREATMENT GROUPS:
1. R: High voltage postoperative radiotherapy by external beam to the axilla, supraclavicular fossa and parasternal nodes with a target dose of 48 Gy in 20 fractions over 48 days, and to the thoracic wall with a dose of 38 Gy in 20 fractions over 48 days. In cases of periglandular growth in the axilla, the dose to the axilla and supraclavicular fossa was elevated to 60 Gy in 25 fractions over 55 days.
2. R + Tam: Postoperative radiotherapy, as above, plus tamoxifen 30 mg, daily for 1 year.
3. Tam: Adjuvant tamoxifen, as above, without postoperative radiotherapy.

ELIGIBILITY CRITERIA FOR ENTRY:
1. AGE: ≤70 years.
2. CLINICAL STAGES: T1b,N0,M0; T2a-b,N0-1,M0.
3. PRIMARY TREATMENT: Modified radical mastectomy with no residual tumor.
4. AXILLARY NODE DISSECTION: Required.
5. NUMBER OF NODES REMOVED: Not specified.
6. MENSTRUAL STATUS: Postmenopausal (>5 years after menopause).
7. RECEPTOR STATUS: Not required.

METHOD OF RANDOMIZATION:
By telephone to the trial secretariat where randomization was performed with closed envelopes and permuted blocks of 6 for each stratum.

RANDOMIZATION STRATIFICATIONS:
By hospital; tumor size (≤2 cm, >2 cm); number of positive nodes (0, 1-3, ≥4).

YEARS OF ACCRUAL: 1978-1983.

TOTAL NUMBER OF PATIENTS RANDOMIZED: 718.

ADDITIONAL ELIGIBILITY OR STRATIFICATION CRITERIA:
Criteria for ineligibility included history of other malignant disease; and previous radiotherapy or cytotoxic drugs.

FOLLOW-UP, COMPLIANCE AND TOXICITY:
Clinical assessment with full blood count and liver function tests were done monthly during chemotherapy. Patients not receiving chemotherapy were checked by clinical examination 5 times during the first year. Clinical assessment was done every 3 months during the second and third year, every 4 months during the fourth year, every 6 months during the fifth year, and annually thereafter. Bone scans were performed every 6 months during the first 2 years, and then annually until the fifth year. Mammography of the remaining breast was performed annually. Blood counts and liver function tests were performed every 6 months during the first 2 years and then annually until the fifth year.

COMPLIANCE BEFORE FIRST RECURRENCE:
All patients received radiotherapy in accordance with the allocation. Tamoxifen treatment was discontinued due to side effects in 3 patients.

DISCUSSION:
The trial did not reveal any statistically significant differences in survival between the three treatment groups. The combined treatment of radiotherapy and tamoxifen, however, resulted in a better recurrence-free survival than either radiotherapy alone or tamoxifen alone. These differences were most pronounced among patients with histologically positive axillary nodes.

PUBLICATIONS:
Ryden S, Möller T, Hafström L, Ranstam J, Westrup C, Wiklander O for the Southern Swedish Breast Cancer Group. Adjuvant therapy of breast cancer: Compliance and data validity in a multicenter trial. Controlled Clin. Trials 1986; 7:290-305.

Ryden S, Aspegren K, Borgström S, Hafström L, Killander D, Landberg T, Möller T, Risholm L, Wiklander O. Adjuvant therapy for premenopausal patients with stage II breast cancer. ECCO 3 Congress, Stockholm, 1985, 174. (Abstract)

Table 1. Outcome by allocated treatment for trial 78A2 (S Swedish BCG)
Key: T = total free of relevant event at start of relevant year(s)

(i) Death (D) in each separate year, subdivided by age when randomized (D <50 & D 50+ / T <50 & T 50+)

Year	1. R	2. TamR	3. Tam
1	0+ 7/ 2+ 234	0+ 5/ 1+ 238	0+ 8/ 0+ 243
2	0+ 15/ 2+ 212	0+ 14/ 1+ 221	0+ 8/ 0+ 225
3	0+ 11/ 2+ 167	0+ 8/ 0+ 185	0+ 17/ 0+ 176
4	0+ 6/ 2+ 135	0+ 5/ 0+ 153	0+ 6/ 0+ 127
5+	0+ 9/ 2+ 99	0+ 11/ 0+ 122	0+ 12/ 0+ 106

(ii) First recurrence or prior death (F) in each separate year, subdivided by age when randomized (F <50 & F 50+ / T <50 & T 50+)

Year	1. R	2. TamR	3. Tam
1	0+ 24/ 2+ 234	0+ 16/ 1+ 238	0+ 26/ 0+ 243
2	0+ 24/ 2+ 195	0+ 15/ 1+ 212	0+ 25/ 0+ 210
3	0+ 8/ 2+ 144	0+ 6/ 0+ 176	0+ 15/ 0+ 149
4	0+ 8/ 2+ 119	0+ 10/ 0+ 144	0+ 7/ 0+ 110
5+	0+ 8/ 2+ 81	0+ 11/ 0+ 108	0+ 5/ 0+ 84

(iii) Recurrence with survival (R) and Death (D) in all years together, subdivided by various characteristics when randomized (R & D / T)

Age at Entry		1. R	2. TamR	3. Tam
40 - 49		0+0/2	0+0/1	0+0/0
50 - 59		8+20/62	4+11/62	9+13/75
60 - 69		15+23/158	10+28/161	18+31/156
70 +		1+5/14	1+4/15	0+7/12
Any age (including "not known")		24+48/236	15+43/239	27+51/243

	Nodal Status (N0, N1-3, N4+)			
<50	N0 (by axillary	0+0/1	0+0/0	0+0/0
50 +	clearance)	2+16/96	6+13/96	6+11/102
<50	N1-N3 (by axillary	0+0/1	0+0/1	0+0/0
50 +	clearance)	10+18/94	4+13/89	14+23/96
50 +	N4+ (by axillary sample/clearance)	12+14/44	5+17/53	7+17/45

	Estrogen Receptors (ER, fmol/mg)			
<50	ER poor or ER < 10	0+0/1	0+0/0	0+0/0
50 +		5+17/58	3+12/47	11+18/68
50 +	ER+ or ER 10-99	6+9/50	3+6/51	7+7/39
50 +	ER++ or ER 100+	8+10/53	6+13/70	4+12/69

TITLE: A RANDOMIZED TRIAL OF ADJUVANT TAMOXIFEN IN POST-MENOPAUSAL WOMEN WITH AXILLARY NODE POSITIVE BREAST CANCER

STUDY NUMBER: 78B
STUDY TITLE: TORONTO-EDMONTON

ABBREVIATED TRIAL NAME: TORONTO-EDMONTON
STUDY NUMBER: 78B

INVESTIGATORS' NAMES AND AFFILIATIONS:
KI Pritchard (1), JW Meakin (2), NF Boyd (3), AHG Paterson (4), U Ambus (5), G DeBoer (3), DJA Sutherland (1), AJ Dembo (1), RH Wilkinson (6), AA Bassett (7), WK Evans (8), H Wiezel (5), and other investigators for the Toronto-Edmonton Breast Cancer Study Group.

(1) Toronto-Bayview Regional Cancer Centre, and Sunnybrook Medical Centre, University of Toronto, Toronto.
(2) Ontario Cancer Foundation, & University of Toronto.
(3) The Ontario Cancer Institute, & University of Toronto.
(4) The Cross Cancer Institute, & University of Alberta, Edmonton.
(5) The Toronto Hospital, & University of Toronto.
(6) North York General Hospital, & University of Toronto.
(7) Mount Sinai Hospital, & University of Toronto.
(8) Ottawa Regional Cancer Centre, & University of Ottawa.

TRIAL DESIGN: TREATMENT GROUPS:
1. Tam: Tamoxifen, 10 mg, three times daily for 2 years.
2. Nil: No further systemic therapy.

ELIGIBILITY CRITERIA FOR ENTRY:
1. AGE: No restriction
2. CLINICAL STAGES: T1-3, N1-2, M0 (Stages I-III) UICC staging.
3. PRIMARY TREATMENT: Complete mastectomy and axillary node dissection.
4. NODE DISSECTION: Complete axillary dissection (including at least level 1 & 2 nodes) mandatory.
5. NUMBER OF NODES REMOVED: Not specified.
6. MENSTRUAL STATUS: Postmenopausal (at least 6 months from last menses, or over 50 if prior hysterectomy without oophorectomy done).
7. ESTROGEN AND PROGESTERONE RECEPTORS: Not required.

METHOD OF RANDOMIZATION:
Telephone to a central office in the Biostatistics Department of the Princess Margaret Hospital where allocations had been pre-arranged using random number tables. No clinician had access to allocations.

RANDOMIZATION STRATIFICATIONS:
Optional loco-regional post-operative radiotherapy: (yes, no); tumor size: (<2 cm, 2-5 cm, >5 cm, unknown); number of positive axillary nodes: (1-3, >4, uncertain); time from menopause: (<5 years, >5 years).

YEARS OF ACCRUAL: 1978-1984

TOTAL NUMBER OF PATIENTS RANDOMIZED: 400.

AIM, RATIONALE, SPECIAL FEATURES:
Patients who have operable breast cancer, but histologically involved axillary lymph nodes, have a ten year survival of only 25%. When this study was designed in 1978, although trials had suggested that both hormonal and chemotherapy were useful in the adjuvant setting in premenopausal women, there was no good evidence for the use of either modality in postmenopausal women. In postmenopausal women with metastatic disease, however, tamoxifen had been clearly shown to have an overall response rate of 30% in unselected postmenopausal women, accompanied by the ease of oral administration and absence of major toxicity. In addition, it had been shown to be useful in preventing tumor formation in the DMBA rat model when given in the subclinical phase of tumor growth, following DMBA administration, suggesting potential utility as an adjuvant.

ADDITIONAL ELIGIBILITY CRITERIA OR STRATIFICATION METHODS:
Patients were excluded from study for the following additional reasons: not randomized within 12 weeks following mastectomy; history of prior malignancy (excluding treated basal or squamous cell carcinoma of skin); long term steroid or chemotherapy use for other illness; receiving non-protocol anti-tumor therapy; residual tumor in axilla or on chest wall. Staging included liver function tests, serum calcium determination, chest roentgenogram (PA+lateral), and radionuclide bone scan.

Estrogen receptor was measured in 299 and progesterone receptor in 293 patients.

Number of histologically positive nodes was: N1-3=237; N≥4=155; unknown=8.

COMPLIANCE AND TOXICITY:
Although all 400 patients are included in the results as randomized, nine were ineligible at randomization, because of previous breast cancer (4), other previous cancers (2), metastatic disease (2), and because of inadequate surgery (a subcutaneous mastectomy) (1). Follow-up was performed every 3 months for two years, every 6 months for five years and every year thereafter. Follow-up included physical examination, full blood count, liver function tests, serum calcium determination, and yearly chest roentgenogram.

Of 198 patients randomized to receive tamoxifen, 133 received it as per protocol while 29 had it discontinued early, 12 received reduced doses, nine had an interrupted course, two had both reduction and interruption, 11 had more than two years of therapy, and two did not receive any tamoxifen. Of 202 patients randomized to the control arm (nil), two in fact received tamoxifen

DISCUSSION:
Recurrence rates were decreased for the entire group of patients receiving tamoxifen as part of this study. For reasons which are unclear, this beneficial effect was not seen in the subgroup of 125 patients who received optional locoregional radiation therapy. This may relate to the smaller numbers studied, to the generally poor prognostic factors present in this group as a whole (more had >4 positive nodes, etc) or to a documented increase in the time from surgery to starting tamoxifen in this subgroup of women.

Our results demonstrate a significantly reduced relapse rate for patients with 1-3 positive, and >4 positive nodes, as well as a significant reduction in both locoregional and distant metastases in patients receiving tamoxifen. We saw no overall survival benefits for tamoxifen in this study.

PUBLICATIONS:
Pritchard KI, Meakin JW, Boyd NF, Ambus U, DeBoer G, Dembo AJ, Paterson AHG, Sutherland DJA, Wilkinson RH, Bassett AA, Evans WK, Beale FA, Clark RM, Keane TJ, Toronto-Edmonton Breast Cancer Study Group. A randomized trial of adjuvant tamoxifen in postmenopausal women with axillary node positive breast cancer. In Adjuvant Therapy of Cancer IV, SE Jones, SE Salmon (Eds), Grune and Stratton, Orlando, Florida, 1984, pp 339-347.

Pritchard KI, Meakin JW, Boyd NF, DeBoer G, Paterson AHG, Ambus U, Dembo AJ, Sutherland DJA, Wilkinson RH, Bassett AA, Evans WK, Beale FA, Clark RM, Keane TJ for the Toronto-Edmonton Breast Cancer Study Group. Adjuvant tamoxifen in postmenopausal women with axillary node positive breast cancer: An update. In Adjuvant Therapy of Cancer V, SE Salmon (Ed), Grune and Stratton, Orlando, Florida, 1987, pp 391-400.

Table 1. Outcome by allocated treatment for trial 78B (Toronto-Edmont.)
Key: T = total free of relevant event at start of relevant year(s)

(i) Death (D) in each separate year, subdivided by age when randomized (D <50 & D 50+ / T <50 & T 50+)

Year	1. Tam	2. Nil
1	0+ 15/ 4+ 194	0+ 5/ 6+ 196
2	1+ 9/ 4+ 179	1+ 11/ 6+ 191
3	0+ 10/ 3+ 158	1+ 17/ 5+ 164
4	1+ 5/ 3+ 123	0+ 7/ 4+ 121
5+	0+ 6/ 2+ 92	0+ 10/ 3+ 88

(ii) First recurrence or prior death (F) in each separate year, subdivided by age when randomized (F <50 & F 50+ / T <50 & T 50+)

Year	1. Tam	2. Nil
1	1+ 33/ 4+ 194	2+ 47/ 6+ 196
2	1+ 22/ 3+ 151	0+ 24/ 4+ 132
3	0+ 8/ 2+ 103	0+ 18/ 4+ 85
4	0+ 6/ 2+ 72	0+ 5/ 3+ 52
5+	0+ 6/ 2+ 44	0+ 9/ 2+ 36

(iii) Recurrence with survival (R) and Death (D) in all years together, subdivided by various characteristics when randomized (R & D / T)

Age at Entry	1. Tam	2. Nil
40 - 49	0+2/4	0+2/6
50 - 59	13+20/78	19+17/77
60 - 69	16+21/96	22+26/91
70 +	1+4/20	9+7/28
Any age (including "not known")	30+47/198	50+52/202

Nodal Status (N0, N1-3, N4+)

		1. Tam	2. Nil
<50	N1-N3 (by axillary clearance)	0+0/1	0+1/4
50+		14+19/111	29+21/117
<50	N4+ (by axillary sample/clearance)	0+2/3	0+1/2
50+		16+25/76	19+27/74
50+	Remainder (i.e. other N+ or N?)	0+1/7	2+2/5

Estrogen Receptors (ER, fmol/mg)
Breakdown by ER status was not available.

TITLE: PROPHYLACTIC TAMOXIFEN IN THE PRIMARY TREATMENT OF EARLY BREAST CANCER: THE NAPLES (GUN) STUDY

STUDY NUMBER: 78C1-4
STUDY TITLE: GUN NAPLES

ABBREVIATED TRIAL NAME: GUN NAPLES
STUDY NUMBER: 78C1-4

INVESTIGATORS' NAMES AND AFFILIATIONS:
S De Placido (1), C Pagliarulo (1), A Marinelli (1) C Gallo (2), G Petrella (3), G Delrio (4), AR Bianco (1).

(1) Division of Medical Oncology, University of Naples.
(2) Institute of Health Statistics, University of Naples.
(3) Division of Surgery, University of Naples.
(4) Institute of Biology, University of Naples.

TRIAL DESIGN: TREATMENT GROUPS:
1. Tam: Tamoxifen, 30 mg, orally, daily for 2 years.
2. Nil: No adjuvant therapy.

ELIGIBILITY CRITERIA FOR ENTRY:
1. AGE: <80 years.
2. CLINICAL/PATHOLOGICAL STAGES: Histologically confirmed, non-inflammatory, unilateral stage I-III (T3a) breast cancer.
3. PRIMARY TREATMENT: Radical, modified radical or segmental mastectomy; the latter reserved for small (T1) tumors plus radiotherapy to residual breast.
4. NODE DISSECTION: Mandatory.
5. NUMBER OF NODES REMOVED: Minimum of 6.
6. MENSTRUAL/NODAL STATUS: Premenopausal node negative; postmenopausal node negative and positive.
7. RECEPTOR STATUS: Not required.

METHOD OF RANDOMIZATION:
A simple randomization allocation was performed. Randomization lists for groups of 50 patients were prepared from random digit tables. Treatment allocations were then transferred to sealed envelopes.

A parallel study was attempted to determine the hormonal effects of the treatment. In order to have a sufficient number of patients for this study, a biased allocation was performed during 1980 with assignment of patients to the tamoxifen arm in a 2:1 ratio. This was rectified in 1981 and 1982 and, in 1983, the reverse was done with a 2:1 allocation in favor of the control group during the last year of entry in an attempt to rebalance the two arms.

RANDOMIZATION STRATIFICATIONS:
None.

YEARS OF ACCRUAL: 1978-1983.

TOTAL NUMBER OF PATIENTS RANDOMIZED: 308: 146 to the tamoxifen arm and 162 to the control group.

ADDITIONAL ELIGIBILITY OR STRATIFICATION CRITERIA:
The disease was assessed by clinical examination with TNM staging, chest X-ray, liver isotope or ultrasound scan, isotope bone scan, complete hematological and biochemical work-up, including liver function tests. CT scan, fine needle aspiration of suspicious lesions, cytological examination of serous effusions, and peritoneoscopy were added whenever indicated.

Estrogen and progesterone receptors were measured in the majority of primary tumors and were considered receptor positive if they contained ≥10 fmol/mg of cytosol protein.

FOLLOW-UP, COMPLIANCE AND TOXICITY:
Clinical, hematological and biochemical assessment of each patient was done every 3 months for the first 2 years postmastectomy, and then at 6 month intervals. Chest X-rays and liver ultrasound scan were done every 6 months, unless otherwise indicated.

COMPLIANCE BEFORE FIRST RECURRENCE:
Compliance with protocol was good: at least 85% of the planned drug dose was administered and no major protocol violations were seen in 85.4% of the patients. Compliance was fair to poor in 14.6% of the patients.

DISCUSSION:
Overall mortality was 11.6% (17/146) for the tamoxifen group and 11.1% (18/162) for the control group. The overall relapse rate was 22.6% (33/146) for the tamoxifen group and 35.2% (57/162) for the control group. (p=.0005 by the generalized Wilcoxon test.)

PUBLICATIONS:
Bianco AR, Pagliarulo C, De Placido S, D'Istria M, Petrella G, Marinelli A, Contegiacomo A, and Delrio G. Adjuvant tamoxifen, alone or in combination with chemotherapy, in the treatment of early breast cancer. In: Antioestrogens in oncology: Past, present and prospects". Proceedings of the International Symposium of Hormonotherapy. Pannuti (ed), Amsterdam 1985, Excerpta Medica, 131-138.

Bianco AR, De Placido S, Gallo C, Pagliarulo C, Marinelli A, Petrella G, D'Istria M, and Delrio G. Adjuvant therapy with tamoxifen in operable breast cancer. Lancet 1988; ii:1094-99.

Table 1. Outcome by allocated treatment for trial 78C1 (GUN Naples)
Key: T = total free of relevant event at start of relevant year(s)

(i) Death (D) in each separate year,
subdivided by age when randomized (D <50 & D 50+ / T <50 & T 50+)

Year	1. Tam	2. Nil
1	0+ 0/ 3+ 28	0+ 0/ 9+ 20
2	0+ 1/ 3+ 28	0+ 2/ 9+ 20
3	0+ 1/ 3+ 27	0+ 1/ 9+ 18
4	0+ 1/ 3+ 26	0+ 4/ 9+ 17
5+	1+ 4/ 3+ 25	1+ 1/ 9+ 13

(ii) First recurrence or prior death (F) in each separate year,
subdivided by age when randomized (F <50 & F 50+ / T <50 & T 50+)

Year	1. Tam	2. Nil
1	0+ 1/ 3+ 28	0+ 3/ 9+ 20
2	0+ 3/ 3+ 27	1+ 3/ 9+ 17
3	0+ 1/ 3+ 23	2+ 4/ 8+ 14
4	0+ 3/ 3+ 22	0+ 2/ 6+ 10
5+	1+ 2/ 3+ 19	0+ 2/ 6+ 8

(iii) Recurrence with survival (R) and Death (D) in all years together,
subdivided by various characteristics when randomized (R & D / T)

Age at Entry		1. Tam	2. Nil
< 40		0+1/1	0+0/1
40 - 49		0+0/2	2+1/8
50 - 59		1+5/15	2+4/7
60 - 69		2+2/13	3+4/12
70 +		0+0/0	1+0/1
Any age (including "not known")		3+8/31	8+9/29

	Nodal Status (N0, N1-3, N4+)		
< 50	N0 (by axillary	0+1/3	2+1/9
50 +	clearance)	0+1/9	2+0/7
50 +	N1-N3 (by axillary clearance)	2+1/9	3+2/6
50 +	N4+ (by axillary sample/clearance)	1+5/10	1+6/7

	Estrogen Receptors (ER, fmol/mg)		
< 50	ER poor or ER < 10	0+1/1	0+0/1
50 +		2+4/11	1+4/7
< 50	ER+ or ER 10-99	0+0/0	0+1/3
50 +		1+1/6	1+4/7
< 50	ER++ or ER 100+	0+0/0	0+0/1
50 +		0+1/4	4+0/4

Trial continued overleaf

Table 1. Outcome by allocated treatment for trial 78C2 (GUN Naples)
Key: T = total free of relevant event at start of relevant year(s)

(i) Death (D) in each separate year,
subdivided by age when randomized (D <50 & D 50+ / T <50 & T 50+)

Year	1. Tam	2. Nil
1	0+ 0/ 4+ 28	0+ 1/ 3+ 11
2	0+ 1/ 4+ 28	0+ 0/ 3+ 10
3	0+ 0/ 4+ 27	0+ 0/ 3+ 10
4	0+ 1/ 4+ 27	0+ 1/ 3+ 10
5+	0+ 2/ 4+ 26	0+ 1/ 3+ 9

(ii) First recurrence or prior death (F) in each separate year,
subdivided by age when randomized (F <50 & F 50+ / T <50 & T 50+)

Year	1. Tam	2. Nil
1	0+ 2/ 4+ 28	0+ 1/ 3+ 11
2	0+ 1/ 4+ 25	0+ 1/ 3+ 10
3	0+ 3/ 4+ 24	0+ 1/ 3+ 9
4	1+ 0/ 4+ 21	0+ 1/ 3+ 8
5+	0+ 1/ 3+ 20	0+ 0/ 3+ 7

(iii) Recurrence with survival (R) and Death (D) in all years together,
subdivided by various characteristics when randomized (R & D / T)

Age at Entry		1. Tam	2. Nil
40 - 49		1+0/4	0+0/3
50 - 59		2+1/11	0+1/5
60 - 69		1+3/16	1+1/4
70 +		0+0/1	0+1/2
Any age (including "not known")		4+4/32	1+3/14

	Nodal Status (N0, N1-3, N4+)		
<50	N0 (by axillary	0+0/3	0+0/3
50+	clearance)	0+0/10	1+2/7
50+	N1-N3 (by axillary clearance)	1+2/11	0+1/4
<50	N4+ (by axillary	1+0/1	0+0/0
50+	sample/clearance)	2+2/7	0+0/0

	Estrogen Receptors (ER, fmol/mg)		
<50	ER poor or ER < 10	0+0/0	0+0/2
50+		1+2/9	0+1/3
<50	ER+ or ER 10-99	1+0/4	0+0/0
50+		1+1/5	0+2/4
50+	ER++ or ER 100+	1+1/9	0+0/3

Table 1. Outcome by allocated treatment for trial 78C3 (GUN Naples)
Key: T = total free of relevant event at start of relevant year(s)

(i) Death (D) in each separate year,
subdivided by age when randomized (D <50 & D 50+ / T <50 & T 50+)

Year	1. Tam	2. Nil
1	0+ 0/ 15+ 49	0+ 1/ 15+ 57
2	0+ 1/ 15+ 49	0+ 4/ 15+ 56
3	0+ 3/ 15+ 48	0+ 0/ 15+ 52
4	1+ 0/ 15+ 33	0+ 0/ 12+ 44
5+	0+ 0/ 7+ 7	0+ 0/ 3+ 19

(ii) First recurrence or prior death (F) in each separate year,
subdivided by age when randomized (F <50 & F 50+ / T <50 & T 50+)

Year	1. Tam	2. Nil
1	0+ 2/ 15+ 49	1+ 6/ 15+ 57
2	2+ 3/ 15+ 44	1+ 8/ 13+ 49
3	1+ 5/ 13+ 40	3+ 3/ 12+ 36
4	0+ 2/ 10+ 17	0+ 1/ 5+ 25
5+	0+ 0/ 4+ 3	0+ 1/ 1+ 9

(iii) Recurrence with survival (R) and Death (D) in all years together,
subdivided by various characteristics when randomized (R & D / T)

Age at Entry		1. Tam	2. Nil
< 40		0+0/2	2+0/4
40 - 49		2+1/13	3+0/11
50 - 59		7+2/26	10+0/25
60 - 69		1+2/20	4+2/21
70 +		0+0/3	0+3/11
Any age (including "not known")		10+5/64	19+5/72

	Nodal Status (N0, N1-3, N4+)		
<50	N0 (by axillary	1+0/13	4+0/12
50+	clearance)	2+1/25	3+0/28
<50	N1-N3 (by axillary	0+0/0	1+0/3
50+	clearance)	5+1/15	7+4/20
<50	N4+ (by axillary	1+1/2	0+0/0
50+	sample/clearance)	1+2/9	4+1/9

	Estrogen Receptors (ER, fmol/mg)		
<50	ER poor or ER < 10	1+1/6	1+0/2
50+		2+3/13	5+1/20
<50	ER+ or ER 10-99	1+0/6	1+0/5
50+		1+1/14	6+3/17
<50	ER++ or ER 100+	0+0/1	2+0/4
50+		3+0/14	2+0/12

Table 1. Outcome by allocated treatment for trial 78C4 (GUN Naples)
Key: T = total free of relevant event at start of relevant year(s)

(i) Death (D) in each separate year,
subdivided by age when randomized (D <50 & D 50+ / T <50 & T 50+)

Year	1. Tam	2. Nil
1	0+ 0/ 2+ 16	0+ 0/ 14+ 34
2	0+ 0/ 2+ 16	0+ 0/ 14+ 34
3	0+ 0/ 2+ 10	0+ 0/ 13+ 29
4	0+ 0/ 0+ 0	0+ 0/ 0+ 0
5+	0+ 0/ 0+ 0	0+ 0/ 0+ 0

(ii) First recurrence or prior death (F) in each separate year,
subdivided by age when randomized (F <50 & F 50+ / T <50 & T 50+)

Year	1. Tam	2. Nil
1	0+ 0/ 2+ 16	1+ 4/ 14+ 34
2	0+ 1/ 2+ 16	0+ 6/ 12+ 28
3	0+ 0/ 1+ 8	0+ 0/ 7+ 12
4	0+ 0/ 0+ 0	0+ 0/ 0+ 0
5+	0+ 0/ 0+ 0	0+ 0/ 0+ 0

(iii) Recurrence with survival (R) and Death (D) in all years together,
subdivided by various characteristics when randomized (R & D / T)

Age at Entry		1. Tam	2. Nil
< 40		0+0/0	0+0/4
40 - 49		0+0/2	1+0/10
50 - 59		0+0/8	7+0/22
60 - 69		1+0/8	2+0/8
70 +		0+0/0	1+0/4
Any age (including "not known")		1+0/18	11+0/48

	Nodal Status (N0, N1-3, N4+)		
<50	N0 (by axillary	0+0/2	1+0/14
50+	clearance)	0+0/8	2+0/20
50+	N1-N3 (by axillary clearance)	1+0/6	4+0/9
50+	N4+ (by axillary sample/clearance)	0+0/2	4+0/5

	Estrogen Receptors (ER, fmol/mg)		
<50	ER poor or ER < 10	0+0/0	0+0/2
50+		0+0/2	2+0/7
<50	ER+ or ER 10-99	0+0/1	0+0/7
50+		1+0/3	3+0/9
<50	ER++ or ER 100+	0+0/1	0+0/1
50+		0+0/2	2+0/7

TITLE: PROPHYLACTIC CHEMOHORMONAL THERAPY FOR PREMENOPAUSAL PATIENTS WITH OPERABLE BREAST CANCER: THE NAPLES (GUN) STUDY

STUDY NUMBER: 78C5
STUDY TITLE: GUN NAPLES 5

ABBREVIATED TRIAL NAME: GUN NAPLES
STUDY NUMBER: 78C5

INVESTIGATORS' NAMES AND AFFILIATIONS:
S De Placido (1), C Pagliarulo (1), A Marinelli (1) C Gallo (2), G Petrella (3), G Delrio (4), AR Bianco (1).

(1) Division of Medical Oncology, University of Naples.
(2) Institute of Health Statistics, University of Naples.
(3) Division of Surgery, University of Naples.
(4) Institute of Biology, University of Naples.

TRIAL DESIGN: TREATMENT GROUPS:
1. CMF: Cyclophosphamide 100 mg/M^2, orally, days 1 to 14; methotrexate 40 mg/M^2 and 5-fluorouracil 600 mg/M^2, intravenously, on days 1 and 8 for 9 consecutive courses.
2. CMF+Tam: CMF, as above, plus tamoxifen 30 mg, orally, daily, for 2 years.

ELIGIBILITY CRITERIA FOR ENTRY:
1. AGE: <80 years.
2. CLINICAL/PATHOLOGICAL STAGES: Histologically confirmed, non-inflammatory, unilateral stage I-III (T3a) breast cancer.
3. PRIMARY TREATMENT: Radical, modified radical or segmental mastectomy; the latter reserved for small (T1) tumors plus radiotherapy given to residual breast.
4. NODE DISSECTION: Mandatory.
5. NUMBER OF NODES REMOVED: Minimum of 6.
6. MENSTRUAL/NODAL STATUS: Premenopausal node positive.
7. RECEPTOR STATUS: Not required.

METHOD OF RANDOMIZATION:
Simple randomization allocation was performed. Randomization lists, for groups of 50 patients, were prepared from random digit tables. Treatment allocations were then transferred to sealed envelopes.

RANDOMIZATION STRATIFICATIONS:
None

YEARS OF ACCRUAL: 1978-1983.

TOTAL NUMBER OF PATIENTS RANDOMIZED: 125: 60 to the tamoxifen arm and 65 to the CMF only arm.

ADDITIONAL ELIGIBILITY OR STRATIFICATION CRITERIA:
Patients with irregular menses were considered premenopausal if the plasma hormone profile, including estrogen (E1 + E2), FSH, and LH determination was of the premenopausal type. The disease was assessed by clinical examination with TNM staging, chest X-ray, liver isotope or ultrasound scan, isotope bone scan, complete hematological and biochemical work-up, including liver and kidney function tests. CT scan, fine needle aspiration of suspicious lesions, cytological examination of serous effusions, and peritoneoscopy were added whenever indicated.

Estrogen and progesterone receptors were measured in the majority of primary tumors and were considered receptor positive if they contained ≥10 fmol/mg of cytosol protein.

FOLLOW-UP, COMPLIANCE AND TOXICITY:
Clinical, hematological and biochemical assessment of each patient was done every 3 months for the first 2 years postmastectomy, and then at 6 month intervals. Chest X-rays and liver ultrasound scan were done every 6 months, isotope bone scans every 12 months, unless otherwise indicated.

COMPLIANCE BEFORE FIRST RECURRENCE:
Compliance with the protocol was good: at least 85% of the planned drug dose was administered and no major protocol violations were seen in 89.6% of the patients.

DISCUSSION:
Overall mortality was 20% in both treatment arms. Overall relapse rate was 26.7% following CMF+Tam and 46.2% following CMF alone. The difference reaches statistical significance (p=0.02) by the use of the generalized Wilcoxon test (Breslow).

PUBLICATIONS:
Bianco AR, Pagliarulo C, De Placido S, D'Istria M, Petrella G, Marinelli A, Contegiacomo A, and Delrio G. Adjuvant tamoxifen, alone or in combination with chemotherapy, in the treatment of early breast cancer. In: Anti-oestrogens in oncology: Past, present and prospects". Proceedings of the International Symposium of Hormonotherapy. Pannuti (ed), Amsterdam 1985, Excerpta Medica, 131-138.

Bianco AR, De Placido S, Gallo C, Pagliarulo C, Marinelli A, Petrella G, D'Istria M, and Delrio G. Adjuvant therapy with tamoxifen in operable breast cancer. Lancet 1988; ii:1094-99.

Table 1. Outcome by allocated treatment for trial 78C5 (GUN Naples)
Key: T = total free of relevant event at start of relevant year(s)

(i) Death (D) in each separate year,
subdivided by age when randomized (D <50 & D 50+ / T <50 & T 50+)

Year	1. CMF	2. TamCMF
1	2+ 0/ 55+ 10	1+ 1/ 52+ 8
2	3+ 0/ 53+ 10	4+ 0/ 51+ 7
3	2+ 1/ 45+ 10	4+ 0/ 44+ 7
4	4+ 0/ 36+ 4	0+ 0/ 26+ 4
5+	0+ 1/ 21+ 2	2+ 0/ 19+ 2

(ii) First recurrence or prior death (F) in each separate year,
subdivided by age when randomized (F <50 & F 50+ / T <50 & T 50+)

Year	1. CMF	2. TamCMF
1	7+ 0/ 55+ 10	4+ 1/ 52+ 8
2	13+ 4/ 48+ 10	6+ 0/ 45+ 7
3	4+ 1/ 27+ 4	2+ 0/ 32+ 5
4	0+ 0/ 18+ 0	3+ 0/ 23+ 3
5+	1+ 0/ 12+ 0	2+ 0/ 12+ 2

(iii) Recurrence with survival (R) and Death (D) in all years together,
subdivided by various characteristics when randomized (R & D / T)

Age at Entry	1. CMF	2. TamCMF
< 40	2+6/15	2+3/14
40 - 49	12+5/40	4+8/38
50 - 59	3+2/10	0+1/8
Any age (including "not known")	17+13/65	6+12/60

Nodal Status (N0, N1-3, N4+)

		1. CMF	2. TamCMF
< 50	N1-N3 (by axillary	5+2/23	1+1/19
50 +	clearance)	0+0/1	0+0/4
< 50	N4+ (by axillary	9+9/32	5+10/33
50 +	sample/clearance)	3+2/9	0+1/4

Estrogen Receptors (ER, fmol/mg)

		1. CMF	2. TamCMF
< 50	ER poor or ER < 10	3+8/18	1+6/23
50 +		1+1/2	0+0/2
< 50	ER+ or ER 10-99	8+2/19	4+3/18
50 +		2+1/6	0+0/4
< 50	ER++ or ER 100+	2+0/6	0+0/0
50 +		0+0/0	0+1/2

Short reports

TITLE: SCOTTISH ADJUVANT TAMOXIFEN TRIALS

STUDY NUMBER: 78D1-4
STUDY TITLE: SCOTTISH

ABBREVIATED TRIAL NAME: SCOTTISH
STUDY NUMBER: 78D1-4

INVESTIGATORS' NAMES AND AFFILIATIONS:
HJ Stewart, GK White, APM Forrest & Members of The Breast Cancer Trials Committee, Scottish Cancer Trials Office (MRC), Edinburgh, Scotland.

TRIAL DESIGN: TREATMENT GROUPS:
1. Tam: Adjuvant tamoxifen, 20 mg, daily for 5 years. If disease free, re-randomized to stop or continue tamoxifen until relapse.
2. Nil: No adjuvant systemic therapy but tamoxifen, 20 mg, daily, given on confirmation of first relapse, for a minimum of 6 weeks.

ELIGIBILITY CRITERIA FOR ENTRY:
1. AGE: <80 years of age.
2. CLINICAL STAGES: Operable invasive primary breast cancer less advanced than T_4 or N_3 (UICC 1978).
3. PRIMARY TREATMENT: Simple (total) mastectomy followed by routine adjuvant radiotherapy only when node spread confirmed by sampling; if no nodal tissue submitted for histological examination, the need for postoperative radiotherapy determined randomly; no radiotherapy given following an axillary clearance or negative sample.
4. AXILLARY NODE DISSECTION: By stated choice - sample or clearance.
5. NUMBER OF NODES REMOVED: Not specified.
6. MENSTRUAL STATUS: All categories except premenopausal patients with involved axillary nodes.

METHOD OF RANDOMIZATION:
Allocated treatment identified on computer-generated random lists following phone call to trials office and confirmation of eligibility. Options balanced in subgroups in blocks of 6.

RANDOMIZATION STRATIFICATIONS:
1. MENSTRUAL STATUS: Last menstrual period more than or less than 12 months earlier.
2. NODE STATUS: Histology positive; negative or not available.
3. AXILLARY SURGERY: Node sample or axillary clearance.
4. GEOGRAPHICAL REGION OF REFERRAL: North, East, S-E, or West.

YEARS OF ACCRUAL: 1978-1984 (Pilot trial 1978-80; Main trial 1980-84).

TOTAL NUMBER OF PATIENTS RANDOMIZED: 1323 (includes 107 in pilot trial).

PUBLICATIONS:
Report from the Breast Cancer Trials Committee, Scottish Cancer Trials Office (MRC), Edinburgh, Scotland. Adjuvant tamoxifen in the management of operable breast cancer: the Scottish trial. Lancet 1987; ii:171-175.

Table 1. Outcome by allocated treatment for trial 78D1 (Scottish PilotB)
Key: T = total free of relevant event at start of relevant year(s)

(i) Death (D) in each separate year,
subdivided by age when randomized (D <50 & D 50+ / T <50 & T 50+)

Year	1. Tam	2. Nil
1	0+ 0/ 1+ 54	1+ 3/ 4+ 48
2	0+ 3/ 1+ 54	0+ 9/ 3+ 45
3	1+ 5/ 1+ 51	1+ 5/ 3+ 36
4	0+ 2/ 0+ 46	0+ 5/ 2+ 31
5+	0+ 9/ 0+ 44	0+ 5/ 2+ 26

(ii) First recurrence or prior death (F) in each separate year,
subdivided by age when randomized (F <50 & F 50+ / T <50 & T 50+)

Year	1. Tam	2. Nil
1	1+ 1/ 1+ 54	1+ 11/ 4+ 48
2	0+ 4/ 0+ 53	1+ 9/ 3+ 37
3	0+ 5/ 0+ 49	0+ 6/ 2+ 28
4	0+ 1/ 0+ 44	0+ 5/ 2+ 21
5+	0+ 9/ 0+ 43	0+ 5/ 2+ 16

(iii) Recurrence with survival (R) and Death (D) in all years together,
subdivided by various characteristics when randomized (R & D / T)

Age at Entry	1. Tam	2. Nil
40 - 49	0+1/1	0+2/4
50 - 59	0+10/20	5+11/19
60 - 69	1+7/25	3+10/19
70 +	0+2/9	1+6/10
Any age (including "not known")	1+20/55	9+29/52

Nodal Status (N0, N1-3, N4+)

		1. Tam	2. Nil
50 +	N0 (by clinical evidence only)	0+0/3	0+0/3
<50	Remainder (i.e. other N+ or N?)	0+1/1	0+2/4
50 +		1+19/51	9+27/45

Estrogen Receptors (ER, fmol/mg)

		1. Tam	2. Nil
<50	ER poor or ER < 10	0+1/1	0+1/1
50 +		0+1/3	0+4/5
50 +	ER+ or ER 10-99	0+2/8	2+2/5
50 +	ER++ or ER 100+	0+2/8	2+4/6

Table 1. Outcome by allocated treatment for trial 78D2 (Scottish B)
Key: T = total free of relevant event at start of relevant year(s)

(i) Death (D) in each separate year,
subdivided by age when randomized (D <50 & D 50+ / T <50 & T 50+)

Year	1. Tam	2. Nil
1	1+ 10/ 9+ 175	0+ 13/ 2+ 179
2	0+ 12/ 8+ 165	1+ 15/ 2+ 165
3	1+ 11/ 7+ 125	0+ 11/ 1+ 127
4	1+ 4/ 5+ 81	0+ 13/ 1+ 84
5+	0+ 2/ 2+ 46	0+ 3/ 0+ 42

(ii) First recurrence or prior death (F) in each separate year,
subdivided by age when randomized (F <50 & F 50+ / T <50 & T 50+)

Year	1. Tam	2. Nil
1	1+ 21/ 9+ 175	0+ 36/ 2+ 179
2	2+ 19/ 8+ 154	2+ 42/ 2+ 142
3	0+ 8/ 5+ 108	0+ 9/ 0+ 85
4	0+ 4/ 4+ 70	0+ 5/ 0+ 55
5+	0+ 2/ 2+ 41	0+ 2/ 0+ 32

(iii) Recurrence with survival (R) and Death (D) in all years together,
subdivided by various characteristics when randomized (R & D / T)

Age at Entry	1. Tam	2. Nil
40 - 49	0+3/9	1+1/2
50 - 59	8+18/72	17+19/75
60 - 69	6+17/78	15+23/71
70 +	1+4/25	7+13/33
Any age (including "not known")	15+42/184	40+56/181

Nodal Status (N0, N1-3, N4+)

		1. Tam	2. Nil
50 +	N0 (by clinical evidence only)	0+1/1	0+0/0
<50	N1-N3 (by axillary clearance)	0+1/2	0+1/1
50 +		2+3/29	6+5/28
<50	N4+ (by axillary sample/clearance)	0+0/2	0+0/0
50 +		3+6/17	8+8/22
<50	Remainder (i.e. other N+ or N?)	0+2/5	1+0/1
50 +		10+29/128	25+42/129

Estrogen Receptors (ER, fmol/mg)

		1. Tam	2. Nil
<50	ER poor or ER < 10	0+2/2	0+0/0
50 +		1+13/29	7+19/35
<50	ER+ or ER 10-99	0+0/4	0+0/0
50 +		5+9/34	7+4/37
50 +	ER++ or ER 100+	3+1/45	15+6/40

Table 1. Outcome by allocated treatment for trial 78D3 (Scottish C)
Key: T = total free of relevant event at start of relevant year(s)

(i) Death (D) in each separate year,
subdivided by age when randomized (D <50 & D 50+ / T <50 & T 50+)

Year	1. Tam	2. Nil
1	1+ 9/ 97+ 281	0+ 3/ 98+ 282
2	1+ 5/ 95+ 270	4+ 12/ 98+ 279
3	1+ 10/ 81+ 228	1+ 13/ 79+ 233
4	1+ 7/ 61+ 157	3+ 8/ 58+ 161
5+	0+ 4/ 31+ 80	2+ 2/ 32+ 81

(ii) First recurrence or prior death (F) in each separate year,
subdivided by age when randomized (F <50 & F 50+ / T <50 & T 50+)

Year	1. Tam	2. Nil
1	3+ 16/ 97+ 281	13+ 18/ 98+ 282
2	6+ 17/ 93+ 263	7+ 28/ 85+ 263
3	3+ 7/ 74+ 209	6+ 18/ 67+ 205
4	0+ 9/ 53+ 145	1+ 10/ 44+ 138
5+	1+ 4/ 27+ 71	1+ 1/ 23+ 68

(iii) Recurrence with survival (R) and Death (D) in all years together,
subdivided by various characteristics when randomized (R & D / T)

Age at Entry		1. Tam	2. Nil
< 40		4+2/24	3+3/24
40 - 49		5+2/73	15+7/74
50 - 59		10+16/125	15+10/103
60 - 69		7+15/109	13+14/120
70 +		1+4/46	9+14/59
Any age (including "not known")		27+39/378	55+48/380
	Nodal Status (N0, N1-3, N4+)		
<50	N0 (by axillary	0+0/1	0+0/0
50+	clearance)	0+0/1	0+0/1
<50	N0 (by clinical	9+4/96	18+10/98
50+	evidence only)	18+35/279	37+38/281
50+	Remainder (i.e. other N+ or N?)	0+0/1	0+0/0
	Estrogen Receptors (ER, fmol/mg)		
<50	ER poor or ER < 10	5+1/31	3+2/23
50+		5+10/54	6+8/57
<50	ER+ or ER 10-99	0+0/22	7+3/27
50+		3+7/60	7+6/59
<50	ER++ or ER 100+	0+0/4	0+0/6
50+		3+7/60	10+3/44

Table 1. Outcome by allocated treatment for trial 78D4 (Scottish D)
Key: T = total free of relevant event at start of relevant year(s)

(i) Death (D) in each separate year,
subdivided by age when randomized (D <50 & D 50+ / T <50 & T 50+)

Year	1a. Nil	1b. R	2a. Tam	2b. TamR
1	0+ 1/ 7+ 16	0+ 1/ 6+ 17	0+ 1/ 7+ 16	0+ 1/ 8+ 16
2	0+ 0/ 7+ 15	0+ 3/ 6+ 16	0+ 0/ 7+ 15	1+ 2/ 8+ 15
3	0+ 0/ 7+ 15	0+ 2/ 5+ 13	1+ 1/ 6+ 14	1+ 0/ 7+ 12
4	0+ 1/ 4+ 10	0+ 0/ 3+ 10	0+ 0/ 5+ 10	0+ 0/ 5+ 10
5+	0+ 0/ 3+ 5	0+ 0/ 1+ 5	0+ 0/ 4+ 5	0+ 2/ 1+ 7

(ii) First recurrence or prior death (F) in each separate year,
subdivided by age when randomized (F <50 & F 50+ / T <50 & T 50+)

Year	1a. Nil	1b. R	2a. Tam	2b. TamR
1	2+ 1/ 7+ 16	0+ 5/ 6+ 17	1+ 2/ 7+ 16	1+ 2/ 8+ 16
2	1+ 2/ 5+ 15	0+ 2/ 6+ 12	0+ 0/ 6+ 14	1+ 2/ 7+ 14
3	0+ 2/ 4+ 13	0+ 1/ 5+ 10	0+ 1/ 5+ 13	0+ 0/ 6+ 11
4	0+ 0/ 4+ 6	1+ 0/ 3+ 8	0+ 0/ 5+ 9	0+ 0/ 5+ 9
5+	0+ 0/ 3+ 2	0+ 0/ 1+ 3	0+ 0/ 4+ 4	0+ 2/ 1+ 6

(iii) Recurrence with survival (R) and Death (D) in all years together,
subdivided by various characteristics when randomized (R & D / T)

Age at Entry		1a. Nil	1b. R	2a. Tam	2b. TamR
< 40		0+0/0	0+0/1	0+0/1	0+0/1
40 - 49		3+0/7	1+0/5	0+1/6	0+2/7
50 - 59		0+0/4	1+4/8	0+0/4	1+2/6
60 - 69		1+0/6	1+2/6	1+1/6	0+1/6
70 +		2+2/6	0+0/3	0+1/6	0+2/4
Any age (including "not known")		6+2/23	3+6/23	1+3/23	1+7/24
	Nodal Status (N0, N1-3, N4+)				
<50	N0 (by clinical	2+0/5	1+0/6	0+1/6	0+1/7
50+	evidence only)	3+2/13	2+5/16	1+2/15	1+4/13
<50	Remainder (i.e.	1+0/2	0+0/0	0+0/1	0+1/1
50+	other N+ or N?)	0+0/3	0+1/1	0+0/1	0+1/3
	Estrogen Receptors (ER, fmol/mg)				
<50	ER poor or ER < 10	1+0/1	0+0/1	0+0/2	0+1/1
50+		0+1/3	0+1/3	0+0/3	0+1/1
<50	ER+ or ER 10-99	1+0/3	1+0/1	0+0/2	0+0/0
50+		0+0/0	0+1/2	0+1/5	0+0/0
50+	ER++ or ER 100+	0+0/1	1+0/1	0+0/2	0+0/3

Short reports

TITLE: INTERNATIONAL STUDY OF CMF AND TAMOXIFEN IN EARLY POOR RISK BREAST CANCER

STUDY NUMBER: 78E1-4
STUDY TITLE: UK/ASIA COLLAB

ABBREVIATED TRIAL NAME: UK/ASIA COLLAB
STUDY NUMBER: 78E1-4

INVESTIGATORS' NAMES AND AFFILIATIONS:
F Senanayake (1), E Boeson (1), D Choy (2), AIM Cook (3), CD Derry (4), G Deutsch (5), SR Drake (4), RGB Evans (6), J Godwin (7), R Gray (7), B Hafner (7), K Halnan (8), I Hanham (9), S Jayatilake (10), DJ Jussawalla (11), R Khoo (2), L Lim (7), G Mair (12), DS Murrell (5), VR Pai (11), R Peto (7), S Richards (7), PA Shetty (11), D Skeggs (1), R Wilson (13).

(1) Royal Free Hospital, London.
(2) Queen Mary Hospital, Hong Kong.
(3) Dryburn Hospital, Durham.
(4) Kent and Canterbury Hospital.
(5) Royal Sussex Hospital, Brighton.
(6) Royal Victoria Infirmary, Newcastle-upon-Tyne.
(7) Clinical Trial Service Unit, Radcliffe Infirmary, Oxford.
(8) Hammersmith Hospital, London.
(9) Westminster Hospital, London.
(10) Maraharagama Cancer Hospital, Sri Lanka.
(11) Tata Memorial Hospital, Bombay.
(12) The London Hospital, London.
(13) The General Hospital, Newcastle upon Tyne.

TRIAL DESIGN: TREATMENT GROUPS:
1. CMF: Cyclophosphamide 500 mg/M^2 (max. 750 mg), methotrexate 25 mg/M^2 (max. 40 mg) and 5-fluorouracil 600 mg/M^2 (max 1000 mg), intravenously, day 1 of 3 week cycle for 6 months, then cyclophosphamide 100 mg orally days 1-4 of 3 week cycle, methotrexate 25 mg/M^2 orally and 5-fluorouracil 600 mg/M^2, orally, day 1 of 3 week cycle, (all same max. dose) for the next 18 months.
2. Tam: Tamoxifen 20 mg, orally, twice daily, for 2 years.
3. CMF + Tam: CMF, as above, plus tamoxifen, as above.
4. Nil: No adjuvant endocrine or chemotherapy.

ELIGIBILITY CRITERIA FOR ENTRY:
1. AGE: <70.
2. CLINICAL/PATHOLOGICAL STAGES: Stage II breast cancer, with histologically confirmed ipsilateral node involvement.
3. PRIMARY TREATMENT: Surgery + radiotherapy to completely eradicate loco-regional disease.
4. NODE DISSECTION: Some form of surgical staging required.
5. NUMBER OF NODES REMOVED: Not specified.
6. MENSTRUAL STATUS: Not specified.
7. RECEPTOR STATUS: Not specified.

METHOD OF RANDOMIZATION:
By telephone to a central office in UK, and by envelope, in Asia. Allocations were based on random number tables and treatment allocations balanced in blocks of 6.

RANDOMIZATION STRATIFICATIONS:
By centres: 78E1-Hong Kong, 78E2-Sri Lanka, 78E3-Bombay, 78E4-UK.

YEARS OF ACCRUAL: 1978 - 1985.

TOTAL NUMBER OF PATIENTS RANDOMIZED: 478.

AIM, RATIONALE, SPECIAL FEATURES:
This randomized trial involved collaboration between centres not only in Britain but also in India, Hong Kong and Sri Lanka. The primary aim was to study the effects of treatment on mortality among all randomized patients, with no withdrawals of the few women who were entered in slight conflict with the above guidelines. A secondary objective was to pilot the use of these treatments in areas where they are not usually available and thus provide a useful test of their generalisability to certain parts of Asia.

Since two quite different adjuvant treatments required evaluation, the natural trial design was to use a "factorial" one in which 25% received one treatment, 25% the other, 25% both, and 25% neither.

ADDITIONAL ELIGIBILITY OR STRATIFICATION CRITERIA:
Mastectomy, axillary clearance and radiotherapy suggested but not mandatory.

COMPLIANCE AND TOXICITY:

	Allocated CMF	Not Allocated CMF
Late/missing data	20	20
No cytotoxic therapy	17	97
Nonstandard cytotoxic	10	1
Standard cytotoxic	77	6
Per cent fully compliant	77/104 (74%)	97/104 (93%)

	Allocated tamoxifen	Not allocated tamoxifen
Late/missing data	12	25
No hormonal therapy	3	93
Nonstandard hormonal	6	1
Standard hormonal	103	5
Per cent fully compliant	103/109 (94%)	93/98 (95%)

DISCUSSION:
There were no obvious differences between the pattern of recurrence or death of British and Asian patients. Although randomization was not stratified by prognostic factors, no gross imbalances in any important prognostic features were apparent. The results of all three actively treated groups were relatively equal, with poorer results seen among the untreated control groups.

PUBLICATIONS:
Senanayake F. Adjuvant hormonal chemotherapy in early breast cancer: Early results from a controlled trial. Lancet 1984; ii:1148-1149.

Table 1. Outcome by allocated treatment for trial 78E1 (Hong Kong)
Key: T = total free of relevant event at start of relevant year(s)

(i) Death (D) in each separate year,
subdivided by age when randomized (D <50 & D 50+ / T <50 & T 50+)

Year	1. CMF	2. Tam	3. TamCMF	4. Nil
1	1+ 2/ 17+ 26	1+ 2/ 14+ 25	0+ 0/ 29+ 14	0+ 0/ 20+ 23
2	1+ 2/ 16+ 20	0+ 3/ 12+ 22	2+ 1/ 29+ 12	1+ 2/ 16+ 23
3	2+ 1/ 13+ 16	0+ 1/ 8+ 17	2+ 0/ 25+ 9	1+ 2/ 14+ 16
4	0+ 2/ 7+ 12	1+ 2/ 7+ 13	1+ 1/ 16+ 8	2+ 1/ 10+ 12
5+	1+ 2/ 4+ 8	0+ 0/ 4+ 9	1+ 0/ 12+ 5	0+ 0/ 5+ 9

(ii) First recurrence or prior death (F) in each separate year,
subdivided by age when randomized (F <50 & F 50+ / T <50 & T 50+)

Year	1. CMF	2. Tam	3. TamCMF	4. Nil
1	2+ 5/ 17+ 26	2+ 2/ 14+ 25	3+ 1/ 29+ 14	2+ 2/ 20+ 23
2	3+ 6/ 15+ 21	1+ 6/ 12+ 23	1+ 1/ 26+ 12	2+ 2/ 17+ 21
3	0+ 2/ 12+ 12	1+ 0/ 9+ 16	2+ 2/ 23+ 10	2+ 3/ 12+ 19
4	0+ 1/ 9+ 8	1+ 1/ 6+ 13	1+ 0/ 17+ 6	0+ 2/ 10+ 11
5+	0+ 2/ 5+ 5	1+ 1/ 4+ 10	0+ 0/ 11+ 5	0+ 0/ 6+ 7

(iii) Recurrence with survival (R) and Death (D) in all years together,
subdivided by various characteristics when randomized (R & D / T)

Age at Entry	1. CMF	2. Tam	3. TamCMF	4. Nil
< 40	0+2/6	0+0/3	2+2/12	1+0/7
40 - 49	1+2/11	4+2/11	1+4/17	1+4/13
50 - 59	4+4/14	2+5/15	1+2/10	3+3/17
60 - 69	3+4/11	0+3/9	1+0/4	1+2/6
70+	0+1/1	0+0/1	0+0/0	0+0/0
Any age (including "not known")	8+13/43	6+10/39	5+8/43	6+9/43

Nodal Status (N0, N1-3, N4+)

		1. CMF	2. Tam	3. TamCMF	4. Nil
50 +	N0 (by axillary clearance)	0+0/0	0+0/1	0+0/0	0+0/0
<50	N1-N3 (by axillary	1+1/6	2+0/5	0+1/12	2+2/12
50+	clearance)	3+4/10	2+2/7	0+0/5	3+2/9
<50	N4+ (by axillary	0+0/1	2+1/4	0+1/4	0+1/1
50+	sample/clearance)	2+3/7	0+2/6	0+1/2	0+0/4
<50	Remainder (i.e.	0+3/10	0+1/5	3+4/13	0+1/7
50+	other N+ or N?)	2+2/9	0+4/11	2+1/7	1+3/10

Estrogen Receptors (ER, fmol/mg)
Breakdown by ER status was not available.

Trial continued overleaf

Table 1. Outcome by allocated treatment for trial 78E2 (Sri Lanka)
Key: T = total free of relevant event at start of relevant year(s)

(i) Death (D) in each separate year,
subdivided by age when randomized (D <50 & D 50+ / T <50 & T 50+)

Year	1. CMF	2. Tam	3. TamCMF	4. Nil
1	1+ 0/ 21+ 15	1+ 2/ 19+ 13	0+ 3/ 25+ 10	1+ 1/ 17+ 19
2	0+ 0/ 20+ 15	1+ 0/ 18+ 11	0+ 0/ 25+ 7	1+ 0/ 16+ 18
3	0+ 0/ 18+ 14	0+ 0/ 14+ 10	0+ 0/ 22+ 7	0+ 0/ 13+ 17
4	0+ 0/ 0+ 0	0+ 0/ 1+ 0	0+ 0/ 0+ 0	0+ 0/ 0+ 0
5+	0+ 0/ 0+ 0	0+ 0/ 0+ 0	0+ 0/ 0+ 0	0+ 0/ 0+ 0

(ii) First recurrence or prior death (F) in each separate year,
subdivided by age when randomized (F <50 & F 50+ / T <50 & T 50+)

Year	1. CMF	2. Tam	3. TamCMF	4. Nil
1	2+ 0/ 21+ 15	3+ 2/ 19+ 13	0+ 3/ 25+ 10	3+ 1/ 17+ 19
2	2+ 1/ 16+ 11	4+ 0/ 13+ 8	2+ 0/ 17+ 3	0+ 3/ 11+ 13
3	0+ 0/ 7+ 2	0+ 1/ 4+ 5	0+ 0/ 6+ 0	1+ 0/ 6+ 4
4	0+ 0/ 0+ 0	0+ 0/ 1+ 0	0+ 0/ 0+ 0	0+ 0/ 0+ 0
5+	0+ 0/ 0+ 0	0+ 0/ 0+ 0	0+ 0/ 0+ 0	0+ 0/ 0+ 0

(iii) Recurrence with survival (R) and Death (D) in all years together,
subdivided by various characteristics when randomized (R & D / T)

Age at Entry	1. CMF	2. Tam	3. TamCMF	4. Nil
<40	2+0/8	2+1/10	1+0/8	1+1/7
40 - 49	1+1/13	3+1/9	1+0/17	1+1/10
50 - 59	1+0/13	0+2/9	0+2/7	3+1/11
60 - 69	0+0/2	1+0/4	0+1/3	1+0/8
Any age (including "not known")	4+1/36	6+4/32	2+3/35	6+3/36

Nodal Status (N0, N1-3, N4+)

		1. CMF	2. Tam	3. TamCMF	4. Nil
<50	N0 (by axillary	0+0/0	0+0/0	0+0/0	0+0/0
50+	clearance)	0+0/0	0+0/0	0+0/1	0+0/0
<50	N1-N3 (by axillary	3+1/20	5+2/16	2+0/23	1+0/12
50+	clearance)	1+0/12	1+2/13	0+2/7	4+1/19
<50	N4+ (by axillary	0+0/0	0+0/2	0+0/2	1+2/5
50+	sample/clearance)	0+0/2	0+0/0	0+1/2	0+0/0
<50	Remainder (i.e.	0+0/0	0+0/1	0+0/0	0+0/0
50+	other N+ or N?)	0+0/1	0+0/0	0+0/0	0+0/0

Table 1. Outcome by allocated treatment for trial 78E4 (Great Britain)
Key: T = total free of relevant event at start of relevant year(s)

(i) Death (D) in each separate year,
subdivided by age when randomized (D <50 & D 50+ / T <50 & T 50+)

Year	1. CMF	2. Tam	3. TamCMF	4. Nil
1	1+ 3/ 9+ 20	1+ 0/ 7+ 21	1+ 1/ 11+ 20	1+ 1/ 10+ 20
2	1+ 1/ 8+ 17	1+ 1/ 6+ 21	0+ 1/ 10+ 19	1+ 3/ 9+ 19
3	0+ 2/ 7+ 15	0+ 0/ 5+ 18	2+ 1/ 10+ 18	0+ 3/ 8+ 16
4	0+ 4/ 7+ 13	0+ 0/ 5+ 17	1+ 1/ 8+ 15	0+ 1/ 6+ 11
5+	0+ 2/ 6+ 8	1+ 0/ 5+ 16	0+ 1/ 6+ 14	2+ 2/ 6+ 10

(ii) First recurrence or prior death (F) in each separate year,
subdivided by age when randomized (F <50 & F 50+ / T <50 & T 50+)

Year	1. CMF	2. Tam	3. TamCMF	4. Nil
1	1+ 4/ 9+ 20	2+ 0/ 7+ 21	1+ 1/ 11+ 20	3+ 7/ 10+ 20
2	2+ 1/ 8+ 16	1+ 1/ 5+ 21	0+ 4/ 9+ 19	2+ 5/ 7+ 13
3	0+ 5/ 6+ 14	1+ 2/ 4+ 18	0+ 2/ 7+ 14	0+ 3/ 5+ 8
4	0+ 3/ 6+ 9	0+ 0/ 3+ 16	2+ 1/ 7+ 11	1+ 0/ 5+ 4
5+	1+ 0/ 5+ 6	1+ 1/ 3+ 13	1+ 3/ 5+ 10	0+ 2/ 3+ 3

(iii) Recurrence with survival (R) and Death (D) in all years together,
subdivided by various characteristics when randomized (R & D / T)

Age at Entry	1. CMF	2. Tam	3. TamCMF	4. Nil
<40	1+0/4	0+1/1	1+3/4	0+3/3
40 - 49	1+2/5	2+2/6	1+1/7	0+3/7
50 - 59	1+8/12	2+1/10	3+2/8	4+5/10
60 - 69	0+4/8	1+0/11	3+3/12	3+5/10
Any age (including "not known")	3+14/29	5+4/28	8+9/31	7+16/30

Nodal Status (N0, N1-3, N4+)

		1. CMF	2. Tam	3. TamCMF	4. Nil
<50	N0 (by axillary	0+0/0	0+0/0	0+1/1	0+0/0
50+	clearance)				
<50	N1-N3 (by axillary	1+0/2	0+1/1	0+0/1	0+0/0
50+	clearance)	0+1/3	1+0/4	1+0/4	1+1/3
<50	N4+ (by axillary	0+0/1	1+1/2	0+0/0	0+0/0
50+	sample/clearance)	0+0/2	0+0/3	0+2/2	0+2/2
<50	Remainder (i.e.	1+2/6	1+1/4	2+3/9	0+6/10
50+	other N+ or N?)	1+11/15	2+1/14	5+3/14	6+7/15

Estrogen Receptors (ER, fmol/mg)

Breakdown by ER status was not available.

Table 1. Outcome by allocated treatment for trial 78E3 (Bombay)
Key: T = total free of relevant event at start of relevant year(s)

(i) Death (D) in each separate year,
subdivided by age when randomized (D <50 & D 50+ / T <50 & T 50+)

Year	1. CMF	2. Tam	3. TamCMF	4. Nil
1	0+ 0/ 8+ 4	0+ 0/ 6+ 6	0+ 0/ 11+ 1	0+ 1/ 6+ 8
2	2+ 0/ 8+ 4	0+ 0/ 6+ 6	1+ 0/ 11+ 1	0+ 0/ 6+ 7
3	0+ 0/ 6+ 4	0+ 0/ 6+ 6	0+ 1/ 10+ 1	1+ 0/ 5+ 7
4	0+ 0/ 2+ 4	0+ 0/ 4+ 3	0+ 0/ 5+ 0	0+ 0/ 1+ 5
5+	0+ 0/ 0+ 3	0+ 0/ 1+ 2	0+ 0/ 4+ 0	0+ 0/ 1+ 0

(ii) First recurrence or prior death (F) in each separate year,
subdivided by age when randomized (F <50 & F 50+ / T <50 & T 50+)

Year	1. CMF	2. Tam	3. TamCMF	4. Nil
1	1+ 0/ 8+ 4	0+ 0/ 6+ 6	2+ 1/ 11+ 1	2+ 4/ 6+ 8
2	1+ 0/ 7+ 4	1+ 1/ 6+ 5	1+ 0/ 8+ 0	1+ 2/ 2+ 3
3	1+ 1/ 6+ 4	1+ 0/ 5+ 3	2+ 0/ 7+ 0	0+ 1/ 0+ 1
4	0+ 0/ 5+ 3	1+ 1/ 4+ 1	0+ 0/ 5+ 0	0+ 0/ 0+ 0
5+	0+ 0/ 1+ 3	0+ 0/ 2+ 0	0+ 0/ 4+ 0	0+ 0/ 0+ 0

(iii) Recurrence with survival (R) and Death (D) in all years together,
subdivided by various characteristics when randomized (R & D / T)

Age at Entry	1. CMF	2. Tam	3. TamCMF	4. Nil
<40	1+1/2	1+0/3	4+1/10	0+0/2
40 - 49	0+1/6	2+0/3	0+0/1	2+1/4
50 - 59	1+0/3	2+0/4	0+1/1	6+0/6
60 - 69	0+0/1	0+0/2	0+0/0	0+1/2
Any age (including "not known")	2+2/12	5+0/12	4+2/12	8+2/14

Nodal Status (N0, N1-3, N4+)

		1. CMF	2. Tam	3. TamCMF	4. Nil
<50	N1-N3 (by axillary	0+0/3	2+0/4	0+0/1	0+0/0
50+	clearance)	0+0/2	0+0/1	0+0/0	1+0/1
<50	N4+ (by axillary	1+2/5	1+0/2	4+1/10	2+1/6
50+	sample/clearance)	1+0/2	2+0/5	0+1/1	5+1/7

TITLE: ADJUVANT TAMOXIFEN IN NODE POSITIVE POSTMENOPAUSAL BREAST CANCER

STUDY NUMBER: 78F
STUDY TITLE: CRFB CAEN C5

ABBREVIATED TRIAL NAME: CRFB CAEN C5
STUDY NUMBER: 78F

INVESTIGATORS' NAMES AND AFFILIATIONS:
T Delozier (1), JP Julien (2), O Switsers (1), C Veyret (1), JE Couette (1), Y Graic (2), JM Ollivier (1), JY Genot (1).

(1) Centre Francois Baclesse, Caen, France.
(2) Centre Henri Becquerel, Rouen, France.

TRIAL DESIGN: TREATMENT GROUPS:
1. Tam: Tamoxifen 40 mg, orally, daily for 36 months.
2. Nil: No adjuvant hormone therapy.

ELIGIBILITY CRITERIA FOR ENTRY:
1. AGE: <75.
2. CLINICAL STAGES: T1-3, N0-1,M0.
3. PRIMARY TREATMENT: Total mastectomy for tumor >3 cm; otherwise tumorectomy. Postoperative irradiation of axillary, supraclavicular and internal mammary nodes; chest or breast irradiation depending on surgical procedure.
4. NODE DISSECTION: Complete axillary dissection.
5. NUMBER OF NODES REMOVED: 14 on average.
6. MENSTRUAL STATUS: Menostasis for at least one year.

METHOD OF RANDOMIZATION:
By sealed numbered envelopes; allocation based on random number tables.

RANDOMIZATION STRATIFICATIONS:
By center: (Centre Francois Baclesse, Centre Henri Becquerel).

YEARS OF ACCRUAL: 1978-1982.

TOTAL NUMBER OF PATIENTS RANDOMIZED: 179.

ADDITIONAL ELIGIBILITY OR STRATIFICATION CRITERIA:
Disease was initially assessed by clinical examination with TNM staging, mammography, chest X-ray, X-ray of head, lumbar spine and pelvis, liver function studies including gamma GT, and alkaline phosphatase. Estrogen and progesterone receptor assays were performed for 74% of the patients.

FOLLOW-UP, COMPLIANCE AND TOXICITY:
The follow-up examinations for the first 3 years included a physical examination every 3 months, a chest X-ray every 6 months, and annually: a skeletal X-ray or scan, a mammography, and blood tests including gamma-GT and alkaline phosphatase. After 3 years the annual follow-up included a physical examination, a chest X-ray, a mammography, and CEA.

Control patients were to be given tamoxifen as first treatment in case of a recurrence.

DISCUSSION:
The study showed that tamoxifen prolonged disease-free survival for postmenopausal women with axillary node involvement. This effect of tamoxifen did not seem to vary with the number of positive axillary nodes (1-3 versus 4+). It seemed greater for the patients with progesterone receptors than for the patients without. A similar result was observed for estrogen receptors.

PUBLICATIONS:
Delozier T, Julien JP, Juret P, Veyret C, Couette JE, Graic Y, Ollivier JM, de Ranieri E. Adjuvant tamoxifen in postmenopausal breast cancer: preliminary results of a randomized trial. Breast Cancer Res Treat 1986; 7:105-109.

Delozier T, Julien JP. Hormonotherapie adjuvante dans les cancers du sein curables de la femme menopausee. Horm Reprod Metab 1986;3:100-105.

Table 1. Outcome by allocated treatment for trial 78F (CRFB Caen C5)
Key: T = total free of relevant event at start of relevant year(s)

(i) Death (D) in each separate year,
subdivided by age when randomized (D <50 & D 50+ / T <50 & T 50+)

Year	1. Tam	2. Nil
1	0+ 5/ 2+ 87	0+ 9/ 2+ 88
2	1+ 7/ 2+ 82	0+ 9/ 2+ 79
3	0+ 6/ 1+ 75	0+ 6/ 2+ 70
4	0+ 4/ 1+ 69	0+ 6/ 2+ 64
5+	0+ 1/ 1+ 54	0+ 2/ 1+ 51

(ii) First recurrence or prior death (F) in each separate year,
subdivided by age when randomized (F <50 & F 50+ / T <50 & T 50+)

Year	1. Tam	2. Nil
1	1+ 11/ 2+ 87	0+ 22/ 2+ 88
2	0+ 9/ 1+ 76	0+ 16/ 2+ 66
3	0+ 8/ 1+ 66	1+ 5/ 2+ 49
4	0+ 3/ 1+ 50	0+ 4/ 1+ 38
5+	0+ 2/ 0+ 30	0+ 3/ 0+ 22

(iii) Recurrence with survival (R) and Death (D) in all years together,
subdivided by various characteristics when randomized (R & D / T)

Age at Entry		1. Tam	2. Nil
40 - 49		0+1/2	1+0/2
50 - 59		3+5/21	9+13/37
60 - 69		6+11/39	6+13/34
70 +		1+7/27	3+6/17
Any age (including "not known")		10+24/89	19+32/90

	Nodal Status (N0, N1-3, N4+)		
<50	N1-N3 (by axillary	0+0/0	1+0/2
50 +	clearance)	7+5/47	5+14/50
<50	N4+ (by axillary	0+1/2	0+0/0
50 +	sample/clearance)	3+18/40	13+18/38

	Estrogen Receptors (ER, fmol/mg)		
<50	ER poor or ER < 10	0+1/2	0+0/1
50 +		3+9/22	3+11/18
50 +	ER+ or ER 10-99	2+5/21	4+9/28
50 +	ER++ or ER 100+	2+1/21	6+6/20

TITLE: RANDOMIZED STUDY OF ADJUVANT CMF WITH OR WITHOUT RADIOTHERAPY AND WITH OR WITHOUT OVARIAN ABLATION

STUDY NUMBER: 78G1-3
STUDY TITLE: CCABC CANADA

ABBREVIATED TRIAL NAME: CCABC CANADA
STUDY NUMBER: 78G1-3

INVESTIGATORS' NAMES AND AFFILIATIONS:
J Ragaz, SM Jackson IH Plenderleith, V Ng, K Wilson, VE Basco, M Knowling, J Spinelli.
Cancer Control Agency of British Columbia, Vancouver, B.C. Canada.

TRIAL DESIGN: TREATMENT GROUPS:

1. CMF: Cyclophosphamide 100 mg/M^2, orally, days 1-14, methotrexate 40 mg/M^2, and 5-fluorouracil 600 mg/M^2, intravenously, days 1+8, repeated every 4 weeks for 12 cycles.
In 1981, the chemotherapy was altered to cyclophosphamide 600 mg/M^2, methotrexate 40 mg/M^2 and 5-fluorouracil 600 mg/M^2, all intravenously, every 3 weeks for 9 cycles.

2. CMF + R: CMF, as above, plus radiotherapy to chest wall, internal mammary chain, axilla and supraclavicular area with a dose of 4000 Gy to chest wall and 3750 Gy to the lymph nodes daily over 16 fractions in 3 weeks. Radiotherapy was given after 3 months between chemotherapy cycles.

Until 11/84 patients with an ER ≥10 fmol/mg could also be randomized (76G1) to:

3. CMF + Ov Irr + Pr: CMF as in arm 1 plus ovarian ablation by pelvic field irradiation of 1600 Gy over 4 days, plus prednisone 7.5 mg, daily for 2 years.

4. CMF+ R+Ov Irr + Pr: CMF +R as in arm 2 plus Ov Irr and Pr as in arm 3.

ELIGIBILITY CRITERIA FOR ENTRY:
1. AGE: No restrictions.
2. CLINICAL/PATHOLOGICAL STAGES: Pathologically positive nodes and absence of distant metastasis.
3. PRIMARY TREATMENT: Modified radical mastectomy.
4. NODE DISSECTION: Levels I & II required.
5. NUMBER OF NODES REMOVED: Not specified.
6. MENSTRUAL STATUS: Premenopausal (last menses within one year). If previous hysterectomy but not bilateral oophorectomy, menopausal status determined by FSH levels.

METHOD OF RANDOMIZATION:
By telephone to a central office at the CCABC where allocation was done by a random number table. From 1982, pre-randomization was used.

YEARS OF ACCRUAL: 1979-1986.

TOTAL NUMBER OF PATIENTS RANDOMIZED: 318 (27 patients entered after August 1985 are not included in the present overview). By September 1985, 136 had been allocated CMF and 155 CMF plus radiotherapy. Of those randomized to CMF, 67 ER-positive patients (≥10 fmol/mg) were re-randomized, 33 to ovarian ablation plus CMF and 34 to CMF only. Of the 142 randomized to CMF plus radiotherapy, 67 were re-randomized, 35 to ovarian ablation plus CMF plus radiotherapy and 32 to CMF plus radiotherapy only. All ER-positive patients entered after 11/84 and earlier patients who refused the Ov Irr randomization were randomized between arms 1 and 2 only (76G2). ER-negative patients were also randomized between arms 1 and 2 only (76G3).

AIM, RATIONALE, SPECIAL FEATURES:
This study was designed in 1978 (1) to assess the role of locoregional radiotherapy in conjunction with CMF, and (2) to study the role of adjuvant ovarian irradiation and prednisone in conjunction with CMF ± locoregional radiotherapy in women with ER ≥ 10 fmol/mg. Since the accompanying overview deals only with trials of ovarian ablation vs. none (ignoring trials that are potentially confounded by the addition of prednisone) this comparison does not contribute to the present overview.

ADDITIONAL ELIGIBILITY OR STRATIFICATION CRITERIA:
Patients were excluded if they had had any previous malignancy except cured squamous cell or basal cell carcinoma of the skin.

DISCUSSION:
The overall survival and metastasis-free survival of patients treated with CMF chemotherapy alone and chemotherapy with radiation was similar although local recurrence, as the only evidence of recurrent disease, was significantly less common in the radiotherapy arm (6% vs. 15%).

Neither overall survival nor recurrence were substantially affected by the addition of ovarian irradiation and prednisone to CMF chemotherapy ± locoregional radiation in this study. Chemotherapy with CMF, however, may itself change the hormonal milieu substantially, so that the additional use of ovarian ablation and prednisone may be of no additional value. Alternatively, a real effect of adding ovarian ablation and prednisone to CMF may exist but may have been missed due to the relatively small sample size in this study.

PUBLICATIONS:
Ragaz J, Jackson S, Wilson K, Plenderleith IH, Knowling M, Basco V, Ng V. Randomized study of locoregional radiotherapy (XRT) and ovarian ablation (OOPH) in premenopausal patients with breast cancer treated with adjuvant chemotherapy (CT). Proceedings of the American Society of Clinical Oncology 1988; 7:12.

Table 1. Outcome by allocated treatment for trial 78G1 (CCABC Canada)
Key: T = total free of relevant event at start of relevant year(s)

(i) *Death (D) in each separate year,*
subdivided by age when randomized (D <50 & D 50+ / T <50 & T 50+)

Year	1. CMF	2. CMFR	3. CMFOvIrrPr	4. CMFOvIrrPrR
1	0+ 0/ 29+ 6	1+ 0/ 23+ 8	0+ 0/ 27+ 6	0+ 0/ 30+ 5
2	1+ 1/ 28+ 6	0+ 0/ 22+ 7	1+ 0/ 25+ 5	0+ 0/ 29+ 5
3	0+ 1/ 15+ 4	3+ 0/ 18+ 4	0+ 0/ 17+ 4	1+ 0/ 15+ 4
4	2+ 0/ 11+ 3	1+ 0/ 13+ 4	2+ 0/ 12+ 2	1+ 0/ 12+ 2
5+	1+ 0/ 7+ 1	0+ 0/ 7+ 4	0+ 0/ 6+ 1	1+ 0/ 7+ 1

(ii) *First recurrence or prior death (F) in each separate year,*
subdivided by age when randomized (F <50 & F 50+ / T <50 & T 50+)

Year	1. CMF	2. CMFR	3. CMFOvIrrPr	4. CMFOvIrrPrR
1	1+ 1/ 29+ 6	1+ 0/ 23+ 8	1+ 0/ 27+ 6	0+ 0/ 30+ 5
2	2+ 2/ 28+ 5	3+ 0/ 22+ 8	1+ 0/ 26+ 6	3+ 0/ 30+ 5
3	3+ 0/ 26+ 3	3+ 0/ 19+ 7	2+ 1/ 25+ 5	0+ 0/ 26+ 4
4	2+ 0/ 13+ 2	0+ 0/ 14+ 5	0+ 0/ 14+ 4	1+ 0/ 17+ 4
5+	0+ 0/ 6+ 1	0+ 0/ 11+ 4	0+ 0/ 9+ 1	1+ 0/ 10+ 3

(iii) *Recurrence with survival (R) and Death (D) in all years together,*
subdivided by various characteristics when randomized (R & D / T)

Age at Entry		1. CMF	2. CMFR	3. CMFOvIrrPr	4. CMFOvIrrPrR
< 40		1+0/10	1+4/7	1+0/7	1+1/5
40 - 49		3+4/19	0+2/16	1+3/20	2+2/25
50 - 59		1+2/6	0+0/8	1+0/6	0+0/5
Any age (including "not known")		5+6/35	1+6/31	3+3/33	3+3/35

	Nodal Status (N0, N1-3, N4+)				
<50	N1-N3 (by axillary	1+0/15	1+0/12	2+2/18	1+0/14
50+	clearance)	1+0/3	0+0/4	0+0/5	0+0/4
<50	N4+ (by axillary	3+3/10	0+4/9	0+1/9	1+2/11
50+	sample/clearance)	0+1/2	0+0/3	1+0/1	0+0/1
<50	Remainder (i.e.	0+1/4	0+2/2	0+0/0	1+1/5
50+	other N+ or N?)	0+1/1	0+0/1	0+0/0	0+0/0

	Estrogen Receptors (ER, fmol/mg)				
<50	ER poor or ER < 10	0+0/2	0+0/2	0+1/2	0+0/3
<50	ER+ or ER 10-99	3+4/24	1+5/14	2+2/22	2+3/25
50+		1+2/6	0+0/6	0+0/5	0+0/5
<50	ER++ or ER 100+	1+0/3	0+1/7	0+0/3	1+0/2
50+		0+0/0	0+0/1	1+0/1	0+0/0

Trial continued overleaf

Table 1. Outcome by allocated treatment for trial 78G2 (CCABC Canada)
Key: T = total free of relevant event at start of relevant year(s)

(i) Death (D) in each separate year,
subdivided by age when randomized (D <50 & D 50+ / T <50 & T 50+)

Year	1. CMF	2. CMFR
1	0+ 0/ 9+ 0	0+ 0/ 12+ 4
2	1+ 0/ 1+ 0	0+ 0/ 2+ 1
3	0+ 0/ 0+ 0	0+ 0/ 2+ 0
4	0+ 0/ 0+ 0	0+ 0/ 2+ 0
5+	0+ 0/ 0+ 0	0+ 0/ 1+ 0

(ii) First recurrence or prior death (F) in each separate year,
subdivided by age when randomized (F <50 & F 50+ / T <50 & T 50+)

Year	1. CMF	2. CMFR
1	0+ 0/ 9+ 0	0+ 0/ 12+ 4
2	1+ 0/ 9+ 0	0+ 1/ 11+ 3
3	0+ 0/ 1+ 0	0+ 0/ 5+ 1
4	0+ 0/ 0+ 0	0+ 0/ 2+ 0
5+	0+ 0/ 0+ 0	0+ 0/ 1+ 0

(iii) Recurrence with survival (R) and Death (D) in all years together,
subdivided by various characteristics when randomized (R & D / T)

Age at Entry	1. CMF	2. CMFR
<40	0+1/2	0+0/4
40 - 49	0+0/7	0+0/8
50 - 59	0+0/0	1+0/4
Any age (including "not known")	0+1/9	1+0/16

Nodal Status (N0, N1-3, N4+)

		1. CMF	2. CMFR
<50	N1-N3 (by axillary	0+1/5	0+0/7
50+	clearance)	0+0/0	1+0/3
<50	N4+ (by axillary	0+0/4	0+0/4
50+	sample/clearance)	0+0/0	0+0/1
<50	Remainder (i.e.	0+0/0	0+0/1

Estrogen Receptors (ER, fmol/mg)

		1. CMF	2. CMFR
<50	ER+ or ER 10-99	0+1/8	0+0/11
50+		0+0/0	1+0/3
<50	ER++ or ER 100+	0+0/1	0+0/1
50+		0+0/0	0+0/1

Table 1. Outcome by allocated treatment for trial 78G3 (CCABC Canada)
Key: T = total free of relevant event at start of relevant year(s)

(i) Death (D) in each separate year,
subdivided by age when randomized (D <50 & D 50+ / T <50 & T 50+)

Year	1. CMF	2. CMFR
1	2+ 0/ 54+ 6	0+ 0/ 61+ 11
2	5+ 2/ 48+ 6	6+ 0/ 52+ 11
3	3+ 2/ 33+ 4	5+ 0/ 38+ 11
4	0+ 0/ 18+ 2	2+ 0/ 27+ 7
5+	0+ 0/ 15+ 1	1+ 2/ 16+ 4

(ii) First recurrence or prior death (F) in each separate year,
subdivided by age when randomized (F <50 & F 50+ / T <50 & T 50+)

Year	1. CMF	2. CMFR
1	3+ 1/ 54+ 6	9+ 0/ 61+ 11
2	10+ 3/ 50+ 5	5+ 0/ 52+ 11
3	1+ 0/ 37+ 2	2+ 1/ 41+ 11
4	2+ 0/ 25+ 2	0+ 0/ 30+ 9
5+	1+ 0/ 15+ 2	1+ 1/ 22+ 7

(iii) Recurrence with survival (R) and Death (D) in all years together,
subdivided by various characteristics when randomized (R & D / T)

Age at Entry	1. CMF	2. CMFR
<40	3+1/14	1+4/18
40 - 49	4+9/40	2+10/43
50 - 59	0+4/6	0+2/11
Any age (including "not known")	7+14/60	3+16/72

Nodal Status (N0, N1-3, N4+)

		1. CMF	2. CMFR
<50	N1-N3 (by axillary	3+2/30	0+6/35
50+	clearance)	0+2/4	0+2/6
<50	N4+ (by axillary	4+7/22	3+8/24
50+	sample/clearance)	0+1/1	0+0/3
<50	Remainder (i.e.	0+1/2	0+0/2
50+	other N+ or N?)	0+1/1	0+0/2

Estrogen Receptors (ER, fmol/mg)

		1. CMF	2. CMFR
<50	ER poor or ER < 10	7+10/54	3+14/61
50+		0+4/6	0+2/11

TITLE: ADJUVANT TAMOXIFEN THERAPY FOR EARLY BREAST CANCER

ABBREVIATED TRIAL NAME: INNSBRUCK
STUDY NUMBER: 78H

INVESTIGATORS' NAMES AND AFFILIATIONS:
R Margreiter, FER Steindorfer, P Hausmaninger, H Manfreda, D Schulz, R Marth, G Daxenbichler.
Univ. Klinik für Chirurgie, Innsbruck, Austria.

TRIAL DESIGN: TREATMENT GROUPS:
1. Tam: Tamoxifen, 10 mg, twice daily, orally, x 12 months.
2. Nil: No adjuvant systemic therapy.

ELIGIBILITY CRITERIA FOR ENTRY:
1. AGE: ≤70.
2. CLINICAL/PATHOLOGICAL STAGES: Stage I or II.
3. PRIMARY TREATMENT: Patey-type modified radical mastectomy with adjuvant radiotherapy if the tumor was <2.5 cm and located in the inner quadrants or if there was involvement of the subclavicular lymph nodes.
4. AXILLARY NODE DISSECTION: Required.
5. NUMBER OF NODES REMOVED: ?
6. MENSTRUAL STATUS: No restrictions.
7. RECEPTOR STATUS: Only patients with known receptor status were accepted for this study, and only estrogen receptor positive patients were randomized.

METHOD OF RANDOMIZATION:
By telephone call to a central office.

RANDOMIZATION STRATIFICATIONS:
Menstrual status (premenopausal, postmenopausal); and tumor stage (I, II).

YEARS OF ACCRUAL: 1979-1981.

TOTAL NUMBER OF PATIENTS RANDOMIZED: 234.

AIM, RATIONALE, SPECIAL FEATURES:
This study was designed specifically to evaluate adjuvant tamoxifen in patients who were estrogen receptor positive. Although postmenopausal patients randomized to receive adjuvant tamoxifen had a slightly better disease-free survival than those randomized to no treatment, this difference was not statistically significant. The survival curves for the postmenopausal women were almost superimposable. However, among premenopausal women there was a significantly better disease-free survival (p<0.03) for those randomized to receive adjuvant tamoxifen. A trend towards improved survival among the premenopausal women was not statistically significant.

PUBLICATIONS:
Margreiter R, Steindorfer FER, Hausmaninger P, Manfreda H, Shulz D, Marth R, Daxenbichler G. Adjuvant tamoxifen therapy for early breast cancer: a controlled trial. Reviews of Endocrine-Related Cancer. 1985; 17s:117-121.

Table 1. Outcome by allocated treatment for trial 78H (Innsbruck)
Key: T = total free of relevant event at start of relevant year(s)

(i) Death (D) in each separate year,
subdivided by age when randomized (D <50 & D 50+ / T <50 & T 50+)

Year	1. Tam	2. Nil
1	0+ 1/ 31+ 96	0+ 1/ 27+ 80
2	1+ 2/ 31+ 95	0+ 2/ 27+ 79
3	0+ 3/ 30+ 93	1+ 5/ 27+ 77
4	0+ 1/ 30+ 90	3+ 3/ 26+ 72
5+	2+ 1/ 25+ 73	1+ 1/ 23+ 57

(ii) First recurrence or prior death (F) in each separate year,
subdivided by age when randomized (F <50 & F 50+ / T <50 & T 50+)

Year	1. Tam	2. Nil
1	2+ 4/ 31+ 96	1+ 5/ 27+ 80
2	1+ 5/ 27+ 90	3+ 9/ 26+ 72
3	1+ 5/ 26+ 85	5+ 8/ 23+ 63
4	3+ 5/ 21+ 67	0+ 5/ 18+ 50
5+	0+ 3/ 9+ 38	0+ 4/ 8+ 27

(iii) Recurrence with survival (R) and Death (D) in all years together,
subdivided by various characteristics when randomized (R & D / T)

Age at Entry		1. Tam	2. Nil
< 40		1+0/7	1+2/7
40 - 49		3+3/24	3+3/20
50 - 59		6+5/45	9+7/34
60 - 69		5+3/44	9+1/30
70 +		2+0/5	1+3/15
Any age (including "not known")		18+11/127	23+17/107

Nodal Status (N0, N1-3, N4+)

		1. Tam	2. Nil
< 50	N0 (by axillary	4+0/15	0+3/13
50 +	clearance)	4+3/44	7+3/40
< 50	N1-N3 (by axillary	0+1/8	1+2/6
50 +	clearance)	2+2/30	7+6/28
< 50	N4+ (by axillary	0+2/4	3+0/8
50 +	sample/clearance)	8+3/20	5+3/12
< 50	Remainder (i.e.	0+0/4	0+0/0
50 +	other N+ or N?)	0+0/2	0+0/0

Estrogen Receptors (ER, fmol/mg)

		1. Tam	2. Nil
50 +	ER poor or ER < 10	0+1/2	0+0/0
< 50	ER+ or ER 10-99	4+1/23	4+5/24
50 +		10+7/56	6+11/50
< 50	ER++ or ER 100+	0+2/8	0+0/3
50 +		4+0/38	13+1/30

TITLE: ADJUVANT THERAPY WITH TAMOXIFEN VS. PLACEBO IN OLDER (AGE >65) POSTMENOPAUSAL WOMEN WITH STAGE II BREAST CANCER

STUDY NUMBER: 78J
STUDY TITLE: ECOG EST1178

ABBREVIATED TRIAL NAME: ECOG EST1178
STUDY NUMBER: 78J

INVESTIGATORS' NAMES AND AFFILIATIONS:
FJ Cummings (1), R Gray (2), DC Tormey and TE Davis (3), and the Members of the Eastern Cooperative Oncology Group.
(1) Brown University, Providence, RI.
(2) Harvard School of Public Health, Boston, MA.
(3) University of Wisconsin, Madison, WI.

TRIAL DESIGN: TREATMENT GROUPS
(1) Tam: Tamoxifen 10 mg. twice daily for 24 months; if platelet count <50,000/cu mm or leukocytes <2,500/cu mm, stop drug until counts rise above these levels; reduction to a single tablet daily allowed for uncontrolled vomiting.
(2) Placebo: One tablet twice daily for 24 months; same dose modifications as for tamoxifen.

ELIGIBILITY CRITERIA FOR ENTRY:
1. AGE: ≥66 years.
2. CLINICAL/PATHOLOGICAL STAGES: Potentially curable breast cancer with ≥1 histologically positive ipsilateral axillary lymph node; axillary nodes clinically movable with no fixation of primary tumor to skin, underlying muscle, or chest wall.
3. PRIMARY TREATMENT: Modified radical or radical mastectomy within 10 weeks prior to randomization.
4. AXILLARY DISSECTION: Some form of axillary surgery required.
5. NUMBER OF NODES REMOVED: Not specified.
6. MENSTRUAL STATUS: Postmenopausal.
7. RECEPTOR STATUS: Positive, borderline or unknown.

RANDOMIZATION METHOD:
Via phone to ECOG central office for North American patients; via sealed envelope for South African patients. Treatments assigned at random by permuted blocks within strata with block size 2.

RANDOMIZATION STRATIFICATION:
Axillary node involvement: (1-3; 4 or more); estrogen receptor status positive (≥10 fmol/mg); borderline (3-9.9 fmol/mg) or unknown.

YEARS OF ACCRUAL: August 1978 to February 1982.

TOTAL NUMBER RANDOMIZED: 181.

AIM, RATIONALE, SPECIAL FEATURES:
The specific aims of this double blind study were:
1. to compare disease-free interval, recurrence rate, and survival of tamoxifen vs. placebo, and
2. to assess the importance of ER status on the course of disease and to evaluate the relationship of adjuvant hormonal therapy and ER status.

ADDITIONAL ELIGIBILITY OR STRATIFICATION CRITERIA:
Patients were ineligible for the following: inflammatory carcinoma; bilateral malignancy; evidence of distant metastases on chest radiograph, bone scan, or chemistries; preoperative ulceration or infiltration of skin greater than 2 cm; peau d'orange involving >1/3 of the breast; satellite or parasternal nodules; edema of the arm; palpable supraclavicular or infraclavicular nodes; inadequate hematologic, hepatic, or renal function; prior radiation, chemotherapy, or hormone therapy for breast cancer. Patients whose initial biopsy was more than 4 weeks prior to mastectomy and patients with prior or concomitant malignancy were ineligible.

FOLLOW-UP, COMPLIANCE, AND TOXICITY:
All patients were examined every 3 months by their physicians. Blood count, chemistries, and chest X-ray were obtained every 3 months during therapy and every 6 months thereafter. A bone scan was done every 6 months during treatment and annually thereafter.

The placebo and tamoxifen groups had almost identical distributions of hematologic toxicity (22 vs. 25%), infection (12 vs. 10%), gastrointestinal toxicity (12 vs. 16%), neurologic toxicity (12 vs. 12%), pain (15 vs 11%), and genitourinary toxicity (15 vs. 18%). The patients on placebo did experience significantly fewer hot flashes than the patients on tamoxifen (8 vs. 20%). In both treatment arms, 4% of the patients had severe toxicity, 20% had moderate toxicity, 41% had mild toxicity, and 35% had no toxicity.

DISCUSSION:
Of the patients entering this trial, 45% were aged 66-70, 32% aged 71-75, and 22% were older than 75. Estrogen receptor status was classified as positive in 85%, borderline in 4%, and unknown in 12%. 53% of the patients had 1-3 histologically positive nodes, 34% had 4-10, and 13% had more than 10. Tumor size was <2 cm in 40%, 2-5 cm in 52%, and >5 cm in 8%. Median time from surgery to start of treatment was 5 weeks. The percent alive at 5 years is 72% on tamoxifen, 65% on placebo. Of the 57 deaths used in this analysis, 7 of the 33 on placebo and 9 of the 24 on tamoxifen died before documentation of relapse.

PUBLICATIONS:
Cummings FJ, Gray R, Davis TE, Tormey DC, Harris JE, Falkson G, Arsenau J. Adjuvant tamoxifen treatment of elderly women with stage II breast cancer. Ann. Int. Med. 1985; 103:324-329.

Cummings FJ, Gray R, Davis TE, Tormey DC, Harris JE, Falkson G, Arsenau J. Tamoxifen versus placebo: Double-blind adjuvant trial in elderly women with stage II breast cancer. NCI Monogr. 1986; 1:119-123.

Table 1. Outcome by allocated treatment for trial 78J (ECOG EST1178)
Key: T = total free of relevant event at start of relevant year(s)

(i) Death (D) in each separate year,
subdivided by age when randomized (D <50 & D 50+ / T <50 & T 50+)

Year	1. Tam	2. Nil
1	0+ 4/ 0+ 91	0+ 2/ 0+ 90
2	0+ 2/ 0+ 87	0+ 5/ 0+ 88
3	0+ 5/ 0+ 84	0+ 10/ 0+ 83
4	0+ 7/ 0+ 79	0+ 9/ 0+ 73
5+	0+ 6/ 0+ 59	0+ 7/ 0+ 59

(ii) First recurrence or prior death (F) in each separate year,
subdivided by age when randomized (F <50 & F 50+ / T <50 & T 50+)

Year	1. Tam	2. Nil

(recurrence data not available)

(iii) Recurrence with survival (R) and Death (D) in all years together,
subdivided by various characteristics when randomized (R & D / T)

Age at Entry		1. Tam	2. Nil
Breakdown by age was not available.			
Any age (including "not known")		0+24/91	0+33/90
	Nodal Status (N0, N1-3, N4+)		
50 +	N0 (by clinical evidence only)	0+0/0	0+1/2
50 +	N4+ (by axillary sample/clearance)	0+14/40	0+18/43
50 +	Remainder (i.e. other N+. or N?)	0+10/51	0+14/45
	Estrogen Receptors (ER. fmol/mg)		
50 +	ER poor or ER < 10	0+3/3	0+3/5
50 +	ER+ or ER 10-99	0+14/48	0+12/42
50 +	ER++ or ER 100+	0+6/29	0+15/34

TITLE: CHEMOTHERAPY ± OOPHORECTOMY IN HIGH-RISK PREMENOPAUSAL PATIENTS WITH OPERABLE BREAST CANCER

STUDY NUMBER: 78K2
STUDY TITLE: LUDWIG II

ABBREVIATED TRIAL NAME: LUDWIG II
STUDY NUMBER: 78K2

INVESTIGATORS' NAMES AND AFFILIATIONS:
The Ludwig Breast Cancer Study Group

TRIAL DESIGN: TREATMENT GROUPS:
1. CMFPr: Cyclophosphamide 100 mg/M^2, orally, days 1-14; methotrexate 40 mg/M^2 and 5-fluorouracil 600 mg/M^2 intravenously, day 1 and 8; prednisone 7.5 mg, orally, daily for 12 monthly cycles.
2. CMFPr + Ooph: Surgical oophorectomy, followed by CMFPr, as above.

ELIGIBILITY CRITERIA FOR ENTRY:
1. AGE: No restrictions if premenopausal
2. CLINICAL/PATHOLOGICAL STAGES: Histologically confirmed, non-inflammatory unilateral breast carcinoma (T1a-b, T2a-b, or T3a; N0-1; M0) with ≥4 positive axillary lymph nodes.
3. PRIMARY TREATMENT: Total mastectomy and axillary clearance.
4. NODE DISSECTION: Mandatory
5. NUMBER OF NODES REMOVED: Not specified
6. MENOPAUSAL STATUS: Premenopausal or perimenopausal women.

METHOD OF RANDOMIZATION:
By telephone or telex to the Study Coordination Center. Randomization was carried out by means of pseudorandom numbers generated by a congruence method on a DEC 2060 computer.

RANDOMIZATION STRATIFICATIONS:
Participating institutions.

YEARS OF ACCRUAL: 7/78-8/81.

TOTAL NUMBER OF PATIENTS RANDOMIZED: 356.

ADDITIONAL ELIGIBILITY OR STRATIFICATION CRITERIA:
A chest radiograph and bone scan were required for exclusion of detectable metastatic disease. Other prerequisites were a WBC count of ≥4,000, a platelet count of ≥100,000, adequate liver and renal function. Therapy was begun within 6 weeks after surgery.

Patient characteristics were as follows: 27% were <40 years old, 48% had 4-6 positive axillary lymph nodes and 52% had ≥7, 61% had ER determinations done and 54% of these were ER positive.

FOLLOW-UP, COMPLIANCE AND TOXICITY:
Clinical, hematologic and biochemical assessments of each patient were required every 3 months for 2 years and every 6 months thereafter. Chest X-rays and bone scans were required every 6 months. After 2 years, a bone scan was required annually.

2.5% of the CMFPr group had an oophorectomy and 7.8% of the oophorectomy plus CMFPr group did not have ovarian ablation. 2.1% of patients received no chemotherapy. Chemotherapy was stopped early in 4% of the patients due to recurrence, in 14% due to toxicity and/or refusal, and in 1% for miscellaneous reasons. 54% of the CMF cycles in each regimen were modified primarily because of hematologic toxicity (31%), other toxic effects (15%), or other reasons (8%). Infection (in 8 patients) and mucositis (in 10) were the most frequently reported severe complications.

DISCUSSION:
At a 4-year median follow-up, there was no statistically significant difference between the 2 treatments in terms of disease-free survival (54% vs. 51%; p=0.33) or overall survival (71% vs 70%; p=-0.67), even for patients with tumors containing hormone receptors. A high incidence of amenorrhea (89%) due to suppression of ovarian function was observed for the group receiving CMFPr alone. It is hypothesized that the addition of surgical oophorectomy to the adjuvant therapy regimen was rendered superfluous by this effect of cytotoxic treatment.

A recent update at 7-years median follow-up suggests that treatment effect differences favoring CMFPr + oophorectomy may emerge beyond the 4 years previously reported. Ovarian ablation may benefit those premenopausal patients who are destined to relapse later in the course of their disease, and the overview of older trials may provide larger treatment effect differences than could be obtained from trials followed for shorter durations.

PUBLICATIONS:
Ludwig Breast Cancer Study Group. Chemotherapy with or without oophorectomy in high-risk premenopausal patients with operable breast cancer. J. Clin. Oncol. 1985; 3:1059-1067.

Goldhirsch A and Gelber R for the Ludwig Breast Cancer Study Group. Adjuvant treatment for early breast cancer: the Ludwig Breast Cancer Studies. NCI Monographs 1986; 1:55-70.

International Breast Cancer Study Group. Late effects of adjuvant oophorectomy and chemotherapy upon premenopausal breast cancer patients. Ann. Oncol. 1990; 1:30-35.

Table 1. Outcome by allocated treatment for trial 78K2 (Ludwig II)
Key: T = total free of relevant event at start of relevant year(s)
N.B: only crude data (body counts) are currently available on this trial.

(i) Death (D) in each separate year, subdivided by age when randomized (D <50 & D 50+ / T <50 & T 50+)

Year	1. CMFPr	2. CMFPrOoph
Unknown	48+ 0/ 179+ 0	47+ 0/ 177+ 0

(ii) First recurrence or prior death (F) in each separate year, subdivided by age when randomized (F <50 & F 50+ / T <50 & T 50+)

Year	1. CMFPr	2. CMFPrOoph

(recurrence data not available)

(iii) Recurrence with survival (R) and Death (D) in all years together, subdivided by various characteristics when randomized (R & D / T)

Age at Entry	1. CMFPr	2. CMFPrOoph
Breakdown by age was not available.		
Any age (including not known)	0+48/179	0+47/177

Nodal Status (N0, N1-3, N4+)

	1. CMFPr	2. CMFPrOoph
N4+ (by axillary sample/clearance)	0+48/179	0+47/177

Estrogen Receptors (ER, fmol/mg)

Breakdown by ER status was not available.

TITLE: CHEMOENDOCRINE VS ENDOCRINE VS NO ADJUVANT THERAPY FOR POSTMENOPAUSAL WOMEN WITH OPERABLE BREAST CANCER AND AXILLARY NODE METASTASES

STUDY NUMBER: 78K3
STUDY TITLE: LUDWIG III

ABBREVIATED TRIAL NAME: LUDWIG III
STUDY NUMBER: 78K3

INVESTIGATORS' NAMES AND AFFILIATIONS:
The Ludwig Breast Cancer Study Group.

TRIAL DESIGN: TREATMENT GROUPS:
1. PrTam: Prednisone 7.5 mg, orally, and tamoxifen 20 mg, daily, for 12 months.
2. CMFPrTam: Prednisone and tamoxifen as above, plus cyclophosphamide 100 mg/M^2, orally, days 1-14; methotrexate 40 mg/M^2 and 5-fluorouracil 600 mg/M^2, intravenously, on day 1 and 8 for 12 cycles.
3. Nil: No adjuvant therapy.

ELIGIBILITY CRITERIA FOR ENTRY:
1. AGE: ≤65 years.
2. CLINICAL/PATHOLOGICAL STAGES: Histologically confirmed, non-inflammatory unilateral breast carcinoma (T1a-b, T2a-b, or T3a; N0-1; M0) with positive axillary lymph nodes.
3. PRIMARY TREATMENT: Total mastectomy and axillary clearance.
4. NODE DISSECTION: Mandatory.
5. NUMBER OF NODES REMOVED: Not specified.
6. MENOPAUSAL STATUS: Postmenopausal.

METHOD OF RANDOMIZATION:
Telephone or telex to the Study Coordination Center. Randomization was carried out by means of pseudorandom numbers generated by a congruence method on a DEC 2060 computer.

RANDOMIZATION STRATIFICATIONS: Participating institutions.

YEARS OF ACCRUAL: 7/78-8/81.

TOTAL NUMBER OF PATIENTS RANDOMIZED: 503.

ADDITIONAL ELIGIBILITY OR STRATIFICATION CRITERIA:
A chest X-ray and a bone scan were required for exclusion of detectable metastatic disease. Other pre-requisites were a WBC count of ≥4,000, a platelet count of ≥100,000, adequate liver and renal function. Therapy initiated within 6 weeks after surgery.

Characteristics of the patients in this trial were as follows: median age of 59 years (range 40-65), 56% had 1-3 positive axillary lymph nodes, 35% had tumors ≤2 cm, 51% had ER determinations done and 65% of these were ER positive.

FOLLOW-UP, COMPLIANCE AND TOXICITY:
Clinical, hematologic and biochemical assessments were required every 3 months for 2 years and every 6 months thereafter. Chest X-rays and bone scans were required every 6 months. After 2 years, a bone scan was required annually.

2.5% of the observation group received additional treatment for breast cancer without evidence of relapse. 10.4% of the CMFPrTam group and 3.9% of the PrT group received <50% their scheduled treatment.

DISCUSSION:
The following observations were made at a median follow-up of 4 years: the disease-free survival (DFS) was significantly increased for patients who received CMFPrTam, compared with those who received endocrine therapy alone (p=0.01) and those who received no adjuvant treatment after mastectomy (p<0.0001). DFS was significantly longer for CMFPrTam-treated patients compared with the observation group in all subpopulations. For patients with ER-positive tumors, the DFS of the CMFPrTam group was similar to that of the PrTam group. For patients with ER-negative tumors, DFS was significantly longer for CMFPrTam-treated patients than for PrTam and for observation patients whose DFS were similar. CMFPrTam-treated patients had a lower incidence of failure in the mastectomy scar and in regional and distant sites. At 4 years' median follow-up, differences in overall survival (OS) among the treatment groups were not significant for the entire population or for subpopulations with ER-positive or ER-negative tumors. With 37, 50 and 50 deaths in the three treatment groups, respectively, pairwise absolute differences in OS of 14% (e.g., 65%-79% 4-year rates) could have been detected with a power of 80%. Further follow-up is needed to ascertain whether an early DFS difference will translate into an OS benefit.

At a further follow-up of 6 years' median, OS for the CMFPrTam-treated group has become statistically significantly greater than for the observation group (p=0.04).

PUBLICATIONS:
Ludwig Breast Cancer Study Group: Randomized trial of chemo-endocrine therapy, endocrine therapy, and mastectomy alone in postmenopausal patients with operable breast cancer and axillary node metastases. Lancet 1984; 1:1256-1260.

Goldhirsch A and Gelber R for the Ludwig Breast Cancer Study Group: Adjuvant treatment for early breast cancer: the Ludwig Breast Cancer Studies. NCI Monographs 1986; 1:55-70.

Table 1. Outcome by allocated treatment for trial 78K3 (Ludwig III)
Key: T = total free of relevant event at start of relevant year(s)

(i) Death (D) in each separate year,
subdivided by age when randomized (D <50 & D 50+ / T <50 & T 50+)

Year	1. TamPr	2. TamCMFPr	3. Nil
1	0+ 2/ 4+ 160	0+ 9/ 2+ 169	0+ 6/ 5+ 163
2	1+ 18/ 4+ 155	1+ 11/ 2+ 151	0+ 14/ 5+ 152
3	3+ 21/ 3+ 132	0+ 12/ 1+ 137	1+ 13/ 5+ 137
4	0+ 5/ 0+ 95	0+ 3/ 1+ 100	0+ 11/ 4+ 100
5+	0+ 3/ 0+ 54	0+ 3/ 1+ 56	0+ 6/ 3+ 55

(ii) First recurrence or prior death (F) in each separate year,
subdivided by age when randomized (F <50 & F 50+ / T <50 & T 50+)

Year	1. TamPr	2. TamCMFPr	3. Nil

(recurrence data not available)

(iii) Recurrence with survival (R) and Death (D) in all years together,
subdivided by various characteristics when randomized (R & D / T)

Age at Entry	1. TamPr	2. TamCMFPr	3. Nil
40 – 49	0+4/4	0+1/2	0+1/5
50 – 59	0+22/79	0+23/87	0+31/91
60 – 69	0+26/79	0+15/82	0+19/72
70 +	0+1/2	0+0/0	0+0/0
Any age (including "not known")	0+53/164	0+39/171	0+51/168

Nodal Status (N0, N1–3, N4+)

		1. TamPr	2. TamCMFPr	3. Nil
< 50	N1–N3 (by axillary clearance)	0+2/2	0+1/2	0+0/4
50 +		0+19/85	0+17/98	0+20/90
< 50	N4+ (by axillary sample/clearance)	0+2/2	0+0/0	0+1/1
50 +		0+30/75	0+21/71	0+30/73

Estrogen Receptors (ER, fmol/mg)

Breakdown by ER status was not available.

TITLE: PREDNISONE AND TAMOXIFEN AS ADJUVANT ENDOCRINE THERAPY IN POSTMENOPAUSAL WOMEN WITH OPERABLE BREAST CANCER AND AXILLARY NODE METASTASES

STUDY NUMBER: 78K4
STUDY TITLE: LUDWIG IV

ABBREVIATED TRIAL NAME: LUDWIG IV
STUDY NUMBER: 78K4

INVESTIGATORS' NAMES AND AFFILIATIONS:
The Ludwig Breast Cancer Study Group.

TRIAL DESIGN: TREATMENT GROUPS:
1. Tam + Pr: Tamoxifen 20 mg, and prednisone 7.5 mg, orally, daily for 1 year.
2. Nil: No adjuvant therapy.

ELIGIBILITY CRITERIA FOR ENTRY:
1. AGE: 66-80 years.
2. CLINICAL/PATHOLOGICAL STAGES: Histologically confirmed, non-inflammatory unilateral breast carcinoma (T1a-b, T2a-b, or T3a; N0-1; M0) with positive axillary lymph nodes.
3. PRIMARY TREATMENT: Total mastectomy and axillary clearance.
4. NODE DISSECTION: Mandatory.
5. NUMBER OF NODES REMOVED: Not specified.
6. MENSTRUAL STATUS: Postmenopausal.
7. RECEPTOR STATUS: Not specified.

METHOD OF RANDOMIZATION:
Telephone or telex to the Study Coordination Center. Randomization was carried out by means of pseudorandom numbers generated by a congruence method on a DEC 2060 computer.

RANDOMIZATION STRATIFICATIONS:
Participating institutions.

YEARS OF ACCRUAL: 7/78-8/81.

TOTAL NUMBER OF PATIENTS RANDOMIZED: 349.

ADDITIONAL ELIGIBILITY OR STRATIFICATION CRITERIA:
A chest X-ray and a bone scan were required for exclusion of detectable metastatic disease. Other pre-requisites were a WBC count of ≥4,000, a platelet count of ≥100,000, adequate liver and renal function. Therapy initiated within 6 weeks after surgery.

FOLLOW-UP, COMPLIANCE AND TOXICITY:
Clinical, hematologic and biochemical assessments were required every 3 months for 2 years and every 6 months thereafter. Chest X-rays and bone scans were required every 6 months. After 2 years, a bone scan was required annually.

COMPLIANCE BEFORE FIRST RECURRENCE:
About 3% of the observation group received additional treatment for breast cancer without evidence of relapse. About 3.8% of the tamoxifen plus prednisone group received less than half of their scheduled treatments.

DISCUSSION:
At a median follow-up of 4 years, disease-free survival was significantly increased for women who received prednisone and tamoxifen, compared with those who received no adjuvant therapy (51% vs 42% p=0.02).

Analysis of sites of first failure indicated reduction in local, regional, and contralateral breast relapses in the prednisone and tamoxifen group. In a similar analysis confined to the patients with known estrogen receptor status, the reduction of these recurrences was seen exclusively in the group of patients with estrogen receptor positive tumors.

PUBLICATIONS:
Gelber RD. Adjuvant therapy for postmenopausal women with operable breast cancer. Part II. Randomized trials comparing endocrine therapy with surgery alone. In: Adjuvant Therapy of Cancer IV (Jones SE, Salmon SE, Eds.) Orlando: Grune & Stratton, 1984, pp 393-404.

Ludwig Breast Cancer Study Group: Randomized trial of chemo-endocrine therapy, endocrine therapy, and mastectomy alone in postmenopausal patients with operable breast cancer and axillary node metastases. Lancet 1984; 1:1256-1260.

Goldhirsch A and Gelber R for the Ludwig Breast Cancer Study Group: Adjuvant treatment for early breast cancer: the Ludwig Breast Cancer Studies. NCI Monographs 1986; 1:55-70.

Table 1. Outcome by allocated treatment for trial 78K4 (Ludwig IV)
Key: T = total free of relevant event at start of relevant year(s)

(i) Death (D) in each separate year,
subdivided by age when randomized (D <50 & D 50+ /T <50 & T 50+)

Year	1. TamPr	2. Nil
1	0+ 13/ 0+ 182	0+ 6/ 0+ 167
2	0+ 11/ 0+ 163	0+ 10/ 0+ 157
3	0+ 15/ 0+ 151	0+ 18/ 0+ 145
4	0+ 7/ 0+ 110	0+ 6/ 0+ 115
5+	0+ 5/ 0+ 63	0+ 6/ 0+ 68

(ii) First recurrence or prior death (F) in each separate year,
subdivided by age when randomized (F <50 & F 50+ /T <50 & T 50+)

Year	1. TamPr	2. Nil

(recurrence data not available)

(iii) Recurrence with survival (R) and Death (D) in all years together,
subdivided by various characteristics when randomized (R & D /T)

Age at Entry	1. TamPr	2. Nil
50 – 59	0+0/0	0+0/1
60 – 69	0+14/70	0+22/81
70 +	0+37/112	0+24/85
Any age (including "not known")	0+51/182	0+46/167

Nodal Status (N0, N1-3, N4+)

		1. TamPr	2. Nil
50 +	N1–N3 (by axillary clearance)	0+22/114	0+20/103
50 +	N4+ (by axillary sample/clearance)	0+29/68	0+26/64

Estrogen Receptors (ER, fmol/mg)

Breakdown by ER status was not available.

TITLE: GHENT STUDY OF TAMOXIFEN IN NODE NEGATIVE BREAST CANCER

ABBREVIATED TRIAL NAME: GHENT UNIV
STUDY NUMBER: 79A

INVESTIGATORS' NAMES AND AFFILIATIONS:
A De Schryver, J Huys, and L Vakaet of University Hospital, Ghent, Belgium.

TRIAL DESIGN: TREATMENT GROUPS:
1. Tam: Tamoxifen 20 mg, orally, twice daily for 24 months.
2. Nil: No adjuvant therapy.

ELIGIBILITY CRITERIA FOR ENTRY:
1. AGE: < 75.
2. CLINICAL STAGES: All except M1 (distant metastases), N2 (fixed axillary lymph nodes), N3 (supraclavicular nodes), T4, or inflammatory tumors.
3. PRIMARY TREATMENT: For all patients, radical mastectomy (Halsted or Patey procedures); for medial tumors, radiotherapy (50 Gy) to the internal mammary chain and supraclavicular fossa. Oophorectomy for pre-menopausal women (less than two years after last menstrual bleeding).
4. AXILLARY NODE DISSECTION: Complete axillary dissection.
5. NUMBER OF NODES REMOVED: About 14 on average, but not all were examined.
6. AXILLARY NODE INVOLVEMENT: Negative.

METHOD OF RANDOMIZATION:
A list was kept by a secretary. Allocation was based on a table of random numbers.

RANDOMIZATION STRATIFICATIONS:
Premenopausal, postmenopausal.

YEARS OF ACCRUAL: 1978-1984.

TOTAL NUMBER OF PATIENTS RANDOMIZED: 85.

AIM, RATIONALE, SPECIAL FEATURES:
This study is part of another study which also included a comparison of tamoxifen with chemotherapy for axillary node positive patients.

Extent of disease was initially assessed by clinical examination with TNM staging, peripheral blood counts, chest X-ray, axial bone scan, liver (SGOT, SGPT, LDH, Alk.Phos.) and kidney (creatinine, urea) function tests.

ADDITIONAL ELIGIBILITY OR STRATIFICATION CRITERIA:
Men and pregnant or lactating women were excluded. Patients with the following were also excluded: abnormal liver function studies; blood urea nitrogen >25 mg/100 ml; peripheral blood cell counts <100,000 platelets or <4000 leucocytes per mm^3; a history of cancer of another site; another life-threatening disease.

FOLLOW-UP, COMPLIANCE AND TOXICITY:
A physical examination was performed every 3 months for the first 2 years, every 6 months for the next 3 years, and once a year after 5 years.

There was no problem with compliance.

PUBLICATIONS: None.

Table 1. Outcome by allocated treatment for trial 79A (Ghent Univ.)
Key: T = total free of relevant event at start of relevant year(s)

(i) Death (D) in each separate year,
subdivided by age when randomized (D <50 & D 50+ / T <50 & T 50+)

Year	1. Tam	2. Nil
1	0+ 1/ 0+ 39	0+ 1/ 0+ 46
2	0+ 3/ 0+ 35	0+ 3/ 0+ 45
3	0+ 1/ 0+ 29	0+ 1/ 0+ 34
4	0+ 0/ 0+ 23	0+ 1/ 0+ 28
5+	0+ 0/ 0+ 18	0+ 2/ 0+ 20

(ii) First recurrence or prior death (F) in each separate year,
subdivided by age when randomized (F <50 & F 50+ / T <50 & T 50+)

(recurrence data not available)

(iii) Recurrence with survival (R) and Death (D) in all years together,
subdivided by various characteristics when randomized (R & D / T)

Age at Entry		1. Tam	2. Nil
Breakdown by age was not available.			
Any age (including "not known")		0+5/39	0+8/46
	Nodal Status (N0, N1-3, N4+)		
50+	N0 (by clinical evidence only)	0+5/39	0+8/46
	Estrogen Receptors (ER, fmol/mg)		
Breakdown by ER status was not available.			

TITLE: CHEMOENDOCRINE THERAPY IN ER-POSITIVE POST-MENOPAUSAL PATIENTS WITH OPERABLE BREAST CANCER AND POSITIVE AXILLARY NODES: A SOUTHWEST ONCOLOGY GROUP TRIAL

STUDY NUMBER 79B1
STUDY TITLE: SWOG 7827A

ABBREVIATED TRIAL NAME: SWOG 7827A
STUDY NUMBER: 79B1

INVESTIGATOR NAMES AND AFFILIATIONS:
CK Osborne (1), SE Rivkin (2), SJ Green (3), WA Knight (1), R McDivitt (4), T Cruz (5), and D Tesh (2).

  (1) Department of Medicine, University of Texas Health Science Center, San Antonio, TX.
  (2) Tumor Institute of Swedish Hospital Medical Center, Seattle, WA
  (3) SWOG Statistical Center, Fred Hutchinson Cancer Center, Seattle, WA.
  (4) Department of Pathology, Barnes Hospital, St. Louis, MO
  (5) Department of Surgery, University of Texas Health Science Center, San Antonio, TX.

TRIAL DESIGN: TREATMENT GROUPS:
1. Tam: Tamoxifen 10 mg, orally, twice daily for 12 months.
2. CMFVPr: Cyclophosphamide 60 mg/M^2, orally, daily for 12 months; methotrexate 15 mg/M^2 and 5-fluorouracil 400 mg/M^2, intravenously, weekly, for 12 months; and vincristine 0.625 mg/M^2, intravenously, weekly for 10 weeks. Prednisone 30 mg/M^2, orally, days 1-14, 20 mg/M^2 orally, days 15-28, and 10 mg/M^2 orally, days 29-42.
3. Tam + CMFVPr: Tamoxifen, as above, plus CMFVPr, as above.

ELIGIBILITY CRITERIA FOR ENTRY:
1. AGE: No restrictions.
2. CLINICAL STAGES: T1-3a, N1, M0
3. PRIMARY TREATMENT: Radical or modified radical mastectomy with or without postoperative radiation, or partial mastectomy with axillary dissection and radiation.
4. NODE DISSECTION: Some form of axillary surgery required.
5. MENSTRUAL STATUS: Postmenopausal.
6. ESTROGEN RECEPTOR STATUS: Positive.

METHOD OF RANDOMIZATION:
By telephone to the Statistical Center. A computerized dynamic balancing algorithm is used to randomize patients equally to the treatment arms. Treatment assignment is balanced on the stratification factors within major institution groups. Prior to the availability of the computerized system, a permuted block design was used to randomize patients within stratification levels.

RANDOMIZATION STRATIFICATIONS:
Postoperative radiation therapy (yes, no), number of positive axillary nodes (1-3, ≥4), tumor size (<2 cm, 2-5 cm, > 5 cm).

YEARS OF ACCRUAL: 1979 - present.

TOTAL NUMBER OF PATIENTS: Still open to accrual. 639 patients had been recruited at the time of the 1985 EBCTCG overview.

ADDITIONAL ELIGIBILITY OR STRATIFICATION CRITERIA:
Adequate renal and liver function was required prior to entry. It was also required that therapy be initiated within 42 days of mastectomy.

PUBLICATIONS: None.

Table 1. Outcome by allocated treatment for trial 79B1 (SWOG 7827 A)

The results from this trial have been blinded.

TITLE: CHEMOTHERAPY ALONE VERSUS CHEMOTHERAPY PLUS SURGICAL OOPHORECTOMY IN PREMENOPAUSAL ER-POSITIVE PATIENTS WITH OPERABLE BREAST CANCER AND POSITIVE AXILLARY NODES

STUDY NUMBER: 79B2
STUDY TITLE: SWOG 7827B

ABBREVIATED TRIAL NAME: SWOG 7827B
STUDY NUMBER: 79B2

INVESTIGATOR NAMES AND AFFILIATIONS:
CK Osborne (1), SE Rivkin (2), SJ Green (3), WA Knight (1), R McDivitt (4), T Cruz (5), and D Tesh (2).

(1) Department of Medicine, University of Texas Health Center, San Antonio, TX.
(2) Tumor Institute of Swedish Hospital Medical Center, Seattle, WA.
(3) SWOG Statistical Center, Fred Hutchinson Cancer Center, Seattle, WA.
(4) Department of Pathology, Barnes Hospital, St. Louis, MO.
(5) Department of Surgery, University of Texas Health Science Center, San Antonio, TX.

TRIAL DESIGN: TREATMENT GROUPS:

1. CMFVPr: Cyclophosphamide 60 mg/M^2, orally, daily for 12 months; methotrexate 15 mg/M^2 and 5-fluorouracil 400 mg/M^2, intravenously, weekly for 12 months; vincristine 0.625 mg/M^2 intravenously, weekly for 10 weeks. Prednisone 30 mg/M^2 orally, days 1-14; 20 mg/M^2 orally, days 15-28; and 10 mg/M^2 orally, days 29-42.
2. CMFVPr + Ooph: CMFVPr, as above, plus surgical oophorectomy.

ELIGIBILITY CRITERIA FOR ENTRY:
1. AGE: No restrictions.
2. CLINICAL STAGES: T1-3a, N1, M0.
3. PRIMARY TREATMENT: Radical or modified radical mastectomy with or without post-operative radiation or partial mastectomy with axillary dissection and radiation.
4. NODE DISSECTION: Some form of axillary surgery required.
5. MENSTRUAL STATUS: Premenopausal.
6. ESTROGEN RECEPTOR STATUS: Positive.

METHOD OF RANDOMIZATION:
By telephone to the Statistical Center. A computerized dynamic balancing algorithm is used to randomize patients equally to the treatment arms. Treatment assignment is balanced on the stratification factors within major institution groups. Prior to the availability of the computerized system, a permuted block design was used to randomize patients within stratification levels.

RANDOMIZATION STRATIFICATIONS:
Postoperative radiation therapy (yes or no), number of positive axillary nodes (1-3 vs ≥4), tumor size (<2 cm, 2-5 cm, >5 cm).

YEARS OF ACCRUAL: 1979 - study still accruing. 176 patients had been recruited at the time of the 1985 EBCTCG overview.

TOTAL NUMBER OF PATIENTS: Still accruing.

ADDITIONAL ELIGIBILITY OR STRATIFICATION CRITERIA:
Adequate renal and liver function was required prior to entry. It was also required that therapy be initiated within 42 days of mastectomy.

PUBLICATIONS: None.

Table 1. Outcome by allocated treatment for trial 79B2 (SWOG 7827 B)

The results from this trial have been blinded.

TITLE: RANDOMIZED TRIAL OF CMFVP CHEMOTHERAPY FOR EITHER 1 OR 2 YEARS IN ER NEGATIVE PATIENTS WITH OPERABLE BREAST CANCER AND POSITIVE AXILLARY NODES

STUDY NUMBER: 79B3
STUDY TITLE: SWOG 7827C

ABBREVIATED TRIAL NAME: SWOG 7827C
STUDY NUMBER: 79B3

INVESTIGATOR NAMES AND AFFILIATIONS:
CK Osborne (1), SE Rivkin (2), SJ Green (3), WA Knight (1), R McDivitt (4), T Cruz (5), and D Tesh (2).

(1) Department of Medicine, University of Texas Health Science Center, San Antonio, TX.
(2) Tumor Institute of Swedish Hospital Medical Center, Seattle, WA.
(3) SWOG Statistical Center, Fred Hutchinson Cancer Center, Seattle, WA.
(4) Department of Pathology, Barnes Hospital, St. Louis, MO.
(5) Department of Surgery, University of Texas Health Science Center, San Antonio, TX.

TRIAL DESIGN: TREATMENT GROUPS:
1. CMFVPr: Cyclophosphamide 60 mg/M^2, orally, daily for 12 months; methotrexate 15 mg/M^2 and 5-fluorouracil 400 mg/M^2, intravenously, weekly for 12 months; vincristine 0.625 mg/M^2 intravenously, weekly for 10 weeks. Prednisone 30 mg/M^2 orally, days 1-14; 20 mg/M^2 orally, days 15-28; and 10 mg/M^2, orally, days 29-42.
2. CMFVPr: CMFVPr, as above, for 24 months.

ELIGIBILITY CRITERIA FOR ENTRY:
1. AGE: No restrictions.
2. PATHOLOGIC STAGES: T1-3a, N1, M0.
3. PRIMARY TREATMENT: Radical or modified radical mastectomy with or without post-operative radiation, or partial mastectomy with axillary dissection and radiation.
4. NODE DISSECTION: Some form of axillary surgery required.
5. MENOPAUSAL STATUS: Premenopausal or postmenopausal.
6. ESTROGEN RECEPTOR STATUS: ER-negative only.

METHOD OF RANDOMIZATION:
By telephone to the Statistical Center. A permuted block design was used to randomize patients within stratification levels, with randomization assignment overridden when necessary to maintain institution balance.

RANDOMIZATION STRATIFICATIONS:
Postoperative radiation therapy (yes, no); number of positive axillary nodes (1-3, ≥4); tumor size (<2 cm, 2-5 cm, >5 cm).

YEARS OF ACCRUAL: 1979-1984.

TOTAL NUMBER OF PATIENTS: 444.

ADDITIONAL ELIGIBILITY OR STRATIFICATION CRITERIA:
Adequate renal and liver function was required prior to entry. It was also required that therapy be initiated within 42 days of mastectomy.

PUBLICATIONS: None.

Table 1. Outcome by allocated treatment for trial 79B3 (SWOG 7827 C)

The results from this trial have been blinded.

TITLE: CASE WESTERN RESERVE UNIVERSITY MULTI-INSTITUTIONAL STUDY ON EVALUATION OF INTENSE 5-DRUG CHEMOTHERAPY AND/OR ANTIESTROGEN THERAPY OF PREMENOPAUSAL PATIENTS WITH STAGE II, III CARCINOMA OF THE BREAST

STUDY NUMBER: 79C1
STUDY TITLE: CASE WESTERN B

ABBREVIATED TRIAL NAME: CASE WESTERN B
STUDY NUMBER: 79C1

INVESTIGATORS' NAMES AND AFFILIATIONS:
CA Hubay (1), OH Pearson (3), JS Marshall (3), JP Crowe (1), NH Gordon (1,2).

(1) Department of Surgery, Case Western Reserve University, School of Medicine, Cleveland, OH.
(2) Department of Epidemiology and Biostatistics, Case Western Reserve School of Medicine, Cleveland, OH.
(3) Department of Medicine, Case Western Reserve University, School of Medicine, Cleveland, OH.

TRIAL DESIGN: TREATMENT GROUPS:
1. Ooph + Tam: Oophorectomy after mastectomy with tamoxifen 20 mg, twice daily, orally for 3 years.
2. Ooph + Tam + CMFVPr: Oophorectomy and tamoxifen, as above, plus cyclophosphamide cyclophosphamide 1.5 mg/kg, orally, daily for 1 year; prednisone 1.0 mg/kg orally, daily, for 10 days, 0.5 mg/kg orally, daily for 10 days, 0.25 mg/kg orally, daily, for 10 days, then stopped. Vincristine, 15 mg/kg intravenously, weekly for 6 weeks, then every sixth week for 1 year; 5-fluorouracil, 7 mg/kg and methotrexate, 0.36 mg/kg, intravenously, weekly first three months, biweekly second three months, triweekly for last 6 months. Dose of 5-fluorouracil + methotrexate escalated as tolerated until WBC <4000/mm^3.

ELIGIBILITY CRITERIA FOR ENTRY:
1. AGE: Not specified (premenopausal).
2. CLINICAL/PATHOLOGICAL STAGES: All except M1 (distant metastases), and N0.
3. PRIMARY TREATMENT: Radical or modified radical mastectomy.
4. AXILLARY NODE DISSECTION: Yes.
5. NUMBER OF NODES REMOVED: Not specified.
6. MENSTRUAL STATUS: Premenopausal.
7. RECEPTOR STATUS: Positive (>3 fmol/mg).

METHOD OF RANDOMIZATION:
By envelopes, prepared and sealed by the department of biostatistics.

RANDOMIZATION STRATIFICATIONS:
None specified.

YEARS OF ACCRUAL: 1980-1985.

TOTAL NUMBER OF PATIENTS RANDOMIZED: 41.

FOLLOW-UP, COMPLIANCE AND TOXICITY:
Patients underwent physical examination at 3 month intervals, chest X-ray, blood counts and chemistries at 6 month intervals and bone scan and mammography at 12 month intervals. Of the 24 patients randomized to oophorectomy plus tamoxifen, one had poor, one had fair, and 22 had good compliance. Of the 17 patients randomized to endocrine therapy plus chemotherapy, all had good compliance.

PUBLICATIONS: None.

Table 1. Outcome by allocated treatment for trial 79C1 (Case Western B)
Key: T = total free of relevant event at start of relevant year(s)

(i) Death (D) in each separate year,
subdivided by age when randomized (D <50 & D 50+ / T <50 & T 50+)

Year	1. TamOoph	2. TamCMFVPrOoph
1	0+ 0/ 19+ 5	1+ 0/ 17+ 0
2	0+ 0/ 17+ 4	1+ 0/ 16+ 0
3	0+ 0/ 15+ 2	0+ 0/ 12+ 0
4	0+ 0/ 10+ 1	0+ 0/ 6+ 0
5+	0+ 0/ 8+ 1	0+ 0/ 3+ 0

(ii) First recurrence or prior death (F) in each separate year,
subdivided by age when randomized (F <50 & F 50+ / T <50 & T 50+)

Year	1. TamOoph	2. TamCMFVPrOoph
1	1+ 0/ 19+ 5	1+ 0/ 17+ 0
2	2+ 0/ 16+ 3	2+ 0/ 16+ 0
3	0+ 0/ 13+ 2	1+ 0/ 9+ 0
4	1+ 0/ 9+ 1	0+ 0/ 2+ 0
5+	0+ 0/ 5+ 1	0+ 0/ 1+ 0

(iii) Recurrence with survival (R) and Death (D) in all years together,
subdivided by various characteristics when randomized (R & D / T)

Age at Entry	1. TamOoph	2. TamCMFVPrOoph
< 40	0+0/1	1+1/5
40 - 49	4+0/18	1+1/12
50 - 59	0+0/5	0+0/0
Any age (including "not known")	4+0/24	2+2/17

	Nodal Status (N0, N1-3, N4+)		
< 50	N1-N3 (by axillary	2+0/11	0+1/6
50 +	clearance)	0+0/3	0+0/0
< 50	N4+ (by axillary	2+0/8	2+1/11
50 +	sample/clearance)	0+0/2	0+0/0

	Estrogen Receptors (ER, fmol/mg)		
< 50	ER poor or ER < 10	1+0/2	0+1/4
50 +		0+0/2	0+0/0
< 50	ER+ or ER 10-99	3+0/16	2+1/12
50 +		0+0/2	0+0/0
< 50	ER++ or ER 100+	0+0/1	0+0/1
50 +		0+0/1	0+0/0

TITLE: CASE WESTERN RESERVE UNIVERSITY MULTI-INSTITUTIONAL STUDY ON EVALUATION OF INTENSIVE 5-DRUG CHEMOTHERAPY AND/OR ANTIESTROGEN THERAPY OF POSTMENOPAUSAL PATIENTS WITH STAGE II, III CARCINOMA OF THE BREAST

STUDY NUMBER: 79C2
STUDY TITLE: CASE WESTERN B

ABBREVIATED TRIAL NAME: CASE WESTERN B
STUDY NUMBER: 79C2

INVESTIGATORS' NAMES AND AFFILIATIONS:
CA Hubay (1), OH Pearson (3), JS Marshall (3), JP Crowe (1), NH Gordon (1,2).

- (1) Department of Surgery, Case Western Reserve University, School of Medicine, Cleveland, OH.
- (2) Department of Epidemiology and Biostatistics, Case Western Reserve School of Medicine, Cleveland, OH.
- (3) Department of Medicine, Case Western Reserve University, School of Medicine, Cleveland, OH.

TRIAL DESIGN: TREATMENT GROUPS:
1. Tam: Tamoxifen 20 mg, twice daily, orally for 3 years.
2. Tam + CMFVPr: Tamoxifen, as above, plus cyclophosphamide 1.5 mg/kg orally, daily for 1 year; prednisone 1.0 mg/kg orally, daily, for 10 days, 0.5 mg/kg orally, daily for 10 days, 0.25 mg/kg orally, daily, for 10 days, then stopped; vincristine 15 mg/kg, intravenously, weekly for 6 weeks, then every sixth week for 1 year; 5-fluorouracil 7 mg/kg and methotrexate, 0.36 mg/kg, intravenously, weekly first three months, bi-weekly second three months, tri-weekly for last 6 months. Dose of 5-fluorouracil + methotrexate escalated as tolerated until WBC <4000 /mm^3.

ELIGIBILITY CRITERIA FOR ENTRY:
1. AGE: < 76 (postmenopausal).
2. CLINICAL/PATHOLOGICAL STAGES: All except M1 (distant metastases) or N0.
3. PRIMARY TREATMENT: Radical or modified radical mastectomy.
4. AXILLARY NODE DISSECTION: Yes.
5. NUMBER OF NODES REMOVED: Not specified.
6. MENSTRUAL STATUS: Postmenopausal.
7. RECEPTOR STATUS: Estrogen receptor positive (>3 fmol/mg).

METHOD OF RANDOMIZATION:
By envelopes, prepared and sealed by department of biostatistics.

YEARS OF ACCRUAL: 1980-1985.

TOTAL NUMBER OF PATIENTS RANDOMIZED: 92.

FOLLOW-UP, COMPLIANCE AND TOXICITY:
Patients underwent physical examination at 3 month intervals, chest X-ray, blood counts and chemistries at 6 month intervals and bone scan and mammography at 12 month intervals. Determination of death was by patient records and/or death certificate.

Of the 47 patients randomized to tamoxifen alone, two had poor and 45 had good compliance. Of the 45 patients randomized to tamoxifen plus chemotherapy, 7 had poor, 3 had fair, and 35 had good compliance.

PUBLICATIONS:
Pearson OH, Hubay CA, Gordon NH, Marshall JS, Crowe JP, Arafah B, McGuire WL. Endocrine versus endocrine plus 5-drug chemotherapy in postmenopausal women with stage II estrogen receptor positive breast cancer. Cancer 1989; 64:1819-1823.

Table 1. Outcome by allocated treatment for trial 79C2 (Case Western B)
Key: T = total free of relevant event at start of relevant year(s)

(i) Death (D) in each separate year, subdivided by age when randomized (D <50 & D 50+ / T <50 & T 50+)

Year	1. Tam	2. TamCMFVPr
1	0+ 0/ 1+ 46	0+ 2/ 2+ 43
2	0+ 2/ 1+ 43	0+ 1/ 1+ 39
3	0+ 4/ 1+ 34	0+ 1/ 1+ 30
4	0+ 1/ 0+ 18	0+ 3/ 1+ 19
5+	0+ 1/ 0+ 13	0+ 0/ 1+ 9

(ii) First recurrence or prior death (F) in each separate year, subdivided by age when randomized (F <50 & F 50+ / T <50 & T 50+)

Year	1. Tam	2. TamCMFVPr
1	0+ 4/ 1+ 46	0+ 4/ 2+ 43
2	1+ 5/ 1+ 37	0+ 0/ 1+ 33
3	0+ 3/ 0+ 24	0+ 3/ 1+ 26
4	0+ 2/ 0+ 12	0+ 1/ 1+ 16
5+	0+ 2/ 0+ 6	0+ 0/ 1+ 7

(iii) Recurrence with survival (R) and Death (D) in all years together, subdivided by various characteristics when randomized (R & D / T)

Age at Entry		1. Tam	2. TamCMFVPr
40 - 49		1+0/1	0+0/2
50 - 59		4+4/17	1+1/20
60 - 69		4+3/24	0+4/17
70 +		0+1/5	0+2/6
Any age (including "not known")		9+8/47	1+7/45

	Nodal Status (N0, N1-3, N4+)		
< 50	N1-N3 (by axillary	0+0/0	0+0/1
50 +	clearance)	2+4/25	1+1/28
< 50	N4+ (by axillary	1+0/1	0+0/1
50 +	sample/clearance)	6+4/21	0+6/15

	Estrogen Receptors (ER, fmol/mg)		
< 50	ER poor or ER < 10	1+0/1	0+0/2
50 +		1+0/3	0+2/7
50 +	ER+ or ER 10-99	5+5/25	0+4/19
50 +	ER++ or ER 100+	2+3/18	1+1/17

TITLE: ADJUVANT CHEMOTHERAPY ± TAMOXIFEN IN NODAL POSITIVE BREAST CANCER TREATED BY THE GYNECOLOGICAL ADJUVANT BREAST GROUP (GABGI)

STUDY NUMBER: 79D1
STUDY TITLE: GABG WEST GERMANY

ABBREVIATED TRIAL NAME: GABG W GERMANY
STUDY NUMBER: 79D1

INVESTIGATORS' NAMES AND AFFILIATIONS:
M Kaufmann (1), W Jonat (2), F Kubli (1), J Maass (2), H Caffier (3), R Kreienberg (4), J Hilfrich (5), W Kleine (6), K Brunnert (7), M Mahlke (8), U Abel (9).

(1) Dept. Obstet. a. Gynecol., University Hospital Heidelberg.
(2) Dept. Obstet. a Gynecol., University Hospital Hamburg.
(3) Dept. Obstet. a. Gynecol., University Hospital Würzburg.
(4) Dept. Obstet. a. Gynecol., University Hospital Mainz.
(5) Dept. Obstet. a. Gynecol., University Hospital Hannover.
(6) Dept. Obstet. a. Gynecol., University Hospital Freiburg.
(7) Women's Hospital, Community Hospital Karlsruhe.
(8) Women's Hospital, Community Hospital Krefeld.
(9) Inst. Statistics a. Documentation, German Cancer Research Center, Heidelberg.

TRIAL DESIGN: TREATMENT GROUPS:
1. AC: Doxorubicin 30 mg/M^2 and cyclophosphamide 300 mg/M^2, intravenously, day 1 and 8 every 3 weeks for 8 cycles.
2. AC + Tam: AC, as above, plus tamoxifen 30 mg, orally, daily for 2 years.

ELIGIBILITY CRITERIA FOR ENTRY:
1. AGE: ≤65 years.
2. CLINICAL /PATHOLOGICAL STAGES: Histologically proven invasive operable breast carcinoma (T1-3,N1-3,M0) with ≥1 positive axillary lymph node.
3. PRIMARY TREATMENT: Modified radical mastectomy without radiotherapy.
4. AXILLARY NODE DISSECTION: Level I, II, and III mandatory.
5. NUMBER OF NODES REMOVED: Minimum of 10.
6. MENSTRUAL STATUS: No restriction.
7. RECEPTOR STATUS: Negative (<20 fmol/mg of cytosol protein).

METHOD OF RANDOMIZATION:
Randomization was done by envelope and was separate for each hospital. Allocations were based on random number tables and stored in the central office. Treatment allocations were balanced in blocks of 5 for each hospital.

RANDOMIZATION STRATIFICATIONS:
Participating institutions.

YEARS OF ACCRUAL: 1/81-6/85.

TOTAL NUMBER OF PATIENTS RANDOMIZED: 377.

ADDITIONAL ELIGIBILITY OR STRATIFICATION CRITERIA:
The disease was assessed by clinical examination with TNM staging, chest X-ray, bone scintigraphy, liver ultrasound and complete blood chemistry profile.

Biochemical steroid hormone receptor analyses were performed according to EORTC criteria.

No dose reductions were planned. Chemotherapy was initiated within 21 days of surgery.

FOLLOW-UP, COMPLIANCE AND TOXICITY:
Clinical assessments with complete blood chemistry were done every 3 months for the first 2 years. Thereafter follow-up was every 6 months until the fifth year, then annually. Chest X-rays, bone scintigraphy and liver ultrasound were also performed every 6 months for the first 2 years, then annually.

COMPLIANCE BEFORE FIRST RECURRENCE:
About 75% of the patients received the scheduled cytotoxic treatment and more than 95% of the women received tamoxifen for 2 years.

PUBLICATIONS:
Kaufmann M, Jonat W, Caffier H, Hilfrich J, Melchert F, Mahlke M, Abel U, Maass H, Kubli F for the Gynecological Adjuvant Breast Cancer Study Group. Adjuvant chemohormonotherapy selected by axillary node and hormone receptor status in node-positive breast cancer. Rev. on Endocrine-Rel. Cancer 1985; 17s:57-63.

Table 1. Outcome by allocated treatment for trial 79D1 (GABG I Germany)
Key: T = total free of relevant event at start of relevant year(s)

(i) Death (D) in each separate year,
subdivided by age when randomized (D <50 & D 50+ / T <50 & T 50+)

Year	1. AC	2. TamAC
1	2+ 3/ 89+ 99	4+ 3/ 75+ 114
2	4+ 11/ 73+ 76	4+ 2/ 63+ 84
3	3+ 5/ 47+ 38	3+ 5/ 37+ 57
4	3+ 1/ 21+ 12	2+ 3/ 20+ 24
5+	0+ 0/ 9+ 4	0+ 0/ 7+ 5

(ii) First recurrence or prior death (F) in each separate year,
subdivided by age when randomized (F <50 & F 50+ / T <50 & T 50+)

Year	1. AC	2. TamAC
1	7+ 17/ 89+ 99	8+ 12/ 75+ 114
2	12+ 16/ 56+ 51	9+ 9/ 54+ 60
3	4+ 2/ 28+ 11	2+ 4/ 23+ 29
4	0+ 0/ 14+ 4	0+ 3/ 11+ 9
5+	1+ 0/ 5+ 1	0+ 0/ 4+ 1

(iii) Recurrence with survival (R) and Death (D) in all years together,
subdivided by various characteristics when randomized (R & D / T)

Age at Entry		1. AC	2. TamAC
< 40		4+4/24	4+4/27
40 - 49		8+8/65	2+9/48
50 - 59		10+13/61	7+9/80
60 - 69		4+7/36	8+4/34
70 +		1+0/2	0+0/0
Any age (including "not known")		27+32/188	21+26/189

	Nodal Status (N0, N1-3, N4+)		
<50	N4+ (by axillary	10+11/67	5+10/61
50 +	sample/clearance)	13+15/79	11+12/93
<50	Remainder (i.e.	2+1/22	1+3/14
50 +	other N+ or N?)	2+5/20	4+1/21

	Estrogen Receptors (ER, fmol/mg)		
<50	ER poor or ER < 10	9+7/50	6+9/40
50 +		5+13/46	9+7/56
<50	ER+ or ER 10-99	3+5/30	0+4/28
50 +		6+5/27	4+3/33
<50	ER++ or ER 100+	0+0/9	0+0/7
50 +		4+2/26	2+3/25

TITLE: HEIDELBERG STUDY II OF ADJUVANT ENDOCRINE THERAPY IN POTENTIALLY CURABLE POSTMENOPAUSAL BREAST CANCER

STUDY NUMBER: 79D2
STUDY TITLE: HD2 W GERMANY

ABBREVIATED TRIAL NAME: HD2 W GERMANY
STUDY NUMBER: 79D2

INVESTIGATORS' NAMES AND AFFILIATIONS:
M Kaufmann (1), H Schmid (1), K Brunnert (2), F Kubli (1).

(1) Dept. Obstet. a. Gynecol., University Hospital Heidelberg.
(2) Women's Hospital, Community Hospital Karlsruhe.

TRIAL DESIGN: TREATMENT GROUPS:
1. Tam: Tamoxifen 30 mg, orally, daily for 2 years.
2. Nil: No adjuvant therapy.

ELIGIBILITY CRITERIA FOR ENTRY:
1. AGE: All ages.
2. CLINICAL PATHOLOGICAL STAGES: Histologically proven invasive operable breast carcinoma (T1-3,N0,M0).
3. PRIMARY TREATMENT: Modified radical mastectomy without radiotherapy.
4. AXILLARY NODE DISSECTION: Level I, II, and III mandatory.
5. NUMBER OF NODES REMOVED: Minimum of 10.
6. MENSTRUAL STATUS: Postmenopausal.
7. RECEPTOR STATUS: Not specified.

METHOD OF RANDOMIZATION:
Randomization was done by envelope and was separate for each hospital. Allocations were based on random number tables and stored in the central office. Treatment allocations were balanced in blocks of 5 for each hospital.

RANDOMIZATION STRATIFICATIONS:
Participating institutions.

YEARS OF ACCRUAL: 8/79-4/85.

TOTAL NUMBER OF PATIENTS RANDOMIZED: 130.

ADDITIONAL ELIGIBILITY OR STRATIFICATION CRITERIA:
The disease was assessed by clinical examination with TNM staging, chest X-ray, bone scintigraphy, liver ultrasound and complete blood chemistry profile.

Biochemical steroid hormone receptor analyses were performed according to EORTC criteria.

No dose reductions were planned. Chemotherapy was initiated within 21 days of surgery.

FOLLOW-UP, COMPLIANCE AND TOXICITY:
Clinical assessments with complete blood chemistry was done every 3 months for the first 2 years. Thereafter follow-up was every 6 months until the fifth year, then annually. Chest X-rays, bone scintigraphy and liver ultrasound were also performed every 6 months for the first 2 years, then annually.

COMPLIANCE BEFORE FIRST RECURRENCE:
More than 95% of the women received tamoxifen for 2 years.

PUBLICATIONS: None.

Table 1. Outcome by allocated treatment for trial 79D2 (HD 2 W. Germany)
Key: T = total free of relevant event at start of relevant year(s)

(i) Death (D) in each separate year, subdivided by age when randomized (D <50 & D 50+ / T <50 & T 50+)

Year	1. Tam	2. Nil
1	0+ 1/ 2+ 65	0+ 1/ 3+ 60
2	0+ 2/ 2+ 61	0+ 0/ 3+ 57
3	0+ 2/ 2+ 52	0+ 6/ 3+ 53
4	0+ 0/ 2+ 40	0+ 0/ 3+ 36
5+	0+ 0/ 1+ 31	1+ 0/ 2+ 25

(ii) First recurrence or prior death (F) in each separate year, subdivided by age when randomized (F <50 & F 50+ / T <50 & T 50+)

Year	1. Tam	2. Nil
1	0+ 1/ 2+ 65	0+ 2/ 3+ 60
2	1+ 3/ 2+ 58	0+ 3/ 3+ 54
3	0+ 2/ 1+ 45	0+ 2/ 3+ 39
4	0+ 0/ 1+ 30	1+ 1/ 2+ 26
5+	0+ 0/ 1+ 21	0+ 0/ 1+ 13

(iii) Recurrence with survival (R) and Death (D) in all years together, subdivided by various characteristics when randomized (R & D / T)

Age at Entry		1. Tam	2. Nil
40 - 49		1+0/2	0+1/3
50 - 59		1+0/20	0+3/17
60 - 69		0+3/34	1+1/28
70 +		0+2/11	0+3/15
Any age (including "not known")		2+5/67	1+8/63

	Nodal Status (N0, N1-3, N4+)		
< 50	N0 (by clinical	1+0/2	0+1/3
50 +	evidence only)	1+5/65	1+7/60

	Estrogen Receptors (ER, fmol/mg)		
< 50	ER poor or ER < 10	0+0/0	0+1/3
50 +		1+3/27	0+4/22
< 50	ER+ or ER 10-99	1+0/2	0+0/0
50 +		0+2/18	0+3/26
50 +	ER++ or ER 100+	0+0/20	1+0/12

Short reports

TITLE: CONTROLLED TRIAL OF ADJUVANT CHEMOTHERAPY WITH CYCLOPHOSPHAMIDE, METHOTREXATE AND 5-FLUOROURACIL FOR BREAST CANCER

STUDY NUMBER: 79E1-2
STUDY TITLE: MANCHESTER II GUYS CMF

ABBREVIATED TRIAL NAME: MANCHESTER II GUY'S CMF,
STUDY NUMBER: 79E1-2

INVESTIGATORS NAMES AND AFFILIATIONS:
A Howell (2), RD Rubens (1), JL Hayward (1), RA Sellwood (2).
(1) Imperial Cancer Research Fund Breast Cancer Unit, Guy's Hospital, London.
(2) Departments of Medical Oncology and Surgery, University Hospital of South Manchester, Manchester.

TRIALS DESIGN: TREATMENT GROUPS:
1. CMF: Cyclophosphamide 80 mg/M^2 orally, days 1-14; methotrexate 32 mg/M^2 and 5-fluorouracil 480 mg/M^2 intra-venously, day 1 and 8 of a 4 week cycle for 12 months.
2. Nil: No adjuvant chemotherapy.

ELIGIBILITY CRITERIA FOR ENTRY:
1. AGE: <70 Manchester, <75 Guy's.
2. CLINICAL/PATHOLOGICAL STAGES: Involved axillary nodes.
3. PRIMARY TREATMENT: Total mastectomy and axillary clearance. Post-operative radiotherapy not used.
4. NODE DISSECTION: Mandatory.
5. NUMBER OF NODES REMOVED: 23.6 + 10.2.
6. MENSTRUAL STATUS: No restrictions.
7. RECEPTOR STATUS: No restrictions.

METHOD OF RANDOMIZATION:
Separate within each Centre. Telephone to a central office where allocations were based on random number tables and balanced in blocks of 9 (Manchester) and 25 (Guy's).

YEARS OF ACCRUAL: 1979-1981 Manchester; 1979-1983 Guy's.

TOTAL NUMBER OF PATIENTS RANDOMIZED: 268 as of July 1985.

AIM, RATIONALE, SPECIAL FEATURES:
The preliminary results of a trial of adjuvant cyclophosphamide, methotrexate, and 5-fluorouracil (CMF) carried out in the Istituto Nazionale dei Tumori in Milan suggested that CMF could considerably prolong postoperative relapse-free survival (RFS) in all groups of patients with involved axillary lymph nodes. Because of the importance of these findings and the results of the National Surgical Adjuvant Breast Project (NSABP), which suggest a beneficial effect of melphalan as adjuvant therapy, a trial comparing no adjuvant treatment, melphalan, and CMF was started in March, 1976, at the University Hospital of South Manchester (75E2). A trial comparing melphalan with no adjuvant treatment was set up at Guys' Hospital in March, 1975 (75E1), and a trial comparing CMF and no treatment was established in October, 1979 (79E2).

Because the same protocol was used at Guy's and Manchester, the trials were amalgamated. The results of treatment with CMF are reported.

ADDITIONAL ELIGIBILITY OR STRATIFICATION CRITERIA:
Preoperative staging was clinical and confirmed after chest radiography, biochemical screening and isotope bone scan. Chemotherapy was started within two weeks of mastectomy.

At Guy's, only premenopausal patients were entered before 1981. Until November 1979 at Manchester the trial also included a third treatment arm (melphalan) to be compared with the same controls (75E2).

FOLLOW-UP, COMPLIANCE AND TOXICITY:
Follow-up included physical examination on day 1 of each treatment cycle and at six weekly intervals in controls. White blood cell and platelet counts were performed on day 1 and 8 of each cycle in patients randomized to receive adjuvant chemotherapy. For the first two years after mastectomy, biochemical screening and chest radiographs were carried out every 3 months, and isotope bone scans every 6 months. Thereafter, follow-up by physical examination only was carried out every 3 months and other investigations were repeated when indicated. Follow-up was annual after 5 years. The trial was assessed by postoperative relapse free survival, overall survival, pattern of recurrent disease and toxicity due to treatment.

PUBLICATIONS:
Howell A, George WD, Crowther D, Rubens RD, Bulbrook RD, Bush H, Howat JMT, Sellwood RA, Hayward JL, Fentiman IS, Chaudary M. Controlled trial of adjuvant chemotherapy with cyclo-phosphamide, methotrexate, and fluorouracil for breast cancer. Lancet 1984; 1:307-311.

Table 1. Outcome by allocated treatment for trial 79E1 (Guy's CMF)
Key: T = total free of relevant event at start of relevant year(s)

(i) Death (D) in each separate year,
subdivided by age when randomized (D <50 & D 50+ / T <50 & T 50+)

Year	1. CMF	2. Nil
1	0+ 1/ 46+ 51	7+ 3/ 49+ 44
2	0+ 1/ 41+ 44	5+ 3/ 36+ 37
3	3+ 5/ 34+ 37	0+ 1/ 25+ 27
4	2+ 0/ 28+ 20	2+ 0/ 18+ 24
5+	0+ 0/ 21+ 8	0+ 0/ 12+ 11

(ii) First recurrence or prior death (F) in each separate year,
subdivided by age when randomized (F <50 & F 50+ / T <50 & T 50+)

Year	1. CMF	2. Nil
1	0+ 3/ 46+ 51	14+ 9/ 49+ 44
2	5+ 10/ 39+ 37	4+ 1/ 27+ 28
3	1+ 0/ 28+ 22	4+ 2/ 15+ 23
4	2+ 0/ 23+ 14	1+ 1/ 10+ 16
5+	0+ 0/ 13+ 2	0+ 0/ 5+ 2

(iii) Recurrence with survival (R) and Death (D) in all years together,
subdivided by various characteristics when randomized (R & D / T)

Age at Entry	1. CMF	2. Nil
< 40	1+2/16	3+6/16
40 - 49	2+3/30	6+8/33
50 - 59	5+6/41	6+6/33
60 - 69	1+1/10	0+1/11
Any age (including "not known")	9+12/97	15+21/93

Nodal Status (N0, N1-3, N4+)

		1. CMF	2. Nil
<50	N1-N3 (by axillary clearance)	1+1/26	3+4/23
50+		2+1/27	0+1/21
<50	N4+ (by axillary sample/clearance)	2+4/20	6+9/25
50+		4+6/24	6+6/23
<50	Remainder (i.e. other N+ or N?)	0+0/0	0+1/1
\-h			

Estrogen Receptors (ER, fmol/mg)

		1. CMF	2. Nil
<50	ER poor or ER < 10	0+1/3	0+2/3
50+		0+1/2	0+1/3
<50	ER+ or ER 10-99	2+3/30	4+6/24
50+		4+2/23	2+2/16
<50	ER++ or ER 100+	0+0/3	0+0/4
50+		1+0/12	2+2/15

Table 1. Outcome by allocated treatment for trial 79E2 (Manchester II)
Key: T = total free of relevant event at start of relevant year(s)

(i) Death (D) in each separate year,
subdivided by age when randomized (D <50 & D 50+ / T <50 & T 50+)

Year	1. CMF	2. Nil
1	1+ 0/ 17+ 22	3+ 1/ 11+ 28
2	1+ 3/ 16+ 22	0+ 1/ 8+ 27
3	1+ 2/ 15+ 19	1+ 2/ 8+ 26
4	1+ 1/ 14+ 17	1+ 2/ 7+ 24
5+	0+ 0/ 11+ 12	0+ 0/ 6+ 17

(ii) First recurrence or prior death (F) in each separate year,
subdivided by age when randomized (F <50 & F 50+ / T <50 & T 50+)

Year	1. CMF	2. Nil
1	2+ 1/ 17+ 22	4+ 5/ 11+ 28
2	2+ 6/ 15+ 21	2+ 3/ 7+ 23
3	0+ 2/ 13+ 15	0+ 2/ 5+ 20
4	1+ 0/ 11+ 11	0+ 0/ 5+ 13
5+	0+ 0/ 4+ 3	0+ 0/ 1+ 4

(iii) Recurrence with survival (R) and Death (D) in all years together,
subdivided by various characteristics when randomized (R & D / T)

Age at Entry	1. CMF	2. Nil
< 40	0+1/3	0+1/3
40 - 49	1+3/14	1+4/8
50 - 59	1+4/15	2+5/19
60 - 69	2+2/7	2+1/13
Any age (including "not known")	4+10/39	5+11/39

Nodal Status (N0, N1-3, N4+)

		1. CMF	2. Nil
<50	N0 (by axillary clearance)	0+0/0	0+0/1
50+		0+0/0	1+0/1
<50	N1-N3 (by axillary clearance)	0+1/9	1+5/10
50+		1+2/13	1+2/20
<50	N4+ (by axillary sample/clearance)	1+3/8	0+0/0
50+		2+4/9	2+4/7

Estrogen Receptors (ER, fmol/mg)

		1. CMF	2. Nil
<50	ER poor or ER < 10	1+2/7	1+3/5
50+		0+3/11	2+3/14
<50	ER+ or ER 10-99	0+1/6	0+0/3
50+		0+1/3	0+0/5
<50	ER++ or ER 100+	0+1/2	0+0/1
50+		1+2/6	2+2/7

191

TITLE: ADJUVANT CHEMOTHERAPY WITH 5-FLUOROURACIL, DOXORUBICIN, CYCLOPHOSPHAMIDE, VINCRISTINE, PREDNISONE (FACVP), AND TAMOXIFEN WITH OR WITHOUT ADDITIONAL CHEMOTHERAPY WITH METHOTREXATE AND VINBLASTINE IN PATIENTS WITH OPERABLE BREAST CANCER

STUDY NUMBER: 80A
STUDY TITLE: MD ANDERSON 8026

ABBREVIATED TRIAL NAME: MD ANDERSON 8026
STUDY NUMBER: 80A

INVESTIGATORS' NAMES AND AFFILIATIONS:
AU Buzdar and GN Hortobagyi
  Division of Medicine, Medical Breast Service, M.D. Anderson Hospital, Houston.

TRIAL DESIGN: TREATMENT GROUPS:
1. FACVPr + Tam: 5-fluorouracil 400 mg/M^2, intravenously, day 1 and 8; doxorubicin 400 mg/M^2, cyclophosphamide 400 mg/M^2 and vincristine 2 mg (<50 years of age), 1.5 mg (>50 years), intravenously, on day 1; and prednisone 40 mg/M^2 (total dose) orally, daily for the first 5 days of each 3 week cycle, followed by tamoxifen 10 mg twice daily for 6 months. The maximum cumulative dose of doxorubicin was 400 mg/M^2 for all patients (except for those with 1-3 positive nodes, who received no more than 300 mg/M^2.
2. FACVPr + Tam + MV: FACVPr and tamoxifen, as above, plus methotrexate 75 mg/M^2, intramuscularly or intravenously at 2 week intervals for 3 months, then vinblastine 1.7 mg/M^2, intravenously, daily for 5 days at 3 week intervals for 4 cycles.

ELIGIBILITY CRITERIA FOR ENTRY:
1. AGE: All ages.
2. CLINICAL STAGES: Any stage, including T3 and N3, if all disease could be resected.
3. PRIMARY TREATMENT: Total surgical removal of all disease.
4. AXILLARY NODE DISSECTION: Not specified.
5. NUMBER OF NODES REMOVED: Not specified.
6. MENSTRUAL STATUS: No restrictions.
7. RECEPTOR STATUS: Positive or unknown.

METHOD OF RANDOMIZATION:
Randomization was done in the Central Data Management Office.

RANDOMIZATION STRATIFICATIONS: Menopausal status, stage of disease, number of involved nodes (1-3, ≥4).

YEARS OF ACCRUAL: 1980-1982.

TOTAL NUMBER OF PATIENTS RANDOMIZED: 235.

ADDITIONAL ELIGIBILITY OR STRATIFICATION CRITERIA:
The staging work-up included physical examination, CBC, platelet, differential, chest X-ray, skeletal films, bone scans, liver scans, or ultrasound and mammogram of the contralateral breast

Patients were excluded from the study if evidence of metastatic disease was unequivocal.

COMPLIANCE BEFORE FIRST RECURRENCE:
Twenty-four patients did not complete the chemotherapy portion of treatment according to schedule. Chemotherapy was discontinued for 9 patients in the FACVPr + Tam + MV subgroup after a median duration of 4 months (range 3-6 months) and for 15 patients in the FACVPr + Tam arm after a median duration of 4 months (range 1-6 months).

PUBLICATIONS:
Buzdar AU, Hortobagyi GN, Marcus CE, Smith TL, Martin R, Gehan EA. Results of adjuvant chemotherapy trials in breast cancer at M.D. Anderson Hospital and Tumor Institute. NCI Monogr. 1986; 1:81-85.

Buzdar AU, Hortobagyi GN, Smith TL, et al. Adjuvant therapy of breast cancer with or without additional treatment with alternate drugs. Cancer 1988; 62:2098-2104.

Table 1. Outcome by allocated treatment for trial 80A (MD Anderson 8026)
Key: T = total free of relevant event at start of relevant year(s)

(i) Death (D) in each separate year, subdivided by age when randomized (D <50 & D 50+ / T <50 & T 50+)

Year	1. TamFACVPr	2. TamCMFVAPr
1	0+ 1/ 63+ 59	0+ 0/ 51+ 62
2	1+ 2/ 63+ 58	0+ 2/ 51+ 62
3	4+ 2/ 62+ 56	1+ 2/ 51+ 60
4	1+ 0/ 48+ 49	0+ 1/ 41+ 50
5+	0+ 0/ 30+ 25	0+ 0/ 22+ 28

(ii) First recurrence or prior death (F) in each separate year, subdivided by age when randomized (F <50 & F 50+ / T <50 & T 50+)

Year	1. TamFACVPr	2. TamCMFVAPr
1	0+ 4/ 63+ 59	2+ 2/ 51+ 62
2	4+ 6/ 62+ 54	1+ 5/ 48+ 60
3	7+ 6/ 41+ 38	3+ 5/ 33+ 39
4	2+ 1/ 14+ 14	2+ 1/ 18+ 15
5+	0+ 0/ 1+ 2	0+ 0/ 2+ 1

(iii) Recurrence with survival (R) and Death (D) in all years together, subdivided by various characteristics when randomized (R & D / T)

Age at Entry	1. TamFACVPr	2. TamCMFVAPr
< 40	3+1/21	1+0/18
40 - 49	4+5/42	6+1/33
50 - 59	3+3/30	5+3/37
60 - 69	9+2/26	2+2/23
70 +	0+0/3	1+0/2
Any age (including "not known")	19+11/122	15+6/113

Nodal Status (N0, N1-3, N4+)

		1. TamFACVPr	2. TamCMFVAPr
< 50	N0 (by clinical	0+0/5	0+0/4
50 +	evidence only)	1+0/4	0+0/4
< 50	Remainder (i.e.	7+6/58	7+1/47
50 +	other N+ or N?)	11+5/55	8+5/58

Estrogen Receptors (ER, fmol/mg)

Breakdown by ER status was not available.

TITLE: NORTH SWEDEN STUDY OF ADJUVANT HORMONAL AND/OR CYTOTOXIC TREATMENT IN POTENTIALLY CURABLE BREAST CANCER

STUDY NUMBER: 80B1
STUDY TITLE: N SWED BCG 191

ABBREVIATED TRIAL NAME: N SWED BCG 191
STUDY NUMBER: 80B1

INVESTIGATORS' NAMES AND AFFILIATIONS:
L-G Larsson (1), K-A Angqvist (2), H Johansson (3) and collaborators in North Sweden Breast Cancer Group.

(1) Dept. of Oncology, University Hospital, Umeå, Sweden.
(2) Dept. of Surgery, University Hospital, Umeå, Sweden.
(3) Centre of Oncology, University Hospital, Umeå, Sweden.

TRIAL DESIGN: TREATMENT GROUPS:
1. Nil: No adjuvant therapy.
2. Tam + Ov Irr: Tamoxifen 20 mg, twice daily for 2 years plus ovarian irradiation.

ELIGIBILITY CRITERIA FOR ENTRY:
1. AGE: <55 years.
2. CLINICAL STAGES: Patients with primary breast cancer; T3, T4 tumors were included provided they were regarded as potentially curable by mastectomy.
3. PRIMARY TREATMENT: Modified radical, including axillary dissection or total mastectomy plus axillary sampling (removal of axillary tail and palpable nodes).
4. AXILLARY NODE DISSECTION: Mandatory.
5. NUMBER OF NODES REMOVED: Not specified.
6. MENSTRUAL STATUS: Not specified.
7. RECEPTOR STATUS: Not specified.

METHOD OF RANDOMIZATION:
The Centre of Oncology in Umeå was responsible for randomization, which was performed with pre-made protocols based on random numbers.

RANDOMIZATION STRATIFICATIONS:
Axillary surgery (dissection; sampling); axillary histopathology (node positive, node negative); risk group ("high risk", perinodal cancer in axilla and/or T3, T4, "low risk", all other patients).

YEARS OF ACCRUAL: 1980-1986.

TOTAL NUMBER OF PATIENTS RANDOMIZED: 77.

ADDITIONAL ELIGIBILITY OR STRATIFICATION CRITERIA:
The disease was assessed by clinical examination with TNM staging, mammography, chest X-ray, and routine blood and liver function tests. Bone scintigraphy was performed in all cases with positive axillary nodes and/or T3, T4 tumors but was otherwise not mandatory.

FOLLOW-UP, COMPLIANCE AND TOXICITY:
After the treatment period clinical assessment was made every 2nd to 3rd month for the first 2 years and thereafter every 4-6 months. Assessment included routine blood tests and, if clinical symptoms or signs appeared, chest X-ray, X-rays of spine and pelvis or bone scan.

PUBLICATIONS: None.

Table 1. Outcome by allocated treatment for trial 80B1 (N SwedenBCG 191)
Key: T = total free of relevant event at start of relevant year(s)

(i) Death (D) in each separate year,
subdivided by age when randomized (D <50 & D 50+ / T <50 & T 50+)

Year	1. Nil	2. TamOvIrr
1	0+ 0/ 18+ 19	0+ 0/ 27+ 13
2	0+ 0/ 16+ 16	2+ 0/ 24+ 11
3	0+ 0/ 10+ 14	0+ 0/ 12+ 8
4	0+ 0/ 5+ 10	0+ 0/ 10+ 6
5+	2+ 1/ 5+ 5	0+ 0/ 7+ 2

(ii) First recurrence or prior death (F) in each separate year,
subdivided by age when randomized (F <50 & F 50+ / T <50 & T 50+)

Year	1. Nil	2. TamOvIrr
1	1+ 0/ 18+ 19	2+ 0/ 27+ 13
2	1+ 1/ 12+ 16	1+ 0/ 20+ 9
3	1+ 1/ 6+ 12	0+ 0/ 9+ 7
4	0+ 0/ 3+ 7	0+ 0/ 9+ 6
5+	0+ 1/ 1+ 3	0+ 0/ 4+ 1

(iii) Recurrence with survival (R) and Death (D) in all years together,
subdivided by various characteristics when randomized (R & D / T)

Age at Entry	1. Nil	2. TamOvIrr
< 40	1+0/1	1+0/2
40 - 49	0+2/17	0+2/25
50 - 59	2+1/19	0+0/13
Any age (including "not known")	3+3/37	1+2/40

Nodal Status (N0, N1-3, N4+)

		1. Nil	2. TamOvIrr
< 50	N0 (by axillary	0+0/3	1+2/9
50 +	clearance)	0+0/6	0+0/4
< 50	N0 (by clinical	1+0/9	0+0/10
50 +	evidence only)	0+0/2	0+0/2
< 50	Remainder (i.e.	0+2/6	0+0/8
50 +	other N+ or N?)	2+1/11	0+0/7

Estrogen Receptors (ER, fmol/mg)

Breakdown by ER status was not available.

TITLE: NORTH SWEDEN STUDY OF ADJUVANT HORMONAL AND/OR CYTOTOXIC TREATMENT IN POTENTIALLY CURABLE BREAST CANCER

STUDY NUMBER: 80B2
STUDY TITLE: N SWED BCG 192

ABBREVIATED TRIAL NAME: N SWED BCG 192
STUDY NUMBER: 80B2

INVESTIGATORS' NAMES AND AFFILIATIONS:
L-G Larsson (1), K-A Angqvist (2), H Johansson (3) and collaborators in North Sweden Breast Cancer Group.

(1) Dept. of Oncology, University Hospital, Umeå, Sweden.
(2) Dept. of Surgery, University Hospital, Umeå, Sweden.
(3) Centre of Oncology, University Hospital, Umeå, Sweden.

TRIAL DESIGN: TREATMENT GROUPS:
1. Nil: No adjuvant therapy.
2. Tam: Tamoxifen 20 mg, twice daily for 2 years.

ELIGIBILITY CRITERIA FOR ENTRY:
1. AGE: ≥55 years.
2. CLINICAL STAGES: Patients with primary breast cancer; T3, T4 tumors were included provided they were regarded as potentially curable by mastectomy.
3. PRIMARY TREATMENT: Modified radical, including axillary dissection or total mastectomy plus axillary sampling (removal of axillary tail and palpable nodes).
4. AXILLARY NODE DISSECTION: Mandatory.
5. NUMBER OF NODES REMOVED: Not specified.
6. MENSTRUAL STATUS: Not specified.
7. RECEPTOR STATUS: Not specified.

METHOD OF RANDOMIZATION:
The Centre of Oncology in Umeå was responsible for randomization, which was performed with pre-made protocols based on random numbers.

RANDOMIZATION STRATIFICATIONS:
Axillary surgery (dissection; sampling); axillary histopathology (node positive, node negative); risk group ("high risk", perinodal cancer in axilla and/or T3, T4, "low risk", all other patients).

YEARS OF ACCRUAL: 1980-1986.

TOTAL NUMBER OF PATIENTS RANDOMIZED: 178.

ADDITIONAL ELIGIBILITY OR STRATIFICATION CRITERIA:
The disease was assessed by clinical examination with TNM staging, mammography, chest X-ray, and routine blood and liver function tests. Bone scintigraphy was performed in all cases with positive axillary nodes and/or T3,T4 tumors but was otherwise not mandatory.

FOLLOW-UP, COMPLIANCE AND TOXICITY:
After the treatment period clinical assessment was made every 2nd to 3rd month for the first 2 years and thereafter every 4-6 months. Assessment included routine blood tests and, if clinical symptoms or signs appeared, chest X-ray, X-rays of spine and pelvis or bone scan.

PUBLICATIONS: None.

Table 1. Outcome by allocated treatment for trial 80B2 (N SwedenBCG 192)
Key: T = total free of relevant event at start of relevant year(s)

(i) Death (D) in each separate year, subdivided by age when randomized (D <50 & D 50+ / T <50 & T 50+)

Year	1. Nil	2. Tam
1	0+ 2/ 0+ 89	0+ 1/ 0+ 89
2	0+ 2/ 0+ 78	0+ 5/ 0+ 78
3	0+ 5/ 0+ 63	0+ 3/ 0+ 59
4	0+ 1/ 0+ 40	0+ 0/ 0+ 43
5+	0+ 1/ 0+ 21	0+ 0/ 0+ 24

(ii) First recurrence or prior death (F) in each separate year, subdivided by age when randomized (F <50 & F 50+ / T <50 & T 50+)

Year	1. Nil	2. Tam
1	0+ 5/ 0+ 89	0+ 5/ 0+ 89
2	0+ 4/ 0+ 68	0+ 8/ 0+ 69
3	0+ 6/ 0+ 49	0+ 0/ 0+ 46
4	0+ 0/ 0+ 27	0+ 1/ 0+ 32
5+	0+ 2/ 0+ 12	0+ 0/ 0+ 11

(iii) Recurrence with survival (R) and Death (D) in all years together, subdivided by various characteristics when randomized (R & D / T)

Age at Entry		1. Nil	2. Tam
50 - 59		1+3/18	2+2/24
60 - 69		2+4/40	0+4/35
70 +		3+4/31	3+3/30
Any age (including "not known")		6+11/89	5+9/89

Nodal Status (N0, N1-3, N4+)

50 +	N0 (by axillary clearance)	4+5/34	2+3/35
50 +	N0 (by clinical evidence only)	0+0/24	1+0/26
50 +	Remainder (i.e. other N+ or N?)	2+6/31	2+6/28

Estrogen Receptors (ER, fmol/mg)

Breakdown by ER status was not available.

TITLE: NORTH SWEDEN STUDY OF ADJUVANT HORMONAL AND/OR CYTOTOXIC TREATMENT IN POTENTIALLY CURABLE BREAST CANCER

STUDY NUMBER: 80B3
STUDY TITLE: N SWED BCG 193

ABBREVIATED TRIAL NAME: N SWED BCG 193
STUDY NUMBER: 80B3

INVESTIGATORS' NAMES AND AFFILIATIONS:
L-G Larsson (1), K-A Angqvist (2), H Johansson (3) and collaborators in North Sweden Breast Cancer Group.

(1) Dept. of Oncology, University Hospital, Umeå, Sweden.
(2) Dept. of Surgery, University Hospital, Umeå, Sweden.
(3) Centre of Oncology, University Hospital, Umeå, Sweden.

TRIAL DESIGN: TREATMENT GROUPS:
1. Nil: No adjuvant therapy.
2. Tam + Ov Irr: Tamoxifen 20 mg, twice daily for 2 years.
3. AC: Adriamycin 40 mg/M^2 intravenously, day 1, plus cyclophosphamide 100 mg/M^2 orally days 3, 4, 5 and 6 of a 3 week cycle for 8 cycles.
4. AC + Tam + Ov Irr: AC, as above, plus tamoxifen, as above, plus ovarian irradiation.

ELIGIBILITY CRITERIA FOR ENTRY:
1. AGE: <55 years.
2. CLINICAL STAGES: Patients with primary breast cancer; regarded as "high risk" (i.e. perinodal cancer in axilla and/or T3, T4 if regarded as potentially curable by mastectomy).
3. PRIMARY TREATMENT: Modified radical, including axillary dissection or total mastectomy plus axillary sampling (removal of axillary tail and palpable nodes).
4. AXILLARY NODE DISSECTION: Mandatory.
5. NUMBER OF NODES REMOVED: Not specified.
6. MENSTRUAL STATUS: Not specified.
7. RECEPTOR STATUS: Not specified.

METHOD OF RANDOMIZATION:
The Centre of Oncology in Umeå was responsible for randomization, which was performed with pre-made protocols based or random numbers.

RANDOMIZATION STRATIFICATIONS:
Axillary surgery (dissection; sampling); axillary histopathology (node positive, node negative).

YEARS OF ACCRUAL: 1980-1986.

TOTAL NUMBER OF PATIENTS RANDOMIZED: 56.

ADDITIONAL ELIGIBILITY OR STRATIFICATION CRITERIA:
The disease was assessed by clinical examination with TNM staging, mammography, chest X-ray, and routine blood and liver function tests. Bone scintigraphy was performed in all cases.

FOLLOW-UP, COMPLIANCE AND TOXICITY:
After the treatment period clinical assessment was made every 2nd to 3rd month for the first 2 years and thereafter every 4-6 months. Assessment included routine blood tests and, if clinical symptoms or signs appeared, chest X-ray, X-rays of spine and pelvis or bone scan.

PUBLICATIONS: None.

Table 1. Outcome by allocated treatment for trial 80B3 (N SwedenBCG 193)
Key: T = total free of relevant event at start of relevant year(s)

(i) Death (D) in each separate year, subdivided by age when randomized (D <50 & D 50+ / T <50 & T 50+)

Year	1. Nil	2. TamOvIrr	3. AC	4. TamACOvIrr
1	1+ 0/ 11+ 2	0+ 0/ 11+ 3	0+ 0/ 7+ 7	0+ 0/ 9+ 6
2	1+ 0/ 7+ 2	2+ 0/ 11+ 1	0+ 1/ 6+ 6	0+ 0/ 8+ 4
3	1+ 0/ 6+ 2	1+ 0/ 8+ 1	1+ 0/ 4+ 5	0+ 1/ 5+ 3
4	1+ 0/ 4+ 2	0+ 0/ 3+ 1	0+ 0/ 2+ 3	0+ 0/ 5+ 2
5+	0+ 0/ 2+ 2	0+ 0/ 2+ 1	0+ 1/ 1+ 1	0+ 0/ 1+ 1

(ii) First recurrence or prior death (F) in each separate year, subdivided by age when randomized (F <50 & F 50+ / T <50 & T 50+)

Year	1. Nil	2. TamOvIrr	3. AC	4. TamACOvIrr
1	3+ 0/ 11+ 2	1+ 0/ 11+ 3	0+ 1/ 7+ 7	1+ 1/ 9+ 6
2	2+ 0/ 5+ 2	2+ 0/ 10+ 1	2+ 1/ 5+ 6	0+ 2/ 5+ 3
3	1+ 0/ 3+ 2	1+ 0/ 6+ 1	1+ 0/ 3+ 4	0+ 0/ 4+ 1
4	1+ 0/ 2+ 2	0+ 0/ 2+ 1	0+ 1/ 2+ 2	0+ 0/ 2+ 1
5+	0+ 0/ 0+ 1	0+ 0/ 1+ 0	0+ 0/ 0+ 0	0+ 0/ 0+ 0

(iii) Recurrence with survival (R) and Death (D) in all years together, subdivided by various characteristics when randomized (R & D / T)

Age at Entry	1. Nil	2. TamOvIrr	3. AC	4. TamACOvIrr
<40	0+0/0	0+2/3	0+0/0	0+0/2
40 - 49	3+4/11	1+1/8	1+1/7	1+0/7
50 - 59	0+0/2	0+0/3	2+2/7	2+1/6
Any age (including "not known")	3+4/13	1+3/14	3+3/14	3+1/15

Nodal Status (N0, N1-3, N4+)

		1. Nil	2. TamOvIrr	3. AC	4. TamACOvIrr
<50	N0 (by clinical evidence only)	0+0/0	0+0/0	0+0/1	0+0/1
<50	Remainder (i.e. other N+ or N?)	3+4/11	1+3/11	1+1/6	1+0/8
50 +		0+0/2	0+0/3	2+2/7	2+1/6

Estrogen Receptors (ER, fmol/mg)

Breakdown by ER status was not available.

TITLE: NORTH SWEDEN STUDY OF ADJUVANT HORMONAL AND/OR CYTOTOXIC TREATMENT IN POTENTIALLY CURABLE BREAST CANCER

STUDY NUMBER: 80B4
STUDY TITLE: N SWED BCG 194

ABBREVIATED TRIAL NAME: N SWED BCG 194
STUDY NUMBER: 80B4

INVESTIGATORS' NAMES AND AFFILIATIONS:
L-G Larsson (1), K-A Angqvist (2), H Johansson (3) and collaborators in North Sweden Breast Cancer Group.

(1) Dept. of Oncology, University Hospital, Umeå, Sweden
(2) Dept. of Surgery, University Hospital, Umeå, Sweden.
(3) Centre of Oncology, University Hospital, Umeå, Sweden.

TRIAL DESIGN: TREATMENT GROUPS:
1. Nil: No adjuvant therapy.
2. Tam: Tamoxifen 20 mg, twice daily for 2 years.
3. AC: Adriamycin 40 mg/M^2 intravenously, day 1, plus cyclophosphamide 100 mg/M^2 orally days 3, 4, 5 and 6 of a 3 week cycle for 8 cycles.
4. AC + Tam: AC, as above, plus tamoxifen, as above.

ELIGIBILITY CRITERIA FOR ENTRY:
1. AGE: ≥55 years.
2. CLINICAL STAGES: Patients with primary breast cancer; regarded as "high risk" (i.e. perinodal cancer in axilla and/or T3, T4 if regarded as potentially curable by mastectomy).
3. PRIMARY TREATMENT: Modified radical, including axillary dissection or total mastectomy plus axillary sampling (removal of axillary tail and palpable nodes).
4. AXILLARY NODE DISSECTION: Dissection or sampling mandatory.
5. NUMBER OF NODES REMOVED: Not specified.
6. MENSTRUAL STATUS: Not specified.
7. RECEPTOR STATUS: Not specified.

METHOD OF RANDOMIZATION:
The Centre of Oncology in Umeå was responsible for randomization, which was performed with pre-made protocols based on random numbers.

RANDOMIZATION STRATIFICATIONS:
Axillary surgery (dissection; sampling); axillary histopathology (node positive, node negative).

YEARS OF ACCRUAL: 1980-1986.

TOTAL NUMBER OF PATIENTS RANDOMIZED: 96.

ADDITIONAL ELIGIBILITY OR STRATIFICATION CRITERIA:
The disease was assessed by clinical examination with TNM staging, mammography, chest X-ray, and routine blood and liver function tests. Bone scintigraphy was performed in all cases.

FOLLOW-UP, COMPLIANCE AND TOXICITY:
After the treatment period clinical assessment was made every 2nd to 3rd month for the first 2 years and thereafter every 4-6 months. Assessment included routine blood tests and, if clinical symptoms or signs appeared, chest X-ray, X-rays of spine and pelvis or bone scan.

PUBLICATIONS: None.

Table 1. Outcome by allocated treatment for trial 80B4 (N SwedenBCG 194)
Key: T = total free of relevant event at start of relevant year(s)

(i) Death (D) in each separate year, subdivided by age when randomized (D <50 & D 50+ / T <50 & T 50+)

Year	1. Nil	2. Tam	3. AC	4. TamAC
1	0+ 2/ 0+ 25	0+ 5/ 0+ 24	0+ 2/ 0+ 23	0+ 0/ 0+ 24
2	0+ 1/ 0+ 18	0+ 1/ 0+ 18	0+ 0/ 0+ 18	0+ 0/ 0+ 21
3	0+ 1/ 0+ 14	0+ 1/ 0+ 14	0+ 2/ 0+ 15	0+ 2/ 0+ 15
4	0+ 0/ 0+ 9	0+ 1/ 0+ 12	0+ 1/ 0+ 11	0+ 0/ 0+ 10
5+	0+ 1/ 0+ 5	0+ 0/ 0+ 6	0+ 0/ 0+ 6	0+ 0/ 0+ 6

(ii) First recurrence or prior death (F) in each separate year, subdivided by age when randomized (F <50 & F 50+ / T <50 & T 50+)

Year	1. Nil	2. Tam	3. AC	4. TamAC
1	0+ 6/ 0+ 25	0+ 6/ 0+ 24	0+ 4/ 0+ 23	0+ 0/ 0+ 24
2	0+ 1/ 0+ 13	0+ 2/ 0+ 14	0+ 2/ 0+ 15	0+ 2/ 0+ 20
3	0+ 2/ 0+ 9	0+ 0/ 0+ 12	0+ 0/ 0+ 10	0+ 1/ 0+ 11
4	0+ 1/ 0+ 2	0+ 1/ 0+ 10	0+ 2/ 0+ 6	0+ 0/ 0+ 5
5+	0+ 0/ 0+ 0	0+ 0/ 0+ 3	0+ 2/ 0+ 2	0+ 0/ 0+ 2

(iii) Recurrence with survival (R) and Death (D) in all years together, subdivided by various characteristics when randomized (R & D / T)

Age at Entry		1. Nil	2. Tam	3. AC	4. TamAC
50 - 59		2+2/7	1+2/9	3+1/5	1+0/11
60 - 69		3+3/18	1+5/14	4+4/18	0+2/13
70 +		0+0/0	0+0/1	0+0/0	0+0/0
Any age (including "not known")		5+5/25	2+7/24	7+5/23	1+2/24
	Nodal Status (N0, N1-3, N4+)				
50 +	N0 (by axillary clearance)	0+0/0	0+1/1	0+0/0	0+0/0
50 +	N0 (by clinical evidence only)	0+0/1	0+1/1	0+0/0	0+0/0
50 +	Remainder (i.e. other N+ or N?)	5+5/24	2+5/22	7+5/23	1+2/24

Estrogen Receptors (ER, fmol/mg)

Breakdown by ER status was not available.

TITLE: ADJUVANT CHEMOTHERAPY IN NODE POSITIVE STAGE II BREAST CANCER

STUDY NUMBER: 80C
STUDY TITLE: SE SWED BCG B

ABBREVIATED TRIAL NAME: SE SWED BCG B
STUDY NUMBER: 80C

INVESTIGATORS' NAMES AND AFFILIATIONS:
M Söderberg, T Hatschek, A Molde, H Sellström and K Wiegner
South Eastern Breast Cancer Group Oncologic Center, Regionsjukhuset, Linköping.

TRIAL DESIGN: TREATMENT GROUPS:
1. R: Radiotherapy for 6 weeks with split-course technique (10 + 10 fractions) to a target dose of 45 Gy to the regional lymph nodes and 38 Gy to the chest wall, starting 3 weeks after mastectomy.
2. R + AC: Radiotherapy, as above, plus doxorubicin 40 mg/M^2, intravenously, day 1 and cyclophosphamide 200 mg/M^2, orally, days 3-6 every 3 weeks for 8 cycles.

ELIGIBILITY CRITERIA FOR ENTRY:
1. AGE: Not specified.
2. CLINICAL/PATHOLOGICAL STAGES: Stage II disease (pT1-2,N1,M0).
3. PRIMARY TREATMENT: Radical surgery including axillary clearance.
4. AXILLARY NODE DISSECTION: Mandatory.
5. NUMBER OF NODES REMOVED: Not specified.
6. MENSTRUAL STATUS: Postmenopausal.
7. RECEPTOR STATUS: Not specified.

METHOD OF RANDOMIZATION:
Random allocations to treatment groups by telephone to the Regional Tumour Registry.

RANDOMIZATION STRATIFICATIONS:
Participating hospitals; tumor size (<2 cm, ≥2 cm); number of involves axillary nodes (1-3, ≥4).

YEARS OF ACCRUAL: 1980-1983.

TOTAL NUMBER OF PATIENTS RANDOMIZED: 43.

AIM, RATIONALE, SPECIAL FEATURES:
This study was initiated to compare the toxicity and recurrence rates after 6 months postoperative chemotherapy with doxorubicin and cyclophosphamide or radiotherapy alone. A parallel study comparing R+AC with R+CMF in premenopausal breast cancer also recruited 43 patients.

ADDITIONAL ELIGIBILITY OR STRATIFICATION CRITERIA:
Staging was assessed by clinical examination, mammography, chest X-ray, bone scan and liver function tests.

FOLLOW-UP, COMPLIANCE AND TOXICITY:
During the first year the patients were seen every 3 months, the following two years every 4 months, thereafter every 6 months. The follow-up included clinical examination, full blood count and liver function tests. In asymptomatic patients, chest X-ray and bone scan were performed annually, mammography every other year.

COMPLIANCE BEFORE FIRST RECURRENCE:
Five patients finished the chemotherapy at an early stage, 2 of these due to side effects of treatment. Totally, 16 of 21 patients randomized to R + AC were able to receive the chemotherapy on schedule according to protocol. In the parallel study, 40 of 43 premenopausal patients were able to receive the chemotherapy according to protocol.

DISCUSSION:
The results of the postmenopausal study have contributed to the overview results but have not been reported separately. A toxicity study based on the present material will be published in the near future.

PUBLICATIONS:
Hrafukelsson J, Nilsson K, Söderberg M. Tolerance of radiotherapy combined with adjuvant chemotherapy in breast cancer. Acta Oncol. 1987; 26:269-272.

Table 1. Outcome by allocated treatment for trial 80C (SE Sweden BCG B)
Key: T = total free of relevant event at start of relevant year(s)

(i) Death (D) in each separate year,
subdivided by age when randomized (D <50 & D 50+ / T <50 & T 50+)

Year	1. R	2. ACR
1	0+ 0/ 1+ 21	0+ 2/ 0+ 21
2	0+ 2/ 1+ 21	0+ 1/ 0+ 19
3	0+ 3/ 1+ 19	0+ 1/ 0+ 18
4	1+ 2/ 1+ 14	0+ 0/ 0+ 16
5+	0+ 0/ 0+ 7	0+ 0/ 0+ 7

(ii) First recurrence or prior death (F) in each separate year,
subdivided by age when randomized (F <50 & F 50+ / T <50 & T 50+)

1	1+ 3/ 1+ 21	0+ 4/ 0+ 21
2	0+ 3/ 0+ 18	0+ 1/ 0+ 17
3	0+ 1/ 0+ 15	0+ 0/ 0+ 16
4	0+ 1/ 0+ 12	0+ 0/ 0+ 12
5+	0+ 0/ 0+ 4	0+ 0/ 0+ 4

(iii) Recurrence with survival (R) and Death (D) in all years together,
subdivided by various characteristics when randomized (R & D / T)

Age at Entry		1. R	2. ACR
40 - 49		0+1/1	0+0/0
50 - 59		1+3/12	0+1/8
60 - 69		0+4/9	1+3/13
Any age (including "not known")		1+8/22	1+4/21

	Nodal Status (N0, N1-3, N4+)		
50 +	N1-N3 (by axillary clearance)	1+3/10	0+1/13
<50	N4+ (by axillary	0+1/1	0+0/0
50 +	sample/clearance)	0+4/11	1+3/8

	Estrogen Receptors (ER, fmol/mg)		
50 +	ER poor or ER < 10	0+2/4	0+1/2
<50	ER+ or ER 10-99	0+1/1	0+0/0
50 +		1+3/9	1+1/10

TITLE: CRC ADJUVANT BREAST TRIAL

ABBREVIATED TRIAL NAME: CRC 2
STUDY NUMBER: 80D

STUDY NUMBER: 80D
STUDY TITLE: CRC 2

INVESTIGATORS' NAMES AND AFFILIATIONS:
M Baum (1), D Berstock (2), D Brinkley (1), A Lyons (3), T McElwain (4), J MacIntyre (5), N Orr (6), T Powles (4), W Ross (7), I Smith (4), C Elston (8), K MacRae (9), J Cuzick (10), D Roy (11), R Peto (12), W Odling-Smee (11), P Abram (3), J Houghton (13), D Riley (13).

- (1) King's College Hospital, London
- (2) Clatterbridge General Hospital, Wirral.
- (3) Belvoir Park Hospital, Belfast.
- (4) Royal Marsden Hospital, Surrey.
- (5) Perth Royal Infirmary, Perth.
- (6) Colchester General, Essex.
- (7) Newcastle General Hospital.
- (8) City Hospital, Nottingham.
- (9) Charing Cross Hospital, London.
- (10) Imperial Cancer Research Fund, London.
- (11) Royal Victoria Hospital, Belfast.
- (12) Clinical Trial Service Unit, Oxford.
- (13) CRC Clinical Trials Centre, London.

TRIAL DESIGN: TREATMENT GROUPS:
1. Tam: Tamoxifen 10 mg, orally, twice daily for 2 years.
2. C: Cyclophosphamide 5 mg/kg for 6 days starting less than 24 hours after surgery.
3. Tam + C: Tamoxifen, as above, and cyclophosphamide, as above.
4. Nil: No adjuvant therapy.

ELIGIBILITY CRITERIA FOR ENTRY:
1. AGE: < 75.
2. CLINICAL STAGES: Primary breast cancer.
3. PRIMARY TREATMENT: Total mastectomy with axillary clearance; or total mastectomy with axillary node sampling and adjuvant radiotherapy when nodes positive; or local excision and axillary sampling with radiotherapy to remaining breast.
4. AXILLARY NODE DISSECTION: Optional.
5. NUMBER OF NODES REMOVED: Not specified.
6. MENSTRUAL STATUS: No restrictions.
7. RECEPTOR STATUS: No restrictions.

METHOD OF RANDOMIZATION:
Telephone to a central office where allocations were based on a computerized random number generator and balanced in blocks.

RANDOMIZATION STRATIFICATIONS:
By clinician. No patient for whom a treatment was allocated was subsequently withdrawn. Mastectomy and local excision patients were randomized separately.

YEARS OF ACCRUAL: 1980-1985.

TOTAL NUMBER OF PATIENTS RANDOMIZED: 1768.

AIM, RATIONALE, SPECIAL FEATURES:
This trial was set up to repeat both the Scandinavian Adjuvant Chemotherapy Trial and the NATO trial. It was also the aim of the study that it be acceptable to as many clinicians as possible and be easily incorporated into their individual clinical practices.

If it was believed there was a clear indication for tamoxifen, patients could be randomized between arms 1 and 3 in a parallel study.

ADDITIONAL ELIGIBILITY OR STRATIFICATION CRITERIA:
Female patients with primary breast cancer were ineligible if they were pregnant, ≥75 years, of distant domicile or with distant metastases. Chest X-ray and at least an X-ray of lumbar spine and pelvis were required, but bone scans were not mandatory. Blood count and liver function tests, including alkaline phosphatase, were routinely performed. Radiotherapy usually included radiation to the chest wall (or breast for local excision patients) and internal mammary nodes as well as to the axilla.

FOLLOW-UP, COMPLIANCE AND TOXICITY:
Clinical assessment with additional special investigations determined only by symptoms or suspicion of recurrence were carried out every 3 months for the first 2 years, 6 monthly until 5 years, and then annually. Compliance with tamoxifen was excellent based on patient's reports. Over 90% of patients allocated to receive cyclophosphamide completed the 6 day perioperative course.

Toxicity for the tamoxifen patients appeared to be minimal, and at present it is not known how many failed to complete the prescribed course due to putative side effects. Toxicity related to the cyclophosphamide therapy was relatively minor although over 50% of patients did experience nausea. However, since the patients were in hospital for the duration of their course this could be treated accordingly. Also, 30% of patients did experience some alopecia, although less than 8% required a wig.

DISCUSSION:
The results of the present trial are still at an early stage as the median follow-up is 1.5 years (patient accrual is still continuing). The results for the tamoxifen comparison show a similar trend to that obtained by comparable trials and the overview.

PUBLICATIONS:
CRC Adjuvant Working Party. Cyclophosphamide and tamoxifen as adjuvant therapies in the management of breast cancer. Br. J. Cancer 1988; 57:604-607.

Table 1. Outcome by allocated treatment for trial 80D (CRC2)
Key: T = total free of relevant event at start of relevant year(s)

(i) Death (D) in each separate year,
subdivided by age when randomized (D <50 & D 50+ / T <50 & T 50+)

Year	1. Tam	2. C	3. TamC	4. Nil
1	5+ 9/ 142+ 275	1+ 5/ 134+ 315	1+ 2/ 134+ 325	1+ 5/ 131+ 312
2	3+ 4/ 131+ 258	3+ 8/ 130+ 295	4+ 8/ 126+ 303	4+ 12/ 126+ 296
3	0+ 1/ 91+ 203	2+ 6/ 92+ 222	3+ 3/ 96+ 228	1+ 6/ 99+ 234
4	0+ 0/ 57+ 117	1+ 3/ 56+ 126	1+ 1/ 47+ 120	1+ 0/ 57+ 128
5+	0+ 0/ 13+ 36	0+ 0/ 12+ 31	0+ 0/ 15+ 28	0+ 0/ 19+ 31

(ii) First recurrence or prior death (F) in each separate year,
subdivided by age when randomized (F <50 & F 50+ / T <50 & T 50+)

Year	1. Tam	2. C	3. TamC	4. Nil
1	14+ 21/ 142+ 275	12+ 30/ 134+ 315	8+ 13/ 134+ 325	12+ 35/ 131+ 312
2	6+ 14/ 106+ 221	12+ 25/ 102+ 231	13+ 19/ 105+ 259	16+ 27/ 107+ 235
3	3+ 6/ 60+ 125	5+ 11/ 56+ 134	1+ 7/ 55+ 152	6+ 8/ 62+ 142
4	1+ 1/ 28+ 50	0+ 5/ 22+ 53	2+ 2/ 20+ 50	1+ 2/ 25+ 50
5+	1+ 0/ 1+ 3	0+ 0/ 0+ 4	0+ 0/ 2+ 3	0+ 0/ 0+ 5

(iii) Recurrence with survival (R) and Death (D) in all years together,
subdivided by various characteristics when randomized (R & D / T)

Age at Entry	1. Tam	2. C	3. TamC	4. Nil
< 40	8+3/42	10+1/31	6+2/33	6+4/37
40 - 49	9+5/100	12+6/103	9+7/101	22+3/94
50 - 59	22+7/138	25+14/140	13+6/147	27+12/142
60 - 69	6+6/109	19+7/144	13+6/143	15+9/134
70 +	0+1/28	5+1/31	1+2/35	7+2/36
Any age (including "not known")	45+22/417	71+29/449	42+23/459	77+30/443

Nodal Status (N0, N1-3, N4+)

		1. Tam	2. C	3. TamC	4. Nil
< 50	N0 (by axillary	1+0/28	4+1/34	4+1/47	2+1/30
50 +	clearance)	7+2/68	10+1/83	2+0/69	11+1/83
< 50	N0 (by clinical	4+2/51	4+1/46	2+1/41	10+1/49
50 +	evidence only)	3+2/78	7+5/97	6+1/104	11+4/99
< 50	N1-N3 (by axillary	3+3/32	5+1/18	5+2/22	4+0/16
50 +	clearance)	4+4/47	8+5/37	6+5/51	7+5/41
< 50	N4+ (by axillary	4+2/11	2+3/9	0+0/2	5+0/7
50 +	sample/clearance)	4+1/16	10+5/21	4+2/24	9+2/20
< 50	Remainder (i.e.	5+1/20	7+1/27	4+5/22	7+5/29
50 +	other N+ or N?)	10+5/66	14+6/77	9+6/77	11+11/69

Estrogen Receptors (ER, fmol/mg)

Breakdown by ER status was not available.

TITLE: TOULOUSE STUDY OF ADJUVANT TAMOXIFEN IN 251 POSTMENOPAUSAL BREAST CANCER PATIENTS

STUDY NUMBER: 80E
STUDY TITLE: TOULOUSE

ABBREVIATED TRIAL NAME: TOULOUSE
STUDY NUMBER: 80E

INVESTIGATORS NAMES AND AFFILIATIONS:
A Naja (1), JP Armand (1), C Hill (2).

(1) Centre Claudius Regaud, Toulouse, France.
(2) Institut Gustave Roussy, Villejuif, France.

TRIAL DESIGN: TREATMENT GROUPS:
1. Tam: Tamoxifen 30 mg, orally, daily for 24 months.
2. Nil: No adjuvant therapy.

ELIGIBILITY CRITERIA FOR ENTRY:
1. AGE: 50 to 75.
2. CLINICAL/PATHOLOGICAL STAGES: All except M1 (distant metastases) or N3 (supraclavicular nodes), T1 N- with at least 12 verified axillary nodes.
3. PRIMARY TREATMENT: For tumors <3 cm: segmental mastectomy, axillary clearance and 50 Gy to the breast + 15 Gy booster to tumor bed; for all other tumors: radical mastectomy followed by 50 Gy to the thoracic wall if axillary N+. In all cases, if axillary N+ : 50 Gy to axilla and to internal mammary chain; if axillary N- and tumor in internal quadrant: 50 Gy to the internal mammary nodes.
4. NODE DISSECTION: Not required (no surgery for two patients).
5. NUMBER OF NODES REMOVED: 15 ± 7
6. MENSTRUAL STATUS: Postmenopausal (no menses for at least 1 yr). Patients with previous castration were excluded.

METHOD OF RANDOMIZATION:
By telephone call to the principal investigator's office. Allocation was based on a computer programme and balanced in blocks of 4.

YEARS OF ACCRUAL: 1980-1983.

TOTAL NUMBER OF PATIENTS RANDOMIZED: 251.

AIM, RATIONALE, SPECIAL FEATURES:
A large imbalance in the distribution of nodal status between the two treatment groups has been observed in this trial. To explore the possibility of an identifiable pattern to this imbalance, extensive checks were performed on waiting time between patients, by treatment and nodal status. These checks failed to find any pattern in the imbalance, and it is assumed that it arose by a somewhat extreme play of chance.

ADDITIONAL ELIGIBILITY OR STRATIFICATION CRITERIA:
Only female patients with a primary adenocarcinoma of the breast and in complete remission after the initial treatment were considered for entry. Patients with creatinemia over 15 mg, abnormal liver function, inflammatory symptoms, bilateral breast cancer or history of cancer of another site were excluded. Two patients with tumors > 5 cm refused surgery and received only radiotherapy: 50 Gy to the breast, 30 Gy booster to the tumor, 70 Gy to the axillary nodes and 50 Gy to the internal mammary nodes.

FOLLOW-UP, COMPLIANCE AND TOXICITY:
A physical examination was performed every three months for the first 2 years, and every 6 months for the next 3 years. Every year, a chest X-ray, mammography, PAP smear, and CEA serum assay were performed.

Two patients randomized to the control group were given tamoxifen by mistake, and 2 patients in the tamoxifen group stopped the treatment because of nausea and vomiting (one resumed treatment with a dose of 20 mg instead of 30). All patients are included in the results as randomized.

PUBLICATIONS: None.

Table 1. Outcome by allocated treatment for trial 80E (Toulouse)
Key: T = total free of relevant event at start of relevant year(s)

(i) Death (D) in each separate year,
subdivided by age when randomized (D <50 & D 50+ / T <50 & T 50+)

Year	1. Tam	2. Nil
1	0+ 2/ 0+ 125	0+ 4/ 4+ 122
2	0+ 7/ 0+ 123	0+ 7/ 4+ 118
3	0+ 5/ 0+ 116	0+ 5/ 4+ 111
4	0+ 8/ 0+ 69	0+ 4/ 3+ 68
5+	0+ 0/ 0+ 19	0+ 0/ 0+ 28

(ii) First recurrence or prior death (F) in each separate year,
subdivided by age when randomized (F <50 & F 50+ / T <50 & T 50+)

Year	1. Tam	2. Nil
1	0+ 14/ 0+ 125	1+ 9/ 4+ 122
2	0+ 8/ 0+ 111	0+ 12/ 3+ 113
3	0+ 6/ 0+ 100	1+ 6/ 3+ 99
4	0+ 1/ 0+ 70	0+ 3/ 2+ 66
5+	0+ 0/ 0+ 27	0+ 0/ 0+ 27

(iii) Recurrence with survival (R) and Death (D) in all years together,
subdivided by various characteristics when randomized (R & D / T)

Age at Entry	1. Tam	2. Nil
40 - 49	0+0/0	2+0/4
50 - 59	2+10/48	5+8/42
60 - 69	5+8/48	4+5/47
70 +	0+4/29	2+6/33
Any age (including "not known")	7+22/125	13+19/126

Nodal Status (N0, N1-3, N4+)

		1. Tam	2. Nil
<50	N0 (by axillary	0+0/0	0+0/2
50 +	clearance)	1+1/27	5+3/61
50 +	N0 (by clinical evidence only)	0+0/3	0+0/0
<50	N1-N3 (by axillary	0+0/0	1+0/1
50 +	clearance)	1+5/39	2+10/43
<50	N4+ (by axillary	0+0/0	1+0/1
50 +	sample/clearance)	4+15/52	3+6/17
50 +	Remainder (i.e. other N+ or N?)	1+1/4	1+0/1

Estrogen Receptors (ER, fmol/mg)

Breakdown by ER status was not available.

TITLE: ADJUVANT CMF IN NODE NEGATIVE BREAST CANCER

ABBREVIATED TRIAL NAME: INT MILAN 8004
STUDY NUMBER: 80F

INVESTIGATORS' NAMES AND AFFILIATIONS:
G Bonadonna, M Zambetti, P Valagussa, P Bignami, G DiFronzo, R Silvestrini.
Istituto Nazionale Tumori, Milan.

TRIAL DESIGN: TREATMENT GROUPS:
1. CMF: Cyclophosphamide 600 mg/M^2, methotrexate 40 mg/M^2 and 5-fluorouracil 600 mg/M^2 intravenously every 21 days times 12 courses.
2. Nil: No adjuvant chemotherapy.

ELIGIBILITY CRITERIA FOR ENTRY:
1. AGE: ≤ 75 years.
2. CLINICAL/PATHOLOGICAL STAGES: T1-T2 lesions with histologically negative lymph nodes.
3. PRIMARY TREATMENT: Modified radical mastectomy or quadrantectomy followed by radiotherapy to the breast.
4. AXILLARY NODE DISSECTION: Full axillary node dissection was required.
5. NUMBER OF NODES REMOVED: Not specified.
6. MENSTRUAL STATUS: No restriction.
7. RECEPTOR STATUS: Must be ER negative.

METHOD OF RANDOMIZATION:
Block randomisation with a permuted block of length 4 for the two regimens.

RANDOMIZATION STRATIFICATIONS:
By menopausal status.

YEARS OF ACCRUAL: 12/80-9/85.

TOTAL NUMBER OF PATIENTS RANDOMIZED: 96

AIM, RATIONALE, SPECIAL FEATURES:
This study was designed to determine whether the advantages for adjuvant CMF seen in patients who were node positive could also be demonstrated in node negative patients as well.

DISCUSSION:
A statistically significant improvement in disease-free survival and overall survival has been seen for all patients, premenopausal and postmenopausal patients entered in this trial. Although the effects of therapy in the two menopausal subsets appear to be nearly the same, the effects of therapy appear to be the greatest in the 32 patients with a high labelling index.

PUBLICATIONS:
Bonadonna G, Valagussa P, Tancini G, et al. Current status of the Milan adjuvant chemotherapy trials for node positive and node negative breast cancer. NCI Monogr. 1986; 1:45-49.

Table 1. Outcome by allocated treatment for trial 80F (INT Milan 8004)
Key: T = total free of relevant event at start of relevant year(s)

(i) Death (D) in each separate year, subdivided by age when randomized (D <50 & D 50+ / T <50 & T 50+)

Year	1. CMF	2. Nil
1	0+ 0/ 26+ 21	0+ 2/ 29+ 18
2	0+ 0/ 21+ 16	2+ 0/ 27+ 13
3	0+ 0/ 12+ 12	3+ 2/ 15+ 8
4	0+ 0/ 9+ 6	0+ 0/ 8+ 4
5+	0+ 0/ 3+ 2	0+ 0/ 3+ 0

(ii) First recurrence or prior death (F) in each separate year, subdivided by age when randomized (F <50 & F 50+ / T <50 & T 50+)

Year	1. CMF	2. Nil
1	1+ 1/ 26+ 21	4+ 3/ 29+ 18
2	1+ 0/ 25+ 20	5+ 3/ 25+ 14
3	0+ 0/ 18+ 15	1+ 2/ 18+ 9
4	0+ 0/ 13+ 11	0+ 0/ 10+ 5
5+	0+ 0/ 10+ 8	1+ 0/ 6+ 4

(iii) Recurrence with survival (R) and Death (D) in all years together, subdivided by various characteristics when randomized (R & D / T)

Age at Entry		1. CMF	2. Nil
< 40		0+0/5	2+3/11
40 – 49		2+0/21	4+2/18
50 – 59		1+0/11	3+2/11
60 – 69		0+0/10	1+2/7
Any age (including "not known")		3+0/47	10+9/47

Nodal Status (N0, N1–3, N4+)

		1. CMF	2. Nil
< 50	N0 (by axillary clearance)	2+0/26	6+5/29
50 +		1+0/21	4+4/17
50 +	N1–N3 (by axillary clearance)	0+0/0	0+0/1

Estrogen Receptors (ER, fmol/mg)

		1. CMF	2. Nil
< 50	ER poor or ER < 10	2+0/25	6+5/29
50 +		1+0/21	4+4/18

TITLE: MONTPELLIER STUDY OF ADJUVANT TAMOXIFEN IN POSTMENOPAUSAL BREAST CANCER

ABBREVIATED TRIAL NAME: MONTPELLIER
STUDY NUMBER: 81A

INVESTIGATORS NAMES AND AFFILIATIONS:
JB Dubois (1), H Pujol (1), C Hill (2).

(1) Centre Paul Lamarque, Montpellier, France.
(2) Institut Gustave Roussy, Villejuif, France.

TRIAL DESIGN: TREATMENT GROUPS:
1. Tam: Tamoxifen 30 mg, orally daily, for 24 months.
2. Nil: No adjuvant therapy.

ELIGIBILITY CRITERIA FOR ENTRY:
1. AGE: 50 to 75.
2. CLINICAL STAGES: All except M1 (distant metastases), or N3 (supraclavicular nodes) with at least 12 verified axillary nodes.
3. PRIMARY TREATMENT: For tumors > 5 cm or rapidly growing, 45 Gy to the breast and 55 Gy to the tumor followed by surgery; for other tumors, surgery followed by 45 Gy to the thoracic wall and 5 Gy boost to the scar. In all cases the surgery was a mastectomy + ipsilateral axillary clearance.
4. NODE DISSECTION: Axillary clearance for all cases.
5. NUMBER OF NODES REMOVED: 10 ± 3
6. MENSTRUAL STATUS: Post-menopausal (no menses for at least 1 yr). Patients with previous castration were excluded.

METHOD OF RANDOMIZATION:
Randomization was by telephone to the statistical office where allocation was based on a computer programme and balanced in blocks of 8.

YEARS OF ACCRUAL: 1980-1984.

TOTAL NUMBER OF PATIENTS RANDOMIZED: 203.

AIM, RATIONALE, SPECIAL FEATURES:
This study was designed to determine if tamoxifen in postmenopausal breast cancer is able to: 1) increase the disease free interval 2) improve the overall survival.

ADDITIONAL ELIGIBILITY OR STRATIFICATION CRITERIA:
Only female patients with a primary adenocarcinoma of the breast and in complete remission after the initial treatment were considered for entry. Patients with creatinemia over 15 mg, abnormal liver function, inflammatory symptoms, bilateral breast cancer or history of cancer of another site were excluded.

FOLLOW-UP, COMPLIANCE AND TOXICITY:
A physical examination was performed every three months for the first 2 years, and every 6 months for the next 3 years. Every year, a chest X-ray, mammography, PAP smear, and CEA serum assay were performed.

There was no interruption of tamoxifen for toxicity.

PUBLICATIONS: None.

Table 1. Outcome by allocated treatment for trial 81A (Montpellier)
Key: T = total free of relevant event at start of relevant year(s)

(i) Death (D) in each separate year,
subdivided by age when randomized (D <50 & D 50+ / T <50 & T 50+)

Year	1. Tam	2. Nil
1	0+ 5/ 0+ 101	0+ 3/ 0+ 102
2	0+ 1/ 0+ 96	0+ 3/ 0+ 99
3	0+ 0/ 0+ 80	0+ 3/ 0+ 81
4	0+ 1/ 0+ 47	0+ 0/ 0+ 45
5+	0+ 0/ 0+ 9	0+ 0/ 0+ 10

(ii) First recurrence or prior death (F) in each separate year,
subdivided by age when randomized (F <50 & F 50+ / T <50 & T 50+)

Year	1. Tam	2. Nil
1	0+ 8/ 0+ 101	0+ 9/ 0+ 102
2	0+ 4/ 0+ 82	0+ 8/ 0+ 74
3	0+ 1/ 0+ 57	0+ 2/ 0+ 54
4	0+ 3/ 0+ 33	0+ 3/ 0+ 33
5+	0+ 0/ 0+ 17	0+ 0/ 0+ 14

(iii) Recurrence with survival (R) and Death (D) in all years together,
subdivided by various characteristics when randomized (R & D / T)

Age at Entry		1. Tam	2. Nil
50 - 59		2+1/37	6+2/34
60 - 69		3+6/53	5+3/46
70 +		4+0/11	2+4/22
Any age (including "not known")		9+7/101	13+9/102

	Nodal Status (N0, N1-3, N4+)		
50 +	N0 (by axillary clearance)	3+3/65	3+4/62
50 +	N1-N3 (by axillary clearance)	4+2/28	8+4/29
50 +	N4+ (by axillary sample/clearance)	2+2/7	2+1/11
50 +	Remainder (i.e. other N+ or N?)	0+0/1	0+0/0

Estrogen Receptors (ER, fmol/mg)

Breakdown by ER status was not available.

TITLE: HORMONO-CHEMOTHERAPY VERSUS CHEMOTHERAPY IN STAGE II BREAST CARCINOMA

ABBREVIATED TRIAL NAME: FB BORDEAUX
STUDY NUMBER: 81B

STUDY NUMBER: 81B
STUDY TITLE: FB BORDEAUX

INVESTIGATORS' NAMES AND AFFILIATIONS:
L Mauriac, M Durand, F Bonichon, D Maree, A Avril, J Wafflart, P Mage, J Chauvergne
Fondation Bergonie, Bordeaux, France.

TRIAL DESIGN: TREATMENT GROUPS:
1. CMF: Cyclophosphamide 600 mg/M^2, methotrexate 40 mg/M^2, 5-fluorouracil 600 mg/M^2 in one intravenous injection every 3 weeks for 9 injections.
2. CMF + Tam: CMF, as above, + tamoxifen 30 mg daily for 24 months.

ELIGIBILITY CRITERIA FOR ENTRY:
1. AGE: ≤75.
2. CLINICAL/PATHOLOGICAL STAGES: All histologically node-positive patients except M1 (distant metastases), N2 (fixed axillary lymph nodes), N3 (supraclavicular nodes), and rapidly evolving or inflammatory tumors.
3. PRIMARY TREATMENT: For tumors ≥3 cm, total mastectomy with axillary clearance without radiotherapy; for tumors <3 cm, segmental mastectomy with axillary clearance and radiotherapy to the breast, internal mammary and supraclavicular nodes.
4. NODE DISSECTION: Axillary clearance.
5. NUMBER OF NODES REMOVED: 15 ± 3
6. MENSTRUAL STATUS: No restriction.
7. HORMONAL RECEPTORS: ER ≥10 fmol/mg or PR ≥15 fmol/mg.

METHOD OF RANDOMIZATION:
Sealed envelopes were kept in a central office. Allocation was based on a table of random numbers and balanced in blocks of 8.

YEARS OF ACCRUAL: 1981-1984.

TOTAL NUMBER OF PATIENTS RANDOMIZED: 326.

AIM, RATIONALE, SPECIAL FEATURES:
115 patients (62 in the CMF group and 53 in the CMF+Tam group) were also part of a randomized study of levamisole (Durand 1986).

ADDITIONAL ELIGIBILITY OR STRATIFICATION CRITERIA:
Extent of disease was initially assessed by clinical examination with TNM staging, chest X-ray, bone scintigraphy, and liver and renal function studies.

FOLLOW-UP, COMPLIANCE AND TOXICITY:
Follow-up was performed every 3 months for the first year, every 6 months for the second year, and annually afterwards. Follow-up included clinical examination, chest X-ray and mammogram once a year.

Only one patient in the tamoxifen arm of the trial did not receive two years of tamoxifen. There was no difference in the average doses of chemotherapy between the two arms of the study.

Although all patients are included in the results as randomized, 7 patients were ineligible for one of the following reasons: age >75, N3, or N-.

PUBLICATIONS:
Mauriac L, Durand M, Chauvergne J, Bonichon F, Avril A, Mage B, Dilhuydy MH, Le Treut A, Wafflart J, Maree D, Lagarde C. Adjuvant trial for stage II receptor-positive breast cancer: CMF vs. CMF plus tamoxifen in a single centre. Breast Cancer Res. Treat. 1988; 11:179-186.

Durand M, Mauriac L, Chauvergne J, Hoerni B, Bonichon F. Immunotherapie par le levamisole apres chimiotherapie adjuvante dans les cancers du sein a haut risque metastatique. Bull. Cancer 1986;73:462.

Table 1. Outcome by allocated treatment for trial 81B (FB Bordeaux)
Key: T = total free of relevant event at start of relevant year(s)

(i) Death (D) in each separate year, subdivided by age when randomized (D <50 & D 50+ / T <50 & T 50+)

Year	1. CMF	2. TamCMF
1	0+ 1/ 59+ 101	0+ 0/ 67+ 98
2	1+ 4/ 53+ 92	1+ 1/ 66+ 84
3	1+ 2/ 36+ 62	1+ 0/ 45+ 62
4	1+ 1/ 23+ 26	1+ 0/ 24+ 27
5+	0+ 0/ 3+ 2	0+ 0/ 3+ 4

(ii) First recurrence or prior death (F) in each separate year, subdivided by age when randomized (F <50 & F 50+ / T <50 & T 50+)

Year	1. CMF	2. TamCMF
1	4+ 7/ 59+ 101	0+ 2/ 67+ 98
2	1+ 5/ 55+ 93	2+ 2/ 67+ 95
3	2+ 5/ 43+ 72	2+ 2/ 59+ 74
4	2+ 1/ 24+ 40	1+ 1/ 35+ 45
5+	0+ 0/ 7+ 11	0+ 0/ 9+ 10

(iii) Recurrence with survival (R) and Death (D) in all years together, subdivided by various characteristics when randomized (R & D / T)

Age at Entry	1. CMF	2. TamCMF
< 40	2+1/19	1+2/20
40 - 49	4+2/40	1+1/47
50 - 59	5+3/46	4+1/48
60 - 69	4+2/39	2+0/36
70 +	1+3/16	0+0/14
Any age (including "not known")	16+11/160	8+4/165

	Nodal Status (N0, N1-3, N4+)	1. CMF	2. TamCMF
< 50	N1-N3 (by axillary	3+0/40	1+0/43
50 +	clearance)	7+6/75	3+0/61
< 50	N4+ (by axillary	3+3/19	1+3/24
50 +	sample/clearance)	3+2/26	3+1/37

	Estrogen Receptors (ER, fmol/mg)	1. CMF	2. TamCMF
< 50	ER poor or ER < 10	1+2/7	0+0/9
50 +		0+0/2	0+0/7
< 50	ER+ or ER 10-99	5+1/46	2+3/47
50 +		6+3/61	3+0/59
< 50	ER++ or ER 100+	0+0/6	0+0/11
50 +		4+5/38	3+1/32

TITLE: ADJUVANT SYSTEMIC THERAPY WITH CMF OR CMF + RADIOTHERAPY OR CMF + TAMOXIFEN IN PREMENOPAUSAL HIGH-RISK BREAST CANCER PATIENTS

STUDY NUMBER: 82B
STUDY TITLE: DANISH BCG 82b

ABBREVIATED TRIAL NAME: DANISH BCG 82b
STUDY NUMBER: 82B

INVESTIGATORS' NAMES AND AFFILIATIONS:
Danish Breast Cancer Cooperative Group, Finsen Institute, Copenhagen.

TRIAL DESIGN: TREATMENT GROUPS:
1. CMF: Cyclophosphamide 600 mg/M^2, methotrexate 40 mg/M^2 and 5-fluorouracil 600 mg/M^2, intravenously on day 1 of a 4 week cycle x 9 cycles.
2. CMF + R: CMF, as above, x 8 cycles, plus radiotherapy. The first cycle was given 2-4 weeks postoperatively, the second cycle 9-11 weeks after operation (After completion of radiotherapy) and subsequent cycles were repeated every 4 weeks.
Radiotherapy, 50 Gy, delivered in 25 fractions, was started 4-6 weeks postoperatively (1-2 weeks after the first cycle of CMF) and given to the chest wall, the supraclavicular and axillary nodes.
3. CMF + Tam: CMF, as above, x 9 cycles, plus tamoxifen 10 mg, three times daily for 1 year.

ELIGIBILITY CRITERIA FOR ENTRY:
1. AGE: Not specified.
2. CLINICAL/PATHOLOGICAL STAGES: Invasive breast carcinoma with either positive axillary nodes, and/or tumor >5 cm, and/or invasion of skin or deep fascia.
3. PRIMARY TREATMENT: Total mastectomy and axillary sampling.
4. AXILLARY NODE DISSECTION: Not required.
5. NUMBER OF NODES REMOVED: No restriction.
6. MENSTRUAL STATUS: Pre- or perimenopausal (≤5 years of spontaneous menostasia).
7. RECEPTOR STATUS: No restriction.

METHOD OF RANDOMIZATION:
Randomization is decentralized and performed by 20 different departments which receive the randomization cards from the secretariat in closed numbered envelopes.

RANDOMIZATION STRATIFICATIONS: Institution.

YEARS OF ACCRUAL: 1982 - still accruing although, in June 1986, the CMF+Tam arm was discontinued.

TOTAL NUMBER OF PATIENTS RANDOMIZED: 654, as of February 1985.

AIM, RATIONALE, SPECIAL FEATURES
Adjuvant chemotherapy has been shown to prolong the time to recurrence and, in some studies, the survival of premenopausal women. This study was designed to evaluate the potential benefit of adding either adjuvant radiotherapy or adjuvant tamoxifen to chemotherapy in this group.

FOLLOW-UP, COMPLIANCE AND TOXICITY:
Clinical assessment was done every 3 months for the first year, every 6 months for 4 more years, and annually until 10 years after mastectomy thereafter. Chest X-ray was repeated after 6 and 12 months and then annually for another 4 years.

DISCUSSION:
At 4 years (median observation time 2 years), survival is the same in all three groups, but the disease-free survival of patients on the adjuvant radiotherapy arm is significantly better (p<0.002) than that of patients on the other two arms.

PUBLICATIONS:
Dombernowsky P, Brinker H, Hansen M, et al. Adjuvant therapy of premenopausal and menopausal high-risk breast cancer patients. Acta Oncol. 1988; 27:691-697.

Table 1. Outcome by allocated treatment for trial 82B (Danish BCG 82b)
Key: T = total free of relevant event at start of relevant year(s)

(i) Death (D) in each separate year, subdivided by age when randomized (D <50 & D 50+ / T <50 & T 50+)

Year	1. CMF	2. CMFR	3. TamCMF
1	2+ 6/ 144+ 68	3+ 3/ 165+ 59	5+ 3/ 146+ 72
2	4+ 2/ 122+ 50	7+ 3/ 132+ 40	5+ 4/ 118+ 55
3	1+ 0/ 55+ 22	0+ 0/ 56+ 17	1+ 0/ 55+ 26
4	0+ 0/ 0+ 0	0+ 0/ 0+ 0	0+ 0/ 0+ 0
5+	0+ 0/ 0+ 0	0+ 0/ 0+ 0	0+ 0/ 0+ 0

(ii) First recurrence or prior death (F) in each separate year, subdivided by age when randomized (F <50 & F 50+ / T <50 & T 50+)

Year	1. CMF	2. CMFR	3. TamCMF
1	15+ 11/ 144+ 68	9+ 3/ 165+ 59	16+ 8/ 146+ 72
2	10+ 5/ 82+ 29	6+ 6/ 92+ 29	11+ 6/ 79+ 41
3	1+ 0/ 15+ 8	0+ 0/ 13+ 6	1+ 1/ 20+ 8
4	0+ 0/ 0+ 0	0+ 0/ 0+ 0	0+ 0/ 0+ 0
5+	0+ 0/ 0+ 0	0+ 0/ 0+ 0	0+ 0/ 0+ 0

(iii) Recurrence with survival (R) and Death (D) in all years together, subdivided by various characteristics when randomized (R & D / T)

Age at Entry	1. CMF	2. CMFR	3. TamCMF
<40	8+3/45	3+3/45	8+4/39
40 - 49	11+4/99	2+7/120	9+7/107
50 - 59	8+8/68	3+6/59	8+7/72
Any age (including "not known")	27+15/212	8+16/224	25+18/218

	Nodal Status (N0, N1-3, N4+)	1. CMF	2. CMFR	3. TamCMF
<50	N0 (by clinical	2+1/15	0+1/15	1+1/16
50+	evidence only)	0+0/9	0+2/3	0+0/7
<50	Remainder (i.e.	17+6/129	5+9/150	16+10/130
50+	other N+ or N?)	8+8/59	3+4/56	8+7/65

	Estrogen Receptors (ER, fmol/mg)	1. CMF	2. CMFR	3. TamCMF
<50	ER poor or ER < 10	1+2/8	0+2/10	2+3/16
50+		0+1/6	0+0/5	0+3/7
<50	ER+ or ER 10-99	1+0/29	0+0/26	0+0/27
50+		1+2/16	0+1/7	1+1/13
<50	ER++ or ER 100+	1+0/5	0+0/7	0+0/4
50+		2+0/2	1+0/3	1+0/7

TITLE: ADJUVANT SYSTEMIC THERAPY WITH TAMOXIFEN OR TAMOXIFEN + RADIOTHERAPY OR TAMOXIFEN + CMF IN POSTMENOPAUSAL HIGH RISK BREAST CANCER PATIENTS

STUDY NUMBER: 82C
STUDY TITLE: DANISH BCG 82c

ABBREVIATED TRIAL NAME: DANISH BCG 82c
STUDY NUMBER: 82C

INVESTIGATORS' NAMES AND AFFILIATIONS:
Danish Breast Cancer Cooperative Group, Finsen Institute, Copenhagen.

TRIAL DESIGN: TREATMENT GROUPS:
1. Tam: Tamoxifen, 10 mg, three times daily was started 2-4 weeks postoperatively and was continued for 1 year.
2. Tam + R: Tamoxifen, as above, plus radiotherapy, 50 Gy in 25 fractions, to the chest wall and the supraclavicular and axillary nodes, started 2-4 weeks postoperatively.
3. CMFTam: Cyclophosphamide 600 mg/M^2, methotrexate 40 mg/M^2 and 5-fluorouracil 600 mg/M^2, intravenously, on day 1 of a 4 week cycle × 9 cycles.

ELIGIBILITY CRITERIA FOR ENTRY:
1. AGE: <70 years.
2. CLINICAL/PATHOLOGICAL STAGES: Invasive breast carcinoma with positive axillary lymph nodes and/or tumor >5 cm and/or invasion of skin or deep fascia.
3. PRIMARY TREATMENT: Total mastectomy and axillary sampling.
4. AXILLARY NODE DISSECTION: Not required.
5. NUMBER OF NODES REMOVED: Not specified.
6. MENSTRUAL STATUS: Postmenopausal (>5 years of spontaneous menostasia.)
7. RECEPTOR STATUS: Not specified.

METHOD OF RANDOMIZATION:
Randomization is decentralized and performed by 20 different departments which receive the randomization cards from the secretariat in closed numbered envelopes.

RANDOMIZATION STRATIFICATIONS: Institution.

YEARS OF ACCRUAL: 1982 - still accruing.

TOTAL NUMBER OF PATIENTS RANDOMIZED: 699, as of February 1985.

AIM, RATIONALE, SPECIAL FEATURES:
Adjuvant tamoxifen has been shown to prolong the time to recurrence and, in some studies, the survival of postmenopausal women. This study was designed to evaluate the potential benefit of adding either adjuvant radiotherapy or adjuvant radiotherapy plus adjuvant chemotherapy.

FOLLOW-UP, COMPLIANCE AND TOXICITY:
Clinical assessment was done every 3 months for the first year, every 6 months for 4 more years, and annually until 10 years after mastectomy thereafter. Chest X-ray was repeated after 6 and 12 months and then annually for another 4 years.

DISCUSSION:
At 4 years (median observation time 2 years), survival is the same in all three groups, but the disease-free survival of patients on the adjuvant radiotherapy arm is superior to that of patients on the other 2 arms (p=0.03 for the 3-way comparison).

PUBLICATIONS:
Mouridsen HT, Rose C, Overgaard M, et al. Adjuvant treatment of postmenopausal patients with high risk primary breast cancer. Acta Oncol. 1988; 27:699-705.

Table 1. Outcome by allocated treatment for trial 82C (Danish BCG 82c)
Key: T = total free of relevant event at start of relevant year(s)

(i) Death (D) in each separate year, subdivided by age when randomized (D <50 & D 50+ / T <50 & T 50+)

Year	1. Tam	2. TamR	3. TamCMF
1	0+ 11/ 1+ 232	0+ 8/ 3+ 229	0+ 9/ 4+ 230
2	0+ 7/ 1+ 185	0+ 11/ 3+ 184	0+ 6/ 3+ 186
3	0+ 0/ 1+ 69	0+ 0/ 2+ 77	0+ 0/ 0+ 77
4	0+ 0/ 0+ 0	0+ 0/ 0+ 0	0+ 0/ 0+ 0
5+	0+ 0/ 0+ 0	0+ 0/ 0+ 0	0+ 0/ 0+ 0

(ii) First recurrence or prior death (F) in each separate year, subdivided by age when randomized (F <50 & F 50+ / T <50 & T 50+)

Year	1. Tam	2. TamR	3. TamCMF
1	0+ 37/ 1+ 232	0+ 22/ 3+ 229	0+ 19/ 4+ 230
2	0+ 6/ 1+ 105	0+ 9/ 3+ 113	0+ 14/ 2+ 117
3	0+ 0/ 0+ 16	0+ 0/ 1+ 23	0+ 1/ 0+ 24
4	0+ 0/ 0+ 0	0+ 0/ 0+ 0	0+ 0/ 0+ 0
5+	0+ 0/ 0+ 0	0+ 0/ 0+ 0	0+ 0/ 0+ 0

(iii) Recurrence with survival (R) and Death (D) in all years together, subdivided by various characteristics when randomized (R & D / T)

Age at Entry		1. Tam	2. TamR	3. TamCMF
40 - 49		0+0/1	0+0/3	0+0/4
50 - 59		10+5/66	4+6/80	8+5/88
60 - 69		15+13/164	8+12/146	11+9/138
70 +		0+0/2	0+1/3	0+1/4
Any age (including "not known")		25+18/233	12+19/232	19+15/234

	Nodal Status (N0, N1-3, N4+)	1. Tam	2. TamR	3. TamCMF
<50	N0 (by clinical	0+0/0	0+0/1	0+0/0
50 +	evidence only)	0+0/22	1+2/24	2+1/19
<50	Remainder (i.e.	0+0/1	0+0/2	0+0/4
50 +	other N+ or N?)	25+18/210	11+17/205	17+14/211

	Estrogen Receptors (ER, fmol/mg)	1. Tam	2. TamR	3. TamCMF
<50	ER poor or ER < 10	0+0/0	0+0/0	0+0/1
50 +		5+2/16	0+2/11	5+2/13
<50	ER+ or ER 10-99	0+0/0	0+0/1	0+0/1
50 +		3+2/23	1+2/25	0+0/29
50 +	ER++ or ER 100+	2+1/39	0+1/31	2+1/31

Short reports

TITLE: TAMOXIFEN VS CHEMOTHERAPY VS TAMOXIFEN PLUS CHEMOTHERAPY IN PATIENTS WITH ER POSITIVE STAGE II BREAST CANCER

STUDY NUMBER: 83B
STUDY TITLE: GROCTA ITALY

ABBREVIATED TRIAL NAME: GROCTA TRIAL
STUDY NUMBER: 83B

INVESTIGATORS' NAMES AND AFFILIATIONS:
F. Boccardo (1), P Bruzzi(1), M Cappellini (2), G Isola (3), I Nenci (4), A. Piffanelli (5), A Rubagotti (1, 6), L.Santi (1,6), A Scanni (7), P Sismondi (8), and other participants in the GROCTA (Italian Research Group on Adjuvant Chemo-hormonotherapy).

(1) National Cancer Institute, Genova.
(2) Department of Radiotherapy, S. Maria Nuova Hospital, Florence.
(3) CI-Pharma, Milan.
(4) Institute of Pathology, University of Ferrara.
(5) Institute of Radiology, University of Ferrara.
(6) Institute of Experimental and Clinical Oncology, University of Genoa.
(7) Department of Medical Oncology, Fatebenefratelli Hospital, Milan.
(8) Institute of Gynaecology and Obstetrics, University of Turin:

TRIAL DESIGN: TREATMENT GROUPS:
1. Tam: Tamoxifen 10 mg, three times, daily, 5 yrs.
2. CMFE: Cyclophosphamide 500 mg/M^2, methotrexate 40 mg/M^2, fluorouracil 600 mg/M^2, intravenously, day 1 every 3 wks x 6 cycles, followed by 4-epidoxorubicin 75 mg/M^2, intravenously, every 3 wks, x 4 courses.
3. CMFE + Tam: CMFE as above, plus tamoxifen, as above.

ELIGIBILITY CRITERIA FOR ENTRY:
1. AGE: >35 and <65.
2. CLINICAL STAGES: pT1-T3; pN1, MO; no evidence of distant metastases.
3. PRIMARY TREATMENT: Radical or modified radical mastectomy. Modified radical mastectomy was mandatory in patients with clinically Tb tumors. Postoperative radiotherapy was not recommended. In T1a tumors and a clinically negative axilla, quadrantectomy or lumpectomy followed by axillary dissection was allowed. Postoperative radiotherapy was mandatory in these patients.
4. NODE DISSECTION: Mandatory.
5. NUMBER OF NODES REMOVED: ≥10.
6. MENSTRUAL STATUS: No restrictions.
7. RECEPTOR STATUS: ER positive (10 fmol/mg protein.)

METHOD OF RANDOMIZATION:
Randomization was by telephone to a central office in the National Cancer Institute of Genoa, where allocation was based on random number tables. Treatment allocation were balanced in blocks of varying sizes.

RANDOMIZATION STRATIFICATIONS:
By participating centres

YEARS OF ACCRUAL: 1983-1987

TOTAL NUMBER OF PATIENTS RANDOMIZED: 510, (281, as of July 1985)

AIM, RATIONALE, SPECIAL FEATURES
This study was designed to compare the relative efficacy of chemotherapy alone, tamoxifen alone, or a combination of both modalities in both premenopausal and postmenopausal women.

ADDITIONAL ELIGIBILITY AND STRATIFICATION CRITERIA:
All patients were randomized within six weeks of surgery.

FOLLOW-UP, COMPLIANCE, AND TOXICITY:
Patients were clinically examined every three months for the first five years, and every six months from the fifth to the tenth year. Blood counts were repeated every three months, chest X-rays every six months, and bone scans yearly during the first 5 postoperative years.

DISCUSSION:
Tamoxifen was more effective than chemotherapy in reducing recurrences in both premenopausal and postmenopausal women, but statistically significant only in postmenopausal women. The combination was more effective than either modality alone in both patient groups. There are still insufficient data to perform a survival analysis.

PUBLICATIONS:
Boccardo F, Bruzzi P, Cappellini M, et al.: Tamoxifen (T) vs chemotherapy (CT) vs chemotherapy plus tamoxifen (CTT) in patients with ER positive, stage II breast cancer. An interim analysis of a multicentric Italian trial. Breast Cancer Res. and Treat. 1987; 10:85.

Boccardo F, Bruzzi P, Cappellini M, et al.: Tamoxifen (T) vs chemotherapy (CT) vs chemotherapy plus tamoxifen (CTT) in stage II, ER+ breast cancer patients. 4 yrs results of a multicentric Italian study. Proc. ASCO 1988; 7:11.

Table 1. Outcome by allocated treatment for trial 83B (GROCTA I Italy)
Key: T = total free of relevant event at start of relevant year(s)

(i) Death (D) in each separate year,
subdivided by age when randomized (D <50 & D 50+ / T <50 & T 50+)

Year	1. Tam	2. CMFE	3. TamCMFE
1	1+ 0/ 39+ 47	0+ 0/ 35+ 59	1+ 0/ 38+ 63
2	0+ 0/ 16+ 25	0+ 0/ 14+ 23	0+ 0/ 12+ 28
3	0+ 0/ 0+ 0	0+ 0/ 0+ 0	0+ 0/ 0+ 0
4	0+ 0/ 0+ 0	0+ 0/ 0+ 0	0+ 0/ 0+ 0
5+	0+ 0/ 0+ 0	0+ 0/ 0+ 0	0+ 0/ 0+ 0

(ii) First recurrence or prior death (F) in each separate year,
subdivided by age when randomized (F <50 & F 50+ / T <50 & T 50+)

Year	1. Tam	2. CMFE	3. TamCMFE
1	2+ 1/ 39+ 47	1+ 4/ 35+ 59	2+ 3/ 38+ 63
2	0+ 0/ 32+ 41	2+ 2/ 30+ 49	0+ 0/ 35+ 54
3	0+ 0/ 13+ 22	0+ 0/ 9+ 14	0+ 0/ 12+ 21
4	0+ 0/ 0+ 0	0+ 0/ 0+ 0	0+ 0/ 0+ 0
5+	0+ 0/ 0+ 0	0+ 0/ 0+ 0	0+ 0/ 0+ 0

(iii) Recurrence with survival (R) and Death (D) in all years together,
subdivided by various characteristics when randomized (R & D / T)

Age at Entry		1. Tam	2. CMFE	3. TamCMFE
<40		0+0/8	2+0/6	0+0/8
40 - 49		1+1/31	1+0/29	1+1/30
50 - 59		1+0/27	5+0/43	2+0/39
60 - 69		0+0/20	1+0/16	1+0/24
Any age (including "not known")		2+1/86	9+0/94	4+1/101

Nodal Status (N0, N1-3, N4+)

		1. Tam	2. CMFE	3. TamCMFE
<50	N1-N3 (by axillary clearance)	0+0/19	1+0/16	0+0/22
50+		0+0/28	2+0/33	1+0/44
<50	N4+ (by axillary sample/clearance)	1+1/19	2+0/17	1+1/16
50+		1+0/18	4+0/26	1+0/18
<50	Remainder (i.e. other N+ or N?)	0+0/1	0+0/2	0+0/0
50+		0+0/1	0+0/0	1+0/1

Estrogen Receptors (ER, fmol/mg)

		1. Tam	2. CMFE	3. TamCMFE
<50	ER poor or ER < 10	1+0/10	0+0/2	0+0/6
50+		0+0/1	0+0/6	0+0/8
<50	ER+ or ER 10-99	0+0/26	2+0/29	1+1/28
50+		1+0/32	3+0/30	2+0/34
<50	ER++ or ER 100+	0+0/2	0+0/1	0+0/3
50+		0+0/12	2+0/21	1+0/21

TITLE: ADJUVANT CHEMOTHERAPY VERSUS HORMONE THERAPY IN THE TREATMENT OF PRIMARY BREAST CANCER

STUDY NUMBER: 84A1
STUDY TITLE: BMFT 02 GERMANY

ABBREVIATED TRIAL NAME: BMFT 02 GERMANY
STUDY NUMBER: 84A1

INVESTIGATORS' NAMES AND AFFILIATIONS:
G Bastert (1), K Hübner (2), J Bojar (3), M Schumacher (4), W Sauerbrei (4), H Scheurlen* (5).

(1) Universitäts-Frauenklinik, Homburg/Saar.
(2) Zentrum Pathologice ter Universität, Frankfurt.
(3) Institute für Phys. Chemie der Universität, Düsseldorf.
(4) Institut für Med. Biometrie der Universität, Freiburg.
(5) Institut für Med. Dokumentation und Statistik der Universität, Heidelberg.

* Principal investigator.

TRIAL DESIGN: TREATMENT GROUPS:
1. CMF x 3: Cyclophosphamide 500 mg/M^2, methotrexate 40 mg/M^2, and 5-fluorouracil 600 mg/M^2, intravenously, day 1 and 8 of a 4-5 week cycle x 3 cycles.
2. CMF x 6: CMF, as above, x 6 cycles.
3. CMF x 3 + Tam: CMF as above x 3 cycles plus Tamoxifen 10 mg, three times daily, for 2 years.
4. CMF x 6 + Tam: CMF, as above, x 6 cycles, plus tamoxifen, as above.

ELIGIBILITY CRITERIA FOR ENTRY:
1. AGE: ≤65 years.
2. CLINICAL STAGES: T1$_a$–T3$_a$, N1-2, M0.
3. PRIMARY TREATMENT: Modified radical mastectomy.
4. AXILLARY NODE DISSECTION: Required.
5. NUMBER OF NODES REMOVED: ≥6.
6. MENSTRUAL STATUS: No restriction.
7. RECEPTOR STATUS: No restriction.

METHOD OF RANDOMIZATION:
By telephone to the statistical unit after completion of three cycles of CMF.

RANDOMIZATION STRATIFICATIONS:
By participating hospitals.

YEARS OF ACCRUAL: 1984 - still accruing.

TOTAL NUMBER OF PATIENTS RANDOMIZED: 96, as of July 1985.

AIM, RATIONALE, SPECIAL FEATURES:
This study was designed as a 2 x 2 study to compare chemotherapy, 3 versus 6 months, with or without tamoxifen.

ADDITIONAL ELIGIBILITY OR STRATIFICATION CRITERIA:
Patients were not eligible if they were pregnant, living abroad, or had limited disease, locally advanced disease, or distant metastases.

FOLLOW-UP, COMPLIANCE AND TOXICITY:
Clinical assessment with prespecified investigations were performed every 3 months for the first 2 years, every 4 months for years 3-5, semi-annually for years 6 and 7, and annually thereafter. X-rays, bone scintigraphy, liver sonography were performed less often.

COMPLIANCE BEFORE FIRST RECURRENCE:
Average amounts of CMF actually given in percent of targets were: 92% (CMF x 6); 96% (CMF x 3 +Tam) 87% (CMF x 6 + Tam); and 102% (CMF x 3) as assessed in May 1986.

PUBLICATIONS:
Scheurlen H. Randomized multicentre 2 x 2-factorial design study of chemo/endocrine therapy in operable, node-positive breast cancer. Recent Results in Cancer Research 1989; 115:226-235.

Table 1. Outcome by allocated treatment for trial 84A1 (BMFT 02 Germany)
Key: T = total free of relevant event at start of relevant year(s)

(i) Death (D) in each separate year, subdivided by age when randomized (D <50 & D 50+ / T <50 & T 50+)

Year	1. CMF	2. CMF	3. TamCMF	4. TamCMF
1	0+ 0/ 11+ 16	0+ 0/ 8+ 13	0+ 0/ 9+ 12	0+ 1/ 8+ 19
2	0+ 0/ 1+ 4	0+ 0/ 3+ 4	0+ 0/ 2+ 1	0+ 0/ 2+ 5
3	0+ 0/ 0+ 0	0+ 0/ 0+ 0	0+ 0/ 0+ 0	0+ 0/ 0+ 0
4	0+ 0/ 0+ 0	0+ 0/ 0+ 0	0+ 0/ 0+ 0	0+ 0/ 0+ 0
5+	0+ 0/ 0+ 0	0+ 0/ 0+ 0	0+ 0/ 0+ 0	0+ 0/ 0+ 0

(ii) First recurrence or prior death (F) in each separate year, subdivided by age when randomized (F <50 & F 50+ / T <50 & T 50+)

Year	1. CMF	2. CMF	3. TamCMF	4. TamCMF

(recurrence data not available)

(iii) Recurrence with survival (R) and Death (D) in all years together, subdivided by various characteristics when randomized (R & D / T)

Age at Entry	1. CMF	2. CMF	3. TamCMF	4. TamCMF
< 40	0+0/1	0+0/4	0+0/3	0+0/2
40 - 49	0+0/10	0+0/4	0+0/6	0+0/6
50 - 59	0+0/7	0+0/8	0+0/6	0+0/8
60 - 69	0+0/7	0+0/5	0+0/6	0+1/10
70 +	0+0/2	0+0/0	0+0/0	0+0/1
Any age (including "not known")	0+0/27	0+0/21	0+0/21	0+1/27

Nodal Status (N0, N1-3, N4+)

	1. CMF	2. CMF	3. TamCMF	4. TamCMF
50 + N0 (by axillary clearance)	0+0/0	0+0/0	0+0/0	0+0/1
<50 N1-N3 (by axillary clearance)	0+0/5	0+0/3	0+0/3	0+0/3
50 +	0+0/5	0+0/5	0+0/7	0+0/4
<50 N4+ (by axillary sample/clearance)	0+0/1	0+0/2	0+0/2	0+0/2
50 +	0+0/5	0+0/4	0+0/2	0+1/9
<50 Remainder (i.e. other N+ or N?)	0+0/5	0+0/3	0+0/4	0+0/3
50 +	0+0/6	0+0/4	0+0/3	0+0/5

Estrogen Receptors (ER, fmol/mg)

	1. CMF	2. CMF	3. TamCMF	4. TamCMF
<50 ER poor or ER < 10	0+0/2	0+0/0	0+0/2	0+0/1
50 +	0+0/2	0+0/2	0+0/1	0+1/5
<50 ER+ or ER 10-99	0+0/4	0+0/4	0+0/3	0+0/4
50 +	0+0/3	0+0/4	0+0/4	0+0/3
<50 ER++ or ER 100+	0+0/1	0+0/1	0+0/0	0+0/0
50 +	0+0/4	0+0/3	0+0/4	0+0/5

Short reports

TITLE: POSTOPERATIVE LOCAL RADIOTHERAPY AS AN ADJUNCT TO SYSTEMIC CHEMOTHERAPY IN THE TREATMENT OF PRIMARY BREAST CANCER

STUDY NUMBER: 84A2
STUDY TITLE: BMFT 03 GERMANY

ABBREVIATED TRIAL NAME: BMFT 03 GERMANY
STUDY NUMBER: 84A2

INVESTIGATORS' NAMES AND AFFILIATIONS:
G Bastert (1), R Sauer (2), K Hübner (3), J Bojar (4), M Schumacher (5), W Sauerbrei (5), H Scheurlen* (6).

(1) Universitäts-Frauenklinik, Homburg/Saar.
(2) Strahlentherapeutische Klinik der Universität, Erlangen.
(3) Zentrum Pathologie der Universität, Frankfurt.
(4) Institut für Med. Biometrie der Universität, Freiburg.
(5) Klinikum der Albert-Ludwigs-Universität, Freiburg.
(6) Institut für Med. Dokumentation und Statistik der Universität, Heidelberg.

  * Principal investigator.

TRIAL DESIGN: TREATMENT GROUPS:
1. CMF: Cyclophosphamide 500 mg/M^2, methotrexate 40 mg/M^2, and 5-fluorouracil 600 mg/M^2, intravenously, day 1 and 8 of a 4-5 week cycle, x 6 cycles.
2. CMF + R: CMF, as above, for 6 cycles, plus radiotherapy: 50 Gy to the chest wall; 50 Gy to the axilla and approximately 45 Gy to the internal mammary lymph node chain over a 5-6 week course between cycles 2 and 3 of CMF.

ELIGIBILITY CRITERIA FOR ENTRY:
1. AGE: ≤65 years.
2. CLINICAL STAGES: T1$_a$-3$_a$ N1-2 M0
3. PRIMARY TREATMENT: Modified radical mastectomy.
4. AXILLARY NODE DISSECTION: Required.
5. NUMBER OF NODES REMOVED: ≥6.
6. MENSTRUAL STATUS: No restriction.
7. RECEPTOR STATUS: No restriction.

METHOD OF RANDOMIZATION:
By telephone to the statistical unit after completion of 2 cycles of CMF.

RANDOMIZATION STRATIFICATIONS:
By participating hospitals.

YEARS OF ACCRUAL: 1984 - still accruing.

TOTAL NUMBER OF PATIENTS RANDOMIZED: 45, as of July 1985.

AIM, RATIONALE, AND SPECIAL FEATURES:
This study was designed to evaluate the addition of adjuvant radiotherapy to adjuvant chemotherapy.

ADDITIONAL ELIGIBILITY OR STRATIFICATION CRITERIA:
Patients were not eligible if they were pregnant, living abroad, or had limited disease, locally advanced disease, or distant metastases.

FOLLOW-UP, COMPLIANCE AND TOXICITY:
Clinical assessment with prespecified investigations were performed every 3 months for the first 2 years, every 4 months for years 3-5, semi-annually for years 6 and 7, and annually thereafter. X-rays, bone scintigraphy, liver sonography were performed less often.

COMPLIANCE BEFORE FIRST RECURRENCE:
Average amounts of CMF actually given in percent of the targets were: 89% (CMF + R); and 92% (CMF only), as assessed in May 1986.

PUBLICATIONS: None.

Table 1. Outcome by allocated treatment for trial 84A2 (BMFT 03 Germany)
Key: T = total free of relevant event at start of relevant year(s)

(i) Death (D) in each separate year,
subdivided by age when randomized (D <50 & D 50+ / T <50 & T 50+)

Year	1. CMF	2. CMFR
1	0+ 0/ 4+ 19	0+ 0/ 7+ 15
2	0+ 0/ 1+ 3	0+ 0/ 1+ 4
3	0+ 0/ 0+ 0	0+ 0/ 0+ 0
4	0+ 0/ 0+ 0	0+ 0/ 0+ 0
5+	0+ 0/ 0+ 0	0+ 0/ 0+ 0

(ii) First recurrence or prior death (F) in each separate year,
subdivided by age when randomized (F <50 & F 50+ / T <50 & T 50+)

Year	1. CMF	2. CMFR

(recurrence data not available)

(iii) Recurrence with survival (R) and Death (D) in all years together,
subdivided by various characteristics when randomized (R & D / T)

Age at Entry		1. CMF	2. CMFR
< 40		0+0/1	0+0/0
40 - 49		0+0/3	0+0/7
50 - 59		0+0/11	0+0/7
60 - 69		0+0/7	0+0/8
70 +		0+0/1	0+0/0
Any age (including "not known")		0+0/23	0+0/22

	Nodal Status (N0, N1-3, N4+)		
50 +	N0 (by axillary clearance)	0+0/0	0+0/1
< 50	N1-N3 (by axillary	0+0/2	0+0/5
50 +	clearance)	0+0/11	0+0/8
< 50	N4+ (by axillary	0+0/0	0+0/2
50 +	sample/clearance)	0+0/4	0+0/4
< 50	Remainder (i.e.	0+0/2	0+0/0
50 +	other N+ or N?)	0+0/4	0+0/2

	Estrogen Receptors (ER, fmol/mg)		
< 50	ER poor or ER < 10	0+0/3	0+0/2
50 +		0+0/6	0+0/6
< 50	ER+ or ER 10-99	0+0/1	0+0/4
50 +		0+0/3	0+0/3
< 50	ER++ or ER 100+	0+0/0	0+0/1
50 +		0+0/10	0+0/6

A five-yearly cycle of data collection and analysis is envisaged for the Early Breast Cancer Trialists' Collaborative Group. Volume 1 describes not only the first cycle of results but also the principles and the methods, and therefore provides the background that is needed to make appropriate use of subsequent volumes. Volume 2 is planned to be published in 1991, after which there should be an interval of about 5 years between successive volumes. If you would like details of Volume 2 to be sent to you, please send your name and address to Oxford University Press.

To: Science and Medical Department/AFL
     Oxford University Press
     Walton Street
     Oxford
     OX2 6DP

Please send me publication details of *Treatment of Early Breast Cancer Volume 2. Worldwide evidence in the early 1990s* when they are available.

Name: ....

Address: ....

....